Seaweed Phylogeography

Zi-Min Hu · Ceridwen Fraser
Editors

Seaweed Phylogeography

Adaptation and Evolution of Seaweeds under Environmental Change

Springer

Editors
Zi-Min Hu
Institute of Oceanology
Chinese Academy of Sciences
Qingdao
China

Ceridwen Fraser
Fenner School of Environment and Society
Australian National University
Canberra, ACT
Australia

ISBN 978-94-017-7532-8 ISBN 978-94-017-7534-2 (eBook)
DOI 10.1007/978-94-017-7534-2

Library of Congress Control Number: 2015958556

Printed on acid-free paper

This Springer imprint is published by SpringerNature
The registered company is Springer Science+Business Media B.V. Dordrecht

Preface

Why Seaweed Phylogeography, Why Now?

> Although species-level research on seaweeds, at least with regard to climate change, lags behind similar work in terrestrial environments, there is no reason that phycologists could not model a research program based on the successes of terrestrial botanists, foresters, and agricultural scientists.
>
> Harley CDG et al. (J Phycol, 2012, 48:1064–1078)

Seaweeds (marine macroalgae) are a group of photosynthesizing organisms that generally attach to rock or other hard substrata in coastal areas. Ecologically, many species form dense forests to provide protective habitats for a wide range of flora and fauna, maintain coastal community by modifying physical structure, form the base of the marine food chain, and serve as the primary producers and carbon sinks. It is estimated that, globally, kelps can assimilate about 1.8 kg carbon m^{-2} $year^{-1}$, exceeding the primary production of marine phytoplankton by up to ten times. Economically, seaweeds are used by humans for food, feed, fertilizer, cosmetics, mariculture, pharmaceutical industry, and biofuels.

Seaweeds are critical components of marine biodiversity and play vital roles in ecosystem function, yet many species are vulnerable to global environmental change and anthropogenic impacts. Understanding how such impacts have affected the genetic diversity and biogeographic patterns of seaweeds will facilitate predictions of how seaweeds will respond to ongoing global environmental change, and thus inform management and conservation strategies.

Over recent decades, rapidly evolving DNA sequencing technologies and ever-improving analytical frameworks have allowed us to begin to understand broadscale patterns of genetic diversity of seaweeds, and to interpret the processes affecting their evolution and ecosystem structure. In particular, phylogeographic inferences of how seaweeds responded to Pleistocene climate change cycles suggest that many experienced localized extinction and large-scale range contraction. Phylogeographic research has also shed light on how seaweeds have evolved and dispersed, how invasive species have affected marine ecosystems, and how seaweeds have adapted to heterogeneous habitat niches. Such knowledge is crucial for linking the diversification and evolution of seaweeds to various biological, environmental, and climatic factors for marine biodiversity planning and conservation purposes.

The book *Seaweed Phylogeography: Adaptation and Evolution of Seaweeds under Environmental Change* provides a collection of articles summarizing advances in population genetics and the evolutionary biogeography of seaweeds over the past two decades. It is intended for students at the senior undergraduate and graduate levels as well as professional researchers interested in phycology, marine biology, ecology, and evolutionary biology. While not attempting to comprehensively cover all research in seaweed phylogeography, we hope that this book achieves its goal of providing a useful and interesting summary of major recent discoveries and avenues for future research.

China
Australia
November 2015

Zi-Min Hu
Ceridwen Fraser

Acknowledgments

The topic of this book was originally proposed to Dr. Zi-Min Hu by Natalia Manrique-Hoyos, an editor at Springer Publishers. Her proposal was met with great enthusiasm by Zi-Min, and a skeleton outline was subsequently developed with useful feedback from an anonymous external reviewer. Invitations to contribute to the book, sent to around 40 researchers, met with heartening response. Within three weeks, we had a list of chapter authors that would allow us to cover most of the topics we envisioned. This rapid and positive response re-emphasized the great thirst for such a book.

However, as chapter proposals began to come in, the scope of the book expanded dramatically, and Dr. Ceridwen Fraser was invited to join in as a co-editor. The editors' roles were highly complementary, with Zi-Min proposing and developing the book skeleton, and overseeing compilation and formatting, and Ceridwen focusing on managing the review and revision process for individual chapters. The long-term fruitful communication and consultation between both of us not only helped to bring about the finalization of the book, but also enhanced the mutual recognition and cooperation. We are delighted to see the book realized.

The contributing authors made great efforts to keep the content as updated, relevant, and conclusive as possible, and this book would not exist without their dedication. We thank them for making every effort to meet our deadlines, and for assisting with providing reviews of other chapters. We particularly thank Dr. Stewart Grant who not only contributed an important chapter, but also provided in-depth reviews of other book chapters. His rapid, willing and constructive feedback greatly enhanced both the speed of the review process and the quality of the finished work.

We gratefully acknowledge the professionalism and patience of Springer Publishers and, in particular, the editors Abbey Huang and Becky Zhao for their help and support during the creation of this book. Finally, we would like to thank our families, Catherine (Cui), Tony (Hu), Amit, and baby Rishi (who was born shortly before Ceridwen joined the editorial team), for forgiving us for occasionally working at home on the book.

China
Australia
November 2015

Zi-Min Hu
Ceridwen Fraser

Contents

Contributors

Nikos Andreakis Australian Institute of Marine Science, Townsville, QLD, Australia; College of Marine and Environmental Sciences, James Cook University, QLD, Australia

Jorge Assis CCMAR—Centro de Ciências do Mar, Universidade do Algarve, Faro, Portugal

James A. Coyer Shoals Marine Laboratory, Cornell University, Portsmouth, NH, USA

Christophe Destombe CNRS, Sorbonne Universités, UPMC University Paris VI, UMI 3614, Evolutionary Biology and Ecology of Algae, Station Biologique de Roscoff, Roscoff, France

De-Lin Duan Key Laboratory of Experimental Marine Biology, Institute of Oceanology, Chinese Academy of Sciences, Qingdao, China; Qingdao National Laboratory for Marine Science and Technology, Qingdao, China

Sylvain Faugeron CNRS, Sorbonne Universités, UPMC University Paris VI, UMI 3614, Evolutionary Biology and Ecology of Algae, Station Biologique de Roscoff, Roscoff, France; Centro de Conservación Marina and CeBiB, Facultad de Ciencias Biológicas, Pontificia Universidad Católica de Chile, Santiago, Chile

Ceridwen I. Fraser Fenner School of Environment and Society, Australian National University, Acton, ACT, Australia

W. Stewart Grant Alaska Department of Fish and Game Commercial Fisheries Division, Anchorage, AK, USA

Marie-Laure Guillemin Facultad de Ciencias, Instituto de Ciencias Ambientales y Evolutivas, Universidad Austral de Chile, Valdivia, Chile; CNRS, Sorbonne Universités, UPMC University Paris VI, UMI 3614, Evolutionary Biology and Ecology of Algae, Station Biologique de Roscoff, Roscoff, France

Takeaki Hanyuda Kobe University Research Center for Inland Seas, Nadaku, Kobe, Japan

Galice Hoarau Faculty of Biosciences and Aquaculture, University of Nordland, Bodø, Norway

Zi-Min Hu Key Laboratory of Experimental Marine Biology, Institute of Oceanology, Chinese Academy of Sciences, Qingdao, China; Qingdao National Laboratory for Marine Science and Technology, Qingdao, China

Kensuke Ichihara Department of Integrated Biosciences, Graduate School of Frontier Sciences, University of Tokyo, Kashiwa, Chiba, Japan

Riko Kawaguchi Division of the Natural/Applied Sciences, Graduate School of Humanities and Sciences, Ochanomizu University, Bunkyo-ku, Tokyo, Japan

Hiroshi Kawai Kobe University Research Center for Inland Seas, Nadaku, Kobe, Japan

Norio Kikuchi Coastal Branch of Natural History Museum and Institute, Chiba, Katsuura, Chiba, Japan

Tania M. King Department of Zoology, University of Otago, Dunedin, New Zealand

Helena Korpelainen Department of Agricultural Sciences, Viikki Plant Science Centre, University of Helsinki, Helsinki, Finland

Jing-Jing Li Key Laboratory of Experimental Marine Biology, Institute of Oceanology, Chinese Academy of Sciences, Qingdao, China; Qingdao National Laboratory for Marine Science and Technology, Qingdao, China

Boris López Programa de Doctorado en Biología y Ecología Aplicada (BEA), Facultad de Ciencias del Mar, Universidad Católica del Norte, Coquimbo, Chile; Departamento de Acuicultura y Recursos Agroalimentarios, Universidad de Los Lagos, Osorno, Chile

Juan Lopez-Bautista Department of Biological Sciences, The University of Alabama, Tuscaloosa, AL, USA

Erasmo C. Macaya Laboratorio de Estudios Algales (ALGALAB), Departamento de Oceanografía, Universidad de Concepción, Concepción, Chile; Millennium Nucleus Ecology and Sustainable Management of Oceanic Island (ESMOI), Coquimbo, Chile

Yuka Masakiyo Division of the Natural/Applied Sciences, Graduate School of Humanities and Sciences, Ochanomizu University, Bunkyo-ku, Tokyo, Japan

João Neiva CCMAR—Centro de Ciências do Mar, Universidade do Algarve, Faro, Portugal

Jeanine L. Olsen Groningen Institute for Evolutionary Life Sciences, Groningen, The Netherlands

Gareth A. Pearson CCMAR—Centro de Ciências do Mar, Universidade do Algarve, Faro, Portugal

David R. Schiel Marine Ecology Research Group, School of Biological Sciences, University of Canterbury, Christchurch, New Zealand

Ester A. Serrão CCMAR—Centro de Ciências do Mar, Universidade do Algarve, Faro, Portugal

Alison R. Sherwood Department of Botany, University of Hawaii, Hawaii, Honolulu, USA

Satoshi Shimada Division of the Natural/Applied Sciences, Graduate School of Humanities and Sciences, Ochanomizu University, Bunkyo-ku, Tokyo, Japan

Paul M. South Marine Ecology Research Group, School of Biological Sciences, University of Canterbury, Christchurch, New Zealand

Sandra C. Straub UWA Oceans Institute and School of Plant Biology, The University of Western Australia, Crawley, WA, Australia

Fadia Tala Centro de Investigación y Desarrollo Tecnológico en Algas de la Universidad Católica del Norte (CIDTA-UCN), Coquimbo, Chile

Florence Tellier Departamento de Ecología, Facultad de Ciencias, Universidad Católica de la Santísima Concepción, Concepción, Chile

Martin Thiel Facultad Ciencias del Mar, Universidad Católica del Norte, Coquimbo, Chile; Millennium Nucleus Ecology and Sustainable Management of Oceanic Island (ESMOI), Coquimbo, Chile; Centro de Estudios Avanzados en Zonas Áridas (CEAZA), Coquimbo, Chile

Mads Solgaard Thomsen Marine Ecology Research Group, School of Biological Sciences, University of Canterbury, Christchurch, New Zealand; School of Biological Sciences, Centre of Integrative Ecology, University of Canterbury, Christchurch, New Zealand; School of Plant Biology, UWA Oceans Institute, The University of Western Australia, Crawley, WA, Australia

Shinya Uwai Faculty of Science, Niigata University, Nishiku, Niigata, Japan

Myriam Valero Evolutionary Biology and Ecology of Algae, UMI 3614, UPMC University Paris VI, CNRS, Sorbonne Universités, Roscoff, France

Xénia A. Weber Fenner School of Environment and Society, Australian National University, Acton, ACT, Australia

Thomas Wernberg School of Plant Biology, UWA Oceans Institute, The University of Western Australia, Crawley, WA, Australia; Australian Institute of Marine Science, Crawley, WA, Australia

Laura J. Wilson Fenner School of Environment and Society, Australian National University, Acton, ACT, Australia

Marianela Zanolla Departamento de Biología Vegetal (Botánica), Facultad de Ciencias, Universidad de Málaga, Málaga, Spain

Giuseppe C. Zuccarello School of Biological Sciences, Victoria University of Wellington, Wellington, New Zealand

Part I
The Field of Phylogeographic Research on Seaweeds

Chapter 1
Seaweed Phylogeography from 1994 to 2014: An Overview

Zi-Min Hu, De-Lin Duan and Juan Lopez-Bautista

Abstract Molecular phylogeographic approaches employed for studying genetic diversity and evolution of seaweeds experienced noticeable growth since the mid-1990s and have greatly expanded our understanding of factors and processes contributing to biodiversity, adaptation, and population genetic variation of seaweeds. Herein, we present a numerical synthesis of 126 published references on seaweed phylogeography during the past two decades. We summarize the progress, research hotspots, regional distribution of outputs, potential deficiencies, and future tendencies in this field at a global scale. We also highlight the importance of integrating a statistically rigorous and comparative phylogeographic framework with species distribution models (SDM) and model-based phylogeographic inferences, when exploring cryptic speciation and evolution of seaweeds in response to global climate change, environmental shift, and human interference.

Keywords Coastal ecosystem · Ecological adaptation · Habitat heterogeneity · Phylogeography · Population genetics · Seaweed

Z.-M. Hu (✉) · D.-L. Duan
Key Laboratory of Experimental Marine Biology, Institute of Oceanology, Chinese Academy of Sciences, Qingdao 266071, China
e-mail: huzimin9712@163.com

Z.-M. Hu · D.-L. Duan
Qingdao National Laboratory for Marine Science and Technology, Qingdao 266071, China

J. Lopez-Bautista
Department of Biological Sciences, The University of Alabama, Tuscaloosa, AL 35487, USA

Z.-M. Hu and C. Fraser (eds.), *Seaweed Phylogeography*,
DOI 10.1007/978-94-017-7534-2_1

1.1 Introduction

Phylogeography is a relatively young discipline seeking to explore the principles, patterns, and processes contributing to the geographic distributions of genealogical lineages, especially those found within species in evolutionary and ecological contexts (Avise 2000, 2009). Because the genetic variation found in extant populations or species resulted from thousands or millions of years of accumulation, phylogeography is closely allied with historical biogeography and has played a fundamental role in bridging the gaps between micro- and macroevolutionary processes which predominantly shaped modern patterns of biodiversity on earth. Since its inception (Avise et al. 1987), phylogeography has considerably influenced life and earth sciences and has expanded into many other subdisciplines such as ecology, biology, genetics, oceanography, environmental sciences, and geography (Beheregaray 2008). The rapid growth and extension of phylogeography is mainly ascribed to the accelerating integration of genetic data and mathematical modeling, emerging high-throughput DNA sequencing technologies and increasingly powerful computation algorithms and analysis programs.

The rapid growth of phylogeographic studies in less than three decades also promoted methodological and conceptual shifts in some fields such as statistical phylogeography (Knowles and Maddison 2002), comparative phylogeography (Arbogast and Kenagy 2001) and phylogeographic information systems (Kidd and Ritchie 2006). Such a trend in development probably is far beyond the imagination of the "Father of Phylogeography," John C. Avise (the recipient of the 2009 Alfred Russel Wallace award), who initially described phylogeography as a discipline with conceptual and technical roots linked to the incipient field of molecular genetics in the 1970s. A recent global search in Web of Science database using the terms "phylogeography" or "phylogeographic" in abstracts found that there were more than 3000 articles published between 1987 and 2006 (Beheregaray 2008). Nevertheless, phylogeographic studies predominantly focused on terrestrial organisms and only a small proportion (17 %) on marine species. The paucity of phylogeographic studies on marine species is unfortunate because marine ecosystems are strongly influenced by environmental conditions, including complex oceanographic factors, the long-term influence of paleoclimate change, short-term environmental shifts and human activities. These forces have independently or interactively had fundamental impacts on the temporal and spatial distributions of biodiversity, population genetic differentiation, and evolutionary histories of marine organisms.

Seaweeds (marine macroalgae) are a diverse and widespread group of photosynthetic organisms, classified into three broad groups (red, brown, and green algae) based on pigmentation and cell structure. They can be found in almost all aquatic environments, from marine to brackish and freshwater, and from the tropical islands near the equator to polar regions. Seaweeds are an essential component of coastal marine ecosystems and play a significant role as benthic primary producers, providing food, habitat structure, breeding grounds, and shelter for many coastal organisms. For seaweeds, drifting is an important characteristic allowing them to undergo long-distance dispersal driven by oceanic currents (Thiel and Haye 2006), thereby producing complex biogeographic patterns of population genetic differentiation.

In this chapter, we present a broad description of the state of phylogeographic studies of seaweeds, including some early studies using RAPD (Random Amplified Polymorphic DNA), RFLP (Restriction Fragment Length Polymorphism), SSCP (Single Strand Conformation Polymorphism) and microsatellites that explored environmental factors as contributors to population-level genetic connectivity and differentiation. Our synthesis of phylogeographic literature of seaweeds provides a global viewpoint to appreciate which genetic markers are more common and effective for deciphering intraspecific evolutionary histories, which taxonomic groups have intensively been surveyed, and where are the disparities of research productivity between different regions of the world. This information helps us to identify future research priorities that can promote and reinforce our understanding of adaptive genetic variation and evolution of seaweeds.

1.2 Benchmark Progress

In contrast to the flourishing development of phylogeographic studies of terrestrial organisms soon after the discipline received its name in 1987 (Avise et al. 1987), the approach was only applied to seaweeds rather later. In 1994, Oppen and colleagues, led by Drs. Jeanine L. Olsen and Wytze T. Stam at the University of Groningen published an article entitled "Tracking dispersal routes: phylogeography of the Arctic-Antarctic disjunct seaweed *Acrosiphonia arcta* (Chlorophyta)" in the *Journal of Phycology* (Oppen et al. 1994), marking the pioneer phylogeographic study on (green) seaweeds (Fig. 1.1). Based on integrative evidence of RAPD and RFLP of nuclear ribosomal intergenic spacer (nrDNA IGS), these authors found that *A. arcta* populations in the Arctic and North Atlantic originated from the

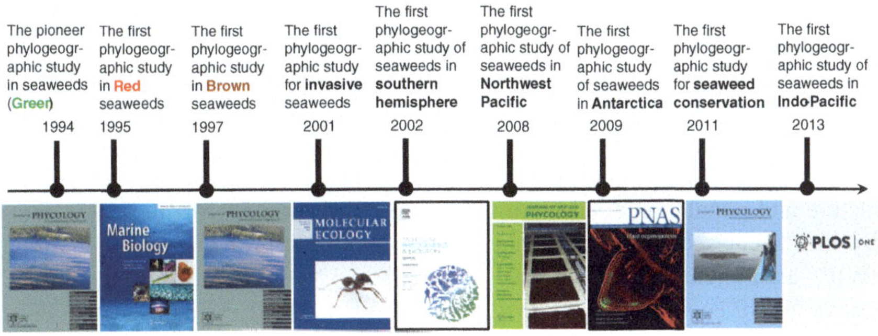

Fig. 1.1 Timeline of phylogeographic researches in seaweeds (key milestones)

Pacific. More importantly, the Arctic Greenland populations underwent independent colonization in contrast to the recolonization of the Antarctic Peninsula populations from southern Chile (Oppen et al. 1994). Afterward, the same research unit published a series of phylogeographic studies for other marine macroalgae (e.g., *Cladophora vagabunda*, Bakker et al. 1995; *Digenea simplex*, Pakker et al. 1996; Palmariaceae, Lindstrom et al. 1996), including the first phylogeographic work on red seaweeds (*Phycodrys rubens*, Oppen et al. 1995a, b) (Fig. 1.1).

In 1997, James Coyer collaborated with his previous colleagues (Olsen and Stam) to investigate genetic variability in the brown seaweed *Postelsia palmaeformis* on spatial scales of 1–250 km by integrating the complementary RAPD and M13 fingerprints (Coyer et al. 1997). They found that RAPDs can easily discriminate *P. palmaeformis* populations isolated by geographic distances of 16–250 km, despite no resolution in discriminating individuals separated by <1–25 m, whereas M13 fingerprinting detected decreased genetic relatedness as geographic distance increased to 25 m (Coyer et al. 1997). This was the first phylogeographic work on brown seaweeds (Fig. 1.1). Olsen and Stam initially introduced phylogeographic approaches to seaweed investigations and opened a window through which phycologists could link environmental variables and life-history features to population genetic differentiation and ecological adaptation of seaweeds over time and space. In 2012, they were awarded the Award of Excellence by the Phycological Society of America for several decades of research achievements in surveying biogeographic histories and evolutionary patterns of benthic seaweeds.

From the beginning of the twenty-first century, phylogeographic approaches have been extensively applied to seaweed research (Fig. 1.1). Based on the geographic distribution of specific chloroplast-encoded large subunit RuBisCo (*rbc*L) haplotypes and the genetic relatedness between source and invasive populations, McIvor et al. (2001) revealed two cryptic invasion routes of the red alga *Polysiphonia harveyi* from Japan to New Zealand and the North Atlantic Ocean, including two separate introductions from Japan (Hokkaido and Honshu) into the northern North Atlantic and a recent introduction from Honshu into at least two areas of California, North Carolina, and New Zealand. Other noteworthy milestones of phylogeographic theory applied to seaweeds include the first phylogeographic studies focused on benthic macroalgae in the Southern Hemisphere (*Halimeda*, Kooistra et al. 2002), the Northwest Pacific (*Ulva*, Shimada et al. 2008), the sub-Antarctic (*Durvillaea antarctica*, Fraser et al. 2009a, b), and the Indo-Pacific (*Sargassum polycystum*, Chan et al. 2013) (Fig. 1.1). In particular, the phylogeographic approach has been demonstrated to be able to provide valuable insights into conservation genetics of endangered seaweed species. Couceiro et al. (2011) used Amplified Fragment Length Polymorphism (AFLP) to investigate the genetic structure and population connectivity of the red alga *Ahnfeltiopsis pusilla*, a naturally rare species categorized as vulnerable in the Northwest Iberian Peninsula (NWIP), at different geographic scales (from <1 to 1200 km). The results indicated that five NWIP enclaves should be designated independently as management units

(MUs), and that the three southernmost sites harboring most of the genetic heritage of *A. pusilla* in NWIP are particularly valuable for conservation as evolutionarily significant units (ESUs).

1.3 Global Glance

Based on a survey of the database ISI Web of Knowledge and the terms "seaweed" and "phylogeography"/"phylogeographic" or "population genetics," a total of 126 papers published from 1994 to 2014 were identified from 27 international journals such as *Proceedings of the National Academy of the Sciences of the United States of America, Proceedings of the Royal Society of Biology B: Biological Sciences, BMC Evolutionary Biology, Aquatic Botany* and *Journal of Phycology*. In general, phylogeographic studies on seaweeds experienced remarkable growth as measured by the number of papers published each year since 2000 (Fig. 1.2), which probably was due to the contribution of the book by Avise (2000) dealing with the history and formation of species, enabling more researchers to become acquainted with this emerging discipline. This tendency, as expected, is in line with the growth of phylogeographic studies at a global scale (Beheregaray 2008).

The far-reaching contribution of phylogeographic approaches to seaweed research is evidenced in Table 1.1, which lists 10 subject categories substantially influenced by this scientific discipline. These studies can be primarily classified in the fields of ecology and evolution (56 %) and population genetics (19 %). The phylogeographic approach has also been extended to address some long-standing questions in seaweed ecology (e.g., hybridization and speciation). For example, Coyer et al. (2002) employed SSCP analyses of nuclear, chloroplast, and mitochondrial markers and confirmed the occurrence of hybridization between *Fucus*

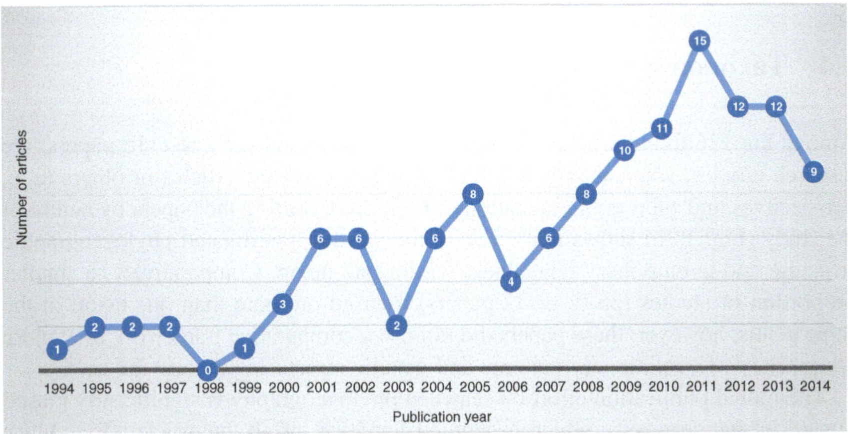

Fig. 1.2 Number of phylogeographic articles published between 1994 and 2014 in seaweeds

Table 1.1 The top 10 scientific journals published articles related to seaweed phylogeography and the contributions of phylogeographic approaches to relevant research areas in seaweed

Journal	No.	Subject category	Representative reference
Journal of Phycology	32	Population genetic structure	Faugeron et al. (2001)
Molecular Ecology	22	Evolutionary biology	Olsen et al. (2010)
Marine Biology	10	Cryptic/genetic diversity	Lee et al. (2013)
Journal of Applied Phycology	7	Environmental adaptation	Kostamo et al. (2012)
Journal of Biogeography	6	Invasive biology	McIvor et al. (2001)
Phycologia	5	Hybridization/speciation	Coyer et al. (2002)
Molecular Phylogenetics and Evolution	4	Biodiversity conservation	Couceiro et al. (2011)
BMC Evolutionary Biology	4	Heredity	Zuccarello et al. (1999)
Botanica Marina	4	Seascape genetics	Krueger-Hadfield et al. (2013)
Aquatic Botany	4	Molecular systematics	Kooistra et al. (2002)

serratus (dioecious) and *F. evanescens* (monoecious) in the Kattegat Sea, Denmark, when densities of *F. serratus* (*Fs*) and *F. evanescens* (*Fe*) exceeded 2 and 14 m^{-2}, respectively. However, the hybridization was asymmetrical since only the *Fe* egg × *Fs* sperm cross was successful and the reciprocal cross was ineffective. Moreover, Zuccarello et al. (1999) used SSCP analysis of the chloroplast-encoded RuBisCo spacer and revealed maternal inheritance of plastids in crosses between isolates of the red alga *Bostrychia radicans* and *B. moritziana*. Listed in Table 1.1 are journals representing the large proportion of published papers dealing with seaweed phylogeography (e.g., *Journal of Phycology, Molecular Ecology*, and *Marine Biology*).

1.4 Taxonomic Coverage

Among the 126 found articles on seaweed phylogeography, 92 % (115 papers) are research articles, whereas only 8 % (11 papers) are review articles or others (e.g., perspectives and short communications) (Fig. 1.3a). Sorting the papers by nature of the study (Fig. 1.3b) shows that 75 % of the articles investigated phylogeographic structure and evolutionary relatedness within one taxon. Comparatively, a smaller proportion of studies (25 % or 32 papers) focused on more than one taxon in the same genus; however, these papers did not use a comparative framework to explore congruency of evolutionary patterns and genetic structuring among the taxa.

The global publication effort on seaweed phylogeography is significantly biased toward brown seaweeds, which accounted for 50 % of all the articles (Fig. 1.4a).

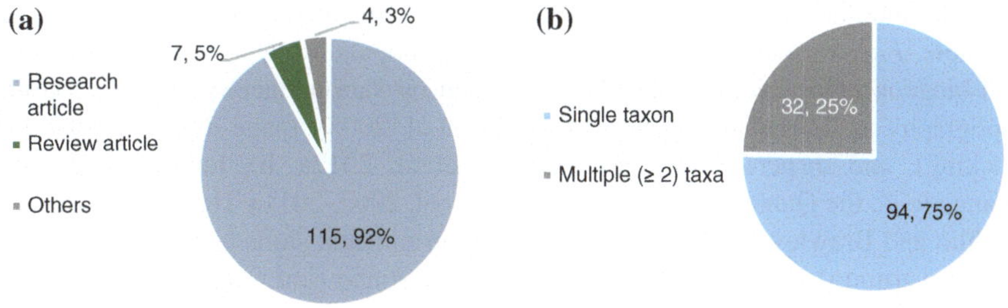

Fig. 1.3 Proportion of articles in seaweed phylogeography published between 1994 and 2014 according to type of articles (**a**) and nature of study (**b**)

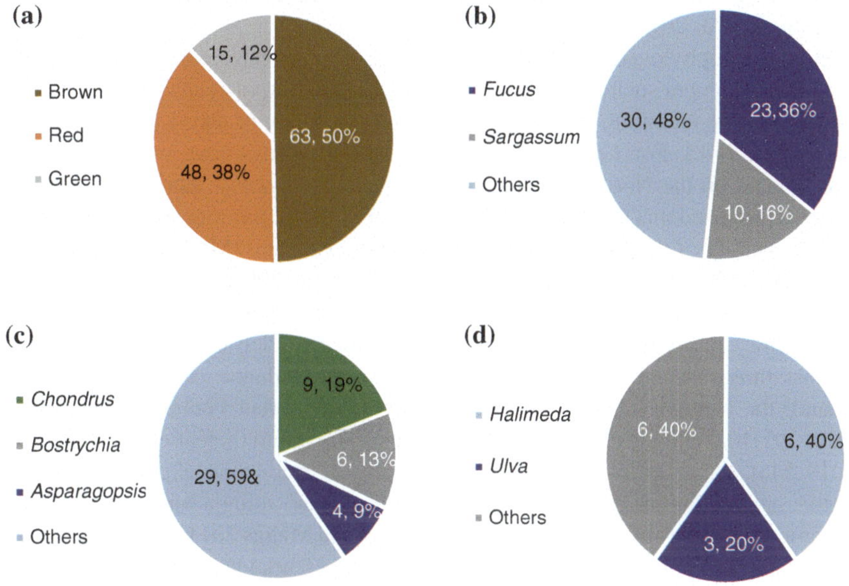

Fig. 1.4 Proportion of articles in seaweed phylogeography published between 1994 and 2014 according to taxonomic catalogs (**a**), the most common surveyed taxa in *brown* (**b**), *red* (**c**) and *green* (**d**)

Studies focusing on red seaweeds represented 38% (48 papers), and a much smaller proportion (12 % or 15 papers) were focused on green seaweeds. Of the 63 articles published on brown seaweeds, the genera *Fucus* and *Sargassum* were the most common taxa surveyed to date, which comprised 23 and 10 articles, respectively (Fig. 1.4b). This taxonomically biased publication activity is probably due to two main reasons: First, the genera *Fucus* and *Sargassum* are characterized by wide geographic distributions from northern cold temperate to southern warm and tropical regimes in the ocean enabling them to be representative models. For example, *Fucus* species' demographic histories, population genetic differentiation, and biogeographic evolution can be influenced by various factors, such as coastal topography and circulation patterns (Muhlin et al. 2008), range dynamics (habitat tracking), and dispersal restrictions (Neiva et al. 2012a, b), long-term climate change (e.g., the Quaternary ice ages) (Coyer et al. 2003, 2011a; Hoarau et al. 2007; Muhlin and Brawley 2009) and short-term environmental shifts (several decades of global warming) (Nicastro et al. 2013). Second, *Fucus* and *Sargassum* have biologically well-defined features, for example, some *Fucus* species can be found in highly heterogeneous marine coastal habitats, such as those with salinity gradients ranging from 2.7 psu to 33.0 psu on a 12 km geographic scale in the North Atlantic (Coyer et al. 2011b). Species of *Sargassum* can form drifting mats after being detached from the substratum and be transported for hundreds of kilometers driven by oceanic currents (Komatsu et al. 2008; Uwai et al. 2009; Cheang et al. 2010a; Hu et al. 2011a, 2013).

In contrast, phylogeographic studies on red seaweeds have expanded into more areas. Some genera such as *Chondrus*, *Bostrychia* and *Asparagopsis* have received relatively more attention, with each genus comprising 9, 6, and 4 papers, respectively (Fig. 1.4c). For *Chondrus*, phylogeographic surveys have overwhelmingly concentrated on the North Atlantic endemic species *C. crispus*. Early phylogenetic analyses proposed that the North Pacific ancestor of *C. crispus* migrated to the Arctic via the opening of the Bering Strait *c.* 3.5 million years ago (Ma) and subsequently colonized both sides of the North Atlantic (Hu et al. 2007a). Recent integrative phylogeographic evidence showed that *C. crispus* experienced large-scale population decline caused by marine glaciations during the late Pleistocene and survived in at least three scattered refugia in the northeastern Atlantic (e.g., southwestern Ireland, the English Channel and the northwestern Iberian Peninsula) (Hu et al. 2010, 2011b). With the retreat of the Last Glacial Maximum (hereafter LGM, 0.026–0.019 Ma), *C. crispus* underwent multidirectional step-stone expansions in the northeastern Atlantic centering on the late Pleistocene refugia and trans-Atlantic migration from Europe to North America (Provan and Maggs 2011; Hu et al. 2011b). More importantly, *C. crispus* exhibits an excellent potential as a research model for seascape genetics, e.g., to explore the ecological role of microgeographic environmental variables in shaping phylogeographic dynamics along coastal communities. A recent study revealed that tidal heights not only contributed to genetic differentiation between high- and low-shore stands, but also restricted genetic exchange within the high-shore stand, extending our understanding of the population genetic structure and evolutionary patterns of *C. crispus* from macrogeographic scale (>500 km) to microgeographic scale (<100–200 m) (Krueger-Hadfield et al. 2013; Hu 2013). The feature of intraspecific hybridization, taxonomic complexity, and highly differentiated lineages in *Bostrychia* (Zuccarello et al 1999, 2006, 2011; Zuccarello and West 2003; Muangmai et al. 2015), and the nature of rapid global invasion in *Asparagopsis* (Andreakis et al. 2007, 2009; Sherwood 2008), make these red algal taxa amenable to becoming research foci concerning cryptic genetic diversity, evolutionary history and invasive processes.

The genera *Halimeda* and *Ulva* are the two top-ranked taxa in limited phylogeographic studies on green seaweeds (Fig. 1.4d). *Halimeda* species consist of phenotypically complex segmented, calcified thalli and are common in reefs and lagoons throughout the tropics (Kooistra et al. 2002). After reproduction, the calcified *Halimeda* segments became the most important contributors of aragonite sediments (up to 90 %) in modern tropical and subtropical carbonate environments (Freile et al. 1995). Historically, *Halimeda* was an important member of the Late Miocene Mediterranean carbonate factory after the Tethyan Seaway was closed during the middle Miocene (Braga et al. 1996). Currently, *Halimeda* species are globally distributed and flourish in variable environments at depths ranging from <1 to 150 m (Kooistra et al. 2002; Verbruggen et al. 2009). These factors enable *Halimeda* to be an ideal model to survey genus-level biogeographic history associated with paleobiogeographical and macroecological processes, and to characterize evolutionary niche dynamics in the ocean (Verbruggen et al. 2005, 2009; Reuter et al. 2012). The green seaweed *Ulva* exhibits considerable morphological plasticity and can grow under variable salinities. Previous molecular research has demonstrated that *Ulva* species have some important phylogeographic imprints of climate change-induced historical reduction of genetic diversity in the Northern Hemisphere (Leskinen et al. 2004; Shimada et al. 2008). Recent genetic surveys indicated that *Ulva* species may be used as models to resolve how intertidal seaweeds adapt to thermal stress and salinity gradients and why some species exhibit more extensive spatial and temporal distribution ranges than others (Ogawa et al. 2015).

1.5 Regions and Countries

Our literature survey shows that the Northeast Atlantic was the most intensively studied region with 23.5 % (32 papers) of all the articles (Fig. 1.5), also see chapters by Neiva et al. (2016) and Li et al. (2016) in this volume. A good portion of studies were conducted for systems from the Northwest Pacific (13.5 % or 17 papers) and from the Northern Hemisphere oceans (14.3 % or 18 papers) in which sampling covered both the North Pacific and the North Atlantic. On the other hand, a relatively small proportion of references (4.8 % or 6 papers) used samples from the Southern Hemisphere oceans, Australia and New Zealand. A smaller proportion of studies (5–7 papers) were focused on the Southeast Atlantic, Indo-Pacific, Antarctic and Sub-Antarctic. There was only one paper each published on seaweed phylogeography in the Mediterranean and South Africa waters. This was surprising since recent evidence indicates that these phylogeographically unexplored regions are important marine biodiversity hot spots (Coll et al. 2010; Griffiths et al. 2010). Collectively, phylogeographic studies on seaweeds were predominantly performed in the Northern Hemisphere oceans but the research effort was not equally distributed among regions (e.g., Northwest Atlantic vs. Northeast Atlantic, North Atlantic vs. North Pacific).

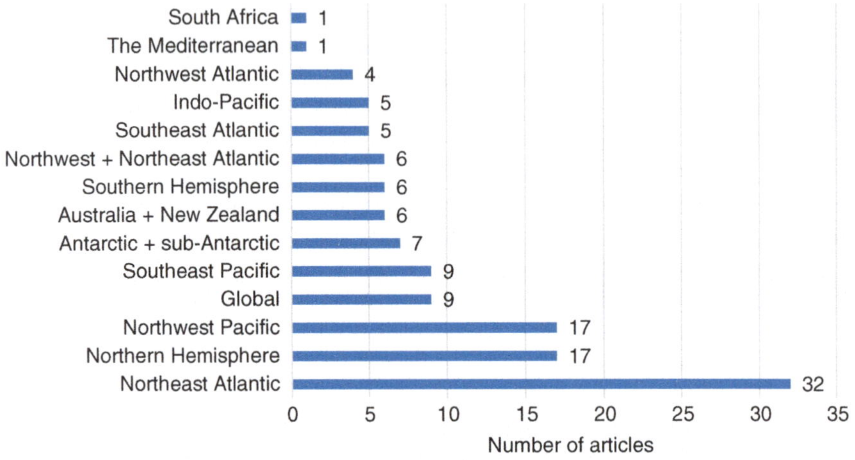

Fig. 1.5 Regional distribution of articles in seaweed phylogeography published between 1994 and 2014 based on the location of organisms studied

When references were classified based on the nationalities of the corresponding authors of the articles (Fig. 1.6), the Netherlands was the most productive country with 14.3 % of all articles (18 papers), followed closely by China with 11.9 % (15 papers) and the USA with 11.1 % (14 papers). The large proportion of phylogeographic studies on seaweeds from the Netherlands was mainly due to the prolific contributions of Drs. Olsen and Stam at the Groningen University and their decades of continuous endeavor in this field (Oppen et al. 1994; Bakker et al. 1995; Peters et al. 1997; Miller et al. 2000; Coyer et al. 2002, 2003, 2006, 2011a, b; Hoarau et al. 2007; Olsen et al. 2010). China's productivity was largely thanks to Dr. Put O. Ang Jr. at the Chinese University of Hong Kong and Dr. Zi-Min Hu at the Institute of Oceanology, Chinese Academy of Sciences (Cheang et al. 2008, 2010a, b; Chan et al. 2013, 2014; Hu et al. 2007b, 2010, 2011a, b, 2013; Wang et al. 2008; Li et al. 2015). There were more than five institutions involved in population genetics and phylogeography of seaweeds in the USA with Drs. Susan H. Browley (University of Maine) and Walter H. Adey (National Museum of Natural History) producing a considerable number of publications (Muhlin et al. 2008; Muhlin and Brawley 2009; Brawley et al. 2009; Adey and Steneck 2001; Adey et al. 2008; Adey and Hayek 2011). The Korean research unit leaded by Dr. Sung Min Boo published 11 papers on population genetic diversity and phylogeography of seaweeds, with the geographic sampling mostly restricted to the Korean Peninsula and adjoining areas (Yang et al. 2008, 2009; Bae et al. 2013; Lee et al. 2013; Kim et al. 2010, 2012, 2014). The research unit at the University of Algarve, Portugal leaded by Dr. Ester A. Serrão made important contributions to phylogeographic studies of seaweeds with nine publications (Moalic et al. 2011; Nicastro et al. 2013; Neiva et al. 2010, 2012a, b, 2014). Dr. Giuseppe C. Zuccarello at Victoria University of Wellington and Dr. Ceridwen I. Fraser at the University of Otago, New Zealand led

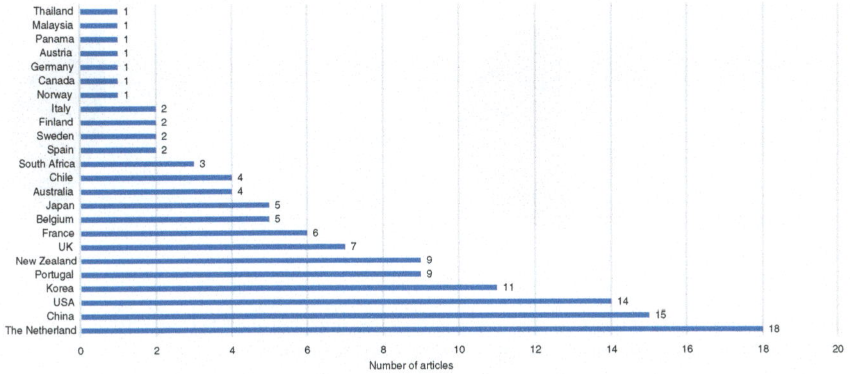

Fig. 1.6 Country distribution of articles in seaweed phylogeography published between 1994 and 2014 based on the location of organisms studied

the phylogeographic studies of seaweeds around Antarctic and sub-Antarctic areas (Zuccarello et al. 1999, 2003, 2006, 2011; Buchanan and Zuccarello 2012; Fraser et al. 2009a, b, 2011, 2013). In the United Kingdom, Drs. Jim Provan and Christine A. Maggs at the Queen's University of Belfast have also demonstrated much endeavor in phylogeographic studies of seaweeds (Provan et al. 2005a, b, 2008, 2011, 2013). Dr. Myriam Valero at the Station Biologique de Roscoff, France and Dr. Marie-Laure Guillemin at the University of Austral de Chile, Chile have published important insights into the phylogeographies of seaweeds in the southeast Pacific (Montecinos et al. 2012; Guillemin et al. 2014; Krueger-Hadfield et al. 2011, 2013; Robuchon et al. 2014). Finally, the Phycology Research Group and Center for Molecular Phylogenetics and Evolution, Ghent University led by Dr. Olivier De Clerck have developed a broad interest in diversification and evolutionary dynamics of seaweeds through the integration of phylogenetic techniques and niche modeling into a Geographic Information System (GIS) framework (Verbruggen 2005, 2009), which may enable us to broadly understand how seaweeds shift their geographic distributions in response to global climate change. Although there are also many other excellent researchers working on seaweed phylogeography, the above examples highlight a few of the most productive groups in the field.

1.6 Genetic Markers

In the survey, we divided the genetic markers employed into three major classes: Class I, mitochondrial DNA (mtDNA) and chloroplast DNA (cpDNA); Class II, nuclear ribosomal DNA (nrDNA) sequences, RAPD, AFLP, SSCP and single nucleotide polymorphism (SNP); Class III, microsatellites. Phylogeographic information and population genetic diversity derived from Class I molecular markers represented 82 studies (51 %), whereas Class II and III molecular markers were used

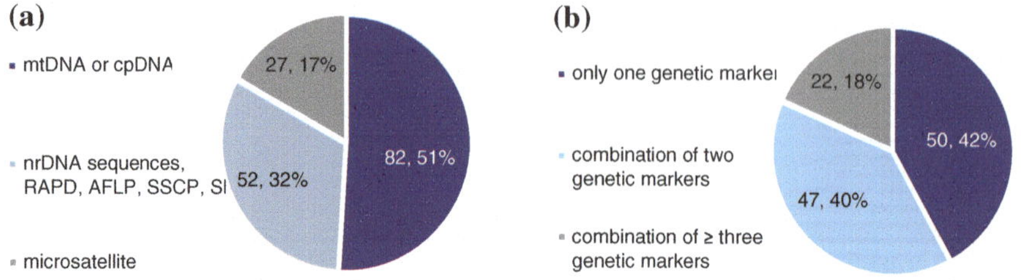

Fig. 1.7 Proportion of articles in seaweed phylogeography published between 1994 and 2014 sorted by classes of genetic marker(s) (**a**) or marker combinations (**b**)

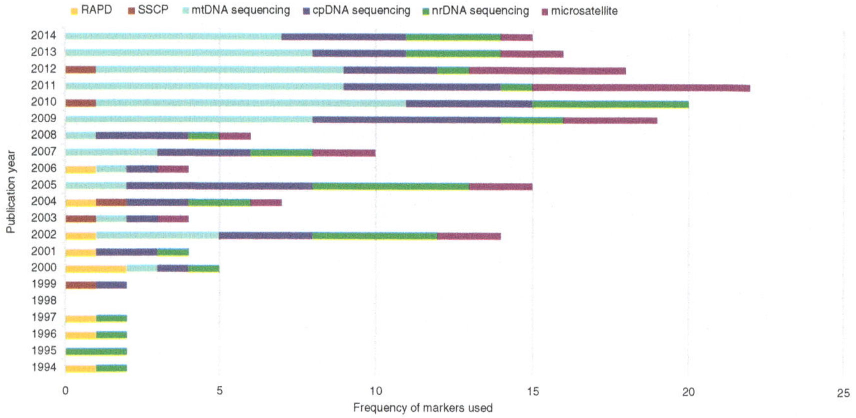

Fig. 1.8 Distribution of genetic marker(s) used for data collection in seaweed phylogeography published between 1994 and 2014

in 52 (32 %) and 27 studies (17 %), respectively (Fig. 1.7a[1]). Among the 126 papers surveyed, 50 papers (42 %) used only one genetic marker, and 47 papers used two genetic markers, whereas 22 papers used ≥3 genetic markers to investigate population-level phylogeographic patterns and genetic differentiation (Fig. 1.7b).

Several interesting trends can be identified when each of the six representative genetic markers (RAPD, SSCP, mtDNA, cpDNA, nrDNA, and microsatellite) are sorted by year (Fig. 1.8). RAPD and nrDNA sequencing (e.g., internal transcribed spacer) were the predominant markers used for seaweeds before 2000. Afterward, nrDNA sequencing showed an intermittently increasing frequency until 2010, whereas the frequency of RAPD decreased significantly and was discontinued after 2006. SSCP was occasionally used for seaweeds from 1999 but gradually lost its advantages when high-throughput sequencing techniques emerged. On the other

[1]The paper which employed more than one genetic markers will be counted for multiple times.

hand, the uniparentally inherited mtDNA and cpDNA markers were used about ≤5 times during the 2000s but have had a continuously increasing popularity in the last 6 years. Microsatellite technology, which was first used for seaweed phylogeography in 2002 as complementary evidence to sequencing data sets, has also shown a trend of increasing frequency since 2007.

1.7 Concluding Remarks and Perspectives

Phylogeographic studies on seaweeds have experienced a noticeable increase over the past two decades, although seaweeds still account for only a small proportion of several thousands of phylogeographic studies. Importantly, these empirical studies have not only expanded our knowledge of how seaweeds respond to severe climate change and environmental shifts via derived adaptive biological characteristics, but have also made valuable contributions to our knowledge of basic biology, intertidal ecology, conservation genetics and the adaptive evolution of seaweeds. Nevertheless, seaweed phylogeography is still confronted with several key challenges.

There is an evident bias of taxonomic coverage in terms of research objectives as brown and red seaweeds were relatively well represented in the phylogeographic literature (Fig. 1.4a). Publications on green seaweeds represented only 12 % of all publications, despite their important biogeographic and evolutionary roles. In fact, green seaweeds (e.g., *Ulva*) occur globally from the cold temperate to tropical regions and the poles, and under strong salinity gradients. These taxa are therefore ideal models for studying environmentally induced adaptive evolution and the interactions between the complex life-histories of seaweeds and their geographic distributions at both vertical and horizontal scales. In addition, there is a wealth of phylogeographic studies on seaweeds from the Northern Hemisphere oceans, but relatively few from the Southern Hemisphere (Fig. 1.5). The disproportional number of surveys of seaweed phylogeography between the Northern and Southern hemispheres is due in part to the vast majority of scientists who live in the Northern Hemisphere, and there are more research institutions there. Empirically, the competitive scenarios of glacial survival versus postglacial recolonization in the Northern Hemisphere appear to receive much more scientists' attention. Coastal areas of North Atlantic and North Pacific were covered by ice sheets during Quaternary glaciations and were recolonized after the ice retreated. Seaweed species in previously glaciated areas show similar phylogeographic patterns (Hewitt 2000; Beheregaray 2008), which are informative about the influence of paleoclimate change on population genetic structure, demographic history and range shifts of seaweeds. However, explanations of the generalized patterns observed in seaweeds in the Northern Hemisphere may not be applicable to the Southern Hemisphere populations or extended to other parts of the world because considerable differences in geomorphologic, environmental, and paleoclimatic history exist among these areas. Currently, phylogeographic knowledge of seaweeds is either inadequate or simply nonexisting for seaweeds inhabiting some key regions on earth, such as southern Africa, the Mediterranean, the Antarctic and sub-Antarctic, Australia, Indo-Pacific (the Coral Triangle) and the Arctic and sub-Arctic. Some of these regions harbor high levels of seaweed endemism and species richness (Lindstrom 2001, 2009; Kerswell 2006; Adey et al. 2008; Coll et al. 2010; Griffiths et al. 2010), providing excellent opportunities to investigate the mechanisms and processes contributing to diversification and evolution of seaweeds at a global scale.

Seaweed phylogeography studies have predominantly been made at macrogeographic scales (>100 km) to test for the influence of long-term environmental factors on population genetic differentiation and demographic modes from the perspective of historical biogeography. Some studies were elaborately designed at moderate geographic scales and revealed significant insights into cryptic genetic diversity, genetic structure, and the influence of ecological variables on morphological variations of seaweeds (Coyer et al. 1997; Bergström et al. 2005; Tatarenkov et al. 2007; Fraser et al. 2009b). However, the knowledge of whether mating systems and microhabitats contribute to genetic differentiation and diversity of coastal seaweeds is still limited. This is surprising since the various physical conditions (e.g., turbidity, wave exposure gradients, and tidal excursion distances) in the intertidal zone make it an ideal ecosystem to explore the genetically and evolutionarily interactive patterns and processes of seaweeds (Hu 2013). Recent surveys have highlighted the significant role of sexual reproduction, inbreeding, and tidal height in substructuring population genetic differentiation in the red seaweed *Chondrus crispus* (Krueger-Hadfield et al. 2011, 2013), opening up an exciting avenue to investigate the relative roles of life-history and microhabitats in shaping phylogeographic connectivity and adaptive evolution of seaweeds.

The integrative comparison of phylogeographic data from multiple co-occurring taxonomic groups can help to discover prevalent patterns and previously unrecognized cryptic biogeographic imprints, enhancing our understanding of why co-distributed organisms have different levels of biodiversity and distributional ranges (Arbogast and Kenagy 2001). Maggs et al. (2008) reanalyzed genetic data for eight benthic marine taxa, including two seaweeds species *Palmaria palmata* and *F. serratus*, identified nine potential marine glacial refugia and refined concordant, nonconcordant and indeterminate biogeographic patterns in North Atlantic. Thereafter, many phylogeographic studies were performed for seaweeds from North Atlantic and North Pacific. We can expect that comparative analysis based on these patterns may help to understand some cryptic and basic information on general evolutionary histories for seaweed phylogeography.

Model-based phylogeograhic inferences, which estimate parameters under different assumed models using likelihood, Bayesian or approximate Bayesian computation (ABC) approaches, have been greatly strengthened as the computational algorithms progressed. They have shown excellent promises to examine the relative support for various hypotheses about demographic histories of marine organisms when a wide range of plausible models are included (Hickerson and Meyer 2008; Ilves et al. 2010). These phylogeographic inferences may enable us to test the proposed hypotheses inferred from molecular data on seaweeds. Finally, integrating phylogeography with species distribution models (SDM) or ecological niche models (ENM) can help us to explicitly elucidate how the species distribution range is affected by climate change, vicariance or dispersal, genetic introgression, and natural selection (Kozak et al. 2008). Based on the species' distribution range shifts between ancestral populations (obtained from paleo-ENM) and current populations, the historical and environmental parameters associated with population demography can be inferred (Richards et al. 2007). A few recent studies have been conducted in this area for intertidal seaweeds (Verbruggen et al. 2009; Neiva et al. 2014) and have provided important phylogeographic insights into range shifts of seaweeds with lower dispersal capability in response to climate change since the LGM. There is still an urgent need to asses, through integrative ecological and evolutionary approaches, the unknown aspects and specific details concerning adaptation and distribution of seaweeds.

Acknowledgements We would like to thank Ceridwen Fraser and an anonymous reviewer for valuable comments on this chapter. This study was partially supported by National Natural Science Foundation of China (31370264) granted to Z.M. Hu.

References

Adey WH, Hayek LA. Elucidating marine biogeography with macrophytes: quantitative analysis of the North Atlantic supports the thermogeographic model and demonstrates a distinct Sub-Arctic region in the northwestern Atlantic. Northeast Nat. 2011;18:1–125.

Adey WH, Steneck RS. Thermogeography over time creates biogeographic regions: a temperature/space/time-integrated model and an abundance-weighted test for benthic marine algal. J Phycol. 2001;37:677–98.

Adey WH, Lindstrom SC, Hommersand MH, Müller KM. The biogeographic origin of Arctic endemic seaweeds: a thermogeographic view. J Phycol. 2008;44:1384–94.

Andreakis N, Procaccini G, Maggs CA, Kooisra WHCF. Phylogeography of the invasive seaweed *Asparagopsis* (Bonnemaisoniales, Rhodophyta) reveals cryptic diversity. Mol Ecol. 2007;16:2285–99.

Andreakis N, Kooistra WHCF, Procaccini G. High genetic diversity and connectivity in the polyploid invasive seaweed *Asparagopsis taxiformis* (Bonnemaisoniales) in the Mediterranean, explored with microsatellite alleles and multilocus genotypes. Mol Ecol. 2009;18:212–26.

Arbogast BS, Kenagy GJ. Comparative phylogeography as an integrative approach to historical biogeography. J Biogeogr. 2001;28(7):819–25.

Avise JC. Phylogeography: the history and formation of species. Cambridge: Harvard University Press; 2000.

Avise JC. Phylogeography: retrospect and prospect. J Biogeogr. 2009;36:3–15.

Avise JC, Arnold J, Ball RM, Bermingham E, Lamb T, Neigel JE, Reeb CA, Saunders NC. Intraspecific phylogeography: the mitochondrial DNA bridge between population genetics and systematics. Ann Rev Ecol Syst. 1987;18:489–522.

Bae DY, Ang PO Jr, Boo SM. Mitochondrial *cox*3 and *trn*W-I sequence diversity of *Sagassum mticum*. Aquat Bot. 2013;104:220–3.

Bakker FT, Olsen JL, Stam WT. Global phylogeography in the cosmopolitan species *Cladophora vagabunda* (Chlorophyta) based on nuclear rDNA internal transcribed spacer sequences. Eur J Phycol. 1995;30:197–208.

Beheregaray LB. Twenty years of phylogeography: the state of the field and the challenges for the Southern Hemisphere. Mol Ecol. 2008;17:3754–74.

Bergström L, Tatarenkov A, Johannesson K, Jönsson RB, Kautsky L. Genetic and morphological identification of *Fucus radicans* sp. nov. (Fucales, Phaeophyceae) in the brackish Baltic Sea. J Phycol. 2005;41(5):1025–38.

Braga JC, Martin JM, Riding R. Internal structure of segment reefs: *Halimeda* algal mounds in the Mediterranean Miocene. Geology. 1996;24:35–8.

Brawley SH, Coyer JA, Blakeslee AMH, Hoarau G, Johnson L, Byers JE, Stam WT, Olsen JL. Historical invasions of the intertidal zone of Atlantic North America associated with distinctive patterns of trade and emigration. Proc Nat Acad Sci USA. 2009;106(20):8239–44.

Buchanan J, Zuccarello GC. Decoupling of short- and long-distance dispersal pathways in the endemic New Zealand seaweed *Carpophyllum maschalocarpum* (Phaeophyceae, Fucales). J Phycol. 2012;48:518–29.

Chan SW, Cheang CC, Chirapart A, Gerung G, Tharith C, Ang PO Jr. Homogeneous population of the brown alga *Sargassum polycystum* in Southeast Asia: possible role of recent expansion and asexual propagation. PLoS ONE. 2013;8(10):e77662.

Chan SW, Cheang CC, Yeung CW, Chirapat A, Gerung G, Ang PO Jr. Recent expansion led to the lack of genetic structure of *Sargassum aquifolium* populations in Southeast Asia. Mar Biol. 2014;161:785–95.

Cheang CC, Chu KH, Ang PO Jr. Morphological and genetic variation in the populations of *Sargassum hemiphyllum* (Phaeophyceae) in the northwestern Pacific. J Phycol. 2008;44:855–65.

Cheang CC, Chu KH, Ang PO Jr. Phylogeography of the marine marcoalga *Sargassum hemiphyullum* (Phaeophyceae, Heterokontophyta) in the Northwestern Pacific. Mol Ecol. 2010a;19:2933–48.

Cheang CC, Chu KH, Fujita D, Yoshida G, Hiraoka M, Critchley AT, Choi HG, Duan DL, Serisawa Y, Ang PO Jr. Low genetic variability of *Sargassum muticum* (Phaeophyceae) revealed by a global analysis of native and introduced populations. J Phycol. 2010b;46: 1063–74.

Coll M, Piroddi C, Steenbeek J, Kaschner K, Lasram FBR, Aguzzi J, Ballesteros E, Bianchi CN, Corbera J, Dailianis T, Danovaro R, Estrada M, Froglia C, Galil BS, Gasol JM, Gertwagen R, Gil J, Guihaumon F, Kesner-Reyes K, Kitsos M-S, Koukouras A, Lampadariou N, Laxamana E, Lopez-Fe CM, Lotze HK, Martin D, Mouillot D, Oro D, Raicevich S, Rius-Barile J, Saiz-Salinas JL, Vicente CS, Somot S, Templado J, Turon X, Vafidis D, Villanueva R, Voultsiadou E. The biodiversity of the Mediterranean Sea: estimates, patterns, and threats. PLoS ONE. 2010;5(8):e11842.

Couceiro L, Maneiro I, Ruiz JM, Barreiro R. Multiscale genetic structure of an endangered seaweed *Ahnfeltiopsis pusilla* (Rhodophyta): implications for its conservation. J Phycol. 2011;47(2):259–68.

Coyer JA, Olsen JL, Stam WT. Genetic variability and spatial separation of the sea palm kelp, *Postelsia palmaeformis* (Phaeophyceae) as assessed with M13 fingerprints and RAPDs. J Phycol. 1997;33:561–8.

Coyer JA, Peters AF, Hoarau G, Stam WT, Olsen JL. Hybridization of the marine seaweeds, *Fucus serratus* and *Fucus evanescens* (Heterokontophyta: Phaeophyceae) in a 100-year-old zone of secondary contact. Proc R Soc B. 2002;269:1829–34.

Coyer JA, Peters AF, Stam WT, Olsen JL. Post-ice age recolonization and differentiation of *Fucus serratus* L. (Phaeophyceae; Fucaceae) populations in Northern Europe. Mol Ecol. 2003;12:1817–29.

Coyer JA, Hoarau G, Costa JF, Hogerdijk B, Serrão EA, Billard E, Valero M, Pearson GA, Olsen JL. Evolution and diversification within the intertidal brown macroalgae *Fucus spiralis*/*F. vesiculosus* species complex in the North Atlantic. Mol Phylogenet Evol. 2011a;58:283–96.

Coyer JA, Hoarau G, Pearson G, Mota C, Jüterbock A, Alpermann T, John U, Olsen JL. Genomic scans detect signatures of selection along a salinity gradient in populations of the intertidal seaweed *Fucus serratus* on a 12 km scale. Mar Genom. 2011b;4:41–9.

Coyer JA, Hoarau G, Skage M, Stam WT, Olsen JL. Origin of *Fucus serratus* (Heterokontophyta; Fucaceae) populations in Iceland and the Faroes: a microsatellite-based assessment. Eur J Phycol. 2006;41:235–46.

Faugeron S, Valero M, Destombe C, Martínez EA, Correa JA. Hierarchical spatial structure and discriminant analysis of genetic diversity in the red alga *Mazzaella laminarioides* (Gigartinales, Rhodophyta). J Phycol. 2001;37:705–16.

Fraser CI, Nikula R, Spencer HG, Waters JM. Kelp genes reveal effects of subantarctic sea ice during the last glacial maximum. Proc Nat Acad Sci USA. 2009a;106(9):3249–53.

Fraser CI, Hay CH, Spencer GH, Waters JM. Genetic and morphological analyses of the southern bull kelp *Durvillaea antarctic* (Phaeophyceae: Durvillaeales) in New Zealand reveal cryptic species. J Phycol. 2009b;45:436–43.

Fraser CI, Nikula R, Waters JM. Oceanic rafting by a coastal community. Proc R Soc B. 2011;278:649–55.

Fraser CI, Zuccarello GC, Spencer HG, Salvatore LC, Garcia GR, Waters JM. Genetic affinities between trans-oceanic populations of non-buoyant macroalgae in the high latitudes of the southern hemisphere. PLoS ONE. 2013;8(7):e69138.

Freile D, Milliman JD, Hillis L. Leeward bank margin *Halimeda* meadows and draperies and their sedimentary importance on the western Great Bahama bank slope. Coral Reefs. 1995;14:27–33.

Griffiths CL, Robinson TB, Lange L, Mead A. Marine biodiversity in South Africa: an evaluation of current states of knowledge. PLoS ONE. 2010;5(8):e12008.

Guillemin M, Valero M, Faugeron S, Nelson W, Destombe C. Tracing the trans-Pacific evolutionary history of a domesticated seaweed (Gracilaria chilensis) with Archaeological and genetic data. PLoS ONE. 2014;9(12):e114039.

Hewitt GM. The genetic legacy of the Quaternary ice ages. Nature. 2000;40:907–13.

Hickerson MJ, Meyer C. Testing comparative phylogeographic models of marine vicariance and dispersal using a hierarchical Bayesian approach. BMC Evol Biol. 2008;8:322.

Hoarau G, Coyer JA, Veldsink JH, Stam WT, Olsen JL. Glacial refugia and recolonization pathways in the brown seaweed *Fucus serratus*. Mol Ecol. 2007;16:3606–16.

Hu ZM. Intertidal population genetic dynamics at a microgeographic seascape scale. Mol Ecol. 2013;22:3191–4.

Hu ZM, Critchley AT, Gao TX, Zeng XQ, Morrell SL, Duan DL. Delineation of *Chondrus* (Gigartinales, Florideophyceae) in China and the origin of *C. crispus* inferred from molecular data. Mar Biol Res. 2007a;3:145–54.

Hu ZM, Zeng XQ, Critchley AT, Morrell SL, Duan DL. Phylogeography of the Northern Atlantic species *Chondrus crispus* (Gigartinales, Rhodophyta) inferred from nuclear rDNA internal transcribed spacer sequences. Hydrobiologia. 2007b;575(1):315–27.

Hu ZM, Guiry MD, Critchley AT, Duan DL. Phylogeographic patterns indicate trans-Atlantic migration from Europe to North America in the red seaweed *Chondrus crispus* (Gigartinales, Rhodophyta). J Phycol. 2010;46:889–900.

Hu ZM, Uwai S, Yu SH, Komatsu T, Ajisaka T, Duan DL. Phylogeographic heterogeneity of the brown macroalga *Sargassum horneri* (Fucaceae) in the northwestern Pacific in relation to late Pleistocene glaciation and tectonic configurations. Mol Ecol. 2011a;20:3894–909.

Hu ZM, Li W, Li JJ, Duan DL. Post-Pleistocene demographic history of the North Atlantic endemic Irish moss *Chondrus crispus*: glacial survival, spatial expansion and gene flow. J Evol Biol. 2011b;24:505–17.

Hu ZM, Zhang J, Lopez-Bautista J, Duan DL. Asymmetric genetic exchange in the brown seaweed *Sargassum fusiforme* (Phaeophyceae) driven by oceanic currents. Mar Biol. 2013;160(6):1407–14.

Ilves KL, Huang W, Wares JP, Hickerson MJ. Colonization and/or mitochondrial selective sweeps across the North Atlantic intertidal assemblage revealed by multi-taxa approximate Bayesian computation. Mol Ecol. 2010;19:4505–19.

Kerswell AP. Global biodiversity patterns of benthic marine algae. Ecology. 2006;87:2479–88.

Kidd KM, Ritchie MG. Phylogeographic information systems: putting the geography into phylogeography. J Biogeogr. 2006;33(11):1851–65.

Kim SY, Weinberer F, Boo SM. Genetic data hint at a common donor region for invasive Atlantic and Pacific populations of *Gracilaria vermiculophylla* (Gracilariales, Rhodophyta). J Phycol. 2010;46:1346–349.

Kim KM, Hoarau G, Boo SM. Genetic structure and distribution of *Gelidium elegans* (Gelidiales, Rhodophyta) in Korea based on mitochondrial *cox*1 sequence data. Aquat Bot. 2012;98:27–33.

Kim SY, Manghisi A, Morabito M, Yang EC, Yoon HS, Miller KA, Boo SM. Genetic diversity and haplotype distribution of *Pachymeniopsis gargiuli* sp. Nov. and *P. lanceolata* (Halymeniales, Rhodophyta) in Korea, with notes on their non-native distributions. J Phycol. 2014;50:885–96.

Knowles LL, Maddison WP. Statistical phylogeography. Mol Ecol. 2002;11:2623–35.

Komatsu T, Matsunaga D, Mikami A, Sagawa T, Boisnier E, Tatsukawa K, Aoki M, Ajisaka T, Uwai S, Tanaka K, Ishida K, Tanoue H, Sugimoto T. Abundance of drifting seaweeds in eastern East China Sea. J Appl Phycol. 2008;20:801–9.

Kostamo K, Korpelainen H, Olsson S. Comparative study on the population genetics of the red algae *Furcellaria lumbricalis* occupying different salinity conditions. J Appl Phycol. 2012;159:561–71.

Kooistra WHCF, Coppejans EGG, Payri C. Molecular systematics, historical ecology, and phylogeography of *Halimeda* (Bryopsidales). Mol Phylogenet Evol. 2002;24:121–38.

Kozak KH, Graham CH, Wiens JJ. Integrating GIS-based environmental data into evolutionary biology. Trends Ecol Evol. 2008;23:141–8.

Krueger-Hadfield SA, Collén J, Daguin-Thiebaut C, Valero M. Genetic population structure and mating system in *Chondrus crispus* (Rhodophyta). J Phycol. 2011;47:440–50.

Krueger-Hadfield SA, Roze D, Mauger S, Valero M. Intergametophytic selfing and microgeographic genetic structure shape populations of the intertidal red seaweed *Chondrus crispus*. Mol Ecol. 2013;22:3242–60.

Lee KM, Boo SM, Kain (Jones) JM, Sherwood AR. Cryptic diversity and biogeography of the widespread brown alga *Colpomenia sinuosa* (Ectocarpales, Phaeophyceae). Bot Mar. 2013;56:15–25.

Leskinen E, Alström-Rapaport C, Pamilo P. Phylogeographical structure, distribution and genetic variation of the green algae *Ulva intestinalis* and *U. compressa* (Chlorophyta) in the Baltic Sea area. Mol Ecol. 2004;13:2257–65.

Li JJ, Hu ZM, Duan DL. Genetic data from the red alga *Palmaria palmata* reveal a mid-Pleistocene deep genetic split in the North Atlantic. J Biogeogr. 2015;42(5):902–13.

Li JJ, Hu ZM, Duan DL. Survival in glacial refugia vs. postglacial dispersal in the North Atlantic: the cases of red seaweeds. In: Hu ZM, Fraser CI editors. Seaweed phylogeography: adaptation and evolution of seaweeds under environmental change. Berlin, Heidelberg: Springer; 2016.

Lindstrom SC. The Bering Strait connection: dispersal and speciation in boreal macroalgae. J Biogeogr. 2001;28:243–51.

Lindstrom SC. The biogeography of seaweeds in Southeast Alaska. J Biogeogr. 2009;36:401–9.

Lindstrom SC, Olsen JL, Stam WT. Recent radiation of the Palmariaceae (Rhodophyta). J Phycol. 1996;32:457–68.

Maggs CA, Castilho R, Foltz D, Henzler C, Jolly MT, Kelly J, Olsen JL, Perez KE, Stam WT, Väinölä R, Viard F, Wares J. Evaluating signatures of glacial refugia for North Atlantic benthic marine taxa. Ecology. 2008;89:S108–22.

McIvor L, Maggs CA, Provan J, Stanhope MJ. *rbc*L sequences reveal multiple cryptic introductions of the Japanese red alga *Polysiphonia harveyi*. Mol Ecol. 2001;10:911–9.

Miller KA, Olsen JL, Stam WT. Genetic divergence correlates with morphological and ecological subdivision in the deep-water elk kelp *Pelagophycus porra* (Laminariales, Phaeophyceae). J Phycol. 2000;36:862–70.

Moalic Y, Arnaud-Haond S, Perrin C, Pearson GA, Serrão EA. Travelling in time with networks: revealing present day hybridization versus ancestral polymorphism between two species of brown algae, *Fucus vesiculosus* and *F. spiralis*. BMC Evol Biol. 2011;11:33.

Montecinos A, Broitman BR, Faugeron S, Haye PA, Tellier F, Guillemin M. Species replacement along a linear coastal habitat: phylogeography and speciation in the red alga *Mazzaella laminarioides* along the southeast Pacific. BMC Evol Biol. 2012;12:97.

Muangmai N, Fraser CI, Zuccarello GC. Contrasting patterns of population structure and demographic history in cryptic species of (Rhodomelaceae, Rhodophyta) from New Zealand. J Phcyol. 2015;51:574–85.

Muhlin JF, Brawley SH. Recent versus relic: discerning the genetic signature of *Fucus vesiculosus* (Heterokontophyta; Phaeophyceae) in the Northwestern Atlantic. J Phycol. 2009;45:828–37.

Muhlin JF, Engel CR, Stessel R, Weatherbee RA, Brawley SH. The influence of coastal topography, circulation patterns, and rafting in structuring populations of an intertidal alga. Mol Ecol. 2008;17:1198–210.

Neiva J, Pearson GA, Valero M, Serrão EA. Surfing the wave on a borrowed board: range expansion and spread of introgressed organellar genomes in the seaweed *Fucus ceranoides* L. Mol Ecol. 2010;19:4812–22.

Neiva J, Pearson GA, Valero M, Serrão EA. Drifting fronds and drifting alleles: range dynamics, local dispersal and habitat isolation shape the population structure of the estuarine seaweed *Fucus ceranoides*. J Biogeogr. 2012a;39:1167–78.

Neiva J, Pearson GA, Valero M, Serrão EA. Fine-scale genetic breaks driven by historical range dynamics and ongoing density-barrier effects in the estuarine seaweed *Fucus ceranoides* L. BMC Evol Biol. 2012b;12:78.

Neiva J, Assis J, Fernandes F, Pearson GA, Serrão EA. Species distribution models and mitochondrial DNA phylogeography suggest an extensive biogeographical shift in the high-intertidal seaweed *Pelvetia canaliculata*. J Biogeogr. 2014;41:1137–48.

Neiva J, Serrão EA, Assis J, Pearson GA, Coyer JA, Olsen JL, Hoarau G, Valero M. Climate oscillations, range shifts and phylogeographic patterns of North Atlantic Fucaceae. In: Hu ZM, Fraser CI editors. Seaweed phylogeography: adaptation and evolution of seaweeds under environmental change. Berlin, Heidelberg: Springer; 2016.

Nicastro KR, Zardi GI, Teixeira S, Neiva J, Serrão EA, Pearson GA. Shift happens: trailing edge contraction associated with recent warming trends threatens a distinct genetic lineage in the marine macroalga *Fucus vesiculosus*. BMC Biol. 2013;11:6.

Ogawa T, Ohki K, Kamiya M. High heterozygosity and phenotypic variation of zoids in apomictic *Ulva prolifera* (Ulvophyceae) from brackish environments. Aquat Bot. 2015;120:185–92.

Olsen JL, Zechman FW, Hoarau G, Coyer JA, Stam WT, Valero M, Nilsson P, Åberg P. The phylogeographic architecture of the fucoid seaweed *Ascophyllum nodosum*: an intertidal "marine tree" and survivor of more than one glacial-interglacial cycle. J Biogeogr. 2010;37:842–56.

Pakker H, Klerk H, van Campen JH, Olsen JL, Breeman AM. Evolutionary and ecological differentiation in the pantropical to warm-temperate seaweed *Digenea simplex* (Rhodophyta). J Phycol. 1996;32:250–7.

Peters AF, van Oppen MJH, Wiencke C, Stam WT, Olsen JL. Phylogeny and historical ecology of the Desmarestiaceae (Phaeophyceae) support a southern hemisphere origin. J Phycol. 1997;33:294–309.

Provan J, Murphy S, Maggs CA. Tracking the invasive history of the green alga *Codium fragile* ssp. *tomentosoides*. Mol Ecol. 2005a;14:189–94.

Provan J, Wattier RA, Maggs CA. Phylogeographic analysis of the red seaweed *Palmaria palmata* reveals a Pleistocene marine glacial refugium in the English Channel. Mol Ecol. 2005b;14:793–803.

Provan J, Maggs CA. Unique genetic variation at a species' rear edge is under threat from global climate change. Proc R Soc B. 2011;279:39–47.

Provan J, Booth D, Todd NP, Beatty GE, Maggs CA. Tracking biological invasions in space and time: elucidating the invasive history of the green alga *Codium fragile* using old DNA. Divers Distrib. 2008;14:343–54.

Provan J, Glendinning K, Kelly R, Maggs CA. Levels and patterns of population genetic diversity in the red seaweed *Chondrus crispus* (Florideophyceae): a direct comparison of single nucleotide polymorphisms and microsatellites. Biol J Linn Soc. 2013;108:251–62.

Richards CL, Carstens BC, Knowles LL. Distribution modeling and statistical phylogeography: an integrative framework for generating and testing alternative biogeographic hypotheses. J Biogeogr. 2007;34:1833–45.

Reuter M, Piller WE, Richoz S. The dispersal of *Halimeda* in northern hemisphere mid-latitudes: paleobiogeographical insights. Perspec Plant Ecol Evol Syst. 2012;14:303–9.

Robuchon M, Gall LL, Mauger S, Valero M. Contrasting genetic diversity patterns in two sister kelp species co-distributed along the coast of Brittany, France. Mol Ecol. 2014;23:2669–85.

Shimada S, Yokoyama N, Arai S, Hiraoka M. Phylogeography of the genus *Ulva* (Ulvophyceae, Chlorophyta), with special reference to the Japanese freshwater and brackish taxa. J Appl Phycol. 2008;20:979–89.

Sherwood AR. Phylogeography of *Asparagopsis taxiformis* (Bonnemaisoniales, Rhodophyta) in the Hawaiian Islands: two mtDNA markers support three separate introductions. Phycologia. 2008;47(1):79–88.

Tatarenkov A, Jönsson RB, Kautsky L, Johannesson K. Genetic structure in populations of *Fucus vesiculosus* (Phaeophyceae) over spatial scales from 10 m to 800 km. J Phycol. 2007;43 (4):675–85.

Thiel M, Haye PA. The ecology of rafting in the marine environment. III. Biogeographical and evolutionary consequences. Oceanogr Mar Biol Ann Rev. 2006;44:323–429.

Uwai S, Kogame K, Yoshida G, Kawai H, Ajisaka T. Geographical genetic structure and phylogeography of the *Sargassum horneri/filicinum* complex in Japan, based on the mitochondrial *cox3* haplotype. Mar Biol. 2009;156:901–11.

van Oppen MJH, Diekmann OE, Wiencke C, Stam WT, Olsen JL. Tracking dispersal routes: phylogeography of the Arctic-Antarctic disjunct seaweed *Acrosiphonia arcta* (Chlorophyta). J Phycol. 1994;30(4):67–80.

van Oppen MJH, Draisma SGA, Olsen JL, Stam WT. Multiple trans-Arctic passages in the red alga *Phycodrys rubens*: evidence from nuclear rDNA ITS sequences. Mar Biol. 1995a;123:179–88.

van Oppen MJH, Olsen JL, Stam WT. Genetic variation within and among North Atlantic and Baltic populations of the benthic alga *Phycodrys rubens* (Rhodophyta). Eur J Phycol. 1995b;30:251–60.

Verbruggen H, De Clerck O, Kooistra WHCF, Coppejans E. Molecular and morphometric data pinpoint species boundaries in *Halimeda* section *Rhipsalis* (Bryopsidales, Chlorophyta). J Phycol. 2005;41:606–21.

Verbruggen H, Tyberghein L, Pauly K, Vlaeminck C, Nieuwenhuyze KV, Kooistra WHCF, Leliaert F, De Clerck O. Macroecology meets macroevolution: evolutionary niche dynamics in the seaweed *Halimeda*. Global Ecol Biogeogr. 2009;18:393–405.

Wang XL, Zhao FJ, Hu ZM, Critchley AT, Morrell SL, Duan DL. Inter-simple sequence repeat (ISSR) analysis of genetic variation of *Chondrus crispus* populations from North Atlantic. Aquat Bot. 2008;88:154–9.

Yang EC, Cho GY, Kogame K, Carlile AL, Boo SM. RuBisCo cistron sequence variation and phylogeography of *Ceramium kondoi* (Ceramiaceae, Rhodophyta). Bot Mar 2008;51:370–77.

Yang EC, Lee SY, Lee WJ, Boo SM. Molecular evidence for recolonization of *Ceramium japonicum* (Ceramiaceae, Rhodophyta) on the west coast of Korea after the last glacial maximum. Bot Mar 2009;52:307–15.

Zuccarello GC, West JA. Multiple cryptic species: molecular diversity and reproductive isolation in the *Bostrychia radicans/B. moritziana* complex (Rhodomelaceae, Rhodophyta) with focus on North American isolates. J Phycol. 2003;39:948–59.

Zuccarello GC, West JA, Kamiya M, King RJ. A rapid method to score plastid haplotypes in red seaweeds and its use in determining parental inheritance of plastids in the red alga *Bostrychia* (Ceramiales). Hydrobiologia. 1999;401:207–14.

Zuccarello GC, Buchanan J, West JA. Increased sample for inferring phylogeographic patterns in *Bostrychia radicans/B. moritziana* (Rhodomelaceae, Rhodophyta) in the Eastern USA. J Phycol. 2006;42:1349–52.

Zuccarello GC, Buchanan J, West JA, Pedroche FF. Genetic diversity of the mangrove-associated alga *Bostychia radicans/B. moritziana* (Ceramiales, Rhodophyta) from southern Central America. J Appl Phycol 2011;59:98–104.

Chapter 2
Paradigm Shifts in the Phylogeographic Analysis of Seaweeds

W. Stewart Grant

Abstract The phylogeographic analysis of seaweeds faces numerous biological, methodological, and conceptual challenges. One challenge is to understand how life-history phases, mating systems, dispersal mechanisms, and physiological tolerances influence the distributions and persistence of local populations. A second challenge has been to develop genetic assay methods that allow us to trace unbroken gene lineages so they can be tested with models using coalescence theory. An integral part of this challenge has been to identify unrecombining sections of DNA with known modes of inheritance. For the most part, seaweed phylogeographic studies have used unrecombining organellar and plastid DNAs, but new methods of phasing haplotypes of nuclear DNA can provide an enormous amount of data to fine-tune hypotheses to distinguish the effects of demography and natural selection. Finally, phylogeography faces conceptual challenges, as we learn ever more about palaeoclimates, the historical shapes of shorelines, and the dynamics of ocean currents. A major advance is the marriage between demographic models and models of natural selection, because both interact to mold the shapes of gene genealogies.

Keywords Historical demography • Modes of inheritance • Modes of reproduction • Molecular markers • Natural selection • Phylogeography • Pleistocene • Population genetics

---

This chapter is dedicated to Bob Vadas (University of Maine) who, by example, taught me to ask questions.

W.S. Grant (✉)
Alaska Department of Fish and Game Commercial Fisheries Division, 333 Raspberry Road, Anchorage, AK 99518, USA
e-mail: phylogeo@gmail.com

Z.-M. Hu and C. Fraser (eds.), *Seaweed Phylogeography*,
DOI 10.1007/978-94-017-7534-2_2

2.1 Introduction

The central concept in phylogeography is the notion that the geographic distributions of gene lineages can be used to understand the 'spatial and temporal dimensions of microevolution' (Avise 2009). Monophyletic lineages are the primary units of analysis and have historically been estimated from nonrecombining, uniparentally inherited cytoplasmic DNAs, such as those in mitochondria and plastids (Avise et al. 1987). The origins of Phylogeographic patterns of genetic diversity among populations can be fully understood only by bringing together ideas from several disciplines, including not only genetics and molecular biology, but also ecology, reproduction and developmental biology, physiology, dispersal biology, paleoclimatology, geology, and oceanography (Dawson et al. 2013; Marske et al. 2013; Habel et al. 2015). This interdisciplinary approach can be seen in recent studies of seaweeds and is embodied in a long-standing research paradigm described by Huxley and Tinbergen, known as *explanatory pluralism* (Box 1) (Huxley 1916; Mayr 1961; Tinbergen 1963). Pluralistic explanations can lead to major breakthroughs in understanding the patterns of biological diversity.

Box 1. Phylogeography and Explanatory Pluralism

Most biologists specialize in a particular area of research because time and funding restrict the use of a wide range of research methods. Unfortunately, such limitations can preclude broad understanding of a biological system. Phylogeography intersects with several disciplines that help to provide explanations for the patterns of genetic diversity among populations. These explanations fall into two broad categories reflecting *proximate* and *ultimate* causes (Tinbergen 1963; Dewsbury 1999). The distinction between these categories is whether the expression of a trait results from an event during the life of an individual, and by extension within the recent history of a population, or is the legacy of an individual's, or population's, evolutionary history.

Proximate causes include *mechanistic* explanations of how a phenomenon occurs. For example, why do some populations have less genetic diversity than other populations? One explanation might be that the amount of diversity in a population is proportional to the population's effective size. Larger populations have higher levels of diversity because the effects random drift are less than in small populations. Proximate causes also include *ontological* explanations that in our example of explaining levels of genetic diversity might include a population's response to climate variability or to the recent introduction of a predator into its ecosystem.

Ultimate causes can be rooted in the *phylogenetic* origins of an individual or population trait. Again in our example, genetic diversity in a population, or species, might reflect its evolutionary history, which we can reconstruct with a large toolbox of genetic methods and concepts. Finally, we can ask whether genetic diversity has *adaptive value*. Are populations with high levels of genetic variability better able to survive short-term environmental challenges, or to adapt to long-term habitat changes?

Phylogeographic explanations typically include both *proximate* and *ultimate* factors to understand patterns in nature. Proximate mechanisms operate within the lifetime of an individual (or population), and *ultimate* mechanisms operate over long timescales. An understanding of how phylogeographic patterns arise depends on incorporating both kinds of factors from other disciplines. Proximate factors include life-cycle variation, mating strategies, environment–biological interactions, physiology, predation, and demographic processes. Ultimate factors include the phylogenetic and genetic legacies arising from a species' evolutionary history and adaptive traits that give rise to phylogeographic patterns. Importantly, phylogeographic explanations have to include information about the interaction of a species' biology with geologic, paleoclimatic, and oceanographic histories in a region.

The goals of this chapter are to explore the fundamental concepts of phylogeography and to outline biological variables in seaweeds that influence phylogeographic patterns. An overview of the literature shows that phylogeographic concepts have shifted in response to the development of new methods for surveying genetic variation and the creation of new theories and statistical models to interpret genetic patterns. These developments have led to the expansion of phylogeography beyond its original formulation. Phylogeographic studies now often include the analysis of nuclear genes, in addition to organellar genes and invoke ideas from numerous disciplines to explain phylogeographic patterns. A major future direction in phylogeography will be to weave demography and natural selection into more holistic explanations of microevolutionary processes.

2.2 Shifting Paradigms in Phylogeography

The term *phylogeography* first appeared in an article by Avise et al. (1987) to describe the association between the lineages in a gene genealogy and geography. The hallmark of phylogeographic inference has been the use of nonrecombining organellar DNA, chiefly from mitochondria and chloroplasts, to reconstruct gene genealogies. However, the development of new molecular markers has prompted an expansion of the methods and concepts used in phylogeography (Hickerson et al. 2010; Marske et al. 2013; Bowen et al. 2014). Phylogeographic studies now draw on a diversity of methods and concepts to infer contemporary and historical processes that influence population distributions and abundances (Hickerson et al. 2010; Marske et al. 2013). Population genetic models are used to interpret genetic variability in terms of equilibrium processes such as genetic drift, gene flow, and mutation (Crow and Kimura 1970). Seascape concepts borrow from work on land plants (Sork et al. 1999) to understand the effects of currents, tides, shoreline topography, and temperature on the dispersal and settlement of spores and gametes (Galindo et al. 2006; Selkoe et al. 2008; Liggins et al. 2013). Species' distribution modeling (SDM), or ecological niche modeling (ENM), (Elith and Leathwick 2009) can predict both historical (Bigg et al. 2007) and contemporary (Jueterbock et al. 2013) geographic distributions of marine organisms. These and many other concepts provide a framework to test hypotheses of historical and contemporary population structures with molecular datasets.

2.3 Methodological Considerations

A seminal challenge in phylogeography is the use of genetic profiles in contemporary populations to understand the roles of historical demography and selection on microevolutionary processes. The chief genetic approach to understand the effects of historical climate change is coalescence analysis of gene genealogies, because historical population size, extirpations and colonizations, and founder events influence the depth of coalescence to a common ancestral haplotype in a genealogy (Fu and Li 1999). The shape of a gene genealogy can provide a window into the past demography of a population (Rogers and Harpending 1992; Hey and Nielsen 2004; Drummond and Rambaut 2007; Beaumont 2010).

A vexing problem in the phylogeographic analysis of seaweeds can be their different morphological forms, depending on environmental conditions. Related species sometimes differ little in their morphologies, making individual identifications difficult. Hence, a major focus of molecular methods in the past few years has been on systematics (Kraft et al. 2010; Saunders and Kucera 2010; Yotsukura et al. 2010; Boo et al. 2011; Lindstrom et al. 2011; Sutherland et al. 2011; Marins et al. 2012; Kirkendale et al. 2013; Lin et al. 2013), and on phylogenetic relationships among higher taxa based on DNA restriction fragment length polymorphisms (RFLP) (Bhattacharya and Druehl 1990) and DNA sequence markers (Saunders et al. 2004; Clarkson and Saunders 2010; Verbruggen et al. 2010). Several researchers routinely use the mtDNA barcode gene, cytochrome oxidase I (*cox*I) (Saunders 2005; McDevit and Saunders 2010), whereas others favor the chloroplast DNA gene, ribulose-bisphosphate carboxylase (*rbc*L) (Freshwater et al. 1994) for phylogenetic and phylogeographic studies.

2.3.1 Molecular Markers

The goals of phylogeography are first to describe genetic relationships among populations, or among closely related species, scattered across a seascape, then to infer the ultimate and proximal processes responsible for the observed genetic patterns. Phylogeographic studies require a careful choice of molecular markers to be able to detect a particular effect size (magnitude of divergence) and appropriate temporal scale of events. Most importantly, a molecular marker must have a predictable mode of inheritance, which can be tested with breeding studies, with pedigree analysis, or with statistical analyses of population data. A molecular marker must be appropriately polymorphic to test a particular hypothesis and amenable to the analysis of large numbers of individuals to be able to provide enough statistical power to detect population structure (Ryman et al. 2006).

Advances in sequencing technology facilitate the use of large numbers of molecular markers used in phylogeographic studies. New approaches for developing molecular markers with restriction site-associated DNA (RAD) and next-generation sequencing (NGS) methods can now identify thousands to hundreds of thousands of DNA polymorphisms in populations of nonmodel species (Metzker 2010; McCormack et al. 2013; Toonen et al. 2013). Genotyping by sequencing small DNA fragments (50–500 bp), 'amplicon sequencing', promises to yield large numbers of genotypes at a much lower cost than current methods of SNP genotyping (Campbell et al. 2015). However, the massive amounts of genotypic data these methods produce require attention to genotype calling, quality control, and statistical analysis (Davey et al. 2011; Nielsen et al. 2011; Patel and Jain 2012). These technologies survey polymorphisms throughout the genome and open the door to investigate proximal and ultimate mechanisms operating on different timescales and to understand the effects of natural selection.

2.3.2 Levels of Polymorphisms Influence Hypothesis Testing

Genetic diversity ultimately arises through mutation, but the level of diversity in a population is influenced by the confounding effects of demographic history and natural selection. The mutation rate of a molecular marker and its level of polymorphism greatly influence the kinds of hypotheses that can be tested. The silent substitution rate of mtDNA is less than one-third of that of cpDNA in vascular land plants (Wolfe et al. 1987). The substitution rate in coding regions of nDNA is about twice that of cpDNA. However, nonsynonymous rates for mtDNA and cpDNA are similar. Sequences of DNA with higher levels of polymorphism are generally assumed to have larger mutation rates than less polymorphic sequences, and this assumption has been used, for example, to estimate mutation rates for one gene from another. Markers are most useful for phylogeographic analysis when their genealogies show a pattern of shared and unique lineages among populations that captures historical and contemporary events.

DNA sequences with low mutation rates might resolve relationships among species or higher taxa, but may not be useful for detecting fine-scale population structure, or for identifying individuals, to track the genetic effects of mating systems. On the other hand, molecular markers with high mutation rates may not resolve deep structure in a species or relationships among species, because of the loss of information through multiple substitutions (rapidly evolving sequences) or allelic convergence (microsatellites). Nucleotide saturation in a gene leads to underestimates of divergence between lineages in a gene genealogy, and the use of sequences with low levels of polymorphism may not resolve important features of recent phylogeographic structure. For example, a 629 bp portion of the mtDNA COI gene with 14 haplotypes provided greater resolution of population structure and colonization history of the kelp *Durvillaea antarctica* along the coast of Chile, than did a 886 bp portion of the cDNA *rbc*L gene that yielded only two haplotypes (Fraser et al. 2010).

Allelic convergence from the stepwise nature of mutation in microsatellites is especially problematic for comparisons between well-differentiated species, because the same microsatellite allelic state (size) can arise from alleles with different ancestries (Hedrick 1999; Rubinsztein et al. 1999; Ellegren 2004). Hence, the use of microsatellites for phylogenetic inference can be open to serious error. Microsatellite allelic convergence can even be a problem in delineating populations. For example, a survey of variability among populations of the red alga *Chondrus crispus* with biallelic single nucleotide polymorphisms (SNPs) and highly polymorphic microsatellites showed different patterns of diversity (Provan et al. 2013), which appeared to be due in part to the convergence of microsatellite alleles. Microsatellites can be plagued by other problems, including allele- and lineage-specific mutation rates and genotyping artifacts, such as null alleles and allelic dropout (Wattier et al. 1998).

When the goal of a study is to identify individuals, or parent–offspring pairs, or to illuminate the effects of mating systems, or life-history phase on genetic variability, markers with high levels of polymorphism may be needed to provide a unique genotype for each individual. For example, a survey on the Canary Islands of six populations of the green alga *Cladophoropsis membranacea* with highly polymorphic microsatellite markers indicated that intertwined mats of unattached filaments included individuals of both the haploid and diploid phases and that haploid plants significantly outnumbered diploid plants (van der Strate et al. 2002).

2.3.3 Use of Multiple Markers

Phylogeographic analyses of seaweeds have been based largely on sequence variability in chloroplast and mitochondrial DNA, chiefly because of their uniparental inheritance and apparent lack of recombination. Plastid DNA is generally circular and is about 70 kb to over 400 kb in size (Coleman and Goff 2004). Mitochondrial DNA in algae is generally smaller with genomic sizes of about 25–50 kb (e.g., Burger et al. 1999; Turmel et al. 1999). However, the use of multiple markers provides stronger insights into evolutionary processes than the use of a single-gene marker, because multiple markers mitigate for the large evolutionary variance among gene lineages (Edwards and Beerli 2000; Sunnucks 2000; Zhang and Hewitt 2003; Heled and Drummond 2008; Brito and Edwards 2009). Different modes of inheritance in organellar and nuclear genes provide complementary

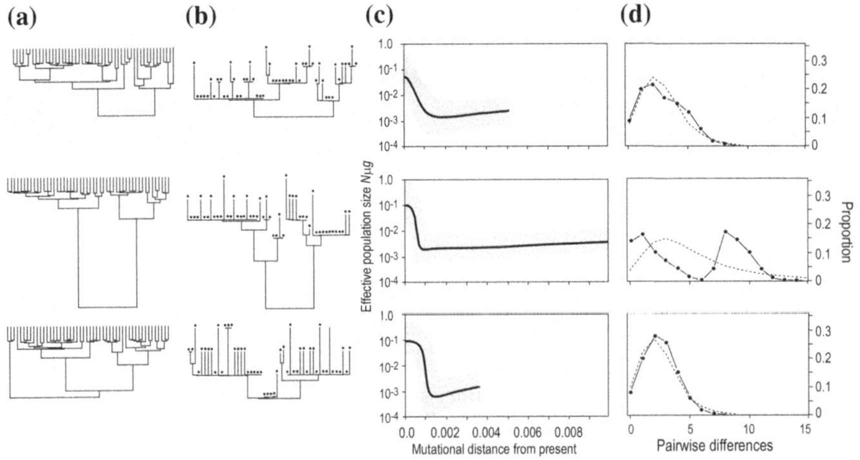

Fig. 2.1 Replicate gene genealogies in simulations of the same population model of a sudden population expansion of three orders of magnitude from $N = 10^3$ to $N = 10^6$ at 0.075 mutation units (μg) in the past. Sample size $n = 50$. **a** Simulated gene genealogies. **b** Genealogies as recovered in neighbor-joining trees of mutations along a 500 bp sequence. **c** Bayesian skyline plots of the simulated sequences. **d** Mismatch distributions (*solid line*) with the expected distributions under a model of population expansion. (reprinted from Grant 2015)

insights into the same population events. Organellar DNA is generally inherited from one parent, generally from the maternal parent in higher plants (e.g., Dumolin et al. 1995), as it is in most animals (Giles et al. 1980), and comparisons with biparentally inherited nuclear DNA allow estimates of sex-biased dispersal, reproductive success in males and females and effective sizes of the total population.

An observed organellar DNA genealogy represents only one of an infinite number of possible evolutionary outcomes for a given population history (Rosenberg and Nordborg 2002). Gene genealogies simulated with the same population history illustrate this point (Grant 2015). Figure 2.1 shows three replicate genealogies in a population experiencing a sudden expansion (Fig. 2.1a). First, a molecular tree does not completely capture the topology of a genealogy (Fig. 2.1b). Second, the Bayesian skyline plots (Fig. 2.1c) and mismatch distributions (Fig. 2.1d) differ considerably among simulations, even though the genealogies were generated with the same population history. The results for each genealogy might provide the basis for constructing conflicting phylogeographic scenarios.

While the use of multiple nuclear loci can mitigate the randomness of individual genealogies, nuclear genes are also beset with the problem of recombination, which can mix gene genealogies during meiosis and violate a fundamental assumption of coalescence analysis. This problem can be overcome to some extent by the use of methods that estimate recombination events in a set of nuclear DNA sequences (Librado and Rozas 2009) and that establish the haplotypic or gametic phases of genotypes (Stephens et al. 2001; Stephens and Donnelly 2003; Scheet and Stephens

2006; Browning and Browning 2011; Garrick et al. 2010). The accuracy of phase estimation is improved by gene genealogies from family data, or by previously phased samples. Incorporating nuclear markers into phylogeographic analyses is especially important in the light of the development of next-generation sequencing that can produce genotypic date for an enormous number of polymorphic nucleotides sites.

2.3.4 Sampling Schemes and Statistical Power

The ability to resolve phylogeographic patterns also depends on sampling a sufficient number of populations. The number of individuals in a sample relative to the *effect size* influences the ability to detect phylogeographic structure. An effect size is the strength of the relationship between two variables (e.g., a correlation), or the size of the difference between means (Nakawaga and Cuthill 2007). When effect sizes are large, sample sizes needed to detect an effect can be relatively small compared to situations in which such signals are weak. Large samples are required to detect small allele- or haplotype frequency differences between populations (Ryman et al. 2006; Bird et al. 2011). Sample sizes needed to detect a given level of divergence between populations, as measured with F_{ST}, can be computed with simulations (Ryman and Palm 2006). Sample sizes also influence the ability of other analyses to detect demographic events. For example, simulations show that small samples from populations undergoing large expansions can fail to detect the expansion in a Bayesian skyline plot analysis (Grant 2015).

Third, an important consideration is the numbers and locations of samples. The numbers and locations of samples, of course, depend on the research problem, and often on the level of funding. However, sample design may be more important for testing a hypothesis than the generation of high-quality data (Meirmans 2015). An investigation of propagule dispersal distances might require collections over small and large distances, while a study of variation on a large geographical scale may require only widely spaced samples along a coast. For example, high and low intertidal populations of the red alga *Chondrus crispus* showed significant differences in microsatellite allele frequencies, and high intertidal populations showed small-scale genetic differentiation, which would not have been detected with coarse sampling (Krueger-Hadfield et al. 2013).

In addition to sampling challenges stemming from the biology of seaweeds and finite research resources, the locations and pooling of samples can influence the outcomes of some analyses (Grant 2015). When gene flow between populations is limited, the demographic analysis of a single population may not reflect the dynamics of a species, or regional group of populations, because of the population-specific histories (Städler et al. 2009). Hence, inadequate or unbalanced sampling among regions may produce misleading inferences about population structure or history. For example, a phylogeographic study of the brown intertidal seaweed *Ishige okamurae* used locations with high levels of contemporary diversity to infer locations of glacial refugia (Lee et al. 2012). However, the conclusions of this study were misleading, because areas of high levels of diversity were also areas that had been disproportionately sampled (Hu and Duan 2013).

On the other hand, the pooling of samples from genetically differentiated populations may violate the assumptions of panmixia in many models. For example, when coalescence theory is used to estimate historical demography (Drummond et al. 2005), or migration and effective population size (Beerli and Felsenstein 2001), the pooling of heterogeneous samples pushes genotypic coalescence time frames deeper into the past, producing misleading results. Another sampling strategy is to pool a few individuals from numerous localities for statistical analyses. This approach is sometimes used when individuals are rare, or when conservation concerns or funding limit the number of individuals that can be analyzed. Simulations show that in some circumstances this 'scattered' sampling can provide a better picture of specieswide demography than the analyses of a few single-population samples (Städler et al. 2009; Chikhi et al. 2010). However, the dynamics of local populations cannot be resolved with this scattered sampling scheme.

2.3.5 *Molecular Clock Calibration*

DNA sequences and the molecular clock hypothesis can be used to date divergences between taxa and demographic histories. The calibration of a molecular clock requires the estimation of the mutation rate for a particular gene. Estimates of mutation rates are most commonly based on nodes in a phylogenetic tree that can be associated with a geological or climatic event. Deep divergences can sometimes be tied to geological plate separations, or to the formation of a geological barrier to dispersal. In marine research, the most recent rise of the Isthmus of Panama about 3.1–3.5 million years from uplift and volcanic accretion initiated divergences between populations in the western tropical Atlantic and eastern Pacific (Knowlton and Weigt 1998; Marko 2002). Comparisons between isolated sister populations, or sister species, yielded mtDNA divergence rate estimates of about 2 % divergence per one million years, and this 'standard' estimate has been used in a large number of studies. Substitution rates have also been estimated with various other geological events (e.g., Ketmaier et al. 2003; Marino et al. 2011), fossils (Gaunt and Miles 2002), climate and sea level history (Crandall et al. 2012; Grant et al. 2012), samples of ancient DNA (aDNA) (Prost et al. 2010), and pedigrees (Santos et al. 2005). Commonly used methods to estimate a mutation rate for a gene include the use of mutation rates in a related species (Bilgin et al. 2009), or in another gene (Qu et al. 2011). For the latter method, mutation rate estimates are adjusted by the relative levels of polymorphism for the two genes, by assuming that higher levels of polymorphism indicate a larger mutation rate. Many of these methods of calibration are not available for seaweeds, and researchers resort to using a standard rate, which appears to be slower in plants than in animals (Wolfe et al. 1987).

Mutation rates derived from ancient divergences generally push the timings of population events too far into the past (Ho et al. 2008, 2011), creating a mismatch between inferred population events and environmental causation. While substitution rates estimated from a phylogenetic tree are theoretically expected to equal mutation rates, contemporary mutation rates appear to be much larger. Mutation rates appear to be time-dependent and decline exponentially over a period of about one million years (Ho et al. 2005). It appears that mutation rates estimated from fossils, or from divergences precipitated by datable geological and oceanographic events, produce estimates of effectively population size and demographical timings that greatly overshoot the timings of the actual events (Ho et al. 2011; Grant 2015). Phylogenetically derived mutation rates often place population expansions before or during the Last Glacial Maximum, implying that glaciations did not impact population demography. However, resistance to major environmental changes during a period of global cooling is unlikely, given the effects that much smaller contemporary climate changes have on the abundances and distributions of seaweed populations (see Grant 2015 for detailed discussion).

2.3.6 Reproductive Skew and Genealogical Models

Many forms of data analysis are based on simulated genealogies using backward-looking coalescence models (Kingman 1982; Hudson 1991). One assumption, not only in these models, but also in forward Fisher–Wright models, is that nodes in a genealogy represent bifurcations. However, most marine species and seaweeds with high fecundities and large early life mortalities (type III life histories) show large differences in reproductive success among families (Hedgecock and Pudovkin 2011). This large variance in successful reproduction (reproductive skew) leads to multifurcations at the nodes in a genealogy, which can influence the phylogeographic analysis of genetic data (Eldon and Wakeley 2006). For example, a star-shaped genealogy is generally interpreted as evidence for a recent population expansion (or a selective sweep, see below) and a molecular clock calibration is used to date the expansion. Mismatch distributions or Bayesian skyline plots are used to make these kinds of demographic explanations and are reported in many phylogeographic studies of seaweeds. Unfortunately, when a genealogy contains numerous multifurcations at a coalescence node, these methods of data analysis tend to overestimate the timings of supposed population expansions and inflate estimates of effective population size. In the extreme, singleton haplotypes in a star-shaped genealogy may have arisen in the previous generation from a single highly successful family. Hence, temporal estimates of demographic events are intermediate between this extreme and the commonly used bifurcation models. The theoretical basis for simulation models that take these factors into account is in early stages of development (Eldon et al. 2015). Better recognition of the effects of reproductive skew on the genetic population structure of marine species will provide important insights into seaweed phylogeography.

2.4 Biological Variables Influencing Phylogeography

Phylogeographic patterns ultimately result from long-term interactions between the biology of an organism and environmental variability. Several features of seaweed biology influence patterns of genetic variability among populations, including the modes of inheritance of particular genes, life-history cycles, extents of sexual or asexual reproduction, self-fertilization, and dispersal ability.

2.4.1 Inheritance of Molecular Marker

The mode of inheritance of a molecular marker greatly influences how phylogeographic patterns can be interpreted. Nuclear genes are passed on to offspring by both parents in diploid organisms, whereas organellar genes are generally passed on to offspring by only one parent, generally the female. However, exceptions have arisen independently in several groups of seaweeds (Bock 2007; Nagasato et al. 2000). In isogamous species (male and female gametes are morphologically similar), organelles may be biparentally inherited but postzygotic mechanisms lead to the occurrence of only one parental organelle in the tissues of the plant. For example, in isogamous brown and green algae, chloroplasts can be biparentally inherited, but only maternal mitochondria remain after embryonic development from the zygote (Miyamura 2010; Nagasato et al. 2010). In the brown alga *Ectocarpus*, maternal inheritance of organelles is correlated with differences in the mobilities of the gametes (Peters et al. 2004). In anisogamous (different morphologies of male and female gametes) and oogamous (large female nonmotile gamete relative to motile male gamete), cpDNA and mtDNA are generally inherited through the female gamete (maternal inheritance) for most groups of seaweeds (Table 2.1) (Motomura et al. 2010).

Several prezygotic and postzygotic mechanisms may be responsible for uniparental inheritance (Table 2.2). These mechanisms can be deterministic, when a cellular mechanism targets one parental organelle or the other, or they can be stochastic, when parental organelles are randomly sorted into cell lineages. In some species, organelles in male gametes may degrade during gametogenesis or after fertilization. This commonly occurs in isogamous green algae (Miyamura 2010). In other groups, organelles of one parent in zygotes may degrade, or the DNA of one parent may be selectively degraded (Miyamura 2010). In the oogamous brown algae *Undaria pinnatifida*, the mtDNA inherited from the male is digested toward the end of the one-cell zygotic stage (Kimura et al. 2010). The degradation of DNA in some gametes can be a source of sustenance to developing zygotes (Sears and Vanwinkle-Swift 1994). Generally, mitochondria of the sex with the higher energy requirements are destroyed in the zygote (Han et al. 2014). Another mechanism leading to effective uniparental inheritance includes the partitioning of one parental organelle or the other into separate cells at early stages of development, or

Table 2.1 Examples of organellar inheritance in algae

Group/species	Mode[a]	Mechanism[b]	References
Chlorophyta			
Bryopsis maxima	cp/mt: M	1d	Kuroiwa and Hori (1986)
Bryopsis plumosa	cp/mt: M	1d	Ogawa (1988)
Caulerpa, 4 species	cp/mt: M	3b	Miyamura and Nagumo (2007)
Chlamydomonas reinhardtii	cp: variable	3b	Sears and Vanwinkle-Swift (1994), Lee and Lemieux (1986)
Spirogyra sp.	cp: M	3a	Smith (1950)
Ulva mutabilis (isogamous)	cp/mt: M	3b	Fjeld and Løvlie (1976), Bråten (Bråten 1971, 1973)
Heterokontophyta			
Alaria esculenta-praelonga hybrids	cp/mt: M		Kraan and Guiry (2000)
Ectocarpus siliculosus (isogamous)	cp/mt: M		Peters et al. (2004)
Fucus vesiculosus	cp/mt: M		Brawley et al. (1976)
Fucus serratus, Fucus evenescens	cp/mt: M		Coyer et al. (2002)
Fucus serratus	mt: M, H	2c	Coyer et al. (2004)
Fucus serratus-evenescens hybrids	mt: H	2c	Hoarau et al. (2009)
Saccarhina (Laminaria) angustata	cp/mt: M	1b	Motomura (1990)
Scytosiphon lomentaria	cp: B	4b	Nagasato et al. (2010), Kato et al. (2006)
	mt: M	3a or 3b	Han et al. (2014)
Undaria pinnatifida	mt: M	3a	Kimura et al. (2010)
Rhodophyta			
Caloglossa leprieurii	cp: M		Zuccarello et al. (1999a)
Bostrychia radicans, B. moritziana	cp: M	1b	Zuccarello et al. (1999a)
Bostrychia moritziana	cp/mt: M		Zuccarello et al. (1999a, b)
Porphya yezoensis	cp & mt: M (86 %), B (11 %), P (2 %)	4b	Choi et al. (2008)

[a]*cp* chloroplast; *mt* mitochondria; *M* maternal; *P* paternal; *B* biparental; *H* heteroplasmy
[b]Explanations of mechanisms listed in Table 2.1

Table 2.2 Mechanisms of uniparental inheritance of chloroplastid and mitochondrial DNA (modified from Birky 1995)

Mechanism
1. *Prezygotic*
(a) Unequal cell division and differential growth: large female gametes, small male gametes
(b) Exclusion of organelles from gametes during meiosis
(c) Degradation of organelles in gamete
(d) Degradation of organellar DNA in gamete
2. *Fertilization*
(a) Exclusion of organelles of one parent from zygote
(b) No organelles exchanged
(c) Paternal leakage (heteroplasmy)
3. *Zygotic: deterministic*
(a) Selective silencing or degradation of organelle
(b) Selective silencing or degradation of organellar DNA
(c) Partitioning of parental organelles into separate cells
(d) Exclusion of organelles from embryonic cells
4. *Zygotic: random sorting among cell lineages*
(a) Exclusion of organelles from embryonic tissue
(b) Random replication and sorting

organelles may be selectively excluded from embryonic daughter cells leading to cell lineages. Random sorting in early embryonic cell division leads to the incorporation of either maternal or paternal cpDNA in the large sporophytes of *Scytosiphon lomentaria* (Kato et al. 2006). In the same species, mtDNA is maternally inherited, because paternal mitochondria are digested in male gametes or in fertilized eggs (Han et al. 2014). The mode of organellar transmission is a legacy of evolutionary processes, but proximate factors can alter how organelles are inherited. For example, individuals of the brown seaweed *Fucus serratus* were heteroplasmic for mtDNA in northeastern Atlantic areas previously covered with glaciers (Coyer et al. 2004).

The identification of the mode of inheritance in an alga is important in phylogeographic analyses using coalescence models. Since modes of inheritance can vary among species, they need to be experimentally verified with the distributions of molecular markers in an experimental family (Zuccarello et al. 1999a). Most analyses of chloroplast or mitochondrial sequences proceed by considering the genetic population structure or demography of only one sex, generally females, as the preponderance of cytoplasmic organelles are maternally inherited. When the abundances of the sexes are similar, inferences can be made about the entire population. Caution is needed, however, when organelles are biparentally inherited. For polymorphic gene markers, biparental inheritance appears as heteroplasmy, and it may be difficult to distinguish between markers passed on by one parent or the other, and hence to make population inferences with genealogical analyses.

2.4.2 *Effect of Life-History Phases on Phylogeography*

Seaweeds generally have multiple life-history stages that alternate between diploid and haploid forms. One exception includes fucoid algae, in which the diploid stage produces gametes and zygotes that develop directly into the diploid form, without cycling through a free-living gametophyte phase. In red algae, the relative abundances of isomorphic gametophytic and sporophytic plants vary among groups, indicating an evolutionary legacy. The gametophytic phase is dominant among genera in the Gigartinales, but tetrasporophytes are dominant in the Gracilariales, Ceramiales, and Gelidiales (Thornber and Gaines 2004; Fierst et al. 2005; Dyck and DeWreede 2006b). In some brown and red algae, the sporophytic stage is large, and the gametophyte stage is small, just visible by eye. In red algae, the sporophytic and gametophytic stages can be isomorphic (e.g., *Mazzaella*) and are sometimes difficult to distinguish in the field (Hannach and Sanelices 1985; Hannach and Waaland 1986; Shaughnessy and DeWreede 1991). In yet other red algae, life histories progress through three stages consisting of a macroscopic diploid tetrasporophyte, haploid male and female macroscopic gametophytes, and a small diploid cystocarpic stage embedded in the female gametophyte (e.g., *Mastocarpus*). These minute cystocarps produce a multitude of mitotic diploid spores that greatly amplify a particular genotype and that develop into genetically identical free-living tetrasporophytic plants. Tetrasporophytes then produce mieotic tetraspores that develop into male and female gametophytes. Most brown seaweeds consist of large sporophytes alternating with microscopic gametophytes.

Inadvertent sampling of different life-history phases may introduce inferential errors, because chromosomal ploidy differs between alternate life-history phases. Mixed-phase sampling of isomorphic plants may introduce small-scale heterogeneity, which may be difficult to interpret in terms of dispersal and population structure. Although mixed-phase sampling is common in most phylogeographical studies of seaweeds, it may not substantially detract from phylogeographic inferences when organellar DNA markers are used. However, when nuclear markers are used, samples of unknown life-history type may produce ambiguous genotypes, because it is difficult to distinguish between homozygous and hemizygous genotypes. In large-scale studies of phylogeographic structure over distances of tens and hundreds of kilometers, mixed-phase sampling may not introduce substantial errors when population heterogeneity is much greater over large geographical distances than within sites (e.g., Wang et al. 2008).

The sampling of one life-history phase at one site and another phase at other sites may confound interpretations of allelic or haplotypic frequency distributions (Engel et al. 2004; Schiel and Foster 2006). The proximate factors regulating the relative abundance of one phase over the other are poorly understood. One possibility is that the propagules or mature plants in the different phases have different survival rates (e.g., Hansen and Doyle 1976; Kain 1982), because of ecological or physiological differences between phases (e.g., *Chondrus crispus*, Mathieson and Burns 1975; Carrington et al. 2001). In species with heteromorphic life-history phases, microscopic gametophytes and large sporophytes inhabit different microenvironments and differences in abundance may be due to selection from ecological variables or to sweepstakes recruitment. Alternatively, differences in gametophyte and sporophyte abundances may be due to bottlenecks in the production of one phase (Santelices 1990). The results of models incorporating several reproductive variables show that fertilization rate and spore output, separately or together, can favor one life-history phase or another (Fierst et al. 2005). When fertilization rates are density dependent, the phase arriving first may then become dominant (Fierst et al. 2005). Phase dominance can shift among seasons, among years, and among sites (Dyck and DeWreede 2006a; Bellgrove and Aoki 2008), so that samples can potentially include individuals of different ploidies and ecologies.

2.4.3 Asexual and Clonal Reproduction

The predominance of sexual or vegetative reproduction in a population can potentially shape phylogeographic patterns by influencing levels of genetic diversity and modes of dispersal between populations. In populations that depend on the recruitment of sexually produced propagules, gamete release and fertilization success are important determinants of abundance. In brown seaweeds, gamete release is influenced by interactions of daylight, concentrations of inorganic carbon, hydrodynamic conditions, and pheromones (Pearson and Brawley 1996; Pearson et al. 1998; Gordon and Brawley 2004; Serrão et al. 1996a, b). In species of the brown seaweed *Fucus,* light and available carbon for photosynthesis limit gamete release to periods of daylight hours during calm weather and minimal tidal currents (Berndt et al. 2002; Pearson and Serrão 2006). For species in the genus *Alaria,* water movement inhibits sperm release by diluting a sperm-releasing pheromone secreted by ripe eggs (Gordon and Brawley 2004). Fertilization success may also be influenced by male–male gamete competition or by female choice (Engel et al. 1999).

Asexual and clonal propagation can occur in different ways and to different extents among species. Reproduction through asexual spores leaves imprints that can be detected with molecular markers. For example, the low frequency of male gametophytes in the red alga *Chondrus crispus* was key to concluding that asexual reproduction was the dominant mode of reproduction in this alga (Prince and Kingsbury 1973; Frederica et al. 1992). In the red alga *Pterocladia capilacea,* asexual propagation occurred only in the northern British Isles, whereas sexual tetrasporic plants were found along the coast of France, and cystocarpic plants were found in northern Spain (Dixon 1965).

Clonal plants can arise by vegetative regeneration from perennial holdfasts, or from fragments of a thallus. In many species of red algae, population abundance is largely maintained by clonal propagation, even though spore release and germination occur. The frequency of sexuality in several seaweeds drops under poor ecological conditions and is especially common in populations at the edge of a species' geographic range (Eckert 2002). A simulation study showed that sexually reproducing individuals tended to cluster in environmentally advantageous areas at lower latitudes, and asexual individuals were more frequent at high latitudes in ecologically marginal and resource-poor areas (Peck et al. 1998). For example, clonal reproduction in the red alga *Ceramium* in the Skagerrak–Baltic was highest in low-salinity areas in the inner Baltic, but absent in high salinity areas toward the open Atlantic (Gabrielsen et al. 2002). A similar increase in clonal reproduction appeared in populations of *Fucus* in the Baltic, where gametes were less viable in low salinities. In these populations, a skewed sex ratio (predominance of females), a short fertilization potential of eggs (about two minutes), and an environmentally driven short reproductive period led to reproductive failures in most years (Serrão et al. 1999). Hence, clonal reproduction was most important in these geographically and ecologically marginal populations (Box 2).

Box 2. The Effect of Asexual Reproduction of Population Structure

Reproduction in brown algae generally occurs by sexually produced zygotes. However, in low-salinity areas in the Baltic Sea, the persistence of *Fucus vesiculosus* populations is facilitated by clonal reproduction. At its northern limit, fertilization success is generally low and fertilized eggs are susceptible to polyspermy, which is lethal (Serrão et al. 1999). At 5 psu, a dwarf form appears among the common larger form. This dwarf form largely reproduces clonally, as indicated by finding 70–80 % of the plants with identical microsatellite genotypes. A larger survey of *F. vesiculosus* (Morocco to Iceland) showed clonality only in Baltic Sea populations. A related species, *F. serratus*, does not show evidence of clonality over its distributional range, even in populations in marginal areas (Coyer et al. 2003).

Fucus is 'diplontic' with newly attached fronds arising from only sexually produced zygotes (Serrão et al. 1999). Clonal individuals can arise through vegetative fragmentation or budding. However, fragmentation in *Fucus* does not generally lead to attached fronds, except for fragments capable of producing adventitious branches. These fragments generate rhizoids from the wounded basal section that facilitate attachment. In addition to wounding, the growth of rhizoidal filaments appears to be stimulated by low salinities (Tatarenkov et al. 2005). Adventitious fragments with the capability of growing rhizoidal filaments have the potential for acting as clonal propagules.

Clonality tends to arise in environments in which sexual reproduction is impaired or impossible. Sexual reproduction is limited in brackish waters by forces that affect longevity and gamete motility (Serrão et al. 1996a, b). Clonality appears to reflect an in situ adaptive response to environmental conditions rather than dispersals in a source–sink system in which selection favors only some immigrant genotypes.

2.4.4 Self-Fertilization

Over 50 % of angiosperms show self-fertilization, at least occasionally (Barrett et al. 1996), and mixed mating systems have evolved in many groups of plants, despite the potentially negative effects of inbreeding (Goowille et al. 2005). The extent of self-fertilization in seaweeds is poorly known. Selfing can occur when an egg produced by a gametophyte is fertilized by spermatia from a male gametophyte originating from the same sporophytic plant. In the giant kelp *Macrocystis*, models of dispersal and field results showed that about half of the meiotic spores from large sporophytes disperse less than 100 m, and this limited dispersal led to aggregations of gametophytic plants from the same sporophyte. 'Selfing' in this case occurred when gametes from sibling gametophytes fused (Raimondi et al. 2004). Inbreeding from self-fertilization may explain periodic local extinctions of *Macrocystis* (Raimondi et al. 2004). These cycles of extinction and colonization produce metapopulation dynamics that affect genetic diversity and genetic population structure along a shoreline (Reed et al. 2000). The incidence of selfing is affected not only by recruitment dynamics along a shore but also by microhabitat differences at a single locality. For example, high intertidal zone plants of *Chondrus* showed greater gametophytic selfing than did lower-shore plants (Krueger-Hadfield et al. 2013). In *Fucus spiralis*, selfing occurred because gametes were released during periods of low water movement, and limited movement promoted fertilizations between gametes from the same individual (Coleman and Brawley 2005; Perrin et al. 2007).

Self-fertilization can be advantageous or detrimental, depending on the ecological circumstances of a population. On the one hand, selfing can lead to inbreeding and the loss of local genetic diversity or the homozygotic exposure of deleterious alleles, which can retard growth, limit fecundity, and lower survival (Charlesworth and Charlesworth 1987). The example of *Macrocystis* appears to demonstrate the negative effects of selfing (Raimondi et al. 2004). While selfing can lead to deleterious effects, it may also be beneficial under some circumstances by boosting reproductive output when plants are rare or when gamete density is low and fertilization is not assured. Under some circumstances, selfing may be the last resort for producing recruits at low population densities in marginal habitats. Many models used by phylogeographers assume that samples are drawn from a randomly mating population. However, the complexity of mating systems in seaweeds may invalidate this assumption. Samples collected along a limited stretch of shore may not be representative of a population, because of the clumped recruitment from limited spore dispersal, or from selfing. Conversely, the sampling of individuals from different patches, genetic mosaics along shore, or between upper and lower tidal levels may produce misleading inferences about local mating dynamics.

2.4.5 Dispersal

Dispersal is the most important ecological variable shaping phylogeographic structure. Dispersals on ecological time frames can be realized by the movements of gametes, spores, fertile fronds, or zygotic propagules and are influenced by complex physical and biological variables, including spore swimming and sinking rates, height of spore release above the sea floor, and length of spore viability (Reed et al. 2004). Gamete and spore longevities define the scope for their movement in currents and depend to some extent on the presence of chloroplasts to provide energy (e.g., Amsler and Neuschul 1990; Reed et al. 1992). Gametes and spores of many seaweeds are viable for only a few hours, or a few days at most, and this generally leads to short-distance dispersals of only a few meters for many seaweeds (Destombe et al. 1992; Williams and Di Fiori 1996; reviewed in Santelices 1990; Forrest et al. 2000; Gaylord et al. 2002; Kusumo et al. 2006; Tellier et al. 2009; among several other studies). In the brown seaweed, *Sargassum muticum*, most spores settled within 2–3 m of the parent, but some germinated as far as 30 m away. Zygotic propagules, when they occur, may remain viable longer than gametes and spores but also do not disperse far (Deysher and Norton 1981; Kendrick and Walker 1995).

For many species, limited dispersals of gametes and zygotic propagules can lead to small genetic neighborhood sizes and to genetic mosaics along the shore (Williams and Di Fiori 1996; Coyer et al. 1997; Coleman and Kelahr 2009; Krueger-Hadfield et al. 2011). In species with long-lived propagules, particle tracking models indicate that spores can move several kilometers in coastal currents (Brennan et al. 2014). For example, spores were found up to 30 km offshore of the east coast of North America (Amsler and Searles 1980; Zechman and Mathieson 1985). Spores of green and bangiophycidean red algae were found at all depths, but spores of brown and florideophycidean red algae occurred only at greater depths in the photic zone. Spores from upper levels tended to be opportunistic species.

A pattern of isolation by distance (IBD) can arise among populations when gene flow is limited by short dispersals to nearby populations. In an unfragmented forest of the kelp *Laminaria digitata*, limited gamete and spore dispersal resulted in genetic differentiation between neighborhoods separated by 10 km without evidence of environmental boundaries (Billot et al. 2003). IBD models generally only include geographic distances between samples, but seascape features may also influence dispersal and their inclusion can improve IBD correlations. For example, Alberto et al. (2010) examined the effect of habitat continuity on IBD in the kelp *Macrocystis pyrifera* along the Santa Barbara Channel and found that genetic distances between samples were positively correlated with geographical distance between samples, but that habitat continuity was negatively correlated to genetic distance (Fig. 2.2). Geographic distance and habitat continuity did not covary, but operated independently to influence genetic distance. Increasing geographic distances between patches produced larger genetic distances, but habitat continuity between patches led to smaller genetic distances, because gene flow was enhanced.

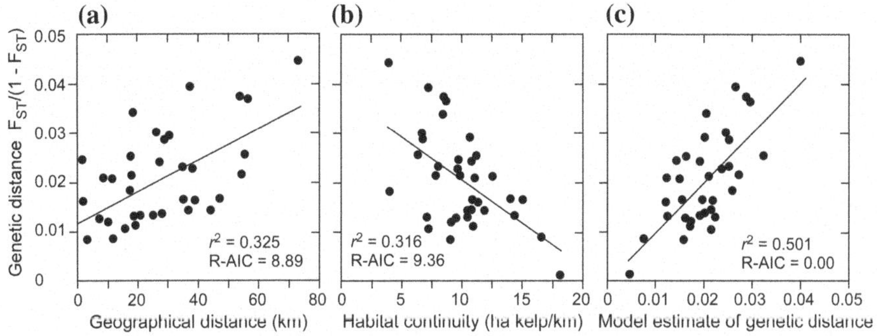

Fig. 2.2 Isolation by distance among populations of *Macrocystis pyrifera* along the coast of California. **a** The effect of geographical distance of genetic distances between populations. **b** The effect of habitat continuity on genetic distance between populations. **c** A two factor model incorporating the effects of geographical distance and habitat continuity on genetic distances. (redrawn from Alberto et al. 2010)

Since spore dispersal is limited, movement across inhospitable habitats is rare. In Chilean populations of the rocky shore bull kelp *Durvillaea antarctica* genetic divergences between populations increased when sandy beaches separated the populations (Fraser et al. 2010). Sandy beaches are inhospitable to the recruitment of these kelps and thus inhibit gene flow between populations.

While small-scale genetic population structure can be attributed to restricted gamete and spore dispersal, some seaweeds show little genetic structure over hundreds or thousands of kilometers. This genetic homogeneity likely reflects rafting of detached reproductive fronds or vegetative fronds that mature during transport in currents. Floating seaweeds and beach-casts of seaweeds are common along seashores, but the effectiveness of drifting plants or fragments to facilitate gene flow varies considerably among species. Detached fragments of the fucoid *Hormosira banksii* produced viable gametes 8 weeks after detachment (McKenzie and Bellgrove 2008). Drifting reproductive fronds of the kelp *Macrocystis pyrifera* produced viable spores for as long as 125 days (Hernández-Carmona et al. 2006). Drifting plants of *Macrocystis pyifera* were found on newly constructed artificial reefs, but kelp germlings appeared to have originated from spores released by a distant stand of kelps, rather than from drifters (Reed et al. 2004). In the bull kelp *Durvillaea antarctica* in New Zealand, the distributions of mtDNA haplotypes in beach-cast samples were largely consistent with the genetic structure of populations, indicating that the current systems shaping the genetic structures of sessile populations also influenced the movement of free-floating plants along the coast (Collins et al. 2010; Bussolini and Waters 2015).

Long-distance dispersals of floating seaweeds have largely been inferred from genetic homogeneity among populations and from particle transport in ocean current models (but see Fraser et al. 2011). The time frames of postulated dispersals vary from contemporary movement in ocean currents to dispersals thousands of years ago. Contemporary movements of detached plants or fragments have been invoked to explain genetic similarity between populations of *Fucus* spp. (Coleman and Brawley 2005; Muhlin et al. 2008) and *Sargassum* spp. (Hu et al. 2011, 2013; Chan et al. 2013, 2014). On longer timescales, rafting in ocean currents may be important for colonizations of distant shores across the open ocean (Waters 2008; Fraser et al. 2010; Hu et al. 2010; among others). Some studies have interpreted genetic homogeneity among populations to indicate postglacial population expansions (*Porphyra umbilicalis*, Teasdale and Klein 2010; *Chondrus crispus*, Hu et al. 2010; *Sargassum* spp., Hu et al. 2011; Chan et al. 2013, 2014), or interhemispheric dispersals (*Macrocystis pyrifera*, Coyer et al. 2001). Ocean currents can enhance or inhibit dispersal (White et al. 2010). Algae entrained in rapid offshore currents can potentially move large distances (Garden et al. 2014, whereas complex near-shore currents and eddies can impede long-distance dispersals (Hoffmann and Ugarte 1985; Hoffmann 1987).

2.5 Reconstructing Historical Population Events

A major goal of marine phylogeography is to construct hypotheses to explain the distributions of genetic diversity across a seascape. Placing historical population events into a time frame can clarify the environmental and biogeographic mechanisms influencing genetic variability and can ultimately reveal the evolutionary mechanisms shaping species' diversity. A wealth of theory can be used to interpret the results of molecular studies. However, several caveats are warranted in the applications of these theories to reconstruct demographic histories, because the assumptions in many theoretical models are not always realized in the sampling of natural populations (Karl et al. 2012; Grant 2015).

Climate variability has an overriding influence on population demography. Climate swings occur on timescales ranging from decades to hundreds of thousands of years, and the challenge is to match genetic signals of demographic change with these climate changes (Fig. 2.3). Global and regional climates have continuously changed over the Holocene and Pleistocene Epochs, and the magnitudes of these changes progressively increase further into the past. Moderate environmental shifts, including El Niños (Philander 1983), Pacific Decadal Oscillations (Mantua et al. 1997) (Fig. 2.3e), and North Atlantic Oscillations (Hurrell 1995), occur in decadal and multidecadal time frames and produce poleward shifts in species' distributions during warm phases and equatorial shifts during cool phases (e.g., Russell et al. 1971; Stebbing et al. 2002). Even small contemporary changes in sea temperature can lead to ecological reorganizations in marine ecosystems (e.g., Hare and Able 2007; Francis et al. 1998; Stenseth et al. 2003). Climate shifts over only a few years can produce changes in sea temperature, nutrients, acidification, grazing, competition with other seaweeds and pathogen exposure, all of which can drive changes in local abundance and distribution (Ling et al. 2009; Wernberg et al. 2011; Wahl et al. 2015). Benthic seaweeds may be particularly vulnerable to environmental changes, because they are sessile and are at the mercy of local environmental

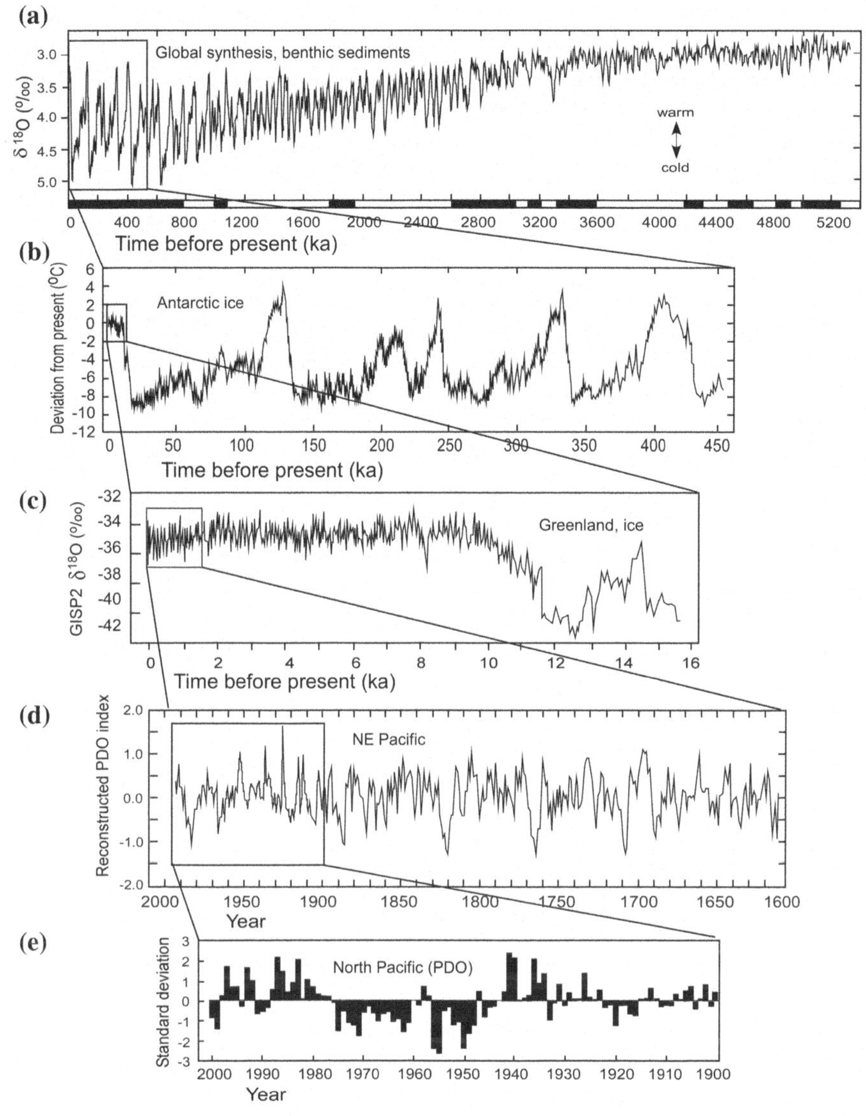

Fig. 2.3 Scales of climate variability over the Quaternary, 2.6 Ma to present. **a** Synthesis of temperature proxy $\delta^{18}O$ records in 57 benthic sediment cores distributed globally (redrawn from Lisiecki and Raymo 2005). **b** Temperature reconstructions from proxy deuterium isotope profiles in Antarctic ice cores (redrawn from Jouzel et al. 2007). **c** Temperature proxy $\delta^{18}O$ records in Greenland ice cores from GISP2 (redrawn from Bond et al. 1997). **d** Extended reconstruction of the Pacific Decadal Oscillation from tree-ring time series along the western coast of North America (redrawn from Gedalof and Smith 2001). **e** Pacific decadal index (redrawn from Mantua and Hare 2002)

conditions (Harley et al. 2012; Koch et al. 2013). For example, shifts in abundances of brown seaweeds around the British Isles have been interpreted to reflect recent climate warming (Yesson et al. 2015).

Greater swings in climate have occurred on centennial timescales, shifts which–in historical times (Fig. 2.3d)–led to Medieval warming (950–1250 AD) and the Little Ice Age (1350–1850 AD) (Mann et al. 2009). These and other changes have been documented in ocean (e.g., Keigwin 1996) and lake sediments (e.g., Gill et al. 2009), tree rings (e.g., Villalba 1994), and historical records of fish landings (Alheit and Hagen 1997). The extent that these periods of regional cooling and warming led to expansions, contractions, or extinctions of seaweed populations is difficult to surmise, because of the lack of fossils or written records. On longer timescales of a few thousand years over the Holocene (Fig. 2.3c), climate swings influenced the structures of plant and animal communities in all of Earth's ecosystems (e.g., Hewitt 1996; Fauvelot et al. 2003; Staubwasser et al. 2003). On even longer timescales, the globe experienced orbitally driven ice ages at 100-thousand-year (kyr) 'Milankovitch' intervals, beginning about 800,000 years ago (Figs. 2.3a, b) (Bond et al. 1993). During these periods of global cooling, glaciers covered large high-latitude areas of North and South America, Europe, and Asia, and coastal areas were reshaped by drops of as much as 120 m in sea level at glacial maxima (Miller et al. 2005; Ludt et al. 2012).

Assessing the effects of Milankovitch climate oscillations on populations has been an enduring focus of phylogeographic research. As a general hypothesis, less genetic diversity within populations and less population structure are expected in high-latitude areas that experienced major disturbances, and greater diversity and population structure are expected in low-latitude areas harboring older diversity (e.g., Fraser et al. 2009). The concept of population contractions into refugia during climate extremes is deeply embedded in the phylogeographical literature and has been used to explain the distributions of mtDNA and cpDNA lineages in several species of seaweeds (e.g., Provan et al. 2005; Hoarau et al. 2007; Provan and Bennett 2008; Lee et al. 2012). However, the basic contraction–expansion model may not be relevant for all marine species. The distributions of some species may have merely been displaced more or less intact by temperature shifts or moved offshore by the exposure of continental shelf areas during drops in sea level without a decline in population size, or without compression into a small area. Hence, in some species, postglacial genetic structure may not show gradients in diversity or population structure. Some responses to historical environmental changes that produce genetic homogeneity may be mistaken for high contemporary levels of gene flow. The expectations of the null diversity model can also be modified by irregular shoreline topologies and ocean currents that create multiple hospitable local habitats during glaciations (Maggs et al. 2008). The genetic structure of populations within a refuge can additionally influence phylogeographic patterns after a postglacial expansion and produce patterns that may be mistaken for high latitude refugia (Gómez and Lunt 2007).

2.6 Comparative Phylogeography

Comparisons of phylogeographic patterns among codistributed species, or species complexes, can reveal general biogeographic patterns that may not be apparent from the analysis of a single species (Arbogast and Kenagy 2001; Hickerson et al. 2010). Phylogeographic concordances among species are especially important for detecting shared responses to historical seascape changes (Avise 2000). However, concordance does not necessarily mean that lineage distributions among species were shaped by the same environmental events, unless the timescales are the same. The careful application of the molecular clock hypothesis provides a means of dating dispersals, population expansions, and vicariances. Comparative phylogeography has been used to test mechanistic hypotheses about the effects of historical and contemporary dispersal barriers to gene flow and of ice age displacements and refugia on sympatric populations of different species.

Phylogeographical comparisons of codistributed seaweeds are already possible for several areas around the globe. In the Northwest Pacific, for instance, marine waters cooled during the LGM and a drop in sea level likely led to offshore and southward shifts in seaweed habitats. Large areas of the continental shelf were exposed, draining the Yellow Sea and largely isolating the Sea of Japan for a brief period during the LGM (Oba et al. 1991; Wang 1999). The rise in sea levels after the LGM led to colonizations over newly submerged areas of the continental shelf. Species of the brown algae *Sargassum* and *Ishige* show star-like haplotype genealogies, sometimes embedded in a complex haplotype network, that indicate recent populations expansions (*S. hornei*, Uwai et al. 2009; Hu et al. 2011; *S. fusiforme*, Hu et al. 2013; *I. okamurae*, Lee et al. 2012). These species also show strong haplotype frequency differences among areas, indicating periods of isolation, but the patterns of differentiation among species are not consistent with one another. Together these results show the profound impact of the last glaciation in displacing and isolating populations. However, the diversity of phylogeographic patterns indicates species-specific responses to climate variability.

In the southwestern Pacific, large areas of the Sunda Shelf were drained, uniting several of the large islands and reducing the extents of available shoreline habitats (Voris 2000). The Shelf was submerged again about 14 600 years ago (Hanebuth et al. 2000), stimulating a massive reoccupation by marine species. Marine species invading shelf habitats show molecular signatures of a recent expansion (Lourie et al. 2005; Crandall et al. 2012), including species of the brown seaweed *Sargassum* (*S. aquifolium* Chan et al. 2014; *S. polycystum*, Chan et al. 2013). In contrast to phylogeographical patterns in the northwestern Pacific, populations across Southeast Asia are largely genetically homogeneous, perhaps because each species was isolated in a single area during global cooling before expanding across the Sunda Shelf.

Another well-studied area is the southeastern Pacific along Chile. During ice age maxima, the Patagonian glacier in the southern part of South America reached to the sea as far north as about 41° S (McCulloch et al. 2000). These shores were most recently open to colonization only after the LGM (0.026–0.019 Ma). A shift in the composition of marine species occurs at about 30°–33° S, but it was uncertain whether the environmental factors that produced the biogeographic transition also produced a genetic discontinuity in species with distributions spanning the biogeographical break. A multilocus survey of populations of the kelp *Lessonia nigrescens* with mtDNA, cpDNA, and nuclear markers resolved three major parapatric lineages, which were concordant with the biogeographical transition. A genetically intermediate lineage was limited to the transition zone. A second study of the red alga *Mazzaella laminarioides* using mtDNA and cpDNA markers revealed three strictly parapatric lineages, whose geographic distributions were not concordant with biogeographic breaks along the coast (Montecinos et al. 2012). It was uncertain, however, whether the geographic boundaries between genetic lineages reflected natural selection or founder events and high-density blocking of immigrants (Waters et al. 2013). A third study surveyed mtDNA and cpDNA markers in populations of the kelp *Durvillaea antarctica* located below the biogeographical break at 30° S (Fraser et al. 2010). One mtDNA lineage appeared along the shores of central Chile, and a second genetic lineage was restricted to the southern shores of Patagonia. This last lineage was closely allied with a lineage in the southern ocean, including New Zealand, and appeared to have colonized Patagonia after the LGM by long-distance rafting. These various comparisons show that species respond to environmental variability in different ways, and that it is difficult to generalize from the studies of only a few species.

2.7 Effects of Natural Selection on Phylogeographic Structure

The construction of demographic hypotheses fundamentally assumes that DNA sequence polymorphisms are 'neutral' to natural selection and that the shape of a genealogy is due solely to genetic drift and gene flow. Unfortunately, genetic imprints from drift and gene flow can look the same as those from various forms of natural selection. For example, background selection prevents slightly deleterious mutations from drifting to higher frequencies, producing an allele frequency spectrum with an excess of low-frequency alleles that resembles a spectrum in a recently expanded population (Charlesworth et al. 1993). Additionally, strong directional selection at a locus can produce a 'selective sweep' as genes linked to a selected locus are carried along to higher frequencies (Smith and Haigh 1974). A selective sweep can lead to the loss of genetic diversity that is indistinguishable from the loss of diversity from a bottleneck in population size (Nei et al. 1975).

Departures from neutrality are generally interpreted only in terms of demographic process, because it is difficult to construct testable hypotheses for the many possible forms of selection. Several statistics can be used to test for departures from neutrality, but they cannot always distinguish the effects of population history from natural selection (Tajima 1989; Fu and Li 1993; Fu 1997; among others). Even so, the potential effects of natural selection cannot be ignored in constructing phylogeographical hypotheses (Gagnaire et al. 2015) (Box 3). Benthic seaweeds may be more affected by climate changes than other marine species, because seaweeds are sessile, are restricted to the photic zone, and are sensitive to physical variables associated with climate change (Harley et al. 2012; Clark et al. 2013; Koch et al. 2013).

Box 3. Confounding Effects of Demography and Natural Selection on Gene Genealogies

The geographical distributions of mtDNA (*Cox*3) haplotypes among populations of the brown seaweed *Sargassum horneri* along the coasts of China and Japan show strong regional differences among five lineages (Fig. 2.4a) (Hu et al. 2011). Haplotype and nucleotide diversities tended to be larger in southern areas and smaller in northern areas, even considering differences in sample sizes. This pattern was interpreted to reflect isolations in multiple refugia during the late Pleistocene, followed by postglacial secondary contact. Contemporary dispersals in ocean currents were further postulated to explain patterns of connectivity between populations. These explanations invoke only demographic processes. However, the pattern also supports an alternative explanation involving selection, and illustrates the potentially confounding effects of demography and selection on phylogeographic structure.

A parsimony network shows a linear arrangement of lineages corresponding more or less to the geography of the samples, indicating a close genetic relationship between lineages in adjoining regions (Fig. 2.4b). This

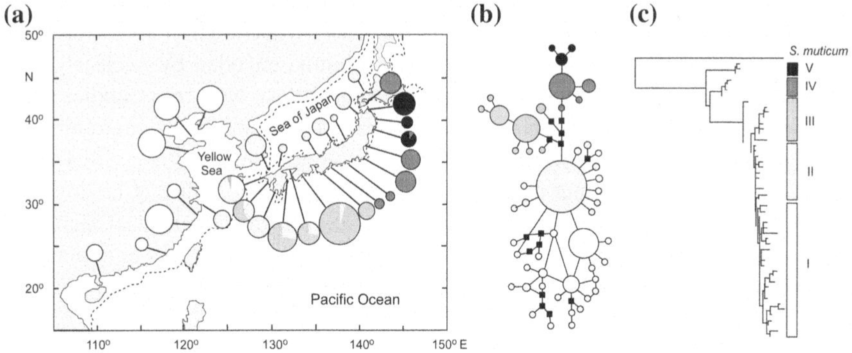

Fig. 2.4 Phylogeographic analysis of mtDNA *Cox3* in *Sargassum horneri* populations in the NW Pacific. **a** Map showing frequencies of haplotypes in lineages I–IV. *Dashed line* indicates shoreline at the LGM about 0.019 Ma. **b** Parsimony haplotype network. *Sizes of circles* are approximately proportional to haplotype frequency, and lines connecting haplotypes represent one mutational step. *Black rectangular blocks* represent hypothetical, but unobserved haplotypes. **c** Neighbor-joining tree of haplotypes. Lineages indicated by *shaded boxes*. (Redrawn from Hu et al. 2011)

arrangement could be interpreted to have resulted from progressive post-glacial colonizations that led to reduced diversity in northern populations, the pioneer mode of colonization (Hewitt 2000). However, an alternative scenario is possible. The shoreline along this coast was warmed by the north-flowing Kuroshio Current during glaciation, so that populations may have not been forced into southern areas, but displaced offshore during lower sea levels. A neighbor-joining tree of the *Cox*3 haplotypes, rooted by the outgroup *S. muticum* (Fig. 2.4c) shows ancestral lineages in the north and derived lineages in the south. This polarization of haplotypes does not support a postglacial expansion from the south.

An alternative scenario invokes both natural selection and demography, in which a cold-adapted, northern lineage expanded into southern waters only after beneficial mutations allowed individuals to complete a life-history cycle in warmer waters. Individuals with beneficial mutations progressively expanded into southern areas, producing a demographic signature of a population expansion with an excess of low-frequency haplotypes and star-shaped haplotype genealogies. The beneficial mutations need not be linked to the sequence that was surveyed in the study, but may be part of the mitochondrial–nuclear unit of selection (Dowling et al. 2008). If this is the case, the effects of selection and demography are confounded and cannot be teased apart by standard phylogeographic analysis. Additional studies of physiology are required.

Seaweeds may be particularly vulnerable to environmental changes because life-history phases may have different ecological requirements (e.g., Mohring et al. 2014). One of the major population markers used in phylogeographical studies, mtDNA may be particularly subject to selection. Mitochondrial DNA encodes several protein subunits, which together with subunits encoded by nuclear genes, form functional respiratory proteins (Dowling et al. 2008). Numerous studies show that the distributions of mtDNA haplotypes are correlated with environmental variables, particularly temperature, in fishes (Consuegra et al. 2015; Silva et al. 2014), and humans (Mishmar et al. 2003), and with altitude in birds (Cheviron and Brumfield 2009) and lizards (Jezkova et al. 2013). The development of technologies that can be used to probe entire genomes will allow researchers to understand the effects of proximate environmental variables on evolutionary responses to short- and long-term climate changes.

2.8 Conclusions

The goals of this chapter have been to explore the myriad life-history and environmental variables that shape phylogeographic structure in seaweeds and to examine the methodological and conceptual challenges to understanding microevolutionary mechanisms. The mechanisms producing phylogeographic patterns can only be understood by holistically examining the variables influencing the abundances and distributions of local populations. The Huxley–Tinbergen research questions (Box 1) provide a framework to investigate proximate and ultimate causes. Phylogeographic patterns in seaweeds are largely interpreted by addressing the first three questions. First, what are the proximate mechanisms controlling vertical and geographic distributions? These include physiological responses to sea temperature and salinity variability, morphological responses to substrate variability and responses to herbivory, among many other variables. Second, how do the ontologies of plants influence their distributions? Numerous studies show that mating systems and life-history variability can influence abundances and distributions of populations. Third, what are the 'phylogenetic' relationships among populations? Phylogeographic research largely falls into this category with a focus on the analysis of gene genealogies. Answering the fourth question–how can we understand the evolutionary significance of variation–is a more difficult research objective. However, the continuing development of next-generation genomic and functional genetic approaches will provide avenues for understanding adaptive responses to natural selection in ever-changing marine environments.

Importantly, a phylogeographic study represents only a snapshot of dynamic processes influenced by continuous climate changes. Attributing phylogeographic patterns to specific climate events is a daunting task, made difficult by large errors in calibrating marker- and species-specific molecular clocks. In the light of the continuously changing climates on all temporal scales around the globe (Fig. 2.3) and frequent miscalibrations of molecular clocks, few reconstructions of historical demography are likely to reflect reality.

Lastly, many phylogeographic studies focus on reconstructing historical demography using models with simplifying assumptions that restrict the construction of broader explanatory hypotheses. The use of neutral models with easily testable hypotheses to explain phylogeographic patterns can lead to a biased view of the nature of microevolutionary processes. Natural selection has doubtlessly shaped, and continues to shape, the patterns of genetic diversity observed in contemporary populations. A growing body of literature demonstrates the influence that natural selection has on respiratory and metabolic genes encoded in the nuclear and organelles. New sequencing technologies capable of producing massive amounts of genomic data are beginning to provide insights into how natural selection targets particular genes (Holsinger 2010). The recognition that natural selection has to be considered in explanations of the origins of phylogeographic patterns are leading to conceptual shifts beyond the original formulation of phylogeography in the 1980s.

Acknowledgments Dr. Sandra Lindstrom kindly commented on an early draft of the manuscript. Drs. Zi-Min Hu and Ceridwen Fraser and an anonymous reviewer provided insightful comments. Thanks to Dr. Zi-Min Hu for the opportunity to articulate my perspective on seaweed phylogeography. The writing of this manuscript was supported by the general funds of the Alaska Department of Fish and Game. This is Professional Publication number PP-280 of the Commercial Fisheries Division of the Alaska Department of Fish and Game, Anchorage Alaska.

References

Alberto F, Raimondi PT, Reed DC, Coelho NC, Leblois R, Whitmer A, Serrão EA. Habitat continuity and geographic distance predict population genetic differentiation in giant kelp. Ecology. 2010;91:49–56.

Alheit J, Hagen E. Long-term climate forcing of European herring and sardine populations. Fish Oceanogr. 1997;6:130–9.

Amsler CD, Neushul M. Nutrient stimulation of spore settlement in kelps *Pterygophora californica* and *Macrocystis pyrifera*. Mar Biol. 1990;107:297–403.

Amsler CD, Searles RB. Vertical distribution of seaweed spores in a water column offshore of North Carolina. J Phycol. 1980;16:617–9.

Arbogast B, Kenagy GJ. Comparative phylogeography as an integrative approach to historical biogeography. J Biogeogr. 2001;28:819–25.

Avise JC. Phylogeography: the history and formation of species. Cambridge: Harvard University Press; 2000.

Avise JC. Phylogeography: retrospect and prospect. J Biogeogr. 2009;36:3–15.

Avise JC, Arnold J, Ball RM Jr, Bermingham E, Lamb T, et al. Intraspecific phylogeography: the mitochondrial DNA bridge between population genetics and systematics. Ann Rev Ecol Syst. 1987;18:489–522.

Barrett SCH, Harder LD, Worley AC. The comparative biology of pollination and mating in flowering plants. Phil Trans R Soc B. 1996;351:1271–80.

Beaumont MA. Approximate Bayesian computation in evolution and ecology. Ann Rev Ecol Evol Syst. 2010;41:379–406.

Beerli P, Felsenstein J. Maximum likelihood estimation of a migration matrix and effective population sizes in *n* subpopulations by using a coalescent approach. Proc Nat Acad Sci USA. 2001;98:4563–8.

Bellgrove A, Aoki MN. Variation in gametophyte dominance in populations of *Chondrus verrucosus* (Gigartinaceae, Rhodophyta). Phycol Res. 2008;56:246–54.

Berndt M-L, Callo JA, Brawley SH. Gamete concentrations and timing and success of fertilization in a rocky shore seaweed. Mar Ecol Prog Ser. 2002;226:273–85.

Bhattacharya D, Druehl LD. Restriction enzyme analysis of variation and taxonomy in the kelp genus *Laminaria* (Laminariales, Phaeophyta). Hydrobiologia. 1990;204:105–10.

Bigg GR, Cunningham CW, Ottersen G, Pogson GH, Wadley MR, Williamson P. Ice-age survival of Atlantic cod: agreement between palaeoecology models and genetics. Proc R Soc B. 2007;275:163–73.

Bilgin R, Çoraman E, Karatas A, Morales JC. Phylogeography of the greater horseshoe bat, *Rhinolophus ferrumequinum* (Chiroptera: Rhinolophidae), in southeastern Europe and Anatolia, with a specific focus on whether the Sea of Marmara is a barrier to gene flow. Act Chiropterologica. 2009;11:53–60.

Billot C, Engel CR, Rousvoal S, Kloareg B, Valero M. Current patterns, habitat discontinuities and population genetic structure: the case of the kelp *Laminaria digitata* in the English Channel. Mar Ecol Prog Ser. 2003;253:111–21.

Bird CE, Karl SA, Smouse PE, Toonen RJ (2011) Detecting and measuring genetic differentiation. In: Held C, Koenemann S, Schubart CD (eds), Phylogeography and population genetics in Crustacea, vol 19. CRC Press, pp 31–55.
Birky CW Jr. Uniparental inheritance of mitochondrial and chloroplast genes: mechanisms and evolution. Proc Nat Acad Sci USA. 1995;92:11331–8.
Bock R (2007) Structure, function, and inheritance of plastid genomes. In: Bock R (ed), Cell and molecular biology of plastids. pp. 29–63 (Topics Curr Genet 19).
Bond G, Broecker W, Johnsen S, McManus J, Labeyrie L, Jouzel J, Bonani G. Correlations between climate records from North Atlantic sediments and Greenland ice. Nature. 1993;365:143–7.
Bond G, Showers W, Cheseby M, Lotti R, Almasi P, de Menocal P, Priore P, Cullen H, Hajdas I, Bonani G. A pervasive millennial-scale cycle in North Altantic Holocene and glacial climates. Science. 1997;278:1257–66.
Boo GH, Lindstrom SC, Klochkova NG, Yotsukura N, Yang EC, Kim HG, Waaland JR, Cho GY, Miller KA, Boo SM. Taxonomy and biogeography of *Agarum* and *Thalassiophyllum* (Laminariales, Phaeophyceae) based on nuclear, mitochondrial and plastid gene sequences. Taxon. 2011;60:831–40.
Bowen BW, Shanker K, Yasuda N, Malay MCD, von der Heyden S, Paulay G, Rocha LA, Selkoe KA, Barber PH, Lessios HA, Crandall ED, Bernardi G, Meyer CP, Carpenter KE, Toonen RJ. Phylogeography unplugged: comparative surveys in the genomic era. Bull Mar Sci. 2014;90:13–46.
Bråten T. The ultrastructure of fertilization and zygote formation in the green alga, *Ulva mutabilis* Føyn. J Cell Sci. 1971;9:621–35.
Bråten T. Autoradiographic evidence for the rapid disintegration of one chloroplast in the zygote of the green alga *Ulva mutabilis*. J Cell Sci. 1973;12:385–9.
Brawley SH, Wetherbeee R, Quatrano RS. Fine-structural studies of the gametes and embryo of *Fucus vesiculosus* L. (Phaeophyta). I. Fertilization and pronuclear fusion. J Cell Sci. 1976;20:233–54.
Brennan G, Kregting L, Beatty GE, Cole C, Elsässer B, Savidge G, Provan J. Understanding macroalgal dispersal in a complex hydrodynamic environment: a combined population genetic and physical modelling approach. J Roy Soc Interface. 2014;11:20140197.
Brito P, Edwards SV. Multilocus phylogeography and phylogenetics using sequence-based markers. Genetica. 2009;135:439–55.
Browning SR, Browning BL. Haplotype phasing: existing methods and new developments. Nat Rev Genet. 2011;12:703–14.
Burger G, Saint-Louis D, Gray MW, Lang BF. Complete sequence of the mitochondrial DNA of the red alga *Porphyra purpurea*: cyanobacterial introns and shared ancestry of red and green algae. Plant Cell. 1999;11:1675–94.
Bussolini LT, Waters JM. Genetic analyses of rafted macroalgae reveal regional oceanographic connectivity patterns. J Biogeogr. 2015;42:1319–26.
Campbell NR, Harmon SA, Narum SR. Genotyping-in-thousands by sequencing (GT-seq): a cost effective SNP genotyping method based on custom amplicon sequencing. Mol Ecol Res. 2015;15:855–67.
Carrington E, Grace SP, Chopin T. Life history phases and the biomechanical properties of the red alga *Chondrus crispus* (Rhodophyta). J Phycol. 2001;37:699–704.
Chan SW, Cheang CC, Chiraart A, Gerung G, Tharith C, Ang P. Homogeneous population of the brown alga *Sargassum polycystum* in Southeast Asia: possible role of recent expansion and asexual propogation. PLoS ONE. 2013;8:e77662.
Chan SW, Heang CC, Yeung CW, Chirapart A, Gerung G, Tharith C, Ang P. Recent expansion led to the lack of genetic structure of *Sargassum aquifolium* populations in Southeast Asia. Mar Biol. 2014;16:785–95.
Charlesworth D, Charlesworth B. Inbreeding depression and its evolutionary consequences. Ann Rev Ecol Syst. 1987;18:237–68.

Charlesworth B, Morgan MT, Charlesworth D. The effect of deleterious mutations on neutral molecular variation. Genetics. 1993;134:1289–303.

Cheviron ZA, Brumfield RT. Migration-selection balance and local adaptation of mitochondrial haplotypes in rufous-collared sparrows (*Zonotrichia capensis*) along an elevational gradient. Ecology. 2009;63:1593–605.

Chikhi L, Sousa VC, Luisi P, Goossens B, Beaumont MA. The confounding effects of population structure, genetic diversity and the sampling scheme on the detection and quantification of population size changes. Genetics. 2010;186:983–95.

Choi SJ, Park EJ, Endo H, Kitade Y, Saga N. Inheritance pattern of chloroplast and mitochondrial genomes in artificial hybrids of *Porphyra yezoensis* (Rhodophyta). Fish Sci. 2008;74:822–9.

Clark JS, Poore AGB, Ralph PJ, Doblin MA. Potential for adaptation in response to thermal stress in an intertidal macroalga. J Phycol. 2013;49:630–9.

Clarkston BE, Saunders GW. A comparison of two DNA barcode markers for species discrimination in the red algal family Kallymeniaceae (Gigartinales, Florideophyceae), with a description of *Euthora timburnonii* sp. nov. Botany. 2010;88:119–31.

Coleman MA, Brawley SH. Are life history characteristics good predictors of genetic diversity and structure? A case study of the intertidal alga *Fucus spiralis* (Heterokontophyta; Phyaeophyceae). J Phycol. 2005;41:753–62.

Coleman AW, Goff LJ. DNA analysis of eukaryotic algal species. J Phycol. 2004;27:463–73.

Coleman AW, Kelaher BP. Connectivity among fragmented populations of a habitat-forming alga, *Phyllospoa comosa* (Phaeophycea, Fucales) on an urbanised coast. Mar Ecol Prog Ser. 2009;381:63–70.

Collins CJ, Freaser CI, Ashcroft A, Waters JM. Asymmetric dispersal of southern bull-kelp (*Duvillaea antarctica*) adults in coastal New Zealand: testing an oceanographic hypothesis. Mol Ecol. 2010;19:4572–80.

Consuegra S, John E, Versppor E, Garcia de Leaniz C. Patterns of natural selection acting on the mitochondrial genome of a locally adapted fish species. Genet Sel Evol. 2015. doi:10.1186/s12711-015-0138-0.

Coyer JA, Olsen JL, Stam WT. Genetic variability and spatial separation in the sea palm kelp *Postelsia palmaeformis* (Phaeophyceae) as assessed with m13 fingerprint and RAPDS. J Phycol. 1997;33:561–8.

Coyer JA, Smith GJ, Andersen RA. Evolution of *Macrocystis* spp. (Phaophyceae) as determined by ITS1 and ITS2 sequences. J Phycol. 2001;37:574–85.

Coyer JA, Peters AF, Hoarau G, Stam WT, Olsen JL. Inheritance patterns of ITS1, chloroplasts, and mitochondria in artificial hybrids of the marine rockweeds, *Fucus serratus* and *F. evenescens* (Heterokontophyta; Fucaceae). Euro J Phycol. 2002;37:173–8.

Coyer JA, Peters AF, Stam WT, Olsen JL. Post-ice age recolonization and differentiation of *Fucus serratus* L. (Phaeophyceae; Fucaceae) populations in Northern Europe. Mol Ecol. 2003;12:1817–29.

Coyer JA, Hoarau G, Stam WT, Olsen JL. Geographically specific heteroplasmy of mitochondrial DNA in the seaweed, *Fucus serratus* (Heterokontophyta: Phaeophyceae, Fucales). Mol Ecol. 2004;13:1323–6.

Crandall ED, Sbrocco EJ, Deboer TS, Barber PH, Carpenter KE. Expansion dating: calibrating molecular clocks in marine species from expansions onto the Sunda Shelf following the Last Glacial Macimum. Mol Biol Evol. 2012;29:707–19.

Crow JF, Kimura M (1970) An introduction to population genetics theory. New York: Harper and Row, xiv + 591 p.

Davey JW, Hohenlohe PA, Etter PD, Boone JQ, Catchen JM, Blaxter ML. Genome-wide genetic marker discovery and genotyping using next-generation sequencing. Nat Rev Genet. 2011;12:499–510.

Dawson MN, Algar AC, Antonelli A, Dávalos LM, Davis E, Early R, Guisan A, Jansson R, Lessard JP, Marske KA, McGuire JL, Stigall AL, Swenson NG, Zimmermann NE, Gavin DG. An horizon scan of biogeography. Front Biogeogr. 2013;5:130–57.

Destombe C, Godin J, Lefebvre C, Dehorter O, Vernet P. Differences in dispersal abilities of haploid and diploid spores of *Gracilaria verrucosa* (Gracilariales, Rhodophyta). Bot Mar. 1992;35:93–98.
Dewsbury DA. The proximate and the ultimate: past, present, and future. Behav Process. 1999;46:189–99.
Deysher L, Norton TA. Dispersal and colonization in *Sargassum muticum* (Yendo) Fensholt. J Exp Mar Biol Ecol. 1981;56:179–95.
Dixon PS. Perennation vegetative propagation and algal life histories, with special reference to *Asparagopsis* and other Rhodophyta. Bot Gothoburg. 1965;3:67–74.
Dowling DK, Friberg U, Lindell J. Evolutionary implications of non-neutral mitochondrial genetic variation. Trends Ecol Evol. 2008;23:546–54.
Drummond AJ, Rambaut A. BEAST: Bayesian evolutionary analysis by sampling trees. BMC Evol Biol. 2007;7:214.
Drummond AJ, Rambaut A, Shapiro B, Pybus OG. Bayesian coalescent inference of past population dynamics from molecular sequences. Mol Biol Evol. 2005;22:1185–92.
Dumolin S, Demesure B, Petit RJ. Inheritance of chloroplast and mitochondrial genomes in pedunculate oak investigated with an efficient PCR method. Theor Appl Genet. 1995;91:1253–6.
Dyck JL, DeWreede RE. Reproduction and survival in *Mazzaella splendens* (Gigartinales, Rhodophyta). Phycologia. 2006a;45:302–10.
Dyck JL, DeWreede RE. Seasonal and spatial patterns of population density in the marine macroalga *Mazzaella splendens* (Gigartinales, Rhodophyta). Phycol Res. 2006b;54:21–31.
Eckert CG. The loss of sex in clonal plants. Evol Ecol. 2002;15:501–20.
Edwards SV, Beerli P. Gene divergence, population divergence, and the variance in coalescence time in phylogeographic studies. Evolution. 2000;54:1839–54.
Eldon B, Wakeley J. Coalescent processes when the distribution of offspring number among individuals is highly skewed. Genetics. 2006;172:2621–33.
Eldon B, Birkner M, Blath J, Freund F. Can the site-frequency spectrum distinguish exponential population growth from multiple-merger coalescents? Genetics. 2015;199:841–56.
Elith J, Leathwick JR. Species distribution models: ecological explanation and prediction across space and time. Ann Rev Ecol Evol Syst. 2009;40:677–97.
Ellegren H. Microsatellites: simple sequences with complex evolution. Nat Rev Genet. 2004;5:435–45.
Engel CR, Wattier R, Destombe C, Valero M. Performance of non-motile male gametes in the sea: analysis of paternity and fertilization success in a natural population of a red seaweed, *Gracilaria gracilis*. Proc R Soc B. 1999;266:1879–86.
Engel CR, Destombe C, Valero M. Mating system and gene flow in the red seaweed *Gracilaria gracilis*: effect of haploid-diploid life history and intertidal rocky shore landscape on fine-scale genetic structure. Heredity. 2004;92:289–98.
Fauvelot C, Bernardi G, Planes S. Reductions in the mitochondrial DNA diversity of coral reef fish provide evidence of population bottlenecks resulting from Holocene sea-level change. Evolution. 2003;57:1571–83.
Fierst J, ter Horst C, Kübler JE, Dudgeon S. Fertilization success can drive patterns of phase dominance in complex life histories. J Phycol. 2005;41:238–49.
Fjeld A, Løvlie A. Genetics of multicellular marine algae. In: Lewin RA, editor. The genetics of algae. Oxford: Blackwell; 1976. p. 219–35.
Forrest BM, Brown SN, Taylor MD, Hurd CL, Hay CH. The role of natural dispersal mechanisms in the spread of *Undaria pinnatifida* (Laminariales, Phaeophyceae). Phycologia. 2000;39:547–53.
Francis RC, Hare SR, Hollowed AB, Wooster WS. Effects of interdecadal climate variability on the oceanic ecosystems of the NE Pacific. Fish Oceanogr. 1998;7:1–21.
Fraser CI, Nikula R, Spencer HG, Waters JM. Kelp genes reveal effects of subantarctic sea ice during the Last Glacial Maximum. Proc Nat Acad Sci USA. 2009;106:3249–53.
Fraser CI, Thiel M, Hamish G, Spencer HG, Waters JM. Contemporary habitat discontinuity and historic glacial ice drive genetic divergence in Chilean kelp. BMC Evol Biol. 2010;10:203.

Fraser CI, Nikula R, Water JM. Ocean rafting by a coastal community. Proc Roy Soc B. 2011;278:649–55.

Frederica S, Brodie J, Hommersand MH. Developmental morphology of *Chondrus crispus* (Gigartinaceae, Rhodophyta). Phycologia. 1992;31:542–63.

Freshwater DW, Fredericq S, Butler BS, Hommersand MH, Chase MW. A gene phylogeny of the red algae (Rhodophyta) based on plastid rbcL. Proc Nat Acad Sci USA. 1994;91:7281–5.

Fu YX. Statistical tests of neutrality of mutations against population growth, hitchhiking and background selection. Genetics. 1997;147:915–25.

Fu YX, Li WH. Statistical tests of neutrality of mutations. Genetics. 1993;133:693–709.

Fu YX, Li WH. Coalescing into the 21st century: an overview and prospects of coalescent theory. Theor Popul Biol. 1999;56:1–10.

Gabrielsen TM, Brochmann C, Rueness J. The Baltic Sea as a model system for studying postglacial colonization and ecological differentiation, exemplified by the red alga *Ceramium tenuicorne*. Mol Ecol. 2002;11:2083–95.

Gagnaire P-A, Broquet T, Aurelle D, Viard F, Souissi A, Bohomme F, Arnaud-Haond S, Bierne N. Using neutral, selected, and hitchhiker loci to assess connectivity of marine populations in the genomic era. Evol Appl. 2015;8:769–86.

Galindo H, Olson D, Palumbi SR. Seascape genetics: a coupled oceanographic-genetic model predicts population structure of Caribbean corals. Curr Biol. 2006;16:1622–6.

Garden CJ, Currie K, Fraser CI, Waters JM. Rafting dispersal constrained by an oceanographic boundary. Mar Ecol Prog Ser. 2014;501:297–302.

Garrick RC, Sunnucks P, Dyer RJ. Nuclear gene phylogeography using PHASE: dealing with unresolved genotypes, lost alleles, and systematic bias in parameter estimation. BMC Evol Biol. 2010;10:118.

Gaunt MW, Miles MA. An insect molecular clock dates the origin of the insects and accords with palaeontological and biogeographic landmarks. Mol Biol Evol. 2002;19:748–61.

Gaylord B, Reed DC, Raimondi PT, Washburn L, McLean TR. A physically based model of macroalgal spore dispersal in the wave and current-dominated nearshore. Ecology. 2002;83:1239–51.

Gedalof Z, Smith DJ. Interdecadal climate variability and regime-scale shifts in Pacific North America. Geophys Res Letts. 2001;28:1515–8.

Giles RE, Blanc H, Cann HM, Wallace DC. Genetics maternal inheritance of human mitochondrial DNA. Proc Nat Acad Sci USA. 1980;77:6715–9.

Gill JL, Williams JW, Jackson ST, Lininger KB, Robinson GS. Pleistocene megafaunal collapse, novel plant communities, and enhanced fire regimes in North America. Science. 2009;326:1100–3.

Gómez A, Lunt DH. Refugia within refugia: patterns of phylogeographic concordance in the Iberian Peninsula. In: Weiss S, Ferrand N, editors. Phylogeography of southern European refugia. Netherlands: Springer; 2007. p. 155–88.

Goowille C, Kalisz S, Eckert CG. The evolutionary enigma of mixed mating systems in plants: occurrence, theoretical explanations, and empirical evidence. Ann Rev Ecol Evol Syst. 2005;36:47–79.

Gordon R, Brawley SH. Effects of water motion on propagule release from algae with complex life histories. Mar Biol. 2004;145:21–9.

Grant WS. Problems and cautions with sequence mismatch analysis and Bayesian skyline plots to infer historical demography. J Hered. 2015. doi:10.1093/jhered/esv020.

Grant WS, Liu M, Gao TX, Yanagimoto T. Limits of Bayesian skyline plot analysis of mtDNA sequences to infer historical demographies in Pacific herring (and other species). Mol Phylogenet Evol. 2012;65:203–12.

Habel JC, Zachos FE, Dapporto L, Rödder D, Radespiel U, Tellier A, Schmitt T. Population genetic revisited–toward a mulidisciplinary research field. Biol J Linn Soc. 2015;115:1–12.

Han WH, Klochkova TA, Chim J, Nagasato C, Motomura T, Kim GH. Identification of three proteins involved in fertilization and parthenogenetic development of a brown alga, *Scytosiphon lomentaria*. Planta. 2014;240:1253–67.

Hanebuth T, Stattegger K, Grootes PM. Rapid flooding of the Sunda Shelf: a late-glacial sea-level record. Science. 2000;288:1033–5.

Hannach G, Santelices B. Ecological differences between the isomorphic reproductive phases of two species if *Iridaea* (Rhodophyta: Gigartinales). Mar Ecol Prog Ser. 1985;22:291–303.

Hannach G, Waaland JR. Environment, distribution and production of *Iridaea*. Aquat Bot. 1986;26:51–78.

Hansen JE, Doyle WT. Ecology and natural history of *Iridaea cordata* (Rhodophyta, Gigartinaceae): population structure. J Phycol. 1976;12:273–8.

Hare JA, Able KW. Mechanistic links between climate and fisheries along the east coast of the United States: explaining population outbursts of Atlantic croaker (*Micropogonias undulates*). Fish Oceanogr. 2007;16:31–45.

Harley CDG, Anderson KM, Demes KW, Jorve JP, Kordas RL, Coyle TA, Graham MH. Effects of climate change on global seaweed communities. J Phycol. 2012;48:1064–78.

Hedgecock D, Pudovkin AI. Sweepstakes reproductive success in high fecund marine fish and shellfish: a review and commentary. Bull Mar Sci. 2011;87:971–1002.

Hedrick PW. Perspective: highly variable loci and their interpretation in evolution and conservation. Evolution. 1999;53:313–8.

Heled J, Drummond AJ. Bayesian inference of population size history from multiple loci. BMC Evol Biol. 2008;8:289.

Hernández-Carmona G, Hughes B, Graham MH. Reproductive longevity of drifting kelp *Macocystis pyrifera* (Phaeophyceae) in Monterey Bay, USA. J Phycol. 2006;42:1199–207.

Hewitt GM. Some genetic consequences of ice ages, and their role in divergence and speciation. Biol J Linn Soc. 1996;58:247–76.

Hewitt GM. The genetic legacy of the Quaternary ice ages. Nature. 2000;405:907–13.

Hey J, Nielsen R. Multilocus methods for estimating populations sizes, migration rates and divergence time with applications to the divergence of *Drosophila pseudoobscura* and *D. persimilis*. Genetics. 2004;167:747–60.

Hickerson MJ, Carstens BC, Cavender-Bares J, Crandall KA, Graham CH, et al. Phylogeography's past, present, and future: 10 years after Avise, 2000. Mol Phylogenet Evol. 2010;54:291–301.

Ho SYW, Phillips MJ, Cooper A, Drummond AJ. Time dependency of molecular rate estimates and systematic overestimation of recent divergence times. Mol Biol Evol. 2005;22:1561–8.

Ho SYW, Saarma U, Barnett R, Haile J, Shaprio B. The effect of inappropriate calibration: three case studies in molecular ecology. PLoS ONE. 2008;3:e1615.

Ho SYW, Lanfear R, Bromham L, Phillips MJ, Soubrier J, Rodrigo AG, Cooper A. Time-dependent rates of molecular evolution. Mol Ecol. 2011;20:3087–101.

Hoarau G, Coyer JA, Veldsink JH, Stam WT, Olsen JL. Glacial refugia and recolonization pathways in the seaweed *Fucus serratus*. Mol Ecol. 2007;16:3606–16.

Hoarau G, Coyer JA, Olsen JL. Paternal leakage of mitochondrial DNA in *Fucus* (Phaeophyceae). J Phycol. 2009;45:621–4.

Hoffmann AJ. The arrival of seaweed propagules at the shore: a review. Bot Mar. 1987;30:151–65.

Hoffmann AJ, Ugarte R. The arrival of propagules of marine macroalgae in the intertidal zone. J Exp Mar Biol Ecol. 1985;92:83–95.

Holsinger KE. Next generation population genetics and phylogeography. Mol Ecol. 2010;1: 2361–3.

Hu ZM, Duan DL. Insufficient geographical sampling could severely influence phylogeographic interpretations. Mar Biol. 2013;160:1521–2.

Hu ZM, Guiry MD, Critchley AT, Duan DL. Phylogeographic patterns indicate trans-Atlantic migration from Europe to North America in the red seaweed *Chondrus crispus* (Gigartinales, Rhodyphyta). J Phycol. 2010;46:889–900.

Hu ZM, Uwai S, Yu SH, Komatsu T, Ajisaka T, Duan DL. Phylogeographic heterogeneity of the brown macroalga *Sargassum horneri* (Fucaceae) in the northwestern Pacific in relation to late Pleistocene glaciation and tectonic configurations. Mol Ecol. 2011;20:3894–909.

Hu ZM, Zhang J, Lopez-Bautista J, Duan DL. Asymmetric genetic exchange in the brown seaweed *Sargassum fusiforme* (Phaeophyceae) driven by oceanic currents. Mar Biol. 2013;160:1407–14.

Hudson RR. Gene genealogies and the coalescent process. Oxford Surv Evol Biol. 1991;7:1–44.

Hurrell JW. Decadal trends in the North Atlantic Oscillation: regional temperatures and precipitation. Science. 1995;269:676–9.

Huxley J. Bird-watching and biological science. Auk. 1916;33:256–70.

Jezkova T, Leal M, Rodríguez-Robles JA. Genetic drift or natural selection? Hybridization and asymmetric mitochondrial introgression in two Caribbean lizards (*Anolis puchellus* and *Anolis krugi*). J Evol Biol. 2013;26:1458–71.

Jouzel J, Masson-Delmotte V, Cattani O, Dreyfus G, Falourd S, Hoffmann G, Minster B, Nouet J, Barnola JM, Chappellaz J, Fischer H, Gallet JC, Johnsen S, Leuenberger M, Loulergue L, Luethi D, Oerter H, Parrenin F, Raisbeck G, Raynaud D, Schilt A, Schwander J, Selmo E, Souchez R, Spahni R, Stauffer B, Steffensen JP, Stenni B, Stocker TF, Tison JL, Werner M, Wolff EW. Orbital and millennial Antarctic climate variability over the past 800,000 years. Science. 2007;317:793–6.

Jueterbock A, Tyberghein L, Verbruggen H, Coyer JA, Olsen JL, Hoarau G. Climate change impact on seaweed meadow distribution in the North Atlantic rocky intertidal. Ecol Evol. 2013;3:1356–73.

Kain JM. The reproductice phenology of nine species of Rhodophyta in the subtidal region of the Isle of Man. Br Phycol J. 1982;17:321–31.

Karl SA, Toonen RJ, Grant WS, Bowen BW. Common misconceptions in molecular ecology: echoes of the modern synthesis. Mol Ecol. 2012;21:4171–89.

Kato Y, Kogame K, Nagasato C, Motomura T. Inheritance of mitochondrial and chloroplast genomes in the isogamous brown alga *Scytosiphon lomentaria* (Phaeophyceae). Phycol Res. 2006;54:65–71.

Keigwin LD. The little ice age and medieval warm period in the Sargasso Sea. Science. 1996;274:1504–8.

Kendrick GA, Walker DI. Dispersal of propagules of *Sargassum* spp. (Sargassaceae: Phaeophyta): observations of local patterns of dispersal and consequences for recruitment and population structure. J Exp Mar Biol Ecol. 1995;192:273–88.

Ketmaier V, Argano R, Caccone A. Phylogeography and molecular rates of subterranean aquatic Stenasellid isopods with a per-Tyrrhenian distribution. Mol Ecol. 2003;12:547–55.

Kimura K, Nagasato C, Uwai S, Motomura T. Sperm mitochondrial DNA elimination in the zygote of the oogamous brown alga *Undaria pinnatifida* (Laminariales, Phaeophyceae). Cytologia. 2010;75:353–61.

Kingman JFC. The coalescent. Stochast Proc Appl. 1982;13:235–48.

Kirkendale L, Saunders GW, Winberg P. A molecular survey of *Ulva* (Chlorophyta) in temperate Australia reveals enhanced levels of cosmopolitanism. J Phycol. 2013;49:69–81.

Knowlton N, Weigt LA. New dates and new rates for divergence across the Isthmus of Panama. Proc R Soc B. 1998;265:2257–63.

Koch M, Bowes G, Ross C, Zhang XH. Climate change and ocean acidification effects on seagrasses and marine macroalgae. Glob Change Biol. 2013;19:103–32.

Kraan S, Guiry M. Molecular and morphologial character inheritance in hybrids of *Alaria esculenta* and *A. praelonga* (Alariaceae, Phaeophyceae). Phycologia. 2000;39:554–9.

Kraft LGK, Kraft GT, Waller RF. Investigation into southern Australian *Ulva* (Ulvophyceae, Chlorophyta) taxonomy and molecular phylogeny indicate both cosmopolitanism and endemic cryptic species. J Phycol. 2010;46:1257–77.

Krueger-Hadfield SA, Collén J, Dauguin-Thiébaut C, Valero M. Genetic population structure and mating system in *Chondrus crispus* (Rhodophyta). J Phycol. 2011;47:440–50.

Krueger-Hadfield SA, Roze D, Mauger S, Valero M. Intergametophytic selfing and microgeographic genetic structure shape populations of the intertidal red seaweed *Chondrus crispus*. Mol Ecol. 2013;22:3242–60.

Kuroiwa T, Hori T. Preferential digestion of male chloroplast nuclei and mitochondrial nuclei during gametogenesis of *Bryopsis maxima* Okamura. Protoplasma. 1986;133:86–7.
Kusumo HT, Pfister CA, Wootton JT. Small-scale genetic structure in the sea palm *Postelsia palmaeformis* Ruprecht (Phaeophyceae). Mar Biol. 2006;149:731–42.
Lee RW, Lemieux C. Biparental inheritance of non-Mendelian gene markers in *Chlamydomonas moewusii*. Genetics. 1986;113:589–600.
Lee KM, Yang EC, Coyer JA, Zuccarello GC, Wang WL, Choi CG, Boo SM. Phylogeography of the seaweed *Ishige okamurae* (Phaeophysceae): evidence for glacial refugia in the northwest Pacific region. Mar Biol. 2012;159:1021–8.
Librado P, Rozas J. DnaSP v5: a software for comprehensive analysis of DNA polymorphism data. Bioinform. 2009;25:1451–2.
Liggins L, Treml EA, Riginos C. Taking the plunge: an introduction to undertaking seascape genetic studies and using biophysical models. Geogr Compass. 2013;7:173–96.
Lin SM, Yang WC, Huisman J, de Clerck O, Lee WJ. Molecular phylogeny of the widespread *Martensia fragilis* complex (Delesseriaceae, Rhodophyta) from the Indo-Pacific region reveals three new species of *Martensia* from Taiwan. Eur J Phycol. 2013;48:173–87.
Lindstrom SC, Hughey JR, Martone PT. New, resurrected and redefined species of *Mastocarpus* (Phllophoraceae, Rhodophyta) from the northeast Pacific. Phycologia. 2011;50:661–83.
Ling S, Johnson CR, Ridgway K, Hobday AJ, Haddon M. Climate-driven range extnsion of sea urchin: inferring future trends by analysis of recent population dynamics. Glob Change Biol. 2009;15:719–31.
Lisiecki LE, Raymo ME. A Pliocene-Pleistocene stack of 57 globally distributed benthic δ^{18} O records. Paleoceanography. 2005;20:PA1003.
Lourie SA, Green DM, Vincent ACJ. Dispersal, habitat differences, and comparative phylogeography of Southeast Asian seahorses (Syngnathidae: *Hippocampus*). Mol Ecol. 2005;14:1073–94.
Ludt WB, Bernal M, Bowen BW, Rocha LA. Living in the past: phylogeography and population histories of Indo-Pacific wrasses (Genus *Halichoeres*) in shallow lagoons versus outer reef slopes. PLoS ONE. 2012;7:e38042.
Maggs CA, Castilho R, Foltz D, Henzler C, Jolly MT, Kelly J, Olsen J, Perez KE, Stam W, Väinölä R, Viard F, Wares J. Evaluating signatures of glacial refugia for North Atlantic Benthic marine taxa. Ecology. 2008;89:S108–22.
Mann ME, Zhang Z, Rutherford S, Bradley RS, Hughes MK, Shindell D, Ammann C, Faluvegi G, Ni F. Global signatures and dynamical origins of the Little Ice Age and Medieval Climate Anomaly. Science. 2009;326:1256–60.
Mantua NJ, Hare SR. The Pacific decadal oscillation. J Oceanogr. 2002;58:35–44.
Mantua NJ, Hare SR, Zhang Y, Wallace JM, Francis RC. A Pacific interdecadal climate oscillations with impacts on salmon production. Bull Am Meteorol Soc. 1997;78:1069–79.
Marino IA, Pujolar JM, Zane L. Reconciling deep calibration and demographic history: Bayesian inference of post glacial colonization patterns in *Carcinus aestuarii* (Nardo, 1847) and *C. maenas* (Linnaeus, 1758). PLoS ONE. 2011;6:e28567.
Marins BV, Amado-Filho GM, Barreto MBB, Longo LL. Taxonomy of the southwestern Atlantic endemic kelp: *Laminaria abyssalis* and *Laminaria brasiliensis* (Phaeophyeae, Laminariales) are not different species. Phycol Res. 2012;60:51–60.
Marko PB. Fossil calibration of molecular clocks and the divergence times of geminate species pairs separated by the Isthmus of Panama. Mol Biol Evol. 2002;19:2005–21.
Marske KA, Rahbek C, Nogués-Bravo D. Phylogeography: spanning the ecology-evolution continuum. Ecography. 2013;36:1169–81.
Mathieson AC, Burns RL. Ecological studies of economic red algae. V. Growth and reproduction of natural and harvested populations of *Chondrus crispus* Stackhouse in New Hampshire. J Mar Exp Mar Biol Ecol. 1975;17:137–56.
Mayr E. Cause and effect in biology. Science. 1961;134:1501–6.
McCormack JE, Hird SM, Zellmer AJ, Carstens BC, Brumfield RT. Applications of next-generation sequencing to phylogeography and phylogenetics. Mol Phylogenet Evol. 2013;66:526–38.

McCulloch RD, Bentley MJ, Purves RS, Hulton NRJ, Sugden DE, Clapperton CM. Climatic inferences from glacial and palaeoecological evidence at the last glacial termination, southern South America. J Quat Sci. 2000;15:409–17.

McDevit DC, Saunders GW. A DNA barcode examination of the Laminariaceae (Phaeophyceae) in Canada reveals novel biogeographical and evolutionary insights. Phycologia. 2010;49:235–48.

McKenzie PF, Bellgrove A. Dispersal of *Hormosira banksii* (Phaeophyceae) via detached fragments: reproductive viability and longevity. J Phycol. 2008;44:1108–15.

Meirmans PG. Seven common mistakes in population genetics and how to avoid them. Mol Ecol. 2015;24:3223–31.

Metzker ML. Sequencing technologies–the next generation. Nat Rev Genet. 2010;11:31–46.

Miller KG, Kominz MA, Browning JV, Wright JD, Mountain GS, Katz ME, Sugarman PJ, Cramer BS, Cristie-Blick N, Pekar SF. The Phanerozoic record of global sea-level change. Science. 2005;310:1293–8.

Mishmar D, Ruiz-Pesin E, Golik P, Macaulay V, Clark AG, Hosseini S, Brandon M, Easley K, Chen E, Brown MD, Sukerni RI, Olckers A, Wallace DC. Natural selection shaped regional mtDNA variation in humans. Proc Nat Acad Sci USA. 2003;100:171–6.

Miyamura S. Cytoplasmic inheritance in green algae: patterns, mechanisms and relation to sex type. J Plant Res. 2010;123:171–84.

Miyamura S, Nagumo T. Inheritance pattern of chloroplast DNA is correlated with gamete types based on sex-specific arrangement of the cell fusion site in *Caulerpa* (Ulvophyceae, Chlorophyta). Phycol Res. 2007;55:47–57.

Mohring MB, Wernberg T, Wright JT, Connell SD, Russell BD. Biogeographic variation in temperature drives performance of kelp gametophytes during warming. Mar Ecol Prog Ser. 2014;513:85–96.

Montecinos A, Broitman BR, Faugeron S, Haye PA, Tellier F, Guillemin M-L. Species replacement along a linear coastal habitat: phylogeography and speciation in the red alga *Mazzaella laminarioides* along the south east Pacific. BMC Evol Biol. 2012;12:97.

Motomura T. Ultrastructure of fertilization in *Laminaria angustata* (Phaeophyta, Laminariales) with emphasis on the behavior of centrioles, mitochondria and chloroplasts of sperm. J Phycol. 1990;26:80–9.

Motomura T, Nagasato C, Kimura K. Cytoplasmic inheritance of organelles in brown algae. J Plant Res. 2010;123:185–92.

Muhlin JF, Engel CR, Stessel R, Weatherbee RA, Brawley SH. The influence of coastal topography, circulation patterns, and rafting in structuring populations of an intertidal alga. Mol Ecol. 2008;17:1198–210.

Nagasato C, Motomura T, Ichimura T. Spindle formation in karyogamy-blocked zygotes of the isogamous brown alga *Scytosiphon lomentaria* (Scytosiponales, Phaeophyceae). Eur J Phycol. 2000;35:339–47.

Nagasato C, Inoue A, Mizuno M, Kanazawa K, Ojima T, Okuda K, Motomura T. Membran fusion process and assembly of cell wall during cytokinesis in the brown alga, *Silvetia babingtonii* (Fucales, Phaeophyceae). Planta. 2010;232:287–298.

Nakagawa S, Cuthill IC. Effect size, confidence interval and statistical significance: a practical guide for biologists. Biol Rev. 2007;82:591–605.

Nei M, Maruyama T, Chakraborty R. The bottleneck effect and genetic variability in populations. Evolution. 1975;29:1–10.

Nielsen R, Paul JS, Albrechtsen A, Song YS. Genotype and SNP calling from next-generation sequencing data. Nat Rev Genet. 2011;12:443–51.

Oba T, Kato M, Kitazato H, Koizumi I, Omura A, Sakai T, Takayama T. Paleoenvironmental changes in the Japan Sea during the last 85,000 years. Paleoceanogaphy. 1991;6:499–518.

Ogawa S. Disappearance of chloroplast nucleoids during male gamete formation in *Bryopsis plumosa* (Hudson) C. Ag. (Chlorophyceae). Bot Gaz. 1988;149:25–9.

Patel RK, Jain M. NGS QC toolkit: a toolkit for quality control of next generation sequencing data. PLoS ONE. 2012;7:e30619.

Pearson GA, Brawley SH. Reproductive ecology of *Fucus distichus* (Phaeophyceae): an intertidal alga with successful external fertilization. Mar Ecol Prog Ser. 1996;143:211–23.
Pearson GA, Serrão EA. Revisiting synchronous gamete release by fucoid algae in the intertidal zone: fertilization success and beyond? Integr Compar Biol. 2006;46:587–97.
Pearson GA, Serrão EA, Brawley SH. Control of gamete release in fucoid algae: sensing hydrodynamic conditions via carbon acquisition. Ecology. 1998;79:1725–39.
Peck JR, Yearsley JM, Waxman D. Explaining the geographic distributions of sexual and asexual populations. Nature. 1998;391:889–92.
Perrin C, Daguin C, van de Sleit M, Engel CR, Pearson GA, Serrão EA. Implications of mating system for genetic diversity of sister algal species: *Fucus spiralis* and *Fucus vesiculosus* (Heterokontophyta, Phaeophyceae). Eur J Phycol. 2007;42:219–30.
Peters AF, Scornet D, Müller DG, Kloareg B, Cock JM. Inheritance of organelles in artificial hybrids of the isogamous multicellular chromist alga *Ectocarpus siliculosus* (Phaeophyceae). Eur J Phycol. 2004;39:235–42.
Philander SGH. El Niño southern oscillation phenomena. Nature. 1983;302:295–301.
Prince JS, Kingsbury JM. The ecology of *Chondrus crispus* at Plymouth, Massachusetts. I. Ontogeny, vegetative anatomy, reproduction and life cycle. Am J Bot. 1973;60:956–63.
Prost S, Smirnov N, Fedorov VB, Sommer RS, Stiller M, Nagel D, Knapp M, Hofreiter M. Influence of climate warming on Arctic mammals? New insights from ancient DNA studies of the collared lemming *Dicrostonyx torquatus*. PLoS ONE. 2010;5:e10447.
Provan J, Bennett KD. Phylogeographic insights into cryptic glacial refugia. Trends Ecol Evol. 2008;23:564–71.
Provan J, Wattier RA, Maggs CA. Phylogeographic analysis of the red seaweed *Palmaria palmata* reveals a Pleistocene marine glacial refugium in the English Channel. Mol Ecol. 2005;14: 793–803.
Provan J, Glendinning K, Kelly R, Maggs CA. Levels and patterns of population genetic diversity in the red seaweed *Chondrus crispus* (Florideophyceae): a direct comparison of single nucleotide polymorphisms and microsatellites. Biol J Linn Soc. 2013;108:251–62.
Qu Y, Luo X, Zhang R, Song G, Zou F, Lei F. Lineage diversification and historical demography of a montane bird *Garrulax elliotii*–implications for the Pleistocene evolutionary history of the eastern Himalayas. BMC Evol Biol. 2011;11:174.
Rainmondi PT, Reed DC, Gaylord B, Washburn L. Effects of self-fertilization of the giant kelp *Macrocystis pyrifera*. Ecology. 2004;85:3267–76.
Reed DC, Amsler CD, Ebeling AW. Dispersal in kelps: factors affecting spore swimming and competency. Ecology. 1992;73:1577–85.
Reed DC, Rainmondi PT, Carr MH, Goldwasser L. The role of dispersal and disturbance in determining spatial heterogeneity in sedentary organisms. Ecology. 2000;81:2011–26.
Reed DC, Schroeter SC, Raimondi PT. Spore supply and habitat availability as sources of recruitment limitation in the giant kelp *Macrocystis pyrifera* (Phaeophyceae). J Phycol. 2004;40:275–84.
Rogers AR, Harpending H. Population growth makes waves in the distribution of pairwise differences. Mol Biol Evol. 1992;9:552–69.
Rosenberg NA, Nordborg M. Genealogical trees, coalescent theory and the analysis of genetic polymorphisms. Nat Rev Genet. 2002;3:380–90.
Rubinsztein DC, Amos B, Cooper G. Microsatellite and trinucleotide-repeat evolution: evidence for mutational bias and different rates of evolution in different lineages. Philos Trans R Soc B. 1999;354:1095–9.
Russell FS, Southward AJ, Boalch GT, Butler EI. Changes in biological conditions in the English Channel off Plymouth during the last half century. Nature. 1971;234:468–70.
Ryman N, Palm S. POWSIM: a computer program for assessing statistical power when testing for genetic differentiation. Mol Ecol Notes. 2006;6:600–2.
Ryman N, Palm S, André C, Carvalho G, Dahlgren TG, Jorde PE, Lairkre L, Larsson LC, Palmé A, Ruzzante DE. Power for detecting divergence: differences between statistical methods and marker loci. Mol Ecol. 2006;15:2031–45.

Santelices B. Patterns of reproduction, dispersal and recruitment in seaweeds. Oceanogr Mar Biol Ann Rev. 1990;28:177–276.

Santos C, Moneiel R, Sierra B, Betencourt C, Fernandez E, Alvarez L, Lima M, Abada A, Aluja MP. Understanding differences between phylogenetic and pedigree-derived mtDNA mutation rate: a model using families from the Azores Islands (Portugal). Mol Biol Evol. 2005;22:1490–505.

Saunders GW. Applying DNA barcoding to red macroalgae: a preliminary appraisal holds promise for future applications. Phil Trans R Soc B. 2005;260:1879–88.

Saunders GW, Kucera H. An evaluation of *rbcL*, *tufA*, UPA, LSU and ITS as DNA barcode markers for the marine green macroalgae. Cryptogam Algol. 2010;31:487–528.

Saunders GW, Chiovitti A, Kraft GT. Small-subunit rNA sequences from representatives of selected families of the Gigartinales and Rhodymeniales (Rhodophyta). 3. Delineating the Gigartinales sensu stricto. Can J Bot. 2004;82:43–74.

Scheet P, Stephens M. A fast and flexible statistical model for large-scale population genotype data: application to inferring missing genotypes and haplotypic phase. Am J Hum Genet. 2006;78:629–44.

Schiel DR, Foster MS. The population biology of large brown seaweeds: ecological consequnces of multiphase life histories in dynamic coastal environments. Annu Rev Ecol Evol Syst. 2006;37:343–72.

Sears BB, Vanwinkle-Swift K. The salvage-turnover-repare (STOR) model for uniparental inheritance in *Chlamydomonas*: DNA as a source of sustenance. J Hered. 1994;85:366–76.

Selkoe KA, Henzler CM, Gaines SD. Seascape genetics and the spatial ecology of marine populations. Fish Fish. 2008;9:363–77.

Serrão EA, Pearson GA, Kautsky L, Brawley SH. Successful external fertilization in turbulent environments. Proc Nat Acad Sci USA. 1996a;93:5286–90.

Serrão EA, Kautsky L, Brawley SH. Distributional success of the marine seaweed *Fucus vesiculosus* L. in the brachish Baltic Sea correlates with osmotic capabilities of Baltic gametes. Oecologia. 1996b;107:1–12.

Serrão EA, Brawley SH, Hedman J, Kautsky L, Samuelsson G. Reproductive success of *Fucus vesiculosus* (Phaeophyceae) in the Baltic Sea. J Phycol. 1999;35:254–69.

Shaughnessy FJ, De Wreede RE. Reliability of the resorcinol method for identifying isomorphic phases in the Gigartinaceae (Rhodophyta). J Appl Phycol. 1991;3:121–7.

Silva G, Lima FP, Martel P, Castilho R. Thermal adaptation and clinal mitochondrial DNA variation of European anchovy. Proc R Soc B. 2014;281:20141093.

Smith GM. The Fresh-water algae of the United States. New York: McGraw-Hill; 1950.

Smith JM, Haigh J. The hitch-hiking effect of a favourable gene. Genet Res. 1974;23:23–35.

Sork VL, Nason J, Campbell DR, Fernandez JF. Landscape approaches to historical and contemporary gene flow in plants. Trends Ecol Evol. 1999;14:219–23.

Städler T, Haubold B, Merino C, Stephan W, Pfaffelhuber P. The impact of sampling schemes on the site frequency spectrum in nonequilibrium subdivided populations. Genetics. 2009;182:205–16.

Staubwasser M, Sirocko F, Grootes PM, Seg M. Climate change at the 4.2 ka BP termination of the Indus valley civilization and Holocene south Asian monsoon variability. Geophys Res Letts. 2003;30:1425.

Stebbing ARD, Turk SMT, Wheeler A, Clarke KR. Immigration of southern fish species to south-west England linked to warming of the North Atlantic (1960–2001). J Mar Biol Assoc UK. 2002;82:177–80.

Stenseth NC, Ottersen G, Hurrell JW, Mysterud A, Lima M, Chan K-S, Yoccoz NG, Ådlandsvik B. Studying climate effects on ecology through the use of climate indices: the North Atlantic Oscillation, El Niño southern oscillation and beyond. Proc R Soc B. 2003;270:2087–96.

Stephens M, Donnelly P. A comparison of Bayesian methods for haplotype reconstruction from population genotype data. Am J Hum Genet. 2003;73:1162–9.

Stephens M, Smith NJ, Donnelly P. A new statistical method for haplotype reconstruction from population data. Am J Hum Genet. 2001;68:978–89.

Sunnucks P. Efficient genetic markers for population biology. Trends Ecol Evol. 2000;15:199–203.

Sutherland JE, Lindstrom SC, Nelson WA, Brodie J, Lynch MDJ, Hwang MS, Choi HG, Miyata M, Kikuchi N, Oliveira MC, Farr T, Neefus C, Mols-Mortensen A, Milstein D, Müller KM. A new look at ancient order: generic revision of the Bangiales (Rhodophyta). J Phycol. 2011;47:1131–51.

Tajima F. Statistical method for testing the neutral mutation hypothesis by DNA polymorphism. Genetics. 1989;123:585–95.

Tatarenkov A, Berström L, Jönsson RB, Serrão EA, Kautsky L, Johannesson K. Intriguing asexual life in marginal populations of the brown seaweed *Fucus vesiculosus*. Mol Ecol. 2005;14:647–51.

Teasdale BW, Klein AS. Genetic variation and biogeographical boundaries within the red alga *Porphyra umbilicalis* (Bangiales, Rhodyphyta). Bot Mar. 2010;53:417–31.

Tellier F, Meynard AP, Correa JA, Faugeron S, Valero M. Phylogeographic analyses of the 30°S south-east Pacific biogeographic transition zone establish the occurrence of a sharp genetic discontinuity in the kelp *Lessonia nigrescens*: vicariance or parapatry? Mol Phylogenet Evol. 2009;53:679–93.

Thornber CS, Gaines SD. Population demographics in species with biphasic life cycles. Ecology. 2004;85:1661–74.

Tinbergen N. On aims and methods of ethology. Z Tier Psychol. 1963;20:410–33.

Toonen RJ, Puritz JB, Forsman ZH, Whitney JL, Fernandez-Silva I, Andrews KR, Bird CE. ezRAD: a simplified method for genomic genotyping in non-model organisms. PeerJ. 2013;1:e203.

Turmel M, Lemieux C, Burger G, Lang F, Otc C, Plate I, Gray MW. The complete mitochondrial DNA sequences of *Nephroselmis olivacea* and *Pedinomonas minor*: two radically different evolutionary patterns within green algae. Plant Cell. 1999;11:1717–29.

Uwai S, Kogame K, Yoshida G, Kawai H, Ajisaka T. Geographical genetic structure and phylogeography of the *Sargassum horneri/filicinum* complex in Japan, based on the mitochondrial *cox3* haplotype. Mar Biol. 2009;156:901–11.

van der Strate HJ, van de Zande L, Stam WT, Olsen JL. The contribution of haploids, diploids and clones to fine-scale population structure in the seaweed *Cladophoropsis membranacea* (Chlorophyta). Mol Ecol. 2002;11:329–45.

Verbruggen H, Maggs CA, Saunders GW, Le Gall L, Yoon HS, De Clerck O. Data mining approach identifies research priorities and data requirements for resolving the red algal tree of life. BMC Evol Biol. 2010;10:16.

Villalba R. Tree-ring and glacial evidence for the medieval warm epoch and the Little Ice Age in southern South America. Clim Chang. 1994;26:183–97.

Voris HK. Maps of Pleistocene sea levels in Southeast Asia: shorelines, river systems and time durations. J Biogeogr. 2000;27:1153–67.

Wahl M, Molis M, Hobday AJ, Dudgeon S, Neumann R, Steinberg P, Campbell AH, Marzinelli E, Connell S. The responses of brown macroalgae to environmental change from local to global scales: direct versus ecologically mediated effects. Perspect Phycol. 2015. doi:10.1127/pip/2015/0019.

Wang PX. Response of the Western Pacific marginal seas to glacial cycles: paleoceanographic and sedimentological features. Mar Geol. 1999;165:5–39.

Wang XL, Zhao FJ, Hu ZM, Critchley AT, Morrell SL, Duan DL. Inter-simple sequence repeat (ISSR) analysis of genetic variation of *Chondrus crispus* populations from North Atlantic. Aquat Bot. 2008;88:154–9.

Waters JM. Driven by the West Wind Drift? A synthesis of southern temperate marine biogeography, with new directions for dispersalism. J Biogeogr. 2008;35:417–27.

Waters JM, Fraser CI, Hewitt GM. Founder takes all: density-dependent processes structure biodiversity. Trends Ecol Evol. 2013;28:78–85.

Wattier R, Engel CR, Saumitou-Laprade P, Valero M. Short allele dominance as a source of heterozygote deficiency at microsatellite loci: experimental evidence at the dinucleotide locus Gv1CT in *Gracilaria gracilis* (Rhodophyta). Mol Eol. 1998;7:1569–73.

Wernberg T, Russell BD, Thomsen MS, Gurgel CFD, Bradshaww CJ, Poloczanska ES, Connell SD. Seaweed communities in retreat from ocean warming. Curr Biol. 2011;21:1828–32.

White C, Selkoe KA, Watson J, Siegel DA, Zacherl DC, Toonen RJ. Ocean currents help explain population genetic structure. Proc R Soc B. 2010;277:1658–94.

Williams SL, Di Fiori RE. Genetic diversity and structure in *Pelvetia fastigiata* (Phaeophyta: Fucales): does a small effective neighborhood size explain fine-scale genetic structure? Mar Biol. 1996;126:371–82.

Wolfe KH, Li WH, Sharp PM. Rates of nucleotide substitution vary greatly among plant mitochondrial, chloroplast, and nuclear DNAs. Proc Nat Acad Sci USA. 1987;84:9054–8.

Yesson C, Bush LE, Davies AJ, Maggs CA, Brodie J. Large brown seaweeds of the British Isles: evidence of changes in abundance over four decades. Estu Coast Shelf Sci. 2015;155:167–75.

Yotsukura N, Shimizu T, Katayama T, Druehl LD. Mitochondrial DNA sequence variation of four *Saccharina* species (Laminariales, Phaeophyceae) growing in Japan. J Appl Phycol. 2010;22:243–51.

Zechman FW, Mathieson AC. The distribution of seaweed propagules in estuarine, coastal and offshore waters of New Hampshire, U.S.A. Bot Mar. 1985;28:283–294.

Zhang DX, Hewitt GM. Nuclear DNA and analyses in genetic studies of populations: practice, problems and prospects. Mol Ecol. 2003;12:563–84.

Zuccarello GC, West JA, Kamiya M, King RJ. A rapid method to score plastid haplotypes in red seaweeds and it use in determining parental inheritance of plastids in the red alga (Ceramiales). Hydrobiologia. 1999a;401:207–14.

Zuccarello GC, Burger G, West JA, King RJ. A mitochondrial marker for red algal intraspecific relationships. Mol Ecol. 1999b;8:1443–7.

Chapter 3
The Dynamic Biogeography of the Anthropocene: The Speed of Recent Range Shifts in Seaweeds

Sandra C. Straub, Mads Solgaard Thomsen and Thomas Wernberg

Abstract The biogeographic boundaries of seaweeds are largely determined by temperature tolerances, physical barriers and limitations to dispersal. Anthropogenic ocean warming and increasing connectivity through human activities are now causing rapid changes in the biogeography of seaweeds. Globally, at least 346 non-native seaweed taxa have been introduced to new regions, and at least 31 species of seaweed have shifted their distributions in response to recent temperature changes. Range-shift speeds were determined for 40 taxa, and compared between three drivers: (I) range expansions caused by introductions, (II) range expansions and (III) contractions caused by climate change (warming/cooling). The speed of change in seaweed biogeography differed between these drivers of change, with expansions significantly faster than contractions, and climate-driven shifts significantly slower than introductions. Some of the best documented introduced species expansions include *Sargassum muticum* (4.4 km/year in Denmark), *Undaria pinnatifida* (35–50 km/year in Argentina) and *Caulerpa cylindracea* (11.9 km/year in the Mediterranean Sea). Examples of seaweeds with recent climate-driven range shifts include *Scytothalia dorycarpa*, a native species in Western Australia, which retracted >100 km poleward as a consequence of a single event (a regional marine heat wave). However, climate-driven range shifts were generally assessed over long time periods (>10 years). *Fucus serratus* (1.7 km/year) and *Himanthalia elongata* (4.4 km/year) have slowly retracted westwards in northern Spain in response to

S.C. Straub · M.S. Thomsen · T. Wernberg (✉)
UWA Oceans Institute and School of Plant Biology,
The University of Western Australia, Crawley 6009, WA, Australia
e-mail: thomas.wernberg@uwa.edu.au

M.S. Thomsen
e-mail: mads.solgaard.thomsen@gmail.com

M.S. Thomsen
Marine Ecology Research Group, School of Biological Sciences,
University of Canterbury, Private Bag 4800, Christchurch 8140, New Zealand

M.S. Thomsen
Centre of Integrative Ecology, School of Biological Sciences,
University of Canterbury, Christchurch, New Zealand

Z.-M. Hu and C. Fraser (eds.), *Seaweed Phylogeography*,
DOI 10.1007/978-94-017-7534-2_3

warming in the Bay of Biscay. In England and South Africa, *Laminaria ochroleuca* (5.4 km/year) and *Ecklonia maxima* (36.5 km/year) have expanded their ranges in response to local warming and cooling, respectively. These changes in seaweed biogeography likely have had substantial implications for biodiversity and ecosystem processes, particularly where the shifting seaweeds have been canopy-forming foundation species. We discuss some of these consequences and different attributes of climate and invasion-driven range shifts in seaweeds.

Keywords Climate change • Dispersal • Invasive species • Range contraction • Range expansion • Seaweed distribution

3.1 Introduction

Species distributions are dynamic, continuously shifting in responses to changes in biological and environmental drivers. In the earlier history of the Earth, large-scale geological events and long-term climate fluctuations, such as continental drift or warming and cooling associated with planetary cycles, were the predominant drivers of changes to species' distributions (Wiens and Donoghue 2005). However, over the past millennium humans have increasingly modified the biological and physical properties of the planet (Worm et al. 2006), and we have now entered the Anthropocene, an era where the human influence on the global Earth system rivals or exceeds natural processes (Karl and Trenberth 2003), speeding up important drivers of species distributions influencing the biogeography of organisms across ecosystems. As a consequence, recent changes in the distribution of many marine taxa have been documented on all continents (Perry et al. 2005; Williams and Smith 2007; Sorte et al. 2010; Wernberg et al. 2011a; Poloczanska et al. 2013).

Seaweeds are dominant organisms on many intertidal and shallow subtidal reefs, where their species-specific distributions often shape local reef communities (Wernberg et al. 2003; Buschbaum et al. 2006; Ingólfsson 2008; Tuya et al. 2009). Although the local effects of biotic interactions can generate continental-scale patterns of species associations (Wootton 2001; Irving and Connell 2006), global biodiversity patterns are not explained by biotic interactions alone but are a consequence of both the biotic and abiotic environments (Lüning 1985; Harley et al. 2006; Tittensor et al. 2010). Two mechanisms have been particularly prevalent in driving recent changes in seaweed distributions: species introductions through the direct relocation of species (transported, deliberate or not, by various vectors) and responses to global climate change (Williams and Smith 2007; Wernberg et al. 2011a; Sorte et al. 2013).

Changes in seaweed distributions include both range extensions, where species colonize new, usually adjacent habitats, and range contractions, where species lose previously occupied areas, going locally extinct at the margins of their distribution range (Fig. 3.1) (Wernberg et al. 2011a; Bartsch et al. 2012; Bates et al. 2014).

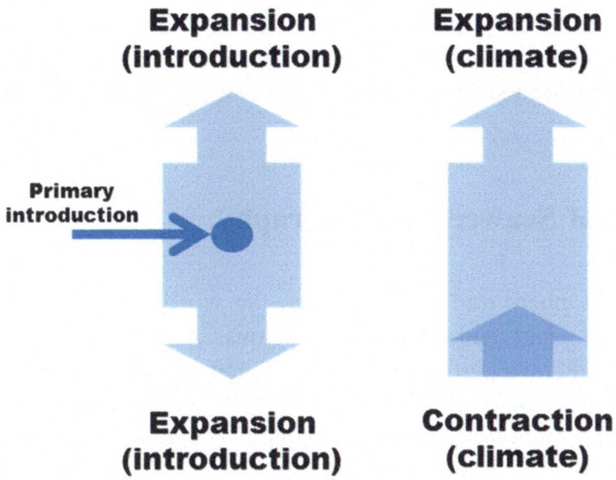

Fig. 3.1 Reconfiguration of seaweed range boundaries takes place as one of three general processes. Either species expand their boundaries in a new area following initial primary introduction to a site where climate conditions are not immediately limiting. Alternatively, native species can expand into previously unoccupied areas, tracking their climate envelope as changing conditions make these suitable. Similarly, species can retract from occupied areas as changing climate makes these unsuitable. The processes, and underlying physical and biological mechanisms, differ between these processes, with expansions driven by dispersal and recruitment dynamics and contractions by performance and mortality (Bates et al. 2014)

For species introductions and climate impacts, range shifts are underpinned by different mechanisms involving dispersal and recruitment (introductions, climate expansion) and attrition and mortality (climate contraction) (Bates et al. 2014). Moreover, whereas range expansion only requires the successful establishment of one or a few individuals in a new location, local extinction and range contraction requires the demise of all individuals and is often preceded by periods of declining abundance and failed recruitment while adult individuals persist in the unfavourable area (Hampe and Petit 2005; Bates et al. 2014). Conversely, environmental conditions are generally not limiting the expansion of introduced species following primary introduction, whereas climate-driven responses track shifts in the climate envelope (Fig. 3.1) (Pinsky et al. 2013; Sunday et al. 2015). Consequently, even if priority effects and other biological (competition, predation) processes can work against the expansion process (Waters et al. 2013), seaweed range shifts are expected to be faster for expansions than contractions and faster for introductions than climate responses (Sorte et al. 2010). The effects on the respective habitats and communities should, however, be of the same magnitude and direction (Sorte et al. 2010).

Here, we first provide a brief overview of natural and anthropogenic factors that shape the biogeography of seaweeds. We then provide a quantitative synthesis of how fast humans are affecting seaweed distributions through an analysis of the speed of reported human-mediated changes in seaweed range boundaries. We also review selected case studies of seaweed range shifts for both native species that have changed their ranges in response to changing environmental conditions, and introduced species, spreading in their new environment. Finally, we discuss the challenges of identifying range shifts and the necessity for monitoring distributions to detect seaweed range shifts.

3.2 Drivers of Seaweed Biogeography

Seaweed biogeographers traditionally group the world's oceans into seven broad regions: the Arctic and Antarctic Polar regions, the cold- and warm-temperate regions of both hemispheres, and the tropical regions of the Atlantic and Indo-Pacific (Lüning 1985; Bartsch et al. 2012). The boundaries between these biogeographic regions are associated with large changes in species composition, maintained by species temperature tolerances (Van den Hoek 1982), natural barriers (Cowman and Bellwood 2013) and species dispersal limits (Wiens 2011). Humans are now assisting seaweeds and other organisms to overcome these geographic boundaries, which previously limited distributions.

3.2.1 Temperature

Seaweeds are confined to the photic zone, where temperature patterns are reasonably well understood, allowing species distributions to be compared to oceanographic patterns (Adey and Steneck 2001). Distribution limits of individual seaweed species typically follow major marine isotherms (Van den Hoek 1982; Lüning 1985), giving rise to strong relationships with the temperature signatures of major ocean currents (Wernberg et al. 2013b).

For seaweeds, these patterns are a product of two key types of temperature boundaries: lethal boundaries, determined by a species' capacity to survive during their unfavourable season; and growth and reproduction boundaries, determined by a species' ability to grow and reproduce during its favourable season (Van den Hoek 1982; Lüning 1985). Seaweeds can be abundant in areas within both boundaries that are within dispersal ranges of the species. However, as thermal windows have changed over geological time (e.g. following ice age cycles), they have biogeographic boundaries and seaweed distributions (Adey and Steneck 2001).

3.2.2 Barriers

If the oceans were a continuous open system, then most species should exhibit cosmopolitan distributions within their respective thermal windows (Myers 1997; Gaylord and Gaines 2000). However, barriers limit dispersal, which leads to discontinuities in species distribution (Myers 1997). These barriers can be 'hard' or 'soft', depending on their underlying mechanism (Luiz et al. 2012; Cowman and Bellwood 2013).

Hard barriers are physical obstacles such as land masses separating marine systems. For example, the final closure of the Tethys seaway around 12 Mya at the northern tip of the Red Sea created a physical barrier which cut off the low-latitude connection between the Indian and Atlantic Oceans (Cowman and Bellwood 2013).

In contrast, soft barriers refer to hydrographical features that disrupt connectivity. Large stretches without suitable substratum, such as deep oceanic basins (Lessios et al. 1998) or extensive beaches (Hidas et al. 2007), can limit the distribution of species with limited dispersal capacity. The greatest example could be the Eastern Pacific Barrier, a 5400-km stretch of deep open ocean between the central and eastern Pacific, likely in existence since the Cenozoic (Grigg and Hey 1992), where only a few marine species are represented on both sides (Lessios et al. 1998). Nearshore gradients in ocean properties, such as the direction and strength of ocean currents, differences in salinity and/or temperature as a result of currents or local upwelling (Luiz et al. 2012; Cowman and Bellwood 2013), can also function as barriers to dispersal. Therefore, many seaweeds show distribution limits concentrated at particular shorelines, often in locations where major currents collide (Gaylord and Gaines 2000; Schils and Wilson 2006; Waters 2008).

Barriers are, however, not permanent especially over geological time scales. Changes in ice cover and sea levels (glaciation, deglaciation, retreating ice caps, historical sea-level alterations) have led to significant alterations in seaweed biogeography. The Baltic Sea, for example, was entirely covered by glaciers during the last ice age, and all present-day seaweeds in the Baltic Sea have colonized following the opening of the Danish Straits about 8000 years ago (Björck 1995). Similarly, recent glacial retreat in the South Shetland Islands has enabled seaweed expansion into newly available habitat in Antarctica (Quartino et al. 2013). Over several glacial cycles, reduced sea levels exposed the Bassian land bridge, a historical barrier between Tasmania and mainland Australia, interrupting connectivity and colonization for several taxa for prolonged periods of time (Burridge et al. 2004; Waters 2008; York et al. 2008). Also, islands emerging due to volcanic activity (e.g. new island formation in Japans Ogasawara Island chain in 2013) create new space for seaweed colonization and can function as stepping stones for long-range dispersers to overcome deep oceanic stretches to reach distant areas (Nogales et al. 2012). Also, dispersal across large sandy stretches can be facilitated by small rocky platforms functioning as intermediate habitats to facilitate dispersal over the barrier (Dethier et al. 2003; Hidas et al. 2007; Mattio et al. 2015).

3.2.3 *Dispersal*

Dispersal is a critical process which allows seaweeds to extend their geographical distribution. Seaweeds employ a broad range of dispersal strategies with some species adapted to short-distance dispersal, typically settling close to their parental populations, and others adapted to long-distance dispersal, typically favouring rapid colonization of new habitats (Santelices 1990).

Most seaweeds disperse by small, largely immotile propagules (zoospores or zygotes) that are transported by waves and currents (Norton 1992; Gaylord et al. 2002). The buoyancy of the propagules, storage components, metabolic rates, and the strength and direction of current flow determine how far these microscopic propagules can disperse (Gaylord et al. 2002), before they have to settle onto hard substrata in the photic zone. In addition to microscopic propagules, seaweeds can also disperse as floating fronds, where a parental thallus is dislodged (breakage of stipes, thallus fragmentation, storms, etc.) and transported by winds and currents (Rothäusler et al. 2012). Many positively buoyant seaweeds can survive, float and disperse for prolonged periods of time (Van den Hoek 1987; Norton 1992; Hobday 2000a; Rothäusler et al. 2012). This dispersal mechanism is particularly efficient for dioecious species that do not rely on concurrent dispersal of male and female thalli (e.g. *Sargassum muticum*) as these species can establish entire new populations from single floating reproductive fronds. Large drifting seaweeds can also function as a raft for smaller negatively buoyant animals and seaweeds (Van den Hoek 1987; Hobday 2000b; Hinojosa et al. 2010; Fraser et al. 2011; Gillespie et al. 2012; Rothäusler et al. 2012; Fraser and Waters 2013). Floating seaweeds can therefore facilitate the colonization of new habitats on remote shores, sometimes by crossing large ocean basins (Fraser et al. 2011; Rothäusler et al. 2012). Dispersal thus depends on both intrinsic seaweed traits such as buoyancy and propagule characteristics, as well as on external factors such as current speed and direction, and environmental conditions that enable survival, settlement and growth (Norton 1992; Hinojosa et al. 2010).

3.2.4 Species Introductions (Human-Assisted Dispersal)

A characteristic feature of the past millennium has been an explosion in travel for trade and colonization, over increasing distances and at decreasing travel times. Through the process of human-assisted dispersal, non-native seaweeds have spread (intentional or not) to habitats far away from their origins (also see Chapter by Neiva et al. (2016) in this volume).

Introduced seaweeds are species that have been relocated beyond their native range by human activities and have become successfully established at a new location. The introduction of seaweeds is a stepwise process, starting with transport and initial arrival through a vector (primary introduction, Fig. 3.1), which is followed by initial survival, establishment and finally successful reproduction and spread (expansion, introduction, Fig. 3.1) to nearby locales (Sakai et al. 2001; Bates et al. 2014). The main vectors responsible for seaweed introductions include hull fouling and aquaculture, but ballast water, breakdown of natural barriers (the Suez canal in particular) and the aquarium trade have also transported seaweed around the world (Williams and Smith 2007) (see Chap. by Neiva et al. (2016) in this volume). Of all introductions, only a small subset establishes permanent populations in their new habitats. It has recently been estimated that at least 346 seaweed taxa have been introduced to, and successfully established populations in, new regions worldwide (many of these taxa having invaded multiple biogeographical regions), breaking down barriers evolved over millennia (see Chap. by Neiva et al. (2016) in this volume). Many of these taxa have also become invasive with significant effects on native species, biodiversity and ecosystem dynamics (Williams and Smith 2007; Thomsen et al. 2009, 2014) (see Chap. by Neiva et al. (2016) in this volume).

For successfully introduced seaweeds, it is implicitly assumed that climate is not the primary limiting constraint on their distribution (or they would not have successfully become established) and that secondary expansion can proceed largely as fast as dispersal allows. Expansion of introduced seaweeds should therefore be rapid relative to climate-induced range changes.

3.2.5 Environmental Change (Human-Induced Climate Change)

Climate and temperature, in particular, play pivotal roles in controlling the global biogeography of seaweeds (Sect. 3.2.1) (Lüning 1985). Consequently, changes in temperature, as for example those associated with anthropogenic greenhouse gas emissions, also alter the distribution of seaweeds (Zachos et al. 2008; Wernberg et al. 2011b; Harley et al. 2012).

On average, anthropogenic emissions of greenhouse gases have caused a decrease in ocean surface seawater pH of ~0.1 since the beginning of the industrial era (IPCC 2014) and ocean warming by ca. 1 °C over the past 4–5 decades, although with substantial local variation (Burrows et al. 2011). While a few regions have cooled due to increased upwelling (e.g., causing kelps to expand their ranges Bolton et al. 2012), most regions have warmed (Lima and Wethey 2012; Hobday and Pecl 2013). Importantly, climate change not only causes gradual and slow increases in temperatures and pH, but also in the frequency and intensity of extreme events (Coumou and Rahmstorf 2012; IPCC 2012). Seaweeds respond to these environmental changes through physiological and morphological acclimations (reversible, phenotypic changes on short timescales), adaptation (irreversible, genotypic changes on medium to long times scales), or migration (changes in distribution on medium timescales) (Bartsch et al. 2012).

Overall, climate change has altered local marine environments leading to changes in distribution and diversity of seaweed communities from local to global scales (Wernberg et al. 2011a; Tanaka et al. 2012; Duarte et al. 2013). As seaweed expansions and contractions follow the external driver of changes in the physical environment, changes in species distributions are expected to be slow relative to introductions (Sorte et al. 2010). Moreover, climate-induced contractions will, in contrast to expansions, typically manifest as repeated recruitment failures and subsequent demise of long-lived populations (Hampe and Petit 2005; Bates et al. 2014). Contractions are therefore expected to be slower than expansions. One obvious exception is the rapid response to extreme events, which can alter local ecosystem structure and functioning abruptly (Wernberg et al. 2013b) and lead to rapid changes in seaweed distributions (Smale and Wernberg 2013).

3.3 Speed of Range Shifts in Seaweeds

In order to determine the rate at which humans have been modifying biogeographic boundaries of seaweeds, we undertook a meta-analysis of the rate of change in distribution limits for recently recorded range shifts for native and introduced seaweeds (range-shift speed). Data bases were searched using key words like 'climate change', 'warming', 'extreme events', 'temperature anomaly' 'heatwaves', 'introduced seaweeds', 'successful invaders', 'shift in distribution', 'shift rates', 'spread rates', 'range shift', 'range expansion' and 'range contraction'. We also backtracked references from relevant reviews and meta-analytical papers (Sorte et al. 2010; Poloczanska et al. 2013; Bates et al. 2014). We included studies that showed data for the directions, distances and time windows of seaweed range shifts, allowing us to calculate annual spread rates. Literature reporting changes in abundance without changes in location were excluded from the dataset, as were studies that did not report a range shift per se. Where rates were not reported directly, but identifiable locations given, rates were calculated (using the Google Earth distance calculator). For introduced species, we did not consider the initial primary introduction distance, only expansion from site of primary introduction into its new environment. Where time was reported as an interval, the midpoint was used. These strict data inclusion criteria limited the number of range-shifting seaweeds included in our analysis, which therefore represents a constrained view of seaweed range shifts.

Range-shift speeds were compared between three drivers of change (cf. Sect. 3.2.5, Fig. 3.1). (I) range expansions following introductions, (II) expansions caused by climate change (typically warm-water species) and (III) range contractions caused by climate change (typically cool-water species). More specifically, we tested (a) whether expansions generally are faster than contractions, and (b) whether introductions are faster than climate-driven changes. Tests were made with permutation-based analysis of variance (Logx + 1 transformed range-shift speeds, 9999 permutations of residuals), followed by two a priori defined planned contrasts (expansion vs. contraction and introduction vs. climate). These analyses did not include range shifts caused by primary introductions or in response to discrete extreme events. These range shifts were excluded due to their artificial and stochastic nature, respectively.

Table 3.1 Summary statistics for human-induced seaweed range shifts

	Expansion (introduction, *n* = 13 taxa)	Expansion (climate, *n* = 22 taxa)	Contraction (climate, *n* = 9 taxa)
Median shift (range)	280 km (40–1450)	192 km (26–593)	116 km (35–1250)
Median time (range)	7 years (1–47)	50 years (2–75)	31 years (1–66)
Taxa (*n* = 41)	*Caulerpa cylindracea* *Caulerpa ollivieri* *Caulerpa taxifoliavar. distichophylla* *Codium fragile* ssp. *fragile* *Codium fragile* ssp. *Tomentosoides* *Fucus serratus* *Grateloupia doryphora* *Grateloupia turuturu* *Heterosiphonia japonica* *Mastocarpus* sp. *Sargassum filicinum* *Sargassum muticum* *Undaria pinnatifida*	*Ahnfeltia plicata* *Bifurcaria bifurcata* *Chondrus crispus* *Codium adhaerens* *Desmarestia aculeata* *Desmarestia ligulata* *Dumontia contorta* *Ecklonia maxima* *Fucus serratus* *Fucus vesiculosus* *Halidrys siliquosa* *Halopithys incurva* *Himanthalia elongata* *Hypnea musciformis* *Laminaria ochroleuca* *Padina pavonica* *Palmaria palmata* *Pelvetia canaliculata* *Sargassum flavifolium* *Sargassum illicifolium* *Turbinaria ornata* *Valonia utricularis*	Assemblage *Durvillea potatorum* *Ecklonia radiata* *Fucus serratus* *Fucus vesiculosus* *Himanthalia elongata* *Sargassum micracanthum* *Sargassum yamamotoi* *Scytothalia dorycarpa*

Details are reported in Appendix

Our literature search returned 71 individual estimates of seaweed range-shift speed (Appendix). In general, of the studies where range-shift speeds could be assessed, expansions following seaweed introductions were detected over larger distances and shorter time periods than expansions and contractions due to climate change (Table 3.1). Studies returning range-shift speeds were reported from all continents except Antarctica (Fig. 3.2), although we found strong geographical biases in what types of range shifts had been recorded. For example, no studies with sufficient information to calculate range-shift speed were reported for climate-driven range expansions or contractions in North and South America, nor expansions and introductions in Australia. Europe had the greatest concentration of range shifts reported with sufficient information for all three categories (Fig. 3.2).

The five areas where climate-induced range-shift speeds are available (SE and SW Australia, Japan, South Africa and SW Europe) are well-known 'temperature hotspots' where the rate of ocean warming since 1950 has been in the top 10 % of observations globally (Hobday and Pecl 2013). Interestingly, the range shift

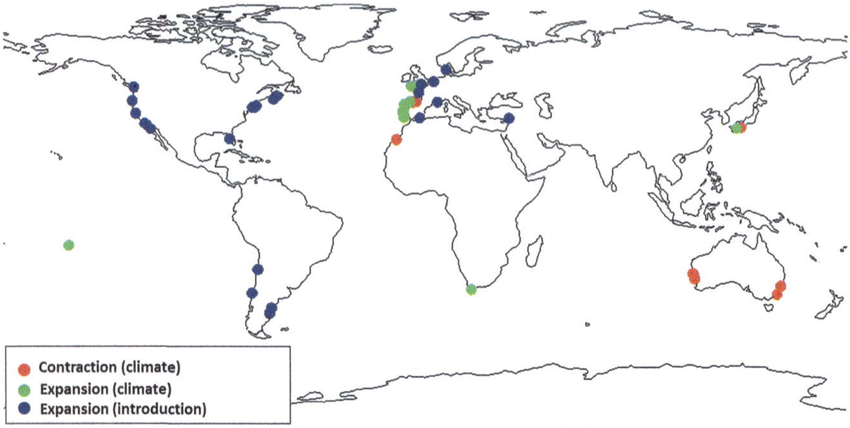

Fig. 3.2 Geographical location of studies reporting range shifts in seaweeds with sufficient information to calculate range-shift speed

reported in the South African warming hotspot was a range expansion of a cool-water kelp (*E. maxima*). However, this expansion was attributed to increased nearshore upwelling (Bolton et al. 2012), a consistent but small-scale phenomenon not captured by global satellite data (Smit et al. 2013). This example highlights that predictions about global range shifts from large-scale satellite images may not capture local distribution patterns, particular where upwelling occurs.

Range-shift speeds were determined for 40 taxa (n = 13, 22, 9 for each of the categories, respectively), with some genera represented by more than one species (Table 3.1). As might be expected, there was little overlap in taxa between categories but one species (*F. serratus*) was represented in all three range-shift categories (Appendix) and another two species (*F. vesiculosus* and *H. elongata*) in both climate change responses (Appendix). These responses highlight the context dependency of range shifts, with the direction of shift presumably determined by a combination of ecological interactions opening/closing opportunities for change as well as the relative position within the species' thermal envelope.

We found support for our range-shift hypotheses (Fig. 3.3). The speed of distributional changes in seaweed range limits differed significantly between the different types of shifts (Fig. 3.3, medians = 35.0, 4.5, and 4.0 km year^{-1}, respectively, $P = 0.0001$, $MS_{2,66} = 26.6$, $pseudo\text{-}F = 31.6$). Expansions were significantly faster than contractions ($P = 0.011$, $MS_{1,2} = 10.1$, $pseudo\text{-}F = 6.9$) and climate-induced shifts were significantly slower than those caused by species introductions ($P = 0.0001$, $MS_{1,2} = 53.1$, $pseudo\text{-}F = 64.1$). However, this test did not include range contractions following a discrete extreme event—a large-scale marine heat wave—where two species of seaweeds were found to contract their ranges by ~100 km in one year (Fig. 3.3, Appendix). These shifts remain some of the fastest observed range changes for any seaweed. When the two heatwave-driven

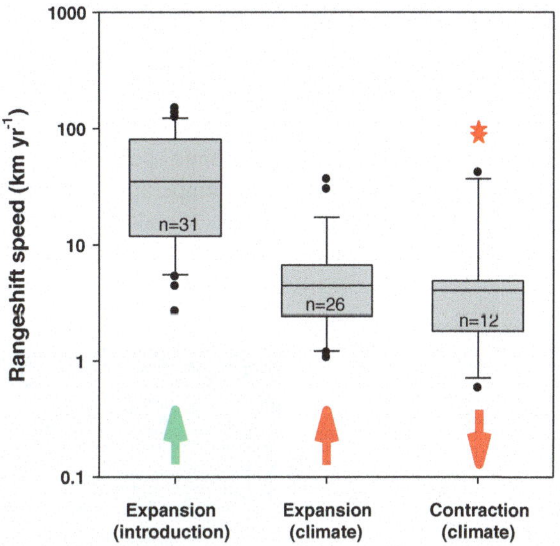

Fig. 3.3 Speed of range shifts in seaweeds. *Arrows* highlight direction (upwards = expansion; downwards = contraction) and underlying cause (*green* = after successful introduction; *red* = climate change) of range shifts. *Red* stars indicate shifts caused by an extreme marine heat wave. The very discrete nature of these shifts differs fundamentally from other reported shifts and consequently these have not been included in the analyses of rates (or the *box* in the plot)

contractions were included in the analysis, the difference between speed of climate-driven contractions and expansions disappeared ($P = 0.112$, $MS_{1,2} = 4.0$, $pseudo\text{-}F = 2.5$) but the difference in speed between introduction and climate-driven range shifts remained ($P = 0.0001$, $MS_{1,2} = 45.9$, $pseudo\text{-}F = 44.8$).

3.4 Case Studies of Seaweed Range Shifts and Ecological Implications

Many range-shifting seaweeds (cf. Table 3.1) are prominent members of their respective communities, where their addition or deletion is likely to have dramatic impacts on ecosystem structure and functioning (Williams and Smith 2007; Thomsen et al. 2010; Wernberg et al. 2013a; Bennett et al. 2015b) (see Chap. by Neiva et al. (2016) in this volume). The scale and nature of these ecological implications depends on the attributes of the shifting species and the impacted habitat (Thomsen et al. 2011). Here, we provide a range of examples of seaweed range shifts and their ecological implications. We also provide an example of a seaweed declining in abundance, a precursor to range contraction(Bates et al. 2014).

3.4.1 Range Contractions (Native Species)

In 2011 an unprecedented marine heat wave off the coast of Western Australia caused dramatic canopy loss of dominant seaweeds, including a 100 km southward range contraction of one of the main canopy-forming species, the fucoid *Scytothalia dorycarpa* (Fig. 3.4). During the heat wave, temperatures exceeded the physiological tolerance of *S. dorycarpa* for many weeks (Smale and Wernberg 2013). The contraction of *S. dorycarpa* co-occurred with a significant decrease in the densities

Fig. 3.4 *Scytothalia dorycarpa* (**a**) is a large (~1 m) fucoid endemic to southern Australia. During an extreme heat wave in 2011, *S. dorycarpa* contracted its range by ~100 km in less than 1 year (Smale and Wernberg 2013). Prior to the heat wave (November 2010, **b**) reefs were covered by a dense mixed canopy of kelp (*Ecklonia radiata*) and *S. dorycarpa*. However, immediately after the heat wave (November 2011, **c**) the canopy had large gaps where *S. dorycarpa* had been extirpated. One year later (November 2012, **d**), the canopy had not recovered and gaps were filling in with foliose and turf algae (*All photos* © T. Wernberg)

of the kelp *Ecklonia radiata* (Wernberg et al. 2013a) and indirectly resulted in changes of the understorey community structure. The net effect was a shift from a dense three-dimensional canopy habitat to reefs with large open patches dominated by much smaller turf forming seaweeds among patches of *E. radiata* (Smale and Wernberg 2013) (Fig. 3.4). Concurrently, with the loss of seaweeds, the biomass and diversity of tropical herbivores increased, facilitating the new canopy-free state by suppressing seaweed reestablishment (Bennett et al. 2015b). The combined effects of the range contraction of *S. dorycarpa* and overall loss of seaweed canopies ultimately resulted in habitat and food loss (Wernberg et al. 2013a; Smale and Wernberg 2013) which are likely to have cascading impacts through altered benthic productivity and food web structure to a variety of higher trophic marine organisms including commercially important crustaceans, fishes and mammals (Lozano-Montes et al. 2011).

In northern Spain, range contractions have been reported for several canopy-forming seaweeds (Appendix), including *Fucus serratus* and *Himanthalia elongata* which have moved westwards in the Bay of Biscay since the late nineteenth century as a response to global warming (Duarte et al. 2013). *H. elongata* changed its range stepwise by 330 km over 120 years, whereas *F. serratus* retracted 197 km over 114 years but also reduced its abundance dramatically in its remaining range, i.e. in the westernmost part of northern Spain (Appendix). For both species the rate of contraction appears to have accelerated in recent years (Duarte et al. 2013). The ecological implications of these two range contractions are largely unknown (Duarte et al. 2013), although both species (and several other large, retreating canopy-forming seaweeds) are important habitat formers for smaller epiphytes and mobile animals (Hawkins and Hartnoll 1985; Lüning 1985; Wernberg et al. 2004; Ingólfsson 2008).

3.4.2 Range Expansions (Native Species)

The warm-water kelp *Laminaria ochroleuca* was first recorded in England in 1948, and subsequently expanded its range eastwards to the Isle of Wright at a rate of 5.4 km per year, as well as expanded northwards to Lundy Island at a rate of 2.5 km per year (Table 3.2). Recent resurveys of the inhabited area suggest that *L. ochroleuca* also expanded from the initially colonized sheltered coastline to moderately wave-exposed open coasts, accompanied by a significant increase in abundance, most likely in response to recent warming (Smale et al. 2014). In the area where *L. ochroleuca* most recently colonized, it competes with the native dominant congener *L. hyperborea*. As both species appear morphologically and functionally similar, it was initially assumed that they would have similar ecosystem function with little impact on the colonized ecosystem (Terazono et al. 2012). However, even small morphological differences may incur large cascading ecosystem effects. For example, Smale et al. (2014) showed that epiphyte cover on the smoother stipe of *L. ochroleuca* was dramatically lower than on the rough stipes

Table 3.2 Overview of the discussed five native and three non-native range shifts as well as the example of abundance change, with their according driver, direction and rate of shift and dispersal means

Species	Division	Driver	Direction	Shift rate (km/year)	Size (cm)	Dispersal means
Fucus serratus	Ochrophyta	Warming	Contraction	1.7	70–100; <200	Negative buoyant, medium production of gametes, short-distance disperser
Himanthalia elongata	Ochrophyta	Warming	Contraction	4.4	300	Rafting of floating receptacles
Scytothalia dorycarpa	Ochrophyta	Heat wave	Contraction	100.0	50–200	Negative buoyant, medium production of gametes
Laminaria ochroleuca	Ochrophyta	Warming	Expansion	2.5–5.4	150	Release of large amounts of spores, short-distance disperser, negative buoyant
Ecklonia maxima	Ochrophyta	Cooling	Expansion	36.5	<1500	Release of very large amounts of spores
Caulerpa cylindracea	Chlorophyta	Introduction	Expansion	11.9	30	Negative buoyant, can regrow from fragments, fragments can re-attach into sediment, clonal spread, medium release of gametes (holocarpy, parental plant dies)
Sargassum muticum	Ochrophyta	Introduction	Expansion	4.4	<1600	Positively buoyant, monocious, selfy, high production of gametes
Undaria pinnatifida	Ochrophyta	Introduction	Expansion	35–50	200	Negative buoyant, local drift of reproductive individuals on dislodged mussels, massive production of gametes
Macrocystis pyrifera	Ochrophyta	Warming	Contraction	95 % cover reduction	<3000	Direct growth on female parental gametophyte; drifting thalli; spores

Fig. 3.5 The kelp *Ecklonia maxima* dominates nearshore reefs in the cold waters around southern Africa west of Cape Agulhas. It is a substantial seaweed which can grow to lengths in excess of 15 m (**a**: Buffels Bay, Cape of Good Hope). Between 2006 and 2008 this species expanded past Cape Agulhas, presumably due to cooling caused by upwelling (**b**: recently colonized intertidal populations at De Hoop Nature Reserve) (*Photos* © T. Wernberg)

of *L. hyperborea*. Thus, a reduction of the epiphytic habitat can be expected if *L. ochroleuca* replaces *L. hyperborea*, potentially with dramatic effects on associated fauna (Christie et al. 2009), trophic interactions (Smale et al. 2014) and biodiversity (Thomsen et al. 2010).

Another example of a recent and unusual range expansion of a native seaweed involves the dominant canopy-forming kelp *Ecklonia maxima* in South Africa (Fig. 3.5). The distribution of *E. maxima* along the southern coastline of South Africa appeared unchanged for ca. 70 years, but suddenly expanded eastwards (between 2006 and 2008) at a rate of 36.5 km per year (Bolton et al. 2012). It is suggested that gradual cooling caused the distribution expansion of *E. maxima*, crossing around Cape Agulhas which is considered a major barrier dividing the western and south coast regions (Anderson et al. 2009). As *E. maxima* is the major kelp along its distributional range, expansion of this species could have substantial ecological consequences (Bolton et al. 2012).

3.4.3 Range Expansion (Introduced Species)

Range expansions of non-native seaweeds can also alter ecosystem functioning after successful establishment. For example, *Caulerpa cylindracea* (Fig. 3.6a) is a highly invasive green seaweed which has spread along the Mediterranean Sea and Canary Islands since the early 1990s at an average rate of 11.9 km per year (Ruitton et al. 2005) (Table 3.2). *C. cylindracea* has invaded both soft and hard substrata and

Fig. 3.6 Some of the most notorious invasive marine species are seaweeds, which have spread rapidly throughout many regions of the world where they have been introduced. *Caulerpa cylindracea* (**a**) growing among turf and foliose seaweeds in its native environment in Western Australia. *Sargassum muticum* (**b**) growing in tide pools in northern Spain, where it is now a dominant element of the seaweed flora. *Undaria pinnatifida* (**c**) growing on tidal platforms in southeastern New Zealand (*Photos* **a**, **c**: © M.S. Thomsen, **b**: © C. Olabarria)

can form dense monospecific stands. The introduction vector is unknown, but where the species has established dense monocultures it has been associated with a decrease in abundance, biodiversity and biotic homogenization of native species (Klein and Verlaque 2008; Verbruggen et al. 2013). By forming multilayered mats that trap sediment, *C. cylindracea* can lead to burial of communities by sediment (Piazzi et al. 2005). Specifically, Baldacconi and Corriero (2009) determined its impacts on sponge assemblages in the Ionian Sea suggesting decreases in sponge cover following the invasion.

The brown, canopy-forming seaweed *Sargassum muticum* (Fig. 3.6b) is also a well-studied invasive seaweed (Engelen et al. 2015). Originating from Asia, *S. muticum* has spread over the last few decades along coastlines in western Europe and western North America (Pedersen et al. 2005; Engelen et al. 2015). Within invaded locations, *S. muticum* can spread rapidly and become a dominant seaweed, sometimes leading to the suppression of local species and alteration of community structure (Stæhr et al. 2000). In 1941, *S. muticum* was first observed outside its native range in the Strait of Georgia (British Columbia, Canada) from where it subsequently spread along the adjacent coastline (Engelen et al. 2015). In 1984 *S. muticum* was sighted in Denmark for the first time in Limfjorden from where it subsequently spread at a rate of 4.4 km per year. *S. muticum* became the most abundant seaweed in Limfjorden, leading to a decrease in cover and abundance of several native canopy-forming seaweeds, including *Halidrys siliquosa*, *Saccharina latissima*, *Fucus vesiculosus* and *Fucus serratus* (Stæhr et al. 2000).

Undaria pinnatifida (Fig. 3.6c) is another high-profile invasive brown seaweed that is native to Japan, Russia and China. In the last 40–50 years, it has invaded Europe (Atlantic and Mediterranean Sea), North America (Pacific coast), south-western Australia, New Zealand and Argentina (Wallentinus 2007). In Argentina, *U. pinnatifida* was first recorded in 1992 and has since extended its range >1000 km southwards from its original site of introduction at a rate averaging between 35 and 50 km per year. While *U. pinnatifida* can have negative impacts on some native seaweeds (Casas et al. 2004), positive effects have also been reported on benthic macrofauna and carbon flow (Dellatorre et al. 2014; Tait et al. 2015).

3.4.4 Abundance Change

Range shifts with a clear change of species distribution at the distribution limits represent extreme transitions from the presence to absence or vice versa. Prior to range contractions, seaweeds will first decrease in abundances within their ranges, where continued reductions in abundance near range limits represent the first steps towards a range shift (Bates et al. 2014). For example, Johnson et al. (2011) documented that previously widespread *Macrocystis pyrifera* (Fig. 3.7) kelp forests decreased drastically in cover at several sites in eastern Tasmania (Table 3.2), likely

Fig. 3.7 The giant kelp (*Macrocystis pyrifera*) is a majestic seaweed often attaining a size of more than 10 m, has declined dramatically in abundance in Tasmania (pictured) over the past couple of decades due to increased warming, nutrient poor water and urchin grazing (*Photo* © T. de Bettignies)

caused by a combination of ocean warming and a strengthening of the East Australian Current (characterized by nutrient poor water) over the past six decades (Johnson et al. 2011). Concurrently, the strengthening of the East Australian Current also led to the range expansion of the sea urchin *Centrostephanus rodgersii* into Tasmanian waters, facilitated by over-fishing of urchin predators (lobsters, Ling et al. 2009). The range-expanding urchins have likely contributed to the decline in *M. pyrifera* through destructive grazing, which has also negatively impacted other native seaweeds (Johnson et al. 2011). With the decrease in abundance of *M. pyrifera*, a fast-growing habitat and food provider, dozens of associated species are losing a unique three-dimensional habitat, resulting in loss of taxonomic diversity and food web complexity (Graham 2004; Ling 2008; Byrnes et al. 2011).

3.5 Perspective and Conclusion: Human Impacts on Seaweed Biogeography

Range shifts caused by species introductions and climate change need close monitoring as they are potentially irreversible and likely to have great ecosystem impacts (Madin et al. 2012). A critical problem, however, is that information on species' range boundaries is scarce and largely qualitative due to lack of baseline information and regular surveys (Wernberg et al. 2011b; Bates et al. 2015; Marcelino and Verbruggen 2015). Ecological niche models can assist to identify areas with suitable habitat, anticipate arrival points and predict the potential extent of range change after a successful introduction (Marcelino and Verbruggen 2015) or environmental change (Molinos et al. 2015; Takao et al. 2015). For example, Takao et al. (2015) found that the present distribution of *Ecklonia cava* around Japan is well represented by SST-based indices. Chronologically observed changes were well in agreement with the projections, and the results further indicated that temperature will be a key factor for distribution *of E. cava* in the future (Takao et al. 2015).

Monitoring laboratory experiments and models projecting future shifts combined will help to identify likely range-shift pathways of seaweeds. In response to range-shifting species, management is necessary and several management tools already in place can be applied through, for example education, raising awareness and protected areas. But existing management tools are not always sufficient, and especially the limited knowledge on range-shift limits adaptive management responses (Madin et al. 2012). Additionally, for successful monitoring, a more widespread use of molecular methods is necessary to determine origin of species to prevent misidentification based on plastic morphology (Bolton 2010) and to identify loss of genetic variability at range edges (Provan and Maggs 2012; Assis et al. 2014; Neiva et al. 2015). Also, more regular surveys are required to be undertaken to determine range edges of populations and identify early range shifts and species introductions.

Future temperature increases are likely to result in more range shifts of seaweeds, especially along north–south orientated coastlines (Wernberg et al. 2011b; Molinos et al. 2015). These range shifts include poleward range extensions of warm-water tropical species, poleward range contractions of cold-water temperate species (Sorte et al. 2010; Wernberg et al. 2011a) and (potentially bidirectional) range expansion of introduced seaweeds (Sorte et al. 2013). Current models for marine species predict that expansions will be more prominent than contractions, leading to an overall increase in the biodiversity of many extratropical regions (Molinos et al. 2015). Globally, however, the narrative is likely to be different, because there will be a net loss of species (Cheung et al. 2009) as extinctions will be far more rapid than the evolution of new species. For seaweeds, this will be exacerbated by the juxtaposition of global patterns of species richness and endemicity (Bolton 1994; Kerswell 2006), hotspots of warming (Hobday and Pecl 2013) and barriers to range shifts (Wernberg et al. 2011a). In particular, southern Australia has the highest species richness and endemicity of seaweeds in the world, as well as some of the fastest warming regions in the world. However, the southern coastline is oriented east–west with very limited landmasses farther south. As seaweeds are pushed poleward towards the edge of the continent, there is great risk that they will 'drop off' to extinction—indeed, it has been estimated that range shifts could result in as much as a 25 % loss of the seaweed flora (Wernberg et al. 2011a).

To determine and model future biogeographic patterns of seaweed distribution, it is also necessary to take into account increasing threats to the coastal environments. Superimposed on temperature increases, increased ocean acidification will also change competitive hierarchies between fleshy, turf and calcifying marine algae, further altering local seaweed communities—and ultimately also range shifts (Hofmann et al. 2012). Also, interactive future effects, especially combined effects of warming and acidification with non-climate stressors, such as reduced water quality, will lower the resilience of communities and single species to perturbations like species invasions and storms (Wernberg et al. 2011a). Concurrently, more frequent and intense discrete events can drive stepwise changes in local environmental structure and cause larger more dramatic range- shifts (Smale and Wernberg 2013). Finally, ecological interactions are influencing the success of introductions and the speed of range shifts, possibly suppressing recovery, enhancing contraction or slowing down expansions (see *M. pyrifera* and *S. dorycarpa* case studies above). The extent to which the ecological context can suppress or enhance range shifts is a question in need of much research effort as we progress from simply detecting change to understanding its underlying drivers and mediators. However, the magnitude of range shifts and biological responses from anthropogenic impacts differ widely among species (Poloczanska et al. 2013).

3.6 Conclusion

There is now substantial evidence that humans have influenced the global biogeography of seaweeds over the last few decades and will continue to do so in the near future. This evidence generally spans timescales of decades, and is unlikely to simply reflect short-term fluctuations such as ENSO events. Humans influence seaweed biogeography through three distinct processes (introductions, climate expansions and climate contractions), which manifest through different processes (dispersal, recruitment and mortality) (Bates et al. 2014) and therefore proceed at different speeds: introduction > expansion > contraction. These changes in seaweed distributions have also been associated with impacts on seaweed-based ecosystems. While we are still to see the long-term ecological and economic consequences, these are likely to be substantial given the ecosystem services derived from seaweed ecosystems (Bennett et al. 2015a).

Acknowledgements This work was supported by the University of Western Australia through an International Postgraduate Research Scholarship to SS and a research collaboration award to TW, the Australian Research Council through a Future Fellowship (FT110100174) to TW. MST was supported by the Marsden Fund Council from Government funding, administered by the Royal Society of New Zealand.

Appendix

Review of published literature and citation searches to compile a global dataset of documented range shifts in native seaweeds or range expansions of successful seaweed invaders. Used key words include climate change, warming, extreme events, heat waves, invasive seaweeds, successful invaders, shift in distribution, range shifts, range expansion and range contraction. Literature was included when data was available for the direction, distance and time window of seaweed shift, so annual spread rates could be calculated. Literature stating a decrease in abundance or not pinpointing location and time window were excluded from the dataset. The two main drivers are Introduction (introduction) and Warming (contraction/ expansion). When unusual driver it is added in brackets

Species	Division	Region	Driver	Annual spread (km/years)	First appearance/absence	Time window (years)	Distance (km)	Reference
Assemblage	Assemblage	SW Australia	Contraction	1.0	1940	50	51	Wernberg et al. (2011a, b)
Assemblage	Assemblage	SE Australia	Contraction	4.2	1940	50	211	Wernberg et al. (2011a, b)
Caulerpa cylindracea	Chlorophyta	Provence, France	Introduction	11.9	1997	7	83	Ruitton et al. (2005)
Caulerpa cylindracea	Chlorophyta	Ligurian Sea	Introduction	44.0	2009	5	220	Altamirano et al. (2014)
Caulerpa ollivieri	Chlorophyta	Mexico	Introduction	19.0	1968	42	800	Ortegón-Aznar et al. (2015)
Caulerpa taxifolia var. distichophylla	Chlorophyta	Mediterranean Sea	Introduction	33.3	2006	6	200	Aplikioti et al. (2016)
Caulerpa taxifolia var. distichophylla	Chlorophyta	Mediterranean Sea	Introduction	87.5	2006	8	700	Aplikioti et al. (2016)
Codium adhaerens	Chlorophyta	Portugal	Expansion	1.2	1955	50	59	Lima et al. (2007)
Codium fragile ssp. *fragile*	Chlorophyta	Nova Scotia	Introduction	11.1	1989	18	200	Watanabe et al. (2010)
Codium fragile ssp. *tomentosoides*	Chlorophyta	NW Atlantic	Introduction	16.0	1955	47	750	Scheibling and Gagnon (2006)
Codium fragile ssp. *tomentosoides*	Chlorophyta	NW Atlantic	Introduction	10.6	1955	47	500	Scheibling and Gagnon (2006)
Codium fragile ssp. *tomentosoides*	Chlorophyta	Northern Chile	Introduction	6.4	2005	7	45	Neill et al. (2006)
Valonia utricularis	Chlorophyta	Portugal	Expansion	3.9	1955	50	197	Lima et al. (2007)
Ahnfeltia plicata	Ochrophyta	Portugal	Expansion	6.6	1955	50	330	Lima et al. (2007)
Bifurcaria bifurcata	Ochrophyta	Britain, Ireland	Expansion	3.1	1964	45	140	Mieszkowska et al. (2006)
Bifurcaria bifurcata	Ochrophyta	Portugal	Expansion	5.1	1955	50	257	Lima et al. (2007)
Chondrus crispus	Ochrophyta	Portugal	Expansion	3.6	1955	50	180	Lima et al. (2007)

(continued)

(continued)

Species	Division	Region	Driver	Annual spread (km/years)	First appearance/absence	Time window (years)	Distance (km)	Reference
Desmarestia aculeata	Ochrophyta	Portugal	Expansion	4.5	1955	50	227	Lima et al. (2007)
Desmarestia ligulata	Ochrophyta	Portugal	Expansion	1.4	1955	50	70	Lima et al. (2007)
Dumontia contorta	Ochrophyta	Portugal	Expansion	1.2	1955	50	62	Lima et al. (2007)
Durvillea potatorum	Ochrophyta	SE Australia	Contraction	0.6	1945	60	35	Millar (2007)
Ecklonia maxima	Ochrophyta	South Africa	Expansion	36.5	2008	2	73	Bolton et al. (2012)
Ecklonia radiata	Ochrophyta	SW Australia	Contraction	88.0	2011	1	88	Wernberg and Bennett, unpublished data 2015
Fucus serratus	Ochrophyta	North Spain	Contraction	1.9	1894	60	116	Duarte et al. (2013)
Fucus serratus	Ochrophyta	North Spain	Contraction	5.1	1955	21	107	Duarte et al. (2013)
Fucus serratus	Ochrophyta	North Spain	Expansion	2.2	1977	12	26	Duarte et al. (2013)
Fucus serratus	Ochrophyta	Spain	Expansion	5.0	1982	20	100	Arrontes (2002)
Fucus serratus	Ochrophyta	North America	Introduction	11.8	1868	17	200	Johnson et al. (2012)
Fucus serratus	Ochrophyta	North America	Introduction	5.3	1868	17	90	Johnson et al. (2012)
Fucus vesiculosus	Ochrophyta	Morocco	Contraction	41.7	1985	30	1250	Nicastro et al. (2013)
Fucus vesiculosus	Ochrophyta	Portugal	Expansion	3.1	1955	50	157	Lima et al. (2007)
Halidrys siliquosa	Ochrophyta	Portugal	Expansion	1.8	1955	50	90	Lima et al. (2007)
Halydris siliquosa	Ochrophyta	Portugal	Expansion	1.1	2006	75	80	Lima et al. (2008)
Himanthalia elongata	Ochrophyta	North Spain	Contraction	1.8	1889	66	116	Duarte et al. (2013)
Himanthalia elongata	Ochrophyta	North Spain	Contraction	4.2	1955	20	84	Duarte et al. (2013)
Himanthalia elongata	Ochrophyta	North Spain	Contraction	26.0	2004	5	130	Duarte et al. (2013)

(continued)

(continued)

Species	Division	Region	Driver	Annual spread (km/years)	First appearance/absence	Time window (years)	Distance (km)	Reference
Himanthalia elongata	Ochrophyta	Portugal	Expansion	4.4	1955	50	219	Lima et al. (2007)
Laminaria ochroleuca	Ochrophyta	SE Atlantic	Expansion	2.5	1948	60	150	Smale et a (2013)
Laminaria ochroleuca	Ochrophyta	SE Atlantic	Expansion	5.4	1948	60	325	Smale et al. (2013)
Padina pavonica	Ochrophyta	Portugal	Expansion	3.7	1955	50	187	Lima et al. (2007)
Pelvetia canaliculata	Ochrophyta	Portugal	Expansion	4.9	1955	50	245	Lima et al. (2007)
Sargassum filicinum	Ochrophyta	Mexico	Introduction	137.5	2003	4	550	Riosmena-Rodriguez (2012)
Sargassum flavifolium	Ochrophyta	Portugal	Expansion	11.9	1955	50	593	Lima et al. (2007)
Sargassum illicifolium	Ochrophyta	Japan	Expansion	10.5	1989	19	200	Tanaka et al. (2012)
Sargassum micracanthum	Ochrophyta	Japan	Contraction	3.9	1977	31	120	Tanaka et al. (2012)
Sargassum muticum	Ochrophyta	Denmark	Introduction	4.4	1984	16	70	Staehr et al. (2000)
Sargassum muticum	Ochrophyta	Canada	Introduction	111.2	1947	6	667	Engelen et al. (2015)
Sargassum muticum	Ochrophyta	northern California	Introduction	80.6	1965	18	1450	Engelen et al. (2015)
Sargassum muticum	Ochrophyta	Mexico	Introduction	35.0	1973	8	280	Engelen et al. (2015)
Sargassum muticum	Ochrophyta	Netherlands	Introduction	50.0	1979	6	300	Engelen et al. (2015)
Sargassum muticum	Ochrophyta	Denmark	Introduction	120.0	1984	5	600	Engelen et al. (2015)
Sargassum muticum	Ochrophyta	France	Introduction	81.8	1983	11	900	Engelen et al. (2015)
Sargassum muticum	Ochrophyta	Mexico	Introduction	2.7	1988	15	40	Espinoza (1990)

(continued)

(continued)

Species	Division	Region	Driver	Annual spread (km/years)	First appearance/absence	Time window (years)	Distance (km)	Reference
Sargassum yamamotoi	Ochrophyta	Japan	Contraction	4.4	1977	31	135	Tanaka et al. (2012)
Scytothalia dorycarpa	Ochrophyta	SW Australia	Contraction	100.0	2011	1	100	Smale and Wernberg (2013)
Scytothalia dorycarpa	Ochrophyta	SW Australia	Contraction	3.2	1961	50	160	Smale and Wernberg (2013)
Turbinaria ornata	Ochrophyta	French Polynesia	Expansion	30.0	1980	10	300	Stewart (2008)
Undaria pinnatifida	Ochrophyta	North America	Introduction	125.0	2000	2	250	Aguilar-Rosas et al. (2004)
Undaria pinnatifida	Ochrophyta	Mexico	Introduction	66.7	2003	3	200	Aguilar-Rosas et al. (2004)
Undaria pinnatifida	Ochrophyta	Argentina	Introduction	35.7	1999	7	250	Dellatorre et al. (2014)
Undaria pinnatifida	Ochrophyta	Argentina	Introduction	50.0	2005	6	300	Dellatorre et al. (2014)
Undaria pinnatifida	Ochrophyta	Argentina	Introduction	50.0	2012	20	1000	Dellatorre et al. (2014)
Grateloupia doryphora	Rhodophyta	Brittany, France	Introduction	150.0	1999	1	150	Simon et al. (2001)
Grateloupia turuturu	Rhodophyta	Gulf of Maine	Introduction	33.0	2007	4	132	Mathieson et al. (2008)
Halopithys incurva	Rhodophyta	Portugal	Expansion	9.5	1955	50	475	Lima et al. (2007)
Heterosiphonia japonica	Rhodophyta	Western North Atlantic	Introduction	66.7	2007	6	400	Newton et al. (2013)
Heterosiphonia japonica	Rhodophyta	Western North Atlantic	Introduction	16.7	2007	6	100	Newton et al. (2013)
Hypnea musciformis	Rhodophyta	Portugal	Expansion	5.4	1955	50	269	Lima et al. (2007)
Mastocarpus sp.	Rhodophyta	Chile	Introduction	18.2	1980	22	400	Macaya et al. (2013)
Mastocarpus sp.	Rhodophyta	Chile	Introduction	31.8	1980	22	700	Macaya et al. (2013)
Palmaria palmata	Rhodophyta	Portugal	Expansion	7.2	1955	50	358	Lima et al. (2007)

References

Adey WH, Steneck RS. Thermogeography over time creates biogeographic regions: a temperature/space/time-integrated model and an abundance-weighted test for benthic marine algae. J Phycol. 2001;37:677–98.

Aguilar-Rosas R, Aguilar-Rosas LE, Avila G, Marcos R. First record of *Undaria pinnatifida* (Harvey) Suringar (Laminariales,Phaeophyta) on the Pacific coast of Mexico. Bot Mar. 2004;47:255–58.

Altamirano M, Andreakis N, Souza-egipsy V, Zanolla M, De la Rosa J. First record of *Caulerpa cylindracea* (Caulerpaceae , Chlorophyta) in Andalusia (Southern Spain). An del Jardín Botánico Madrid. 2014;71:1–9.

Anderson RJ, Bolton JJ, Stegenga H. Using the biogeographical distribution and diversity of seaweed species to test the efficacy of marine protected areas in the warm-temperate Agulhas Marine Province, South Africa. Divers Distrib. 2009;15:1017–27.

Aplikioti M, Louizidou P, Mystikou A, Marcou M, Stavrou P, Kalogirou S, et al. Further expansion of the alien seaweed *Caulerpa taxifolia* var. *distichophylla* (Sonder) Verlaque, Huisman & Procacini (Ulvophyceae, Bryopsidales) in the Eastern Mediterranean Sea. Aquat Invasions. 2016;11:in press.

Arrontes J. Mechanisms of range expansion in the intertidal brown alga *Fucus serratus* in northern Spain. Mar Biol. 2002;141:1059–67.

Assis J, Serrão EA, Claro B, Pearson GA. Climate-driven range shifts explain the distribution of extant gene pools and predict future loss of unique lineages in a marine brown alga. Mol Ecol. 2014;23:2797–810.

Baldacconi R, Corriero G. Effects of the spread of the alga *Caulerpa racemosa* var. *cylindracea* on the sponge assemblage from coralligenous concretions of the Apulian coast (Ionian Sea, Italy). Mar Ecol. 2009;30:337–45.

Bartsch I, Wiencke C, Laepple T. Global seaweed biogeography under a changing climate: the prospected effects of temperature. In: Wiencke C, Bischof K, editors. Seaweed biology. Berlin Heidelberg: Springer; 2012. p. 383–406.

Bates AE, Pecl GT, Frusher S, Hobday AJ, Wernberg T, Smale DA, Sunday JM, Hill NA, Dulvy NK, Colwell RK, Holbrook NJ, Fulton EA, Slawinski D, Feng M, Edgar GJ, Radford BT, Thompson PA, Watson RA. Defining and observing stages of climate-mediated range shifts in marine systems. Glob Environ Change. 2014;26:27–38.

Bates AE, Bird TJ, Stuart-smith RD, Wernberg T, Sunday JM, Barrett NS, Edgar GJ, Frusher S, Hobday AJ, Pecl GT, Smale DA, McCarthy M. Distinguishing geographical range shifts from artefacts of detectability and sampling effort. Divers Distrib. 2015;21:13–22.

Bennett S, Wernberg T, Connell SD, Hobdav AJ, Johnson CR, Poloczanska ES. The "Great Southern Reef": social, ecological and economic value of Australia's neglected kelp forests. Mar Freshwater Res. 2015a. doi:10.1071/MF15232.

Bennett S, Wernberg T, Harvey ES, Santana-Garcon J, Saunders BJ. Tropical herbivores provide resilience to a climate-mediated phase shift on temperate reefs. Ecol Lett. 2015b;18:714–23.

Björck S. A review of the history of the Baltic Sea, 13.0-8.0 ka BP. Quat Int. 1995;27:19–40.

Bolton JJ. Global seaweed diversity: patterns and anomalies. Bot Mar. 1994;37:241–6.

Bolton JJ. The biogeography of kelps (Laminariales, Phaeophyceae): a global analysis with new insights from recent advances in molecular phylogenetics. Helgol Mar Res. 2010;64:263–79.

Bolton J, Anderson R, Smit A, Rothman M. South African kelp moving eastwards: the discovery of *Ecklonia maxima* (Osbeck) Papenfuss at De Hoop Nature Reserve on the south coast of South Africa. African J Mar Sci. 2012;34:147–51.

Burridge CP, Hurt AC, Farrington LW, Coutin PC, Austin CM. Stepping stone gene flow in an estuarine dwelling sparid from south east Australia. J Fish Biol. 2004;64:805–19.

Burrows MT, Schoeman DS, Buckley LB, Moore P, Poloczanska ES, Brander KM, Brown C, Bruno JF, Duarte CM, Halpern BS, Holding J, Kappel CV, Kiessling W, O'Connor MI,

Pandolfi JM, Parmesan C, Schwing FB, Sydeman WJ, Richardson AJ. The pace of shifting climate in marine and terrestrial ecosystems. Science. 2011;334:652–5.
Buschbaum C, Chapman AS, Saier B. How an introduced seaweed can affect epibiota diversity in different coastal systems. Mar Biol. 2006;148:743–54.
Byrnes JE, Reed DC, Cardinale BJ, Cavanaugh KC, Holbrook SJ, Schmitt RJ. Climate-driven increases in storm frequency simplify kelp forest food webs. Glob Change Biol. 2011;17:2513–24.
Casas G, Scrosati R, Luz Piriz M. The invasive kelp *Undaria pinnatifida* (Phaeophyceae, Laminariales) reduces native seaweed diversity in Nuevo Gulf (Patagonia, Argentina). Biol Invasions. 2004;6:411–6.
Cheung WWL, Lam VWY, Sarmiento JL, Kearney K, Watson R, Pauly D. Projecting global marine biodiversity impacts under climate change scenarios. Fish Fish. 2009;10:235–51.
Christie H, Norderhaug KM, Fredriksen S. Macrophytes as habitat for fauna. Mar Ecol Prog Ser. 2009;396:221–33.
Coumou D, Rahmstorf S. A decade of weather extremes. Nat Clim Change. 2012;2:491–6.
Cowman PF, Bellwood DR. Vicariance across major marine biogeographic barriers: temporal concordance and the relative intensity of hard versus soft barriers. Proc R Soc B. 2013;280:20131541.
Dellatorre FG, Amoroso R, Saravia J, Orensanz (Lobo) JM. Rapid expansion and potential range of the invasive kelp *Undaria pinnatifida* in the Southwest Atlantic. Aqua Invasions. 2014;9 (4):467–78.
Dethier MN, McDonald K, Strathmann RR. Colonization and connectivity of habitat patches for coastal marine species distant from source populations. Conserv Biol. 2003;17:1024–35.
Duarte L, Viejo RM, Martínez B, deCastro M, Gomez-Gesteira M, Gallardo T. Recent and historical range shifts of two canopy-forming seaweeds in North Spain and the link with trends in sea surface temperature. Acta Oecologica. 2013;51:1–10.
Engelen AH, Serebryakova A, Ang P, Britton-Simmons K, Mineur F, Pedersen MF, Toth G. Circumglobal invasion by the brown seaweed *Sargassum muticum*. Oceanogr Mar Biol Ann Rev. 2015; 53:81–126.
Espinoza J. The Southern Limit of *Sargassum muticum* (Yendo) Fensholt (Phaeophyta, Fucales) in the Mexican Pacific. Bot Mar. 1990;33:193–96.
Fraser CI, Waters JM. Algal parasite *Herpodiscus durvillaeae* (Phaeophyceae: Sphacelariales) inferred to have traversed the Pacific Ocean with its buoyant host. J Phycol. 2013;49:202–6.
Fraser CI, Nikula R, Waters JM. Oceanic rafting by a coastal community. Proc R Soc B. 2011;278:649–55.
Gaylord B, Gaines SD. Temperature or transport? Range limits in marine species mediated solely by flow. Am Nat. 2000;155:769–89.
Gaylord B, Reed DC, Raimondi PT, Washburn L, McLean SR. A physically based model of macroalgal spore dispersal in the wave and current-dominant nearshore. Ecology. 2002;83:1239–51.
Gillespie RG, Baldwin BG, Waters JM, Fraser CI, Nikula R, Roderick GK. Long-distance dispersal: a framework for hypothesis testing. Trends Ecol Evol. 2012;27:47–55.
Graham MH. Effects of local deforestation on the diversity and structure of southern California giant kelp forest food webs. Ecosystems. 2004;7:341–57.
Grigg R, Hey R. Paleoceanography of the tropical eastern Pacific Ocean. Science. 1992;255:172–8.
Hampe A, Petit R. Conserving biodiversity under climate change: the rear edge matters. Ecol Lett. 2005;8:461–7.
Harley CDG, Hughes AR, Hulgren KM, Miner BG, Sorte CJB, Thornber CS, Rodriguez LF, Tomanek L, Willams SL. The impacts of climate change in coastal marine systems. Ecol Lett. 2006;9:228–41.
Harley CDG, Anderson KM, Demes KW, Jorve JP, Kordas RL, Coyle TA, Graham MH. Effects of climate change on global seaweed communities. J Phycol. 2012;48:1064–78.
Hawkins S, Hartnoll R. Factors determining the upper limits of intertidal canopy-forming algae. Mar Ecol Prog Ser. 1985;20:265–71.

Hidas EZ, Costa TL, Ayre DJ, Minchinton TE. Is the species composition of rocky intertidal invertebrates across a biogeographic barrier in south-eastern Australia related to their potential for dispersal? Mar Freshwater Res. 2007;58:835–42.

Hinojosa IA, Pizarro M, Ramos M, Thiel M. Spatial and temporal distribution of floating kelp in the channels and fjords of southern Chile. Estu Coast Shelf Sci. 2010;87:367–77.

Hobday AJ. Age of drifting *Macrocystis pyrifera* (L.) C. Agardh rafts in the Southern California Bight. J Exp Mar Bio Ecol. 2000a;253:97–114.

Hobday AJ. Persistence and transport of fauna on drifting kelp (*Macrocystis pyrifera* (L.) C. Agardh) rafts in the Southern California Bight. J Exp Mar Bio Ecol. 2000b;253:75–96.

Hobday AJ, Pecl GT. Identification of global marine hotspots: sentinels for change and vanguards for adaptation action. Rev Fish Biol Fish. 2013;24:415–25.

Hofmann LC, Straub S, Bischof K. Competition between calcifying and noncalcifying temperate marine macroalgae under elevated CO2 levels. Mar Ecol Prog Ser. 2012;464:89–105.

Ingólfsson A. The invasion of the intertidal canopy-forming alga *Fucus serratus* L. to southwestern Iceland: possible community effects. Estu Coast Shelf Sci. 2008;77:484–90.

IPCC (2012) Managing the risks of extreme events and disasters to advance climate change adaptation.

IPCC (2014) Climate change 2014 synthesis report summary chapter for policymakers.

Irving AD, Connell SD. Predicting understorey structure from the presence and composition of canopies: an assembly rule for marine algae. Oecologia. 2006;148:491–502.

Johnson L, Brawley S, Adey W. Secondary spread of invasive species: historic patterns and underlying mechanisms of the continuing invasion of the European rockweed *Fucus serratus* in eastern. Biol Invasions. 2012;14:79–97.

Johnson CR, Banks SC, Barrett NS, Cazassus F, Dunstan PK, Edgar GJ, Frusher SD, Gardner C, Haddon M, Helidoniotis F, Hill KL, Holbrook NJ, Hosie GW, Last PR, Ling SD, Melbourne-Thomas J, Miller K, Pecl GT, Richardson AJ, Ridgway KR, Rintoul SR, Ritz DA, Ross DJ, Sanderson JC, Shepherd SA, Slotwinski A, Swadling KM, Taw N. Climate change cascades: shifts in oceanography, species' ranges and subtidal marine community dynamics in eastern Tasmania. J Exp Mar Bio Ecol. 2011;400:17–32.

Karl T, Trenberth K. Modem global climate change. Science. 2003;302:1719–23.

Kerswell AP. Global biodiversity patterns of benthic marine algae. Ecology. 2006;87:2479–88.

Klein J, Verlaque M. The *Caulerpa racemosa* invasion: a critical review. Mar Pollut Bull. 2008;56:205–25.

Lessios HA, Kessing BD, Robertson DR. Massive gene flow across the world's most potent marine biogeographic barrier. Proc R Soc B. 1998;265:583–8.

Lima FP, Wethey DS. Three decades of high-resolution coastal sea surface temperatures reveal more than warming. Nat Commun. 2012;3:1–13.

Lima FP, Ribeiro PA, Queiroz N, Hawkins SJ, Santos AM. Do distributional shifts of northern and southern species of algae match the warming pattern? Glob Chang Biol. 2007;13:2592–604.

Lima FP, Queiroz N, Ribeiro PA, Xavier R, Hawkins SJ, Santos AM. First record of *Halidrys siliquosa* on the Portuguese coast: counter-intuitive range expansion? Mar Biodivers Rec. 2008; doi: 10.1017/S1755267208000018.

Ling SD. Range expansion of a habitat-modifying species leads to loss of taxonomic diversity: a new and impoverished reef state. Oecologia. 2008;156:883–94.

Ling SD, Johnson CR, Frusher SD, Ridgway KR. Overfishing reduces resilience of kelp beds to climate-driven catastrophic phase shift. Proc Nat Acad Sci USA. 2009;106:2234–22345.

Lozano-Montes HM, Loneragan NR, Babcock RC, Jackson K. Using trophic flows and ecosystem structure to model the effects of fishing in the Jurien Bay Marine Park, temperate Western Australia. Mar Freshwater Res. 2011;62(5):421–31.

Luiz OJ, Madin JS, Robertson DR, Rocha LA, Wirtz P, Floeter SR. Ecological traits influencing range expansion across large oceanic dispersal barriers: insights from tropical Atlantic reef fishes. Proc R Soc B. 2012;279:1033–40.

Lüning K. Meeresbotanik: Verbreitung, Ökophysiologie und Nutzung der marinen Makroalgen. In: 1st ed. Stuttgart New York: Georg Thieme Verlag; 1985.

Macaya EC, Pacheco S, Cáceres A, Musleh S. Range extension of the non-indigenous alga *Mastocarpus* sp. along the Southeastern Pacific coast. Rev Biol Mar Oceanogr. 2013; 48: 661–65.

Madin EMP, Ban NC, Doubleday ZA, Holmes TH, Pecl GT, Smith F. Socio-economic and management implications of range-shifting species in marine systems. Glob Environ Change. 2012;22:137–46.

Marcelino VR, Verbruggen H. Ecological niche models of invasive seaweeds. J Phycol. 2015;51:606–20.

Mathieson AC, Dawes CJ, Pederson J, Gladych RA, Carlton JT. The Asian red seaweed *Grateloupia turuturu* (Rhodophyta) invades the Gulf of Maine. Biol Invasions. 2008;10:985–88.

Mattio L, Zubia M, Maneveldt GW, Anderson RJ, Bolton JJ, de Gaillande C, De Clerck O, Payri CE. Marine flora of the Iles Eparses (Scattered Islands): a longitudinal transect through the Mozambique Channel. Acta Oecologica. 2015. doi:10.1016/j.actao.2015.09.001.

Mieszkowska N, Kendall MA, Hawkins SJ, Leaper R, Willamson P, Hardman-Mountford NJ, et al. Changes in the range of some common rocky shore species in Britain - A response to climate change? Hydrobiologia. 2006;555:241–51.

Millar AJ. The Flindersian and Peronian Provinces. Algae Aust Introd. 2007:554–59.

Molinos JG, Halpern BS, Schoeman DS, Brown CJ, Kiessling W, Moore PJ, Pandolfi JM, Poloczanska ES, Richardson AJ, Burrows MT. Climate velocity and the future global redistribution of marine biodiversity. Nat Clim Change. 2015. doi:10.1038/nclimate2769.

Myers AA. Biogeographic barriers and the development of marine biodiversity. Estu Coast Shelf Sci. 1997;44:241–8.

Neill PE, Alcalde O, Faugeron S, Navarrete SA, Correa JA. Invasion of *Codium fragile* ssp. *tomentosoides* in northern Chile: A new threat for *Gracilaria* farming. Aquaculture. 2006;259:202–210.

Neiva J, Assis J, Coelho NC, Fernandes F, Pearson GA, Serrão EA. Genes left behind: climate change threatens cryptic genetic diversity in the canopy-forming seaweed *Bifurcaria bifurcata*. PLoS One. 2015;10:e0131530.

Neiva J, Serrão EA, Assis J, Pearson GA, Coyer JA, Olsen JL, Hoarau G, Valero M. Climate oscillations, range shifts and phylogeographic patterns of North Atlantic Fucaceae. In: Hu ZM, Fraser CI, editors. Seaweed phylogeography: adaptation and evolution of seaweeds under environmental change. Berlin Heidelberg: Springer; 2016.

Newton C, Bracken MES, McConville M, Rodrigue K, Thornber CS. Invasion of the red seaweed *Heterosiphonia japonica* spans biogeographic provinces in the Western North Atlantic Ocean. PLoS One. 2013;8(4):e62261.

Nicastro KR, Zardi GI, Teixeira S, Neiva J, Serrao EA, Pearson GA. Shift happens : trailing edge contraction associated with recent warming trends threatens a distinct genetic lineage in the marine macroalga *Fucus vesiculosus*. BMC Biol. 2013;11:6.

Nogales M, Heleno R, Traveset A, Vargas P. Evidence for overlooked mechanisms of long-distance seed dispersal to and between oceanic islands. New Phytol. 2012;194:313–7.

Norton TA. Dispersal by macroalgae. Br Phycol J. 1992;27:293–301.

Ortegón-aznar I, Rosado-espinosa LA, Aguilar-perera A. Occurrence of the introduced alga *Caulerpa ollivieri* Dostál, 1929 (Caulerpaceae, Chlorophyta) in the Southern Gulf of Mexico. BioInvasions Rec. 2015;4:17–21.

Pedersen MF, Staehr PA, Wernberg T, Thomsen MS. Biomass dynamics of exotic *Sargassum muticum* and native *Halidrys siliquosa* in Limfjorden, Denmark—implications of species replacements on turnover rates. Aquat Bot. 2005;83:31–47.

Perry AL, Low PJ, Ellis JR, Reynolds JD. Climate change and distribution shifts in marine fishes. Science. 2005;308:1912–5.

Piazzi L, Balata D, Ceccherelli G, Cinelli F. Interactive effect of sedimentation and *Caulerpa racemosa* var. *cylindracea* invasion on macroalgal assemblages in the Mediterranean Sea. Estu Coast Shelf Sci. 2005;64:467–74.

Pinsky ML, Worm B, Fogarty MJ, Sarmiento JL, Levin SA. Marine taxa track local climate velocities. Science. 2013;341:1239–42.

Poloczanska ES, Brown CJ, Sydeman WJ, Kiessling W, Schoeman DS, Moore PJ, Brander K, Bruno JF, Buckley LB, Burrows MT, Duarte CM, Halpern BS, Holding J, Kappel CV, O'Connor MI, Pandolfi JM, Parmesan C, Schwing F, Thompson SA, Richardson AJ. Global imprint of climate change on marine life. Nat Clim Change. 2013;3:919–25.

Provan J, Maggs CA. Unique genetic variation at a species' rear edge is under threat from global climate change. Proc R Soc B. 2012;279:39–47.

Quartino ML, Deregibus D, Campana GL, Juan Latorre GE, Momo FR. Evidence of macroalgal colonization on newly ice-free areas following glacial retreat in Potter Cove (South Shetland Islands), Antarctica. PLoS One. 2013;8(3):e58223.

Riosmena-Rodríguez R, Boo GH, López-Vivas JM, Hernandez-Velasco A, Saenz-Arroyo A, Boo SM. The invasive seaweed *Sargassum filicinum* (Fucales, Phaeophyceae) is on the move along the Mexican Pacific coastline. Bot Mar. 2012;55:547–51.

Rothäusler E, Gutow L, Thiel M. Floating seaweeds and their communities. In: Wiencke C, Bischof K, editors. Seaweed biology. Berlin: Springer; 2012. p. 359–80.

Ruitton S, Javel F, Culioli J-M, Meinesz A, Pergent G, Verlaque M. First assessment of the *Caulerpa racemosa* (Caulerpales, Chlorophyta) invasion along the French Mediterranean coast. Mar Pollut Bull. 2005;50:1061–8.

Sakai AK, Allendorf FW, Holt JS, Lodge DM, Molofsky J, With KA, Baughman S, Cabin RJ, Cohen JE, Ellstrand NC, McCauley DE, O'Neil P, Marker IM, Thompson JN, Weller SG. The population biology of invasive species. Ann Rev Ecol Syst. 2001;32:305–32.

Santelices B. Patterns of reproduction, dispersal and recruitment in seaweeds. Oceanogr Mar Biol Ann Rev. 1990;28:177–276.

Scheibling RE, Gagnon P. Competitive interactions between the invasive green alga *Codium fragile* ssp. *tomentosoides* and native canopy-forming seaweeds in Nova Scotia (Canada). Mar Ecol Prog Ser. 2006;325:1–14.

Schils T, Wilson SC. Temperature threshold as a biogeographic barrier in northern Indian ocean macroalgae. J Phycol. 2006;42:749–56.

Simon C, Ar GE, Deslandes E. Expansion of the red alga *Grateloupia doryphora* along the coasts of Brittany (France). Hydrobiologia. 2001;443:23–29.

Smale DA, Wernberg T. Extreme climatic event drives range contraction of a habitat-forming species. Proc R Soc B. 2013;280:20122829.

Smale DA, Wernberg T, Yunnie ALE, Vance T. The rise of *Laminaria ochroleuca* in the Western English Channel (UK) and comparisons with its competitor and assemblage dominant *Laminaria hyperborea*. Mar Ecol. 2014. doi:10.1111/maec.12199.

Smit AJ, Roberts M, Anderson RJ, Dufois F, Dudley SFJ, Bornan TG, Olbers J, Bolton JJ. A coastal seawater temperature dataset for biogeographical studies: large biases between in situ and remotely-sensed data sets around the coast of South Africa. PLoS One. 2013.

Sorte CJB, Williams SL, Carlton JT. Marine range shifts and species introductions: comparative spread rates and community impacts. Glob Ecol Biogeogr. 2010;19:303–16.

Sorte CJB, Ibáñez I, Blumenthal DM, Molinri NA, Miller LP, Grosholz ED, Diez JM, D'Antonio CM, Olden JD, Jones SJ, Dukes JS. Poised to prosper? A cross-system comparison of climate change effects on native and non-native species performance. Ecol Lett. 2013;16:261–70.

Stæhr PA, Pedersen MF, Thomsen MS, Wernberg T, Krause-Jensen D. Invasion of *Sargassum muticum* in Limfjorden (Denmark) and its possible impact on the indigenous macroalgal community. Mar Ecol Prog Ser. 2000;207:79–88.

Stewart HL. The role of spatial and ontogenetic morphological variation in the expansion of the geographic range of the tropical brown alga, *Turbinaria ornata*. Integr Comp Biol. 2008;48:713–19.

Sunday JM, Pecl GT, Frusher S, Hobday AJ, Hill N, Holbrook NJ, Edgar GJ, Stuart-Smith R, Barrett N, Wernberg T, Watson RA, Smale DA, Fulton EA, Slawinski D, Feng M, Radford BT, Thompson PA, Bates AE. Species traits and climate velocity explain geographic range shifts in an ocean-warming hotspot. Ecol Lett. 2015;18:944–53.

Tait LW, South PM, Lilley SA, Thompsen MS, Schiel DR. Assemblage and understory carbon production of native and invasive canopy-forming macroalgae. J Exp Mar Bio Ecol. 2015;469:10–7.

Takao S, Kumagai NH, Yamano H, Fujii M, Yamanaka Y. Projecting the impacts of rising seawater temperatures on the distribution of seaweeds around Japan under multiple climate change scenarios. Ecol Evol. 2015;5:213–23.

Tanaka K, Taino S, Haraguchi H, Prendergast G, Hiraoka M. Warming off southwestern japan linked to distributional shifts of subtidal canopy-forming seaweeds. Ecol Evol. 2012;2:2854–65.

Terazono Y, Nakamura Y, Imoto Z, Hiraoka M. Fish response to expanding tropical *Sargassum* beds on the temperate coasts of Japan. Mar Ecol Prog Ser. 2012;464:209–20.

Thomsen MS, Wernberg T, Tuya F, Silliman BR. Evidence for impacts of nonindigenous macroalgae: a meta-analysis of experimental field studies. J Phycol. 2009;45:812–9.

Thomsen MS, Wernberg T, Altieri A, Tuya F, Gulbransen D, McGlathery KJ, Holmer M, Silliman BR. Habitat cascades: the conceptual context and global relevance of facilitation cascades via habitat formation and modification. Integr Comp Biol. 2010;50(2):158–75.

Thomsen MS, Wernberg T, Olden JD, Griffin GN, Silliman BR. A framework to study the context-dependent impacts of marine invasions. J Exp Mar Bio Ecol. 2011;400:322–7.

Thomsen MS, Wernberg T, Olden JD, Byers JE, Bruno JF, Silliman BR, Schiel DR. Forty years of experiments on aquatic invasive species: are study biases limiting our understanding of impacts? NeoBiota. 2014;22:1–22.

Tittensor DP, Mora C, Jetz W, Lotze HK, Ricard D, Vanden Berghe E, Worm B. Global patterns and predictors of marine biodiversity across taxa. Nature. 2010;466:1098–101.

Tuya F, Wernberg T, Thomsen MS. Habitat structure affect abundances of labrid fishes across temperate reefs in south-western Australia. Environ Biol Fishes. 2009;86:311–9.

Van den Hoek C. The distribution of benthic marine algae in relation to the temperature regulation of their life histories. Biol J Linn Soc. 1982;18:81–144.

Van den Hoek C. The possible significance of long-range dispersal for the biogeography of seaweeds. Helgol Meeresun. 1987;41:261–72.

Verbruggen H, Tyberghein L, Belton GS, Mineur F, Jueterbock A, Hoarau G, Gurgel CFD, De Clerk O. Improving transferability of introduced species' distribution models: new tools to forecast the spread of a highly invasive seaweed. PLoS One. 2013;8:e68337.

Wallentinus I. Alien species alert: *Undaria pinnatifida* (Wakame or japanese kelp). ICES Cooperation Research Report No. 283. Copenhagen: International Council for the Exploration of the Sea; 2007.

Watanabe S, Scheibling RE, Metaxas A. Contrasting patterns of spread in interacting invasive species: Membranipora membranacea and Codium fragile off Nova Scotia. Biol Invasions. 2010;12:2329–342.

Waters JM. Marine biogeographical disjunction in temperate Australia: historical landbridge, contemporary currents, or both? Divers Distrib. 2008;14:692–700.

Waters JM, Fraser CI, Hewitt GM. Founder takes all: density-dependent processes structure biodiversity. Trends Ecol Evol. 2013;28:78–85.

Wernberg T, Kendrick GA, Phillips JC. Regional differences in kelp-associated algal assemblages on temperate limestone reefs in south-western Australia. Divers Distrib. 2003;9:427–41.

Wernberg T, Thomsen MS, Staehr PA, Pedersen MF. Epibiota communities of the introduced and indigenous macroalgal relatives *Sargassum muticum* and *Halidrys siliquosa* in Limfjorden (Denmark). Helgol Mar Res. 2004;58:154–61.

Wernberg T, Russell BD, Thomsen MS, Gurgel CFD, Bradshaw CJA, Poloczanska ES, Connell SD. Seaweed communities in retreat from ocean warming. Curr Biol. 2011a;21:1828–32.

Wernberg T, Russell BDBD, Moore PJPJ, Ling SD, Smale DA, Campbell A, Coleman MA, Steinberg PD, Kendrick GA, Connell SD. Impacts of climate change in a global hotspot for temperate marine biodiversity and ocean warming. J Exp Mar Bio Ecol. 2011b;400:7–16.

Wernberg T, Smale DA, Tuya F, Thomsen MS, Langlois TJ, de Bettgnies T, Bennett S, Rousseaux CS. An extreme climatic event alters marine ecosystem structure in a global biodiversity hotspot. Nat Clim Change. 2013a;3:78–82.

Wernberg T, Thomsen MS, Connell SD, Russell BD, Waters JM, Zuccarello GC, Kraft GT, Sanderson C, West JA, Gurgel CFD. The footprint of continental-scale ocean currents on the biogeography of seaweeds. PLoS One. 2013b;8(11):e80186.

Wiens JJ. The niche, biogeography and species interactions. Phil Trans R Soc B. 2011;366:2336–50.

Wiens JJ, Donoghue MJ. Historical biogeography, ecology and species richness. Trends Ecol Evol. 2005;19(12):639–44.

Williams SL, Smith JE. A global review of the distribution, taxonomy, and impacts of introduced seaweeds. Ann Rev Ecol Evol Syst. 2007;38:327–59.

Wootton JT. Local interactions predict large-scale pattern in empirically derived cellular automata. Nature. 2001;413:841–4.

Worm B, Barbier EB, Beaumont N, Duffy JE, Folke C, Halpern BS, Jackson JBC, Lotze HK, Palumbi SR, Sala E, Selkoe KA, Stachowicz JJ, Watson R. Impacts of biodiversity loss on ocean ecosystem services. Science. 2006;314:787–90.

York KL, Blacket MJ, Appleton BR. The Bassian Isthmus and the major ocean currents of southeast Australia influence the phylogeography and population structure of a southern Australian intertidal barnacle *Catomerus polymerus* (Darwin). Mol Ecol. 2008;17:1948–61.

Zachos JC, Dickens GR, Zeebe RE. An early Cenozoic perspective on greenhouse warming and carbon-cycle dynamics. Nature. 2008;451:279–83.

Part II
Seaweed at Sea: Floating as a Dispersal Mechanism

Chapter 4
Float and Raft: Role of Buoyant Seaweeds in the Phylogeography and Genetic Structure of Non-buoyant Associated Flora

Erasmo C. Macaya, Boris López, Fadia Tala, Florence Tellier and Martin Thiel

> *There is convincing circumstantial evidence that [floating] algae could be an important agent in long-range dispersal of benthic seaweeds. However, the majority of algae cannot float, and they can only be transported by other floating algae or other floating objects.*
>
> van den Hoek (1987)

Abstract Many seaweed species (primary rafters) float at the sea surface and travel with marine currents after detachment from benthic habitats. Various studies have confirmed that dispersal via floating sporophytes and/or gametophytes influences the phylogeography and genetic population structure of these buoyant seaweeds. In

E.C. Macaya (✉)
Laboratorio de Estudios Algales (ALGALAB), Departamento de Oceanografía, Universidad de Concepción, Casilla 160-C, Concepción, Chile
e-mail: emacaya@oceanografia.udec.cl

B. López
Programa de Doctorado en Biología y Ecología Aplicada (BEA), Facultad de Ciencias del Mar, Universidad Católica del Norte, Larrondo 1281, Coquimbo, Chile
e-mail: borislop@gmail.com

B. López
Departamento de Acuicultura y Recursos Agroalimentarios, Universidad de Los Lagos, Avenida Fuchslocher 1305, Osorno, Chile

F. Tala
Centro de Investigación y Desarrollo Tecnológico en Algas de la Universidad Católica del Norte (CIDTA-UCN), Larrondo 1281, Coquimbo, Chile
e-mail: ftala@ucn.cl

F. Tellier
Departamento de Ecología, Facultad de Ciencias, Universidad Católica de la Santísima Concepción, Casilla 297, Concepción, Chile
e-mail: ftellier@gmail.com

Z.-M. Hu and C. Fraser (eds.), *Seaweed Phylogeography*,
DOI 10.1007/978-94-017-7534-2_4

addition, non-buoyant seaweeds (secondary rafters) that grow attached to or intermingled with these primary floaters may also become dispersed by rafting on their floating hosts. Here, we examine reports of non-buoyant seaweed species associated with buoyant seaweeds and discuss potential consequences for their phylogeography and/or genetic population structure. We found that mostly red and brown algae have been reported with floating seaweed rafts, most of them growing as epiphytes and some as obligate parasites (e.g. endophytes) that travel with their hosts. Molecular evidence suggests dispersal associated with primary floaters in 16 non-buoyant seaweeds, although colonization of distant sites could also have occurred via other floating substrata such as wood, buoys, and other man-made materials. Transoceanic dispersal has been inferred for non-buoyant seaweeds (for example, *Gracilaria chilensis* and *Capreolia implexa*) based on low levels of genetic structure and shared haplotypes among populations separated over vast distances of open ocean (e.g. New Zealand–Chile). Some non-buoyant species suspected or shown to be dispersed by rafting are from intertidal habitats, and these algae can resist physiologically stressful conditions during long trips at the sea surface. However, subtidal and low intertidal non-buoyant species have higher potential to be transported because they cohabit with common raft-forming kelps, often growing on them as epiphytes. We conclude that buoyant seaweeds play an important role in driving the phylogeography, evolution, connectivity and distribution of non-buoyant associated seaweeds. Dispersal of non-buoyant seaweeds via these floating seaweeds may have been underestimated in the past.

Keywords Connectivity · Floating alga · Long-distance dispersal · Non-buoyant species · Rafting

4.1 Introduction

At the writing of the above, influential text by van den Hoek (1987), there were no genetic studies available that had demonstrated gene flow mediated by floating algae. Since then, advances in molecular research have led to numerous studies

M. Thiel
Facultad Ciencias del Mar, Universidad Católica del Norte, Larrondo 1281, Coquimbo, Chile
e-mail: thiel@ucn.cl

E.C. Macaya · M. Thiel
Millennium Nucleus Ecology and Sustainable Management of Oceanic Island (ESMOI), Coquimbo, Chile

M. Thiel
Centro de Estudios Avanzados en Zonas Áridas (CEAZA), Coquimbo, Chile

demonstrating that rafting transport of buoyant seaweeds is an important dispersal mechanism for these seaweeds themselves and for their associated biota. Distant populations that have been connected or seeded by long-distance dispersal can show genetic similarities. This has been confirmed for numerous species with floating structures (pneumatocysts) or thalli, such as *Macrocystis pyrifera* (Macaya and Zuccarello 2010a, b), *Carpophyllum maschalocarpum* (Buchanan and Zuccarello 2012), *Sargassum horneri* (Hu et al. 2011), *S. fusiforme* (Hu et al. 2013), *S. polycystum* (Chan et al. 2013), *S. aquifolium* (Chan et al. 2014), *Phyllospora comosa* (Coleman and Kelaher 2009), *Ascophyllum nodosum* (Olsen et al. 2010), *Fucus vesiculosus* (Muhlin et al. 2008; Coyer et al. 2011a), *F. distichus* (Coyer et al. 2011b), *F. ceranoides* (Neiva et al. 2012), and *Durvillaea antarctica* (Fraser et al. 2009, 2010). These buoyant seaweeds can float over extensive distances after detachment from the primary substratum, and occasionally even cross entire ocean basins (Fraser et al. 2009; Coyer et al. 2011a). Particularly, ocean currents such as the Antarctic Circumpolar Current have been inferred to facilitate connectivity among separate populations in positively buoyant kelps (Waters 2008; Fraser et al. 2009; Macaya and Zuccarello 2010b). Long-distance dispersal appears to be most likely at high latitudes where conditions for floating seaweeds are most favourable (Rothäusler et al. 2012). This suggestion is not only supported by the extensive distribution of single haplotypes of *M. pyrifera* and *D. antarctica* in the subantarctic region (Fraser et al. 2009; Macaya and Zuccarello 2010b), but also analysis on *F. vesiculosus* indicate that a single mtDNA-IGS haplotype has crossed the northern North Atlantic (north of 60°N) after the last glaciation and then expanded southward along the entire north-west Atlantic coast (Coyer et al. 2011a) (Fig. 4.1).

Many of the positively buoyant seaweeds are host to a wide diversity of other organisms, including non-buoyant seaweeds (e.g. Gutow et al. 2015). Thus, rafting may not only facilitate gene flow among floating seaweeds but also of their epibiont communities. For example, using mtDNA and AFLPs, Nikula et al. (2013) demonstrated that regular gene flow occurs between populations of invertebrate species that commonly inhabit holdfasts of the bull kelp *D. antarctica.* Passive transport on floating substrata allows relatively immobile (including fully sessile) species with poor dispersal capabilities to travel large distances, including transoceanic journeys (Nikula et al. 2010; Cumming et al. 2014). This also includes many different epiphytic seaweed species lacking buoyant structures. In this chapter we concentrate on these non-buoyant seaweeds, but also include information about (i) the distribution of positively buoyant seaweeds, (ii) the roles of environmental factors that determine floating time and (iii) physiological and reproductive traits of buoyant and non-buoyant seaweeds that permit successful long-distance dispersal, enhancing gene flow.

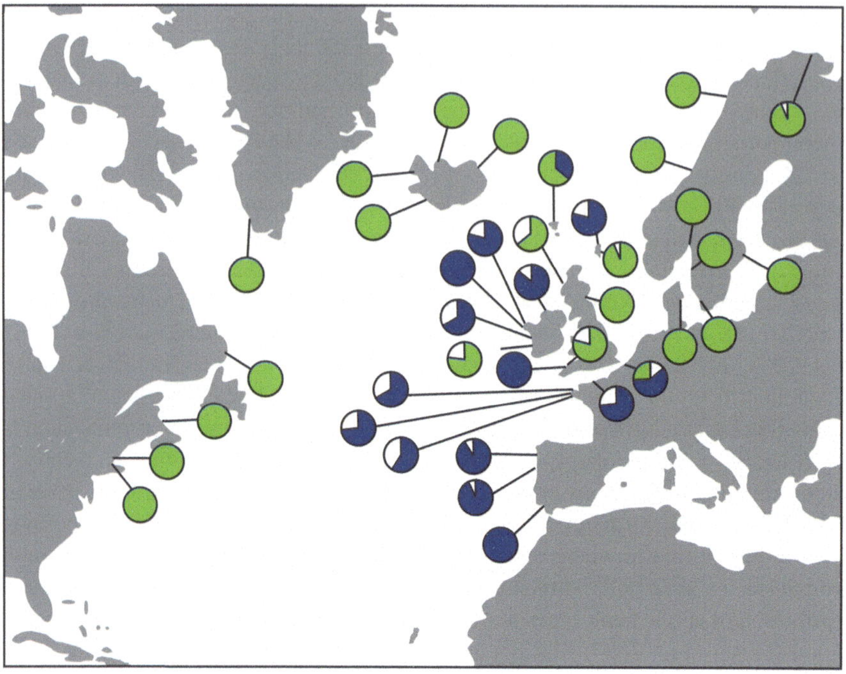

Fig. 4.1 Distribution of mtDNA haplotypes of *Fucus vesiculosus* in the north Atlantic based on microsatellite data. Only the 2 more frequent haplotypes are depicted (*blue* and *yellow*). Modified from Coyer et al. (2011a)

4.2 Buoyant Seaweeds: Distribution and Latitudinal Patterns

Buoyant seaweeds are found throughout all major oceans, even though the total abundance and survival of rafts may show strong regional differences (see Thiel and Gutow 2005a; Rothäusler et al. 2012). Since their source populations are in benthic habitats, high abundances of positively buoyant seaweeds are often found in coastal waters (Hinojosa et al. 2011). Exceptions are the Sargasso Sea and the Central Atlantic Ocean where the two species *Sargassum fluitans* and *S. natans*, not known from benthic habitats, form extensive holopelagic populations (e.g. Gower et al. 2013). In some regions, strong offshore currents move seaweed rafts to the open sea: this occurs frequently in SE Asia where the Kuroshio Current moves into the open North Pacific, and where large amounts of floating seaweeds can be observed at large distances from the nearest shores (Ohno 1984; Komatsu et al. 2014). Also, within the West Wind Drift large amounts of floating kelps can be observed downstream of abundant source populations (Helmuth et al. 1994; Hobday 2000a; Smith 2002; Garden et al. 2014).

The largest abundances of positively buoyant seaweeds are found in cold-temperate waters within latitudes 40–60 in both hemispheres (Ingólfsson 1995; Thiel and Gutow 2005a; Rothäusler et al. 2012). For example, along the southern coast of the South-east Pacific, abundances of two common buoyant kelp species (*D. antarctica* and *M. pyrifera*) increase substantially toward higher latitudes (Wichmann et al. 2012). Also, frequent observations of rafts (Ingólfsson 1995) and reports of high densities of floating seaweeds at 65°N (Khalaman and Berger 2006) underscore this pattern, which most likely is related to higher survival of floating kelps at the sea surface and therefore continuous accumulation of kelp rafts at higher latitudes (Rothäusler et al. 2012).

There are also many other floating substrata that can serve as transport vehicles for non-buoyant seaweeds (Thiel and Gutow 2005a). Several seaweeds grow on floating wood, even though to our knowledge no systematic survey of the epibiont community from driftwood is available. A variety of abiotic substrata, including volcanic pumice and plastics are also known (Bravo et al. 2011; Bryan et al. 2012). In particular, plastics are increasingly reported as transport vehicles for non-buoyant or weakly buoyant seaweeds (Astudillo et al. 2009; Bravo et al. 2011; Kiessling et al. 2015). These abiotic substrata are especially resistant to biological decay and can therefore float for very long time periods over long distances.

4.3 Observations of Non-buoyant Seaweeds Associated with Buoyant Seaweeds

A total of 107 taxa of non-buoyant seaweeds—12 Chlorophyta, 30 Phaeophyceae, 65 Rhodophyta—have been recorded as epiphytes or intermingled with floating substrata (Table 4.1), most of them associated with buoyant seaweeds (84.1 %), but also growing on floating wood and plastic. Some non-buoyant seaweeds have been found on floating substrata, for example the filamentous brown alga *Hincksia granulosa* was reported on buoyant seaweeds (Gutow et al. 2015) and on floating plastics (Astudillo et al. 2009). Among Rhodophyta, the order Ceramiales dominated the number of species (21), followed by Corallinales (18) and Gigartinales (9). Similarly, red algae have been found as major components of the epiphyte community on benthic seaweeds and seagrasses (Whittick 1983; D'Antonio 1985; González et al. 1993; Vairappan 2006; Won et al. 2010; Buza-Jacobucci and Pereira-Leite 2014). For many non-buoyant species (44), their geographic distribution and/or molecular data suggest dispersal on floating substrata, even though they have not yet been recorded with buoyant seaweeds (Table 4.1). We have not included seaweeds reported on the holopelagic species *S. fluitans* and *S. natans*, although Woelkerling (1975) described 68 seaweed species associated with floating samples collected in the western Sargasso Sea.

Table 4.1 Non-buoyant seaweed species reported as epiphytes or intermingled with buoyant seaweeds (S), wood (W) and plastic (P)

Order	Species	Floating substratum	References
Chlorophyta			
Bryopsidales	*Bryopsis rhizopora*	P	Astudillo et al. (2009)
	Bryopsis sp.	S	Gutow et al. (2015)
	Caulerpa flexilis	S	Edgar (1987)
	Caulerpa simpliciuscula	S	Edgar (1987)
	Caulerpa trifaria	S	Edgar (1987)
	Codium fragile	P	Astudillo et al. (2009)
	Derbesia sp.[a]	S	Saunders (2014)
	Wittrockiella lyalli[a]	W	Boedeker et al. (2010)
Cladophorales	*Chaetomorpha* sp.	S	Gutow et al. (2015)
Ulvales	*Ulva rigida*	P	Morton and Britton (2000a, b)
	Ulva spp.[a]	S, P	Edgar (1987), Thiel et al. (2003), Astudillo et al. (2009), Clarkin et al. (2012), Saunders (2014), Gutow et al. (2015)
	Ulvaria sp.[a]	S	Saunders (2014)
Phaeophyceae			
Dictyotales	*Dictyota dichotoma*	S	Edgar (1987)
Ectocarpales	*Adenocystis utricularis*[a]	S	Fraser et al. (2013)
	Asperococcus sp.	S	Clarkin et al. (2012)
	Colpomenia sinuosa	S	Gutow et al. (2015)
	Colpomenia peregrina[a]	S	Lee et al. (2014)
	Ectocarpus acutus	P	Astudillo et al. (2009)
	Ectocarpus sp.	S	Gutow et al. (2015)
	Ectocarpoid sp.[a]	S	Saunders (2014)
	Elachista fucicola	S	Gutow et al. (2015)
	Hincksia granulosa	S, P	Astudillo et al. (2009), Gutow et al. (2015)
	Leathesia marina	S	Clarkin et al. (2012)
	Petalonia sp.[a]	S, P	Bravo et al. (2011), Saunders (2014)
	Pilayella sp.	S	Gutow et al. (2015)
	Scytosiphon lomentaria	P	Astudillo et al. (2009), Bravo et al. (2011), Clarkin et al. (2012)
Fucales	*Acrocarpia paniculata*	S	Edgar (1987)
	Carpoglossum confluens	S	Edgar (1987)
	Cystophora retroflexa	S	Edgar (1987)
	Cystoseira amentacea var. *stricta*[a]	S	Susini et al. (2007)
	Cystophora sp.	S	Edgar (1987)
	Cystoseira sp.	S	Clarkin et al. (2012)
	Fucus spiralis[a]	S	Coleman and Brawley (2005)
	Himanthalia elongata	S	Clarkin et al. (2012)
	Sargassum sp.	S, P	Edgar (1987), Winston et al. (1997)
	Xiphophora gladiata	S	Edgar (1987)

(continued)

Table 4.1 (continued)

Order	Species	Floating substratum	References
Laminariales	*Chorda filum*	S	Clarkin et al. (2012)
	Ecklonia radiata	S	Edgar (1987)
	Laminaria spp.	S	Clarkin et al. (2012)
Sphacelariales	*Herpodiscus durvilleae*[a]	S	Fraser and Waters (2013)
	Halopteris sp.	S	Edgar (1987)
Ralfsiales	Ralfsioid sp.[a]	S	Saunders (2014)
Rhodophyta			
Balliales	*Ballia callitricha*	S	Edgar (1987)
	Ballia scoparia	S	Edgar (1987)
Bonnemaisoniales	*Delisea hypneoides*	S	Edgar (1987)
	Delisea pulchra	S	Edgar (1987)
Ceramiales	*Antithamnion densum*	P	Astudillo et al. (2009)
	Antithamnion sp.	S, P	Bravo et al. (2011), Gutow et al. (2015)
	Bostrychia intricata[a]	S, W	Fraser et al. (2013), Muangmai et al. (2014)
	Bostrychia vaga[a]	S	Muangmai et al. (2014)
	Bostrychia arbuscula[a]	S	Muangmai et al. 2014
	Bostrychia gracilis[a]	S	Muangmai et al. (2014)
	Ceramium virgatum	S	Gutow et al. (2015)
	Ceramium sp.[a]	S	Edgar (1987), Saunders (2014)
	Dasyclonium incisum	S	Edgar (1987)
	Euptilota articulata	S	Edgar (1987)
	Hemineura frondosa	S	Edgar (1987)
	Hymenena sp.[a]	S	Saunders (2014)
	Lenormandia marginata	S	Edgar (1987)
	Microcladia sp.[a]	S	Saunders (2014)
	Neoptilota sp. [a]	S	Saunders (2014)
	Osmundea pinnatifida	S	Clarkin et al. (2012)
	Phycodrys isabellae[a]	S	Saunders (2014)
	Polysiphonia pacifica var. *gracilis*[a]	S	Saunders (2014)
	Polysiphonia mollis	P	Astudillo et al. (2009)
	Polysiphonia sp.	S, P	Edgar (1987), Bravo et al. (2011), Gutow et al. (2015)
	Vertebrata lanosa	S	Ingolfsson (1998), Clarkin et al. (2012)

(continued)

Table 4.1 (continued)

Order	Species	Floating substratum	References
Corallinales	*Amphiroa* sp.	P	Winston et al. (1997)
	Bossiella heteroforma[a]	S	Saunders (2014)
	Bossiella orbigniana[a]	S	Saunders (2014)
	Bossiella sp.[a]	S	Saunders (2014)
	Calliarthron cheilosporioides[a]	S	Saunders (2014)
	Calliarthron sp.[a]	S	Saunders (2014)
	Chiharaea rhododactyla[a]	S	Saunders (2014)
	Clathromorphum parcum[a]	S	Saunders (2014)
	Corallina officinalis	P	Astudillo et al. (2009)
	Corallina sp.[a]	S	Saunders (2014)
	Fosliella sp.	P	Gregory (1983), Winston et al. (1997)
	Hydrolithon farinosum	P	Aliani and Molcard (2003)
	Jania sp.	P	Winston et al. (1997)
	Lithophyllum sp.	P	Winston et al. (1997)
	Lithothamnion sp.[a]	S	Saunders (2014)
	Mesophyllum conchatum[a]	S	Saunders (2014)
	Mesophyllum sp.	P	Winston et al. (1997)
	Pneophyllum sp.[a]	S	Saunders (2014)
Erythropeltidales	*Erythrotrichia* sp.	S	Gutow et al. (2015)
	Smithora sp.[a]	S	Saunders (2014)
Gelidiales	*Capreolia implexa*[a]	S, W	Boo et al. (2014)
	Gelidium sp.	P	Astudillo et al. (2009)
Gigartinales	*Callophyllis variegata*[a]	S	Saunders (2014)
	Callophyllis rangiferinus	S	Edgar (1987)
	Callophyllis dissecta[a]	S	Saunders (2014)
	Chondracanthus harveyanus[a]	S	Saunders (2014)
	Hypnea episcopalis	S	Edgar (1987)
	Mastocarpus californianus[a]	S	Saunders (2014)
	Mastocarpus stellatus	S	Saunders (2014)
	Mazzaella flaccida[a]	S	Saunders (2014)
	Phacelocarpus labillardieri	S	Edgar (1987)
Goniotrichales	*Stylonema alsidii*	S	Gutow et al. (2015)
Gracilariales	*Gracilaria chilensis*[a]	S	Guillemin et al. (2014)
Halymeniales	*Prionitis filiformis*[a]	S	Saunders (2014)
Hildenbrandiales	*Hildenbrandia* sp.[a]	S	Saunders (2014)
Plocamiales	*Plocamium angustum*	S	Edgar (1987)
	Plocamium dilatatum	S	Edgar (1987)
Rhodymeniales	*Champia* sp.	S	Edgar (1987)
	Rhodymenia rhizoides[a]	S	Saunders (2014)
	Rhodymenia sp.	P	Astudillo et al. (2009)

[a]Species not directly observed in buoyant substrata but their geographic distribution and/or molecular data suggest dispersal by rafting

Fig. 4.2 The non-buoyant seaweeds. **a** *Gelidium lingulatum*, **b** *Corallina officinallis* var. *chilensis* (*green arrows*) and *Lessonia spicata* (*red arrows*), **c** *G. rex* growing in holdfast of the buoyant bull kelp *Durvillaea antarctica*

Benthic seaweeds with floating structures commonly grow in close association with other seaweeds in their natural habitats on the seafloor. For example, the buoyant species *D. antarctica* has a compact holdfast that overgrows turf algae growing in the low intertidal zone, the typical habitat of *D. antarctica*. These turf algae become firmly incorporated into the holdfast (Fig. 4.2), and a large number of seaweed species that share the habitat with *D. antarctica* have been found in the holdfasts of this large buoyant kelp (Table 4.2). The species found in holdfasts of the bull kelp range from small seaweeds such as *Bostrychia intricata* to large non-buoyant kelps such as *Lessonia spicata* (Fig. 4.2b). Forty-seven non-buoyant seaweeds have been found attached to or incorporated in their holdfasts during an extensive survey of stranded bull kelps along the continental coast of Chile: 3 Chlorophyta, 6 Phaeophyceae, 37 Rhodophyta and several unidentified crustose algae have been recorded. Some of the most common species associated with the holdfasts of *D. antarctica* are *L. spicata*, *Corallina officinalis* var. *chilensis*, and *Gelidium lingulatum* (Table 4.2).

Some non-buoyant seaweed species are obligate epiphytes such as the red alga *Vertebrata lanosa* that grows almost exclusively on the buoyant species *A. nodosum* and *F. vesiculosus* (Garbary et al. 1991; Garbary and Deckert 2001); this species mostly grows on injured host tissues or lateral pits (Pearson and Evans

Table 4.2 Macroalgae species found in holdfasts of stranded bull kelp *Durvillaea antarctica* on sandy beaches of southern Chile (28°S–42°S; winter 2013—winter 2015)

Species	Presence		
	<1.0 %	1.0–10 %	>10 %
Chlorophyta			
Chaetomorpha firma	+		
Codium sp.	+		
Ulva sp.	+		
Phaeophyceae	+		
Adenocystis utricularis	+		
Colpomenia sp.	+		
Desmarestia ligulata	+		
Ectocarpus sp.	+		
Lessonia spicata			+
Macrocystis pyrifera	+		
Rhodophyta	+		
Ahnfeltiopsis durvillei	+		
Ahnfeltiopsis furcellata	+		
Antithamnionella ternifolia	+		
Asparagopsis armata	+		
Bostrychia intricata	+		
Branchioglossum bipinnatifidum	+		
Camontagnea oxyclada	+		
Centroceras clavulatum	+		
Ceramium virgatum	+		
Chondria secundata	+		
Chondrus canaliculatus	+		
Corallina caespitosa	+		
Corallina officinalis var. *chilensis*			+
Crustose algae			+
Delesseria crassinerva	+		
Gelidium chilense		+	
Gelidium lingulatum			+
Gelidium rex		+	
Gigartina skottsbergii	+		
Grateloupia sp.	+		
Griffithsia chilensis	+		
Laurencia sp.	+		
Mastocarpus latissimus	+		
Mazzaella laminarioides	+		
Mazzaella membranacea	+		
Montemaria horridula	+		
Nothogenia spp.	+		

(continued)

Table 4.2 (continued)

Species	Presence		
	<1.0 %	1.0–10 %	>10 %
Plocamium cartilagineum	+		
Polysiphonia morrowii	+		
Polysiphonia pacifica	+		
Prionitis sp.	+		
Pterosiphonia dendroidea	+		
Pyropia sp.		+	
Rhodoglossum sp.	+		
Rhodymenia skottsbergii	+		
Sarcothalia crispata	+		
Schimmelmannia plumosa	+		
Schottera nicaeensis		+	

1990). The brown alga *Herpodiscus durvilleae* is an obligate endophyte of the buoyant *D. antarctica* and epiphytic filaments forming red-brown patches on the host surface (Heesch et al. 2008). A recent study also found *H. durvilleae* on *D. poha* and *D. willana* (Fraser and Waters 2013).

4.4 Phylogeography/Genetic Structure of Non-buoyant Seaweeds Associated with Buoyant Species

Increasingly, molecular evidence of long-distance dispersal through genetic similarity of populations separated by thousands of kilometres has been reported for buoyant seaweeds (e.g. *D. antarctica*, Fraser et al. 2009; *M. pyrifera*, Macaya and Zuccarello 2010a, b). Genetic analyses of non-buoyant seaweeds have also demonstrated low genetic structure and high levels of connectivity; in some cases, rafting on floating substrata (mostly on buoyant seaweeds) has been invoked to explain this pattern. For most studies, long-distance dispersal is inferred and buoyant seaweed species are suggested as the most likely vector (Table 4.1), but other substrata cannot be ruled out (floating wood, plastic, tar lumps or pumice).

For 16 non-buoyant seaweeds, molecular data suggest that transport attached to or intermingled with buoyant seaweeds might contribute to increased dispersal and shaping of the genetic structure with high levels of population connectivity. Most of these studies are recent (since 2005), and most used more than one molecular marker from different cell compartments (Table 4.3). The most common are red algae (62.5 %) followed by brown (31.3 %) and one green seaweed species (6.2 %), ranging from small filaments (e.g. *Bostrychia*) to thick, leathery species (e.g. *Cystoseira*).

Table 4.3 Summary of molecular studies of non-buoyant seaweed species for which genetic evidence suggest dispersal associated with buoyant seaweeds

Order	Species	Habitat	Substrata	Sampled area	Molecular markers	References
Chlorophyta						
Bryopsidales	*Wittrockiella lyalli*	High intertidal	Rock	NZ/Chile	SSU [N]/LSU [N]/ ITS2 [N]	Boedeker et al. (2010)
Phaeophyceae						
Ectocarpales	*Adenocystis utricularis*	Middle intertidal	Rock-epiphyte	NZ/Chile/Sub Antarctic Is.	COI [M]/*rbc*L [C]/ LSU [N]/28S [N]	Fraser et al. (2013)
	Colpomenia peregrina	Intertidal–subtidal	Rock-epiphyte	Widely sampled (NW–NE Pacific, SW Pacific, N Atlantic)	*cox*3 [M]/*atp*6 [M]/RuBisCo spacer [C]	Lee et al. (2014)
Fucales	*Cystoseira amentacea* var. *stricta*	Intertidal	Rock	France	RAPD	Susini et al. (2007)
	Fucus spiralis	Intertidal	Rock	USA	Microsatellite (5 loci)	Coleman and Brawley (2005)
Sphacelariales	*Herpodiscus durvilleae*	Low intertidal	Endophyte-epiphyte of *D. antarctica*	New Zealand/Falklands Is.	COI [M]/*rbc*L [C]/ LSU [N]	Fraser and Waters (2013)
Rhodophyta						
Ceramiales	*Bostrychia arbuscula*	Upper intertidal	Rock-epiphyte	NZ	COI [M]/*rbc*L [C]/ LSU [N]	Muangmai et al. (2014)
	B. gracilis	Upper intertidal	Rock-epiphyte	NZ	COI [M]/*rbc*L [C]/ LSU [N]	Muangmai et al. (2014)
	B. intricata	High intertidal	Rock-epiphyte	Australia/NZ/Chile/South Africa	COI [M]/*rbc*L [C]/ LSU [N]/28S [N]	Fraser et al. (2013), Muangmai et al. (2014)
	B. vaga	Upper intertidal	Rock-epiphyte	Australia/NZ/Sub Antarctic Is.	COI [M]/*rbc*L [C]/ LSU [N]	Muangmai et al. (2014)

(continued)

Table 4.3 (continued)

Order	Species	Habitat	Substrata	Sampled area	Molecular markers	References
	Phycodrys isabellae	Low intertidal–subtidal	Rock (and epizoic)	Canada/USA	COI [M]/*tuf*A [M]	Saunders (2014)
Gelidiales	*Capreolia implexa*	Intertidal	Rock/epiphyte	Australia/NZ/Chile	*cox*1 [M]/*rbc*L [C]	Boo et al. (2014)
Gigartinales	*Callophyllis variegata*	Low intertidal subtidal	Rock	Canada/USA	COI [M]/*tuf*A [M]	Saunders (2014)
	Chondracanthus harveyanus	Mid–low intertidal	Rock	Canada/USA	COI [M]/*tuf*A [M]	Saunders (2014)
Gracilariales	*Gracilaria chilensis*	Subtidal	Sand	NZ/Chile	ITS2 [N]/ Microsatellites (5 loci)	Guillemin et al. (2014)
Nemaliales	*Nothogenia fastigiata*	Intertidal	Rock	Campbell Is./Chile	COI [M]/*rbc*L [C]/ ITS [N]	Lindstrom et al. (2015)

Records from Saunders (2014) where COI data are consistent with the kelp conveyor hypothesis were included
Notes [N]: nuclear, [M]: mitochondrial, and [C]: chloroplastic markers; NZ: New Zealand

4.4.1 Role of the West Wind Drift (WWD) in Dispersal and Connectivity of Non-buoyant Seaweeds

Transoceanic dispersal achieved through the strong, eastward-moving WWD has shaped the phylogeographic structure of several non-buoyant seaweeds. The present distribution of several *Bostrychia* species in the Southern Hemisphere is suggested by long-distance dispersal followed by allopatric speciation (Muangmai et al. 2014). Based on pairwise F_{ST} analyses, Fraser et al. (2013) showed that populations of *B. intricata* from the Falkland Islands and Chile were not significantly differentiated, albeit their sample sizes were low. They also found shared COI and *rbc*L haplotypes for samples from Falkland, Marion Island, New Zealand and southern Chile. A similar pattern was described for the brown alga *Adenocystis utricularis* (Fraser et al. 2013), with genetic similarities among populations from the Falkland Islands, South Georgia and southern Chile; for example, individuals from Macquarie Island and southern Chile were separated by only 2 bp for COI sequences. The association with buoyant seaweeds is the most likely explanation for these findings, since recolonization of the subantarctic region after the Last Glacial Maximum by rafting dispersal via the WWD has been demonstrated for the buoyant kelps *M. pyrifera* (Macaya and Zuccarello 2010a, b) and *D. antarctica* with its associated fauna (Nikula et al. 2010, 2011, 2013).

Similarly, shared haplotypes and low genetic structure have been found in distant populations for the non-buoyant red alga *Capreolia implexa* (Boo et al. 2014); this species was previously described as endemic for Australia and New Zealand, but COI and *rbc*L sequences from specimens collected in three locations from southern Chile revealed its presence in South America. Pairwise divergence was 0.0 % to 0.3 % between individuals from New Zealand and Chile. For COI sequences, pairwise divergence of 0.2 % was found between southern Chile and Stewart Island (NZ). Recent dispersal through rafting on buoyant kelps was inferred, and the phylogeny and haplotype network suggest Stewart Island as the most likely origin for samples from southern Chile (Fig. 4.3).

More variable molecular markers have also demonstrated transoceanic dispersal, for example in the non-buoyant red seaweed *Gracilaria chilensis*. A subset of the microsatellite diversity from eastern New Zealand populations was found in all sampled Chilean populations (Guillemin et al. 2014). Additional analyses using *rbc*L and ITS2 sequences confirmed the relatively recent arrival of *G. chilensis* in southern Chile: shared haplotypes were found in samples collected on both sides of the Pacific. High genetic diversity in the Biobío region of Chile (36–37°S) suggests this area as the most likely introduction point. According to Guillemin et al. (2014), *G. chilensis* reached the Chilean coast from eastern NZ, through association with buoyant seaweeds rafting along the WWD (Fig. 4.3).

Analysis of three molecular markers, LSU, *rbc*L and COI, revealed identical haplotypes between samples of the obligate parasite *H. durvilleae* from Falkland Islands and the subantarctic Campbell Island (Fraser and Waters 2013), and

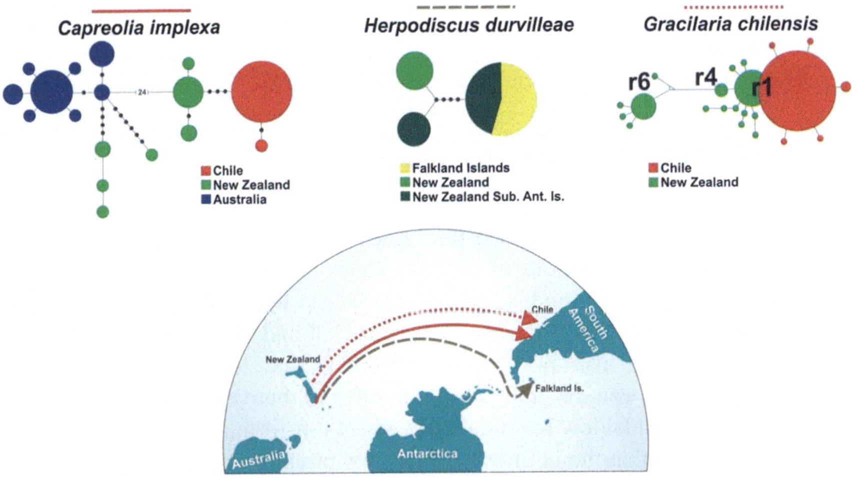

Fig. 4.3 Schematized haplotype networks for the non-buoyant red algae *Capreolia implexa* and *Gracilaria chilensis* and the brown alga *Herpodiscus durvilleae* obligate endo/epiphyte of the bull kelp *Durvillaea antarctica*. Direction of transoceanic dispersal is also depicted for each species. Each *circle* represents one haplotype, circle size is proportional to the haplotype frequency (not comparable among studies). *Small black circle* represents haplotypes that are extinct or were not found. For more information see Guillemin et al. (2014) for *G. chilensis;* Fraser et al. (2013) for *H. durvillaeae* and Boo et al. (2014) for *C. implexa*

consequently recent transoceanic dispersal of the buoyant host *D. antarctica* along the WWD is suggested (Fig. 4.3).

4.4.2 Non-buoyant Fucus and buoyant Ascophyllum, an Example from the Northern Hemisphere

An example of inferred rafting of a non-buoyant seaweed with a buoyant species comes from the association of *Fucus* and *Ascophyllum*. Microsatellite and mtDNA analysis of the non-buoyant alga *Fucus spiralis* in North Atlantic and North Pacific revealed low levels of genetic structure and a shared haplotype (Coyer et al. 2011a). Coleman and Brawley (2005) suggested that this species might disperse in mixed algal rafts associated with the buoyant seaweed *A. nodosum*. Long-distance dispersal might be possible in *A. nodosum*: for example, thalli of this species were found off the coast of West Africa (John 1974), and according to van den Hoek (1987), the specimens travelled at least 5500 km in approximately 430 days. Additionally, microsatellites and mtDNA indicated that there is only a weak present-day population differentiation of *A. nodosum* across the North Atlantic (Olsen et al. 2010).

4.4.3 The Kelp Conveyor Hypothesis

Recently, Saunders (2014) proposed the "Kelp Conveyor Hypothesis" (KCH) to explain the disjunct distributions of several species in California and Haida Gwaii (HG, north-west British Columbia). Many macroalgae were thought to be endemic to HG, but molecular DNA barcode analyses of COI and *tuf*A sequences revealed that 33 species were also present in California. The KCH suggests (i) that non-buoyant species can be transported in rafts of buoyant kelps, particularly species from the subtidal/low intertidal zone where raft-forming kelps usually grow (Table 4.3), and (ii) that the northward Davison Current transports kelp rafts from California to HG. Saunders (2014) found molecular evidence to support both predictions (shared haplotypes and genetic diversity), suggesting that buoyant kelps such as *M. pyrifera* and *Nereocystis luetkeana* are important vectors for non-buoyant seaweeds from California to HG.

4.4.4 Genetic Analysis of Stranded Kelps

A recent study by Bussolini and Waters (2015) analyzed beach-cast rafts of the bull kelp *D. antarctica* in New Zealand. They reported a small proportion of samples having subantarctic origin, suggesting that rafts might travel northward from Auckland Is., Snares Is. or even Macquarie Is. (1400 km distant). To our knowledge no molecular analyses have been carried out on associated non-buoyant seaweeds from beach-cast individuals of buoyant species, although as we showed in section above (see: Observations of non-buoyant seaweeds associated with buoyant seaweeds) many non-buoyant species are associated with holdfasts of *D. antarctica*. Future studies are necessary to demonstrate or confirm that passive dispersal on primary floating seaweeds is possible for non-buoyant species.

4.5 Environmental Factors Determining a Successful Rafting Journey

Successful transport of non-buoyant seaweeds attached to floating substrata depends on factors that can affect various stages of the journey (Thiel 2003; Thiel and Gutow 2005b; Macreadie et al. 2011) (Fig. 4.4). The availability of floating substrata within the geographic range of the non-buoyant seaweeds is one important aspect to be

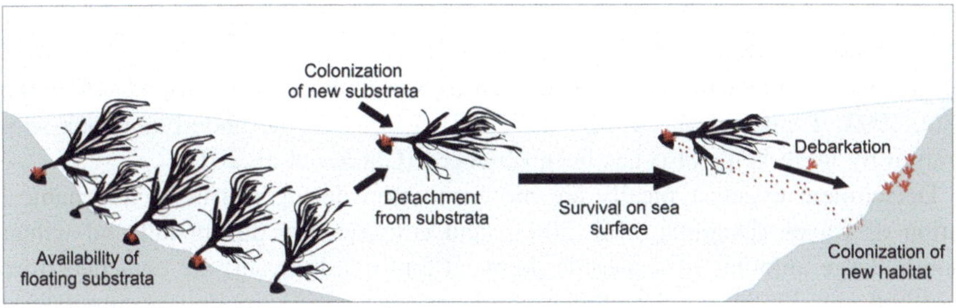

Fig. 4.4 Conceptual model of the steps involved rafting of non-buoyant seaweeds attached to primary floater (e.g. large kelps). Modified from Thiel and Gutow (2005b)

considered. Various biotic and abiotic factors can cause detachment of buoyant seaweeds. Attachment site and strength of travellers on buoyant substrata is also relevant: non-buoyant species weakly attached to their hosts face a high risk of being lost during or shortly after detachment. Further, it is relevant to know whether the non-buoyant seaweeds already grow on the primary floating substrata (e.g. epiphytes) or whether colonization occurs after detachment. Survival of non-buoyant species on the floating substrata depends on a variety of intrinsic and extrinsic factors (physiological tolerance, competitive ability, herbivore resistance, reproductive traits). Finally, colonisations of new habitats depend on the reproductive biology of non-buoyant species and more importantly on substratum availability.

4.5.1 Availability of Floating Substrata

The availability of positively buoyant substrata will depend on the abundance of benthic populations (e.g. kelp beds, mangrove forests), freshwater inputs, coastal topography and/or the ability of non-buoyant seaweeds to colonize and persist on these substrata (Thiel and Gutow 2005a). Among naturally floating substrata, seaweeds probably represent the quantitatively most important substratum for non-buoyant seaweeds (Thiel and Gutow 2005b). Most seaweeds are negatively buoyant and sink to the seafloor when detached from the primary substratum, but some species possess high buoyancy, particularly kelps and other brown seaweeds (Norton and Mathieson 1983) that thrive in temperate environments. In tropical areas, mangroves may provide additional floating substrata, ranging from positively buoyant wood to floating seeds (Kohlmeyer et al. 1995; Filho et al. 2008).

4.5.2 Detachment

Among the mechanisms that explain detachment of seaweeds from the underlying hard bottoms are breakage of the stipe, detachment of the holdfast and lifting of the attachment substratum. Breakage of the stipe can occur regularly during the life cycle of some macroalgae (senescence stage) (Arenas et al. 1995) or can be caused by grazers or burrowing organisms, such as sea urchins or gastropods (Klumpp et al. 1993; Tegner et al. 1995; Thiel 2003). Lifting of the underlying rock substratum by large kelps also has been observed (Garden et al. 2011).

Detachment events typically are more common during storms by mechanical action of waves (Duggins et al. 2003), and consequently during fall and winter months large amounts of seaweeds (buoyant and non-buoyant) are detached from hard bottoms. Following detachment, buoyant seaweeds commonly accumulate in convergence zones where they form large patches consisting of several seaweed species (Hirata et al. 2001; Hinojosa et al. 2010).

4.5.3 Colonization of Floating Substrata

Substrata available in coastal habitats before detachment may be colonized by diverse assemblages of seaweeds that live attached to their structures (e.g. holdfasts and blades of large kelps) (Oliveira et al. 1979; Edgar 1987; Jokiel 1989). The abundance and richness of epiphytic seaweeds depends on the size of the primary raft. In general, large rafts not only accommodate more species (Fig. 4.5) (e.g. Clarkin et al. 2012; Gutow et al. 2015), but also persist for longer periods at the sea surface (Thiel and Gutow 2005b). Larger rafts also offer a greater availability of microhabitats, which favour the presence of epibionts (Macreadie et al. 2011). On large rafts with a lot of shaded surfaces and areas, travellers are also less exposed to abiotic factors such as solar irradiance and temperature (see below).

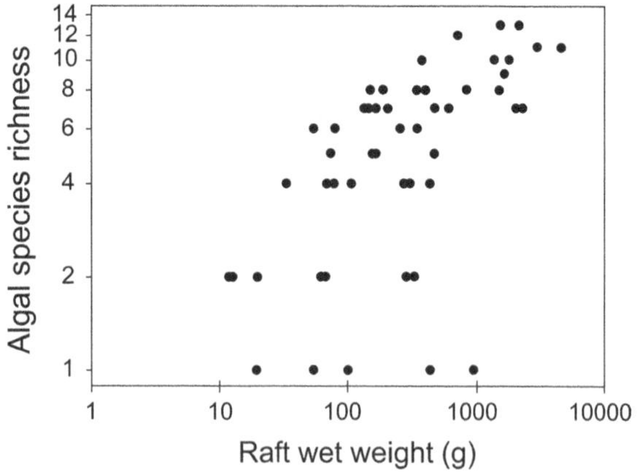

Fig. 4.5 Relationship between algal species richness and total raft wet weight. Relationship was significant (Spearman's correlations $p < 0.01$). Modified from Clarkin et al. (2012)

4.5.4 Attachment Site and Strength

Most non-buoyant seaweeds grow on the holdfasts of positively buoyant seaweeds, but some may also colonize the blades and upper stipe regions of their hosts (Ojeda and Santelices 1984; Edgar 1987; Ingólfsson 1995). During the initial phase of rafting, many organisms living on benthic seaweeds drop off, leaving only the most persistent species on the newly formed seaweed raft (Kingsford and Choat 1985; Gutow et al. 2009, 2015); consequently, attached epiphytic seaweeds are frequently found among initial travellers.

Furthermore, the specific tissue of the host alga on which the non-buoyant seaweeds are growing can affect the success of their transport. Non-buoyant seaweeds are often firmly incorporated in the holdfasts of large, buoyant seaweeds (see e.g. Fig. 4.2). In some large buoyant kelps, the blades or other structures have positive buoyancy (e.g. presence of gas-filled pneumatocysts or gas bubbles inside the blades), while the holdfast is immersed (Gilbert 1984; Thiel and Gutow 2005a). Therefore, non-buoyant seaweeds attached to holdfasts would be less affected than those present at the surface where they are exposed to direct solar radiation and higher temperatures.

4.5.5 Survival on Rafts

Transport of positively buoyant seaweeds along the surface can involve several weeks or months at sea (Hobday 2000b; Hernández-Carmona et al. 2006). Raft trajectories are determined by the actions of winds, waves and currents (Thiel and Gutow 2005a). During a rafting journey, both the buoyant seaweeds (primary rafter such as large kelps) and the epiphytic non-buoyant seaweeds are exposed to a variety of environmental factors that differ from those in benthic habitats. Particularly, solar irradiance and high temperatures can diminish the persistence of floating seaweeds (Hobday 2000b; Vandendriessche et al. 2007; Rothäusler et al. 2009, 2011b, c). Nutrient availability may also play an important role in determining the longevity of floating seaweeds (Edgar 1987). Abundant grazers on the floating seaweeds might either destroy the floating seaweeds or preferentially consume the epiphytic seaweeds, but destructive effects of grazers also depend on temperature and availability of alternative food (Gutow and Franke 2003; Vandendriessche et al. 2007).

4.5.6 Colonization of New Habitats

Any journey on a raft only becomes successful if the travellers are capable of colonizing new habitats. Whether successful colonization occurs depends to a large degree on the physiology and life history of the seaweeds. Species that tolerate the extreme conditions at the sea surface will reach new habitats in better states than those that are sensitive to high temperatures and solar radiation. The reproductive traits of non-buoyant seaweeds determine whether they are able to generate viable (sexually or asexually produced) propagules after arriving in new habitats.

The ecological conditions in new habitats also affect colonization potential. If all suitable habitats are colonized, only very strong competitors will be able to colonize (Stachowicz et al. 2002). Furthermore, if extensive populations of the non-buoyant seaweed already occupy that particular ecological niche in new habitats, new arrivals will have little probability of successfully establishing and reproducing with these existing populations because of density-dependent exclusion (Waters et al. 2013; Neiva et al. 2014).

4.6 Physiology of Non-floating Algae and Reproductive Patterns

In contrast to many mobile animals that potentially escape once buoyant seaweeds are detached from the substratum (e.g. Gutow et al. 2009), the association between epiphytic, non-buoyant seaweeds with the buoyant host is much more intimate, at least during the initial phase of rafting. The epiphytic seaweeds may be growing over the photosynthetic surface or at the base of the thallus (holdfast), depending on the species and host morphology (Diez et al. 2013; Hurd et al. 2014). Most non-buoyant seaweeds associated with buoyant seaweeds occur in the holdfasts that overgrow nearby understory seaweeds. If the buoyant seaweeds retain their holdfast after detachment from the substratum, associated non-buoyant seaweeds can be transported with the rafts. However, the composition and abundance of the rafting community can be highly variable in time and space (Thiel and Gutow 2005a). Edgar (1987) observed that several epiphytes were able to survive up to six months or more on holdfasts of *M. pyrifera* suspended at sea. The size of the non-buoyant seaweeds can also influence their potential to persist, with small filaments less able to survive than larger species (Edgar 1987).

The shift from the seafloor to the sea surface generates important changes in environmental conditions, such as temperature, solar radiation and nutrient supply, affecting the performance of buoyant and non-buoyant seaweeds. Depending on season or latitude, environmental pressures affect the survival potential and dispersal capacity of buoyant seaweeds and the associated flora. Consequently, epiphyte species richness is often less in floating than in benthic floating seaweeds, as shown for benthic and pelagic populations of *F. vesiculosus* and *S. muticum* (Gutow et al. 2015).

Physiological and reproductive adjustments are well documented for the most common buoyant seaweeds (e.g. *Macrocystis, Ascophyllum, Fucus, Durvillaea*). Short and long time variations in physiological responses and growth have been studied in mesocosms (Vandendriessche et al. 2007; Rothäusler et al. 2009, 2011b, c) as well as in field conditions (Rothäusler et al. 2011a; Graiff et al. 2013; Tala et al. 2013). Notwithstanding, the physiological responses and reproductive status of non-buoyant seaweeds associated with floating substrata have not been studied. Considering the molecular studies that support the hypothesis of long-distance dispersal of non-buoyant seaweeds via floating substrata (see above), it is expected that similar physiological adjustments as observed in positively buoyant seaweeds occur in these non-buoyant species.

Physiological and life history traits, size and morphology, littoral distribution, and association with positively buoyant seaweeds, all affect the dispersal and colonization potential of non-buoyant seaweeds. Although non-buoyant seaweeds remain competent, they cannot disperse successfully if their rafting vehicle sinks. Since the floating seaweeds act as important vehicle and primary substratum, tissue disintegration rate (Rothäusler et al. 2011b; Graiff et al. 2013) and losses due to herbivory (Thiel and Gutow 2005a; Rothäusler et al. 2011d) can be other factors that determine the effective dispersal potential.

4.6.1 Physiological Performance

Light (photosynthetically active radiation "PAR" and ultraviolet radiation "UV") and temperature are the main factors influencing growth, reproduction and physiology of seaweeds in temperate and polar regions (Lüning 1990; Wiencke et al. 2007; Hurd et al. 2014). Recovery from stress induced by high solar radiation and temperature is achieved by adjustment of pigment concentrations, photosynthetic efficiency, dynamic photoinhibition and/or repair mechanisms (Franklin and Forster 1997; Gómez et al. 2004; Wieters et al. 2013). Physiological acclimation may operate efficiently within the tolerance range of each species (Eggert 2012), thereby enhancing the persistence of buoyant seaweeds at the sea surface as well as the survival of associated non-buoyant seaweeds. Reduction in growth, photosynthesis, pigment levels and increase in photoinhibition and tissue loss suppress the persistence of seaweeds at the sea surface, especially during stressful summer conditions (Wheeler 1980; Hobday 2000b; Karsten et al. 2001; Vandendriessche et al. 2007; Rothäusler et al. 2011c; Graiff et al. 2013). Irregular fluctuations of light due to daily radiation dose and water turbulence influence the photosynthetic efficiency, photoinhibition and recovery capacity of seaweeds (Dromgoole 1987; Wing and Patterson 1993). Depending on the environmental conditions which the seaweeds face during rafting journeys, successful acclimation is fundamental for their persistence at the sea surface.

The physiological capacity to tolerate the new environmental conditions at the sea surface should be directly related to the bathymetric distribution pattern of seaweeds in their benthic habitats. Raft-forming seaweeds typically grow in low intertidal and subtidal zones (Thiel and Gutow 2005b; Rothäusler et al. 2012). Most non-buoyant seaweeds known to raft are turf-forming species that grow in or close to the holdfasts of larger seaweeds, where the impact of solar radiations is often attenuated by the extensive canopies of these seaweeds. Many of these non-buoyant seaweeds feature traits that are typical of "shade plants" with lower saturation intensity, higher capacity to absorb light, higher sensitivity to UV radiation, and lower photoinhibition during the day than the larger seaweeds that are fully exposed to solar radiation (Gómez et al. 2005; Gómez and Huovinen 2011). Consequently, the rapid change between shaded (benthic) and sunny (floating) conditions can severely affect the tissue health of non-buoyant seaweeds if no effective mechanisms of photo-protection are activated. For example, the intertidal turf species *Gelidium chilense* showed a decrease in the ability to recover photosynthesis levels in presence of UV, which was most pronounced in individuals from the lower edge of their intertidal distribution (Wieters et al. 2013). Abiotic stressors can produce a negative performance in buoyant and non-buoyant seaweeds when they occur during summer with unusually calm seas and particularly warm, sunny days. Nutrient enrichment of coastal waters via upwelling might facilitate recovery from negative effects of radiation and temperature (Edgar 1987; Wieters et al. 2013).

The phylogeographic pattern of populations of the red seaweed *G. chilensis* in Chile supports the hypothesis of transoceanic long-distance dispersal from New Zealand (Guillemin et al. 2014). The high phenotypic and physiological plasticity of *G. chilensis* in response to environmental variability (Santelices and Varela 1993; Molina and Montecino 1996; Edding et al. 2006), and in particular its photosynthetic adaptations to estuarine conditions indicate that *G. chilensis* is a shade-adapted species with a high content of photosynthetic pigments (chlorophyll a and phycobiliproteins), high quantum efficiency, and low light saturating point to cope with the changing light regimes in the estuary (Gómez et al. 2005). Thus, *G. chilensis* exhibits a photobiological strategy that promotes long-distance dispersal.

Seaweeds possess various repair and protective mechanisms to cope with environmental pressure and it is expected that these facilitate long-term rafting. The active synthesis and accumulation of photoprotective substances (e.g. mycosporine-like amino acids [MAAs] in red algae, phlorotannins in brown algae; coumarins in green algae) and effective antioxidant systems allow these seaweeds to tolerate high solar radiation and temperature (Franklin and Forster 1997; Pérez-Rodríguez et al. 2001; Huovinen et al. 2004; Karsten 2008). Different kinds of MAAs in various red seaweeds are related to their bathymetric distribution. Also, higher concentrations occur in geographical locations with high intensities of solar irradiation, which is additionally suggestive of their role as photoprotectors (Huovinen et al. 2004). Undetermined species of *Gelidium* showed high concentrations of MAAs while those of *G. chilensis* were rather low (0.9–1.8 mg g^{-1} Dry Weight DW) when compared to other red seaweeds in southern Chile (8–10 mg g^{-1} DW) (Huovinen et al. 2004). In general, there are no consistent patterns of MAA induction by stressors and its accumulation and synthesis appears to be a rather flexible and species-specific mechanism (Hoyer et al. 2002).

Phlorotannins, a special class of polyphenolic compounds, are the most important defense substances in brown seaweeds, functioning as deterrents of herbivores and microbes, supporting adhesion and strengthening of cell walls, and serving as photoprotector and as potent antioxidant (Schoenwaelder and Wiencke 2000; Arnold and Targett 2002; Karsten 2008). Daily and seasonal changes in phlorotannin concentrations have been described in *Cystoseira tamariscifolia*, being positively related to incident irradiance during the seasonal cycle with higher values in summer compared to other seasons (Abdala-Díaz et al. 2006). In summer, the effect of the increase in the irradiation dosage is regulated by the exudation of phenolic compounds from the tissue. Phlorotannin levels in eight non-buoyant Antarctic brown seaweeds ranged from 0.5 to 9 % DW (mean 3.3 %DW; Iken et al. 2007). These values are higher than those of the tropical and temperate North Pacific brown seaweeds but similar to species from cold-temperate regions (Steinberg 1989; Van Alstyne et al. 1999; Connan et al. 2004). Variability in phlorotannin levels in buoyant seaweeds are associated with changes in solar radiation and herbivore pressure, and this variability tends to be larger than in non-buoyant species (Swanson and Druehl 2002; Cruces et al. 2013). Short-term exposure to UV radiation can induce the production of UV-absorbing phlorotannins in blades of benthic *M. pyrifera* (Swanson and Druehl 2002), but it is not known whether floating *M. pyrifera* can adjust their phlorotannin contents during raft journeys. Rapid induction of soluble phlorotannins has been described in three south Pacific kelps (buoyant *Macrocystis* and *Durvillaea*, non-buoyant *L. spicata*) triggered by UV radiation during short-term thermal stress (Cruces et al. 2013). In floating *D. antarctica* a decrease in the phlorotannin concentrations and antioxidant activities was observed compared to benthic algae during winter and summer (Tala et al. 2013). High radiation can also produce oxidized phlorotannins as described from the peripheral tissues of *Hormosira banksii* thalli (Schoenwaelder 2002). Although rapid synthesis of photoprotective compounds can be stimulated by environmental change as a short-term response, loss of photosynthetic tissues during floating conditions in buoyant and non-buoyant seaweeds could further limit their persistence and dispersal potential.

4.6.2 Reproductive Patterns

If physiological accommodation to new environmental conditions during rafting is not successful, vital processes such as growth and reproduction can be altered. In this way, floating time can be negatively related to the reproductive capacities of buoyant and non-buoyant seaweeds. The possibility to maintain and generate new reproductive cells during rafting journeys has important consequences for dispersal and colonization potential (Macaya et al. 2005; Thiel and Gutow 2005a; McKenzie and Bellgrove 2008). Floating transport of reproductively active thalli has been described for *Postelsia palmaeformis* (Dayton 1973), *Sargassum muticum* (Norton 1977; Critchley et al. 1983), *Turbinaria ornata* (Stewart 2006), *H. banksii* (McKenzie and Bellgrove 2008), *M. pyrifera* (Macaya et al. 2005; Hernández-Carmona et al. 2006), and *D. antarctica* (Tala et al. 2013).

Successful dispersal requires that fertile thalli produce viable spores or gametes that can colonize new hard substrata. Species that reproduce continuously may show advantages during floating conditions compared to strictly seasonally reproducing species. Benthic populations of *M. pyrifera* from northern Chile have their main reproductive activity during spring and winter while in the south reproduction occurs throughout the year (Buschmann et al. 2004). This pattern can be understood as a temporal adjustment in response to latitudinal gradients. Biomass changes occurring during floating time showed that sporophylls of *M. pyrifera* disintegrate quickly at high temperatures and intense solar radiation levels occurring at low latitudes (Macaya et al. 2005; Rothäusler et al. 2009, 2011b, c). As for physiological patterns, currently no specific information exists on the seasonal variation in phase or reproductive stage for non-buoyant seaweeds.

In general, when living near the upper temperature limit for gametogenesis temperate seaweeds do not reproduce, which has important implications for their equatorward distribution limits (Peters and Breeman 1993) and probabilities of surviving as rafters. However, microthalli of some non-floating seaweeds such as *A. utricularis* (Wiencke et al. 1994) and *Colpomenia peregrina* (Orfanidis 1993) showed higher upper survival temperatures than macrothalli (Fig. 4.6). In this case,

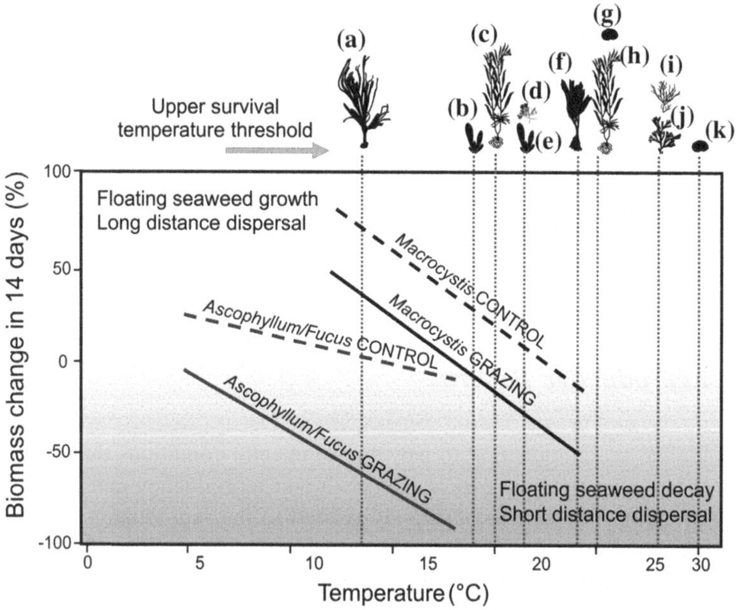

Fig. 4.6 Influence of the combined effects of temperature and grazing (CONTROL vs. GRAZING) on percent biomass change of mainly buoyant seaweeds (modified from Rothäusler et al. 2012) and upper survival temperature (UST) for some non-buoyant and buoyant seaweeds. **a** *Durvillaea antarctica*, **b** *Adenocystis utricularis*, **c** *Macrocystis*, **d** *Bostrychia* sp., **e** *A. utricularis* microthalli, **f** *Lessonia berteroana* microthalli, **g** *Colpomenia peregrina*, **h** *Macrocystis* microthalli, **i** *Gracilaria chilensis*, **j** *Fucus spiralis*, **k** *Colpomenia peregrina* microthalli. UST value from Wiencke et al. (1994) (**a–c**, **e**, **f**, **h**, **j**); Karsten et al. (1996) (**d**); Orfanidis (1993) (**g**, **k**); Macchiavello et al. (1998) (**i**)

vegetative propagation or regrowth from crustose thalli might be important mechanisms facilitating successful dispersal and potential colonization (Gómez and Westermeier 1991; Macchiavello et al. 2003; Cecere et al. 2011). Re-attachment of fragments and regrowth from secondary holdfasts, which are formed when apical regions of lateral branches come in contact with the substratum, have been described in the non-buoyant red alga *Chondracanthus chamissoi* as effective strategies for propagating and maintaining natural populations (Fonck et al. 2008; Sáez et al. 2008).

Reproductive strategies can also be related to seaweed habitats. For example, estuarine populations of *G. chilensis* are characterized by drifting and asexually reproducing individuals, whereas on rocky shores reproduction is sexual with an alternation of haploid and diploid individuals attached to the substratum (Guillemin et al. 2008). Clonal propagation by thallus fragmentation and high tolerance to abiotic stress favours long-term survival, facilitating successful dispersal in non-buoyant species.

Seaweeds with heteromorphic life cycle (macroscopic vs. microscopic; foliose vs. filamentous or crustose thallus) and/or with asexual alternatives of propagation (apomixis, fragmentation) could be the best candidates for long-distance dispersal by rafting. In both cases, the reproductive process has been proposed as an adaptation to environmental variability allowing that the reproductive phase or vegetative structures give rise to a founder individual during initial colonization events (Maggs 1988; Bell 1997; Cecere et al. 2011). *A. utricularis* is a non-buoyant seaweed which has been suggested to be dispersed via the West Wind Drift (Fraser et al. 2013). Its life history comprises diploid macroscopic sporophytes occurring commonly in summer and haploid microscopic filamentous or discoid gametophytes in winter, and it can reproduce either sexually or asexually (Müller 1984). The broad temperature tolerance of filamentous gametophytes (Wiencke et al. 1994) and asexual reproductive potential may be traits that favour long-distance dispersal.

Bostrychia intricata, a turfing non-buoyant red alga, is another example for long-distance dispersal via rafting (Fraser et al. 2013; Muangmai et al. 2014). The dominance of the long-lived tetrasporophyte phase and dispersal primarily by growth and fragmentation of vegetative thalli in *Bostrychia* (West et al. 1996) are traits that also may facilitate long-distance dispersal. Thus, asexual reproductive strategies may explain the lower genetic structure described for many non-buoyant species that are suggested to be dispersed via rafting on floating seaweeds or other buoyant substrata.

The complex life cycles of seaweeds can include the occurrence of monoecious or dioecious stages in the diploid or haploid phases affecting the colonization potential after rafting. Buoyant seaweeds such as *D. antarctica* (Collantes et al. 2002) and *F. vesiculosus* (Serrão et al. 1999) are characterized by dioecious individuals, where the joint voyage of female and male individuals and the synchronization of gamete development can be crucial for colonization of new habitats; fusion of holdfasts can facilitate this joint travel, and can involve different species as well as different sexes (González et al. 2015; Lizée-Prynne et al. 2016). In seaweeds with male and female individuals, differences in ecophysiological traits and reproductive potential between sexes may be important, particularly at the edges of their geographical range compromising their expansion capacity (Tatarenkov et al. 2005; Luthringer et al. 2014).

On the other hand, self-fertilizing hermaphrodites such as *F. spiralis* and *F. distichus* can be especially successful colonizers, because one fertile thallus arriving in a new habitat would be sufficient for successful colonization (Coleman and Brawley 2005; Coyer et al. 2011a). Although vegetative propagation, selfing and inbreeding can lead to decreased local genetic variability as a consequence of single arrivals, these mechanisms can facilitate gene flow and colonization through multiple dispersal events.

4.7 Conclusions and Outlook

This overview shows that rafting dispersal may be of importance not only for primary rafters but also for non-buoyant seaweeds. There are particular biological traits that facilitate successful dispersal of non-buoyant seaweeds, including their acclimation potential and their reproductive biology. Based on this review we predict that many non-buoyant seaweeds that grow in regions with abundant supply of buoyant seaweeds and/or other floating substrata can be efficiently dispersed via rafting. This has important implications for their population connectivity and evolutionary ecology. We suggest that the population dynamics of many non-buoyant seaweeds are strongly affected by the availability and floating persistence of buoyant seaweeds that are host to these secondary rafters. It will be particularly promising to compare the geographic ranges of non-buoyant seaweed species that are commonly found on/in rafts with those of primary rafters. We would expect a high degree of overlap in the geographic ranges of floating seaweeds and common non-buoyant rafters with a high potential for successful transport and colonization (e.g. strong acclimation potential and asexual reproduction).

It may be also important to take into account the frequency of dispersal by buoyant seaweeds, and the resulting consequences for the phylogeographic structure of populations of non-buoyant seaweeds. Dispersal may occur episodically or as single events, while in other species or locations it may occur much more frequently, with important consequences for population connectivity or divergence (see Thiel and Haye 2006).

The currently occurring changes in supply of rafts (e.g. increasing amounts of long-lived floating plastics) and floating persistence of natural rafts (due to climate change—Macreadie et al. 2011) are also expected to affect the dispersal opportunities of non-buoyant seaweeds. Future studies on range extensions or contractions of seaweeds should thus also focus on non-buoyant species, especially those commonly associated with primary rafters.

Acknowledgements We thank the editors for inviting us to contribute this review. Financial support was received through FONDECYT 1131082 to M.T., F. Tellier and F. Tala; FONDECYT 1131023 to F. Tala, M.T. and E.M.; FONDECYT 1110437 to E.M.; PhD-fellowship Becas CONICYT-PCHA/Doctorado Nacional/2014-21140010 to B.L.

References

Abdala-Díaz RT, Cabello-Pasini A, Pérez-Rodríguez E, Ãlvarez RMC, Figueroa FL. Daily and seasonal variations of optimum quantum yield and phenolic compounds in *Cystoseira tamariscifolia* (Phaeophyta). Mar Biol. 2006;148:459–65.

Aliani S, Molcard A. Hitch-hiking on floating marine debris: Macrobenthic species in the western Mediterranean Sea. Hydrobiologia. 2003;503:59–67.

Arenas F, Fernández C, Rico J, Fernández E, Haya D. Growth and reproductive strategies of *Sargassum muticum* (Yendo) Fensholt and *Cystoseira nodicaulis* (Whit.) Roberts. Sci Mar. 1995;59:1–8.

Arnold TM, Targett NM. Marine tannins: the importance of a mechanistic framework for predicting ecological roles. J Chem Ecol. 2002;28:1919–34.

Astudillo JC, Bravo M, Dumont CP, Thiel M. Detached aquaculture buoys in the SE Pacific: potential dispersal vehicles for associated organisms. Aquat Biol. 2009;5:219–31.

Bell G. The evolution of the life cycle of brown seaweeds. Biol J Linn Soc. 1997;60:21–38.

Boedeker C, Ramírez ME, Nelson W. *Cladophoropsis brachyartra* from southern South America is a synonym of *Wittrockiella lyallii* (Cladophorophyceae, Chlorophyta), previously regarded as endemic to New Zealand. Phycologia. 2010;49:525–36.

Boo GH, Mansilla A, Nelson W, Bellgrove A, Boo SM. Genetic connectivity between trans-oceanic populations of *Capreolia implexa* (Gelidiales, Rhodophyta) in cool temperate waters of Australasia and Chile. Aquat Bot. 2014;119:73–9.

Bravo M, Astudillo JC, Lancellotti D, Luna-Jorquera G, Valdivia N, Thiel M. Rafting on abiotic substrata: properties of floating items and their influence on community succession. Mar Ecol Prog Ser. 2011;439:1–17.

Bryan SE, Cook AG, Evans JP, Hebden K, Hurrey L, Colls P, Jell JS, Weatherley D, Firn J. Rapid, long-distance dispersal by pumice rafting. PLoS ONE. 2012;7:e40583.

Buchanan J, Zuccarello GC. Decoupling of short- and long-distance dispersal pathways in the endemic New Zealand seaweed *Carpophyllum maschalocarpum* (Phaeophyceae, Fucales). J Phycol. 2012;48:518–29.

Buschmann AH, Vásquez J, Osorio P, Reyes E, Filun L, Hernandez-Gonzalez MC, Vega JMA. The effect of water movement, temperature and salinity on abundance and reproductive patterns of *Macrocystis* spp. (Phaeophyta) at different latitudes in Chile. Mar Biol. 2004;145:849–62.

Bussolini LT, Waters JM. Genetic analyses of rafted macroalgae reveal regional oceanographic connectivity patterns. J Biogeogr. 2015;42:1319–26.

Buza-Jacobucci G, Pereira-Leite FP. The role of epiphytic algae and different species of *Sargassum* in the distribution and feeding of herbivorous amphipods. Latin Am J Aquat Res. 2014;42:353–63.

Cecere E, Petrocelli A, Verlaque M. Vegetative reproduction by multicellular propagules in Rhodophyta: an overview. Mar Ecol. 2011;32:419–37.

Chan SW, Cheang CC, Chirapart A, Gerung G, Tharith C, Ang P. Homogeneous population of the brown alga *Sargassum polycystum* in Southeast Asia: possible role of recent expansion and asexual propagation. PLoS ONE. 2013;8:e77662.

Chan SW, Cheang CC, Yeung CW, Chirapart A, Gerung G, Ang P. Recent expansion led to the lack of genetic structure of *Sargassum aquifolium* populations in Southeast Asia. Mar Biol. 2014;161:785–95.

Clarkin E, Maggs CA, Allcock AL, Johnson MP. Environment, not characteristics of individual algal rafts, affects composition of rafting invertebrate assemblages in Irish coastal waters. Mar Ecol Prog Ser. 2012;470:31–40.

Coleman MA, Brawley SH. Spatial and temporal variability in dispersal and population genetic structure of a rockpool alga. Mar Ecol Prog Ser. 2005;300:63–77.

Coleman MA, Kelaher BP. Connectivity among fragmented populations of a habitat-forming alga, *Phyllospora comosa* (Phaeophyceae, Fucales) on an urbanised coast. Mar Ecol Prog Ser. 2009;381:63–70.

Collantes G, Merino A, Lagos V. Fenología de la gametogénesis, madurez de conceptáculos, fertilidad y embriogénesis en *Durvillaea antarctica* (Chamisso) Hariot (Phaeophyta, Durvillaeales). Revista de Biología Marina y Oceanografía. 2002;37:83–112.

Connan S, Goulard F, Stiger V, Deslandes E, Ar GE. Interspecific and temporal variation in phlorotannin levels inan assemblage of brown algae. Bot Mar. 2004;47:410–16.

Coyer JA, Hoarau G, Costa JF, Hogerdijk B, Serrão EA, Billard E, Valero M, Pearson GA, Olsen JL. Evolution and diversification within the intertidal brown macroalgae *Fucus spiralis*/ *F. vesiculosus* species complex in the North Atlantic. Mol Phylogenet Evol. 2011a;58:283–96.

Coyer JA, Hoarau G, Van Schaik J, Luijckx P, Olsen JL. Trans-Pacific and trans-Arctic pathways of the intertidal macroalga *Fucus distichus* L. reveal multiple glacial refugia and colonizations from the North Pacific to the North Atlantic. J Biogeogr. 2011b;38:756–71.

Critchley A, Farnham WF, Morrell SL. A chronology of new European sites of attachment for the invasive brown alga, *Sargassum muticum*, 1973–1981. J Mar Biol Ass UK. 1983;63:799–811.

Cruces E, Huovinen P, Gómez I. Interactive effects of UV radiation and enhanced temperature on photosynthesis, phlorotannin induction and antioxidant activities of two sub-Antarctic brown algae. Mar Biol. 2013;160:1–13.

Cumming RA, Nikula R, Spencer HG, Waters JM. Transoceanic genetic similarities of kelp-associated sea slug populations: long-distance dispersal via rafting? J Biogeogr. 2014;41:2357–70.

D'Antonio C. Epiphytes on the rocky intertidal red alga *Rhodomela larix* (Turner) C. Agardh: negative effects on the host and food for herbivores? J Exp Mar Biol Ecol. 1985;86:197–218.

Dayton PK. Dispersion, dispersal, and persistence of the annual intertidal alga, *Postelsia palmaeformis* Ruprecht. Ecology. 1973;54:433–8.

Diez YL, Capote AJ, Suárez AM, Gómez LM, Toyota M. Distribution of epiphytic macroalgae on the thalli of their hosts in Cuba. Acta Bot Bras. 2013;27:815–26.

Dromgoole FI. Photosynthesis of marine algae in fluctuating light. I. Adjustment of rate in constant and fluctuating light regimes. Funct Ecol. 1987;1:377–86.

Duggins DO, Eckman JE, Siddon CE, Klinger T. Population, morphometric and biomechanical studies of three understory kelps along a hydrodynamic gradient. Mar Ecol Prog Ser. 2003;265:57–76.

Edding M, León C, Tala F. Morphological variations of *Gracilaria chilensis* Bird, McLachlan et Oliveira, 1986 (Rhodophyta, Gracilariales) in the southeast Pacific. Gayana. 2006;70:220–7.

Edgar GJ. Dispersal of faunal and floral propagules associated with drifting *Macrocystis pyrifera* plants. Mar Biol. 1987;95:599–610.

Eggert A. Seaweed responses to temperature. In: Wiencke C, Bischof K, editors. Seaweed biology, ecological studies. Berlin Heidelberg: Springer; 2012. p. 47–66.

Filho S, Tagliaro CH, Beasley CR. Seasonal abundance of the shipworm *Neoteredo reynei* (Bivalvia, Teredinidae) in mangrove driftwood from a northern Brazilian beach. Iheringia Série Zoologia. 2008;98:17–23.

Fonck E, Martínez R, Vásquez J, Bulboa C. Factors that affect the re-attachment of *Chondracanthus chamissoi* (Rhodophyta, Gigartinales) thalli. J Appl Phycol. 2008;20:311–4.

Franklin L, Forster R. The changing irradiance environment: consequences for marine macrophyte physiology, productivity and ecology. Eur J Phycol. 1997;32:207–32.

Fraser CI, Nikula R, Spencer HG, Waters JM. Kelp genes reveal effects of subantarctic sea ice during the Last Glacial Maximum. Proc Nat Acad Sci USA. 2009;106:3249–53.

Fraser CI, Thiel M, Spencer HG, Waters JM. Contemporary habitat discontinuity and historic glacial ice drive genetic divergence in Chilean kelp. BMC Evol Biol. 2010;10:203.

Fraser CI, Waters JM. Algal parasite *Herpodiscus durvillaeae* (Phaeophyceae: Sphacelariales) inferred to have traversed the pacific ocean with its buoyant host. J Phycol. 2013;49:202–6.

Fraser CI, Zuccarello GC, Spencer HG, Salvatore LC, Garcia GR, Waters JM. Genetic affinities between trans-oceanic populations of non-buoyant macroalgae in the high latitudes of the Southern Hemisphere. PLoS ONE. 2013;8(7):e69138.

Garbary DJ, Burke J, Lining T. The *Ascophyllum/Polysiphonia/Mycosphaerella* symbiosis. II. Aspects of the ecology and distribution of *Polysiphonia* in Nova Scotia. Bot Mar. 1991;34:391–401.

Garbary DJ, Deckert RJ. Three part harmony-*Ascophyllum* and its symbionts. In: Seckbach J, editor. Symbiosis: mechanisms and model systems. Dortrecht: Kluwer; 2001. p. 309–21.

Garden CJ, Craw D, Waters JM, Smith A. Rafting rocks reveal marine biological dispersal: a case study using clasts from beach-cast macroalgal holdfasts. Estu Coast Shelf Sci. 2011;95:388–94.

Garden CJ, Currie K, Fraser CI, Waters JM. Rafting dispersal constrained by an oceanographic boundary. Mar Ecol Prog Ser. 2014;501:297–302.

Gilbert R. The movement of gravel by the alga *Fucus vesiculosus* (L.) on an arctic intertidal flat. J Sedimen Petrol. 1984;54:463–8.

Gómez I, Figueroa FL, Huovinen P, Ulloa N, Morales V. Photosynthesis of the red alga *Gracilaria chilensis* under natural solar radiation in an estuary in southern Chile. Aquaculture. 2005;244:369–82.

Gómez I, Huovinen P. Morpho-functional patterns and zonation of South Chilean seaweeds: the importance of photosynthetic and bio-optical traits. Mar Ecol Prog Ser. 2011;422:77–91.

Gómez I, López-Figueroa F, Ulloa N, Morales V, Lovengreen C, Huovinen P, Hess S. Patterns of photosynthesis in 18 species of intertidal macroalgae from southern Chile. Mar Ecol Prog Ser. 2004;270:103–16.

Gómez I, Westermeier RC. Frond regrowth from basal disk in *Iridaea laminarioides* (Rhodophyta, Gigartinales) at Mehuin, southern Chile. Mar Ecol Prog Ser. 1991;73:83–91.

González MA, Barrales HL, Candia A, Cid L. Spatial and temporal distribution of dominant epiphytes on *Gracilaria* from a natural subtidal bed in Central-Southern Chile. Aquaculture. 1993;116:135–48.

González AV, Beltrán J, Flores V, Santelices B. Morphological convergence in the inter-holdfast coalescence process among kelp and kelp-like seaweeds (*Lessonia, Macrocystis, Durvillaea*). Phycologia. 2015;54:283–91.

Gower J, Young E, King S. Satellite images suggest a new *Sargassum* source region in 2011. Remote Sens Lett. 2013;4:764–73.

Graiff A, Karsten U, Meyer S, Pfender D, Tala F, Thiel M. Seasonal variation in floating persistence of detached *Durvillaea antarctica* (Chamisso) Hariot thalli. Bot Mar. 2013;56:3–14.

Gregory MR. Virgin plastic granules on some beaches of eastern Canada and Bermuda. Mar Environ Res. 1983;10:73–92.

Guillemin ML, Faugeron S, Destombe C, Viard F, Correa JA, Valero M. Genetic variation in wild and cultivated populations of the haploid- diploid red alga *Gracilaria chilensis*: how farming practices favor asexual reproduction and heterozygosity. Evolution. 2008;62:1500–19.

Guillemin ML, Valero M, Faugeron S, Nelson W, Destombe C. Tracing the trans-pacific evolutionary history of a domesticated seaweed (*Gracilaria chilensis*) with archaeological and genetic data. PLoS ONE. 2014;9(12):e114039.

Gutow L, Beermann J, Buschbaum C, Rivadeneira MM, Thiel M. Castaways can't be choosers—homogenization of rafting assemblages on floating seaweeds. J Sea Res. 2015;95:161–71.

Gutow L, Franke HD. Metapopulation structure of the marine isopod *Idotea metallica*, a species associated with drifting habitat patches. Helgol Mar Res. 2003;56:259–64.

Gutow L, Gimenez L, Boos K, Saborowski R. Rapid changes in the epifaunal community after detachment of buoyant benthic macroalgae. J Mar Biol Ass UK. 2009;89:323–8.

Heesch S, Peters AF, Broom JE, Hurd CL. Affiliation of the parasite *Herpodiscus durvillaeae* (Phaeophyceae) with the Sphacelariales based on DNA sequence comparisons and morphological observations. Eur J Phycol. 2008;43:283–95.

Helmuth B, Veit RR, Holberton R. Long-distance dispersal of a subantarctic brooding bivalve (*Gaimardia trapesina*) by kelp-rafting. Mar Biol. 1994;120:421–6.

Hernández-Carmona G, Hughes B, Graham MH. Reproductive longevity of drifting kelp *Macrocystis pyrifera* (Phaeophyceae) in Monterey Bay, USA. J Phycol. 2006;42:1199–207.

Hinojosa IA, Pizarro M, Ramos M, Thiel M. Spatial and temporal distribution of floating kelp in the channels and fjords of southern Chile. Estu Coast Shelf Sci. 2010;87:367–77.

Hinojosa IA, Rivadeneira MM, Thiel M. Temporal and spatial distribution of floating objects in coastal waters of central-southern Chile and Patagonian fjords. Cont Shelf Res. 2011;31:172–86.

Hirata T, Tanaka J, Iwami T, Ohmi T, Dazai A, Aoki M, Ueda H, Tsuchiya Y, Sato T, Yokohama Y. Ecological studies on the community of drifting seaweeds in the south-eastern coastal waters of Izu Peninsula, central Japan. I. Seasonal changes of plants in species composition, appearance, number of species and size. Phycol Res. 2001;49:215–29.

Hobday AJ. Abundance and dispersal of drifting kelp *Macrocystis pyrifera* rafts in the Southern California Bight. Mar Ecol Prog Ser. 2000a;195:101–16.

Hobday AJ. Persistence and transport of fauna on drifting kelp (*Macrocystis pyrifera* (L.) C. Agardh) rafts in the Southern California Bight. J Exp Mar Biol Ecol. 2000b;253:75–96.

Hoyer K, Karsten U, Wiencke C. Induction of sunscreen compounds in Antarctic macroalgae by different radiation conditions. Mar Biol. 2002;141:619–27.

Hu ZM, Uwai S, Yu SH, Komatsu T, Ajisaka T, Duan D-L. Phylogeographic heterogeneity of the brown macroalga *Sargassum horneri* (Fucaceae) in the northwestern Pacific in relation to late Pleistocene glaciation and tectonic configurations. Mol Ecol. 2011;20:3894–909.

Hu ZM, Zhang J, Lopez-Bautista J, Duan DL. Asymmetric genetic exchange in the brown seaweed *Sargassum fusiforme* (Phaeophyceae) driven by oceanic currents. Mar Biol. 2013;160:1407–14.

Huovinen P, Gómez I, Figueroa FL, Ulloa N, Morales V, Lovengreen CBM. Ultraviolet-absorbing mycosporine-like amino acids in red macroalgae from Chile. Bot Mar. 2004;47:21–9.

Hurd C, Harrison PJ, Bischof K, Lobban CS. Seaweed ecology and physiology. 2nd ed. UK: Cambridge University Press; 2014.

Iken K, Amsler CD, Hubbard JM, McClintock JB, Baker BJ. Allocation patterns of phlorotannins in antarctic brown algae. Phycologia. 2007;46:386–95.

Ingólfsson A. Floating clumps of seaweed around Iceland: natural microcosms and a means of dispersal for shore fauna. Mar Biol. 1995;122:13–21.

Ingólfsson A. Dynamics of macrofaunal communities of floating seaweed clumps off western Iceland: a study of patches on the surface of the sea. J Exp Mar Biol Ecol. 1998;231:119–37.

John DM. New records of *Ascophyllum nodosum* (L.) Le Joli from the warmer parts of the Atlantic Ocean. Phycologia. 1974;10:243–4.

Jokiel PL. Rafting of reef corals and other organisms at Kwajalein Atoll. Mar Biol. 1989;101:483–93.

Karsten U. Defense strategies of algae and cyanobacteria against solar ultraviolet radiation. In: Amsler CD, editor. Algal chemical ecology. Heidelberg: Springer; 2008. p. 273–96.

Karsten U, Bischof K, Wiencke C. Photosynthetic performance of Arctic macroalgae after transplantation from deep to shallow waters. Oecologia. 2001;127:11–20.

Karsten U, Koch S, West JA, Kirst JO. Physiological responses of the eulittoral macroalga *Stictosiphonia hookeri* (Rhodomelaceae, Rhodophyta) from Argentina and Chile: salinity, light and temperature acclimation, Eur J Phycol. 1996;31:361–68.

Khalaman VV, Berger VY. Floating seaweeds and associated fauna in the White Sea. Oceanology. 2006;46:827–33.

Kiessling T, Gutow L, Thiel M. Marine litter as habitat and dispersal vector. In: Bergmann M, Gutow L, Klages M, editors. Marine anthropogenic litter. Berlin: Springer; 2015. p. 141–81.

Kingsford MJ, Choat JH. The fauna associated with drift algae captured with a plankton-mesh purse seine net. Limnol Oceanogr. 1985;30:618–30.

Klumpp DW, Salita-Espinosa JT, Fortes MD. Feeding ecology and trophic role of sea urchins in a tropical seagrass community. Aquat Bot. 1993;45:205–29.

Kohlmeyer J, Bebout B, Vlkmann-Kohlmeyer B. Decomposition of mangrove wood by marine fungi and teredinids in Belize. Mar Ecol. 1995;16:27–39.

Komatsu T, Mizuno S, Natheer A, Kantachumpoo A, Tanaka K, Morimoto A, Hsiao S-T, Rothäusler EA, Shishidou H, Aoki M, Ajisaka T. Unusual distribution of floating seaweeds in the East China Sea in the early spring of 2012. J Appl Phycol. 2014;26:1169–79.

Lee KM, Boo GH, Coyer JA, Nelson WA, Miller KA, Boo SM. Distribution patterns and introduction pathways of the cosmopolitan brown alga *Colpomenia peregrina* using mt cox3 and atp6 sequences. J Appl Phycol. 2014;26:491–504.

Lindstrom SC, Gabrielson PW, Hughey JR, Macaya EC, Nelson WA. Sequencing of historic and modern specimens reveals cryptic diversity in *Nothogenia* (Scinaiaceae, Rhodophyta). Phycologia. 2015;54:97–108.

Lizée-Prynne, Lopez D, Tala B, Thiel FM. No sex-related dispersal limitation in a dioecious, oceanic long-distance traveller: the bull kelp *Durvillaea antarctica*. Bot Mar. 2016;59 (in press).

Lüning K. Seaweeds: their environment, biogeography and ecophysiology. New York: Wiley; 1990.

Luthringer R, Cormier A, Ahmed S, Peters AF, Cock JM. Sexual dimorphism in the brown algae. Persp Phycol. 2014;1:11–25.

Macaya EC, Boltaña S, Hinojosa IA, Macchiavello JE, Valdivia NA, Vásquez NR, Buschmann AH, Vásquez JA, Vega JMA, Thiel M. Presence of sporophylls in floating kelp rafts of *Macrocystis* spp. (Phaeophyceae) along the Chilean pacific coast. J Phycol. 2005;41:913–22.

Macaya EC, Zuccarello GC. DNA barcoding and genetic divergence in the giant kelp *Macrocystis* (Laminariales). J Phycol. 2010a;46:736–42.

Macaya EC, Zuccarello GC. Genetic structure of the giant kelp *Macrocystis pyrifera* along the southeastern Pacific. Mar Ecol Prog Ser. 2010b;420:103–12.

Macchiavello J, Bulboa CR, Edding M. Vegetative propagation and spore-based recruitmentin the carragenophyte *Chondracanthus chamissoi* (Gigartinales, Rhodophyta) in northern Chile. Phycol Res. 2003;51:45–50.

Macchiavello J, De Paula ÉJ, Oliveira EC. Growth rate responses of five commercial strains of *Gracilaria* (Rhodophyta, Gracilariales) to temperature and light. J World Aquacul Soc. 1998;29:259–66.

Macreadie PI, Bishop MJ, Booth DJ. Implications of climate change for macrophytic rafts and their hitchhikers. Mar Ecol Prog Ser. 2011;443:285–92.

Maggs CA. Intraspecific life history variability in the Florideophycidae (Rhodophyta). Bot Mar. 1988;31:465–90.

McKenzie PF, Bellgrove A. Dispersal of *Hormosira banksii* (Phaeophyceae) via detached fragments: reproductive viability and longevity. J Phycol. 2008;44:1108–15.

Molina X, Montecino V. Acclimation to UV irradiance in *Gracilaria chilensis* Bird, McLachlan & Oliveira (Gigartinales, Rhodophyta). Hydrobiologia. 1996;326(327):415–20.

Morton B, Britton JC. Origins of the Azorean intertidal biota: the significance of introduced species, survivors of chance events. Arquipélago Life Mar Sci. 2000a;Suppl 2 (Part A):29–51.

Morton B, Britton JC. The origins of the coastal and marine flora and fauna of the Azores. Oceanogr Mar Biol Ann Rev. 2000;38:3–84.

Muangmai N, West JA, Zuccarello GC. Evolution of four Southern Hemisphere *Bostrychia* (Rhodomelaceae, Rhodophyta) species: phylogeny, species delimitation and divergence times. Phycologia. 2014;53:593–601.

Muhlin JF, Engel CR, Stessel R, Weatherbee RA, Brawley SH. The influence of coastal topography, circulation patterns, and rafting in structuring populations of an intertidal alga. Mol Ecol. 2008;17:1198–210.

Müller DG. Culture studies on the life history of *Adenocystis utricularis* (Phaeophyceae, Dictyosiphonales). Phycologia. 1984;23:87–94.

Neiva J, Assis J, Fernandes F, Pearson GA, Serrão EA. Species distribution models and mitochondrial DNA phylogeography suggest an extensive biogeographical shift in the high-intertidal seaweed *Pelvetia canaliculata*. J Biogeogr. 2014;41:1137–48.

Neiva J, Pearson GA, Valero M, Serrão EA. Drifting fronds and drifting alleles: range dynamics, local dispersal and habitat isolation shape the population structure of the estuarine seaweed *Fucus ceranoides*. J Biogeogr. 2012;39:1167–78.

Nikula R, Fraser CI, Spencer HG, Waters JM. Circumpolar dispersal by rafting in two subantarctic kelp-dwelling crustaceans. Mar Ecol Prog Ser. 2010;405:221–30.

Nikula R, Spencer HG, Waters JM. Evolutionary consequences of microhabitat: population-genetic structuring in kelp- vs. rock-associated chitons. Mol Ecol. 2011;20:4915–24.

Nikula R, Spencer HG, Waters JM. Passive rafting is a powerful driver of transoceanic gene flow. Biol Lett. 2013;20120821.

Norton TA. The growth and development of *Sargassum muticum* (Yendo) Fensholt. J Exp Mar Biol Ecol. 1977;26:41–53.

Norton TA, Mathieson AC. The biology of unattached seaweeds. In: Round FE, Chapman DJ, editors. Progress in phycological research. New York: Elsevier; 1983. p. 336–86.

Ohno M. Observations on the floating seaweeds of near-shore waters of southern Japan. Hydrobiologia. 1984;116(117):408–12.

Ojeda FP, Santelices B. Invertebrate communities in hodfasts of the kelp *Macrocystis pyrifera* from southern Chile. Mar Ecol Prog Ser. 1984;16:65–73.

Oliveira E, Ugadim Y, Paula E. Associated epibiota on *Sargassum* floating on the waters of the Brazilian current-biogeographical remarks. Boletim do Botanica, Universidade de Sao Paulo 1979;7:5–9.

Olsen JL, Zechman FW, Hoarau G, Coyer JA, Stam WT, Valero M, Åberg P. The phylogeographic architecture of the fucoid seaweed *Ascophyllum nodosum*: an intertidal 'marine tree' and survivor of more than one glacial–interglacial cycle. J Biogeogr. 2010;37:842–56.

Orfanidis S. Temperature responses and distribution of several Mediterranean macroalgae belonging to different distribution groups. Bot Mar. 1993;36:359–70.

Pearson GA, Evans LV. Settlement and survival of *Polysiphonia lanosa* (Ceramiales) spores on *Ascophyllum nodosum* and *Fucus vesiculosus* (Fucales). J Phycol. 1990;26:597–603.

Pérez-Rodríguez E, Aguilera J, Gómez I, Figueroa F. Excretion of coumarins by the Mediterranean green alga *Dasycladus vermicularis* in response to environmental stress. Mar Biol. 2001;139:633–9.

Peters AF, Breeman AM. Temperature tolerance and latitudinal range of brown algae from temperate Pacific South America. Mar Biol. 1993;115:143–50.

Rothäusler E, Gómez I, Hinojosa IA, Karsten U, Miranda L, Tala F, Thiel M. Kelp rafts in the Humboldt Current: interplay of abiotic and biotic factors limit their floating persistence and dispersal potential. Limnol Oceanogr. 2011a;56:1751–63.

Rothäusler E, Gómez I, Hinojosa IA, Karsten U, Tala F, Thiel M. Effect of temperature and grazing on growth and reproduction of floating *Macrocystis* spp. (Phaeophyceae) along a latitudinal gradient. J Phycol. 2009;45:547–59.

Rothäusler E, Gómez I, Hinojosa IA, Karsten U, Tala F, Thiel M. Physiological performance of floating giant kelp *Macrocystis pyrifera* (Phaeophyceae): latitudinal variability in the effects of temperature and grazing. J Phycol. 2011b;47:269–81.

Rothäusler E, Gómez I, Karsten U, Tala F, Thiel M. Physiological acclimation of floating *Macrocystis pyrifera* to temperature and irradiance ensures long-term persistence at the sea surface at mid-latitudes. J Exp Mar Biol Ecol. 2011c;405:33–41.

Rothäusler E, Gómez I, Karsten U, Tala F, Thiel M. UV-radiation versus grazing pressure: long-term floating of kelp rafts (*Macrocystis pyrifera*) is facilitated by efficient photoacclimation but undermined by grazing losses. Mar Biol. 2011d;158:127–41.

Rothäusler E, Gutow L, Thiel M. Floating seaweeds and their communities. In: Wiencke C, Bischof K, editors. Seaweed biology. Ecological studies. Heidelberg: Springer; 2012. p. 359–80.

Sáez F, Macchiavello J, Fonck E, Bulboa C. The role of the secondary attachment disc in the vegetative propagation of *Chondracanthus chamissoi* (Gigartinales; Rhodophyta). Aquat Bot. 2008;89:63–5.
Santelices B, Varela D. Intra-clonal variation in the red seaweed *Gracilaria chilensis*. Mar Biol. 1993;116:543–52.
Saunders GW. Long distance kelp rafting impacts seaweed biogeography in the northeast pacific: the kelp conveyor hypothesis. J Phycol. 2014;50:968–74.
Schoenwaelder MEA. Physode distribution and the effect of 'thallus sunburn' in *Hormosira banksii* (Fucales, Phaeophyceae). Bot Mar. 2002;45:262–6.
Schoenwaelder MEA, Wiencke C. Phenolic compounds in the embryo development of several northern hemisphere fucoids. Plant Biol. 2000;2:24–33.
Serrão EA, Brawley SH, Hedman J, Kautsky L, Samuelsson G. Reproductive success of *Fucus vesiculosus* (Phaeophyceae) in the Baltic Sea. J Phycol. 1999;35:254–69.
Smith SDA. Kelp rafts in the Southern Ocean. Glob Ecol Biogeogr. 2002;11:67–9.
Stachowicz JJ, Fried H, Osman RW, Whitlatch RB. Biodiversity, invasion resistance, and marine ecosystem function: reconciling pattern and process. Ecology. 2002;83:2575–90.
Steinberg P. Biogeographical variation in brown algal polyphenolics and other secondary metabolites: comparison between temperate Australasia and North America. Oecologia. 1989;78:373–82.
Stewart HL. Ontogenetic changes in buoyancy, breaking strength, extensibility, and reproductive investment in a drifting macroalga *Turbinaria ornata* (Phaeophyta). J Phycol. 2006;42:43–50.
Susini ML, Thibaut T, Meinesz A, Forcioli D. A preliminary study of genetic diversity in *Cystoseira amentacea* (C. Agardh) Bory var. *stricta* Montagne (Fucales, Phaeophyceae) using random amplified polymorphic DNA. Phycologia. 2007;46:605–11.
Swanson AK, Druehl LD. Induction, exudation and the UV protective role of kelp phlorotannins. Aquat Bot. 2002;73:241–53.
Tala F, Gómez I, Luna-Jorquera G, Thiel M. Morphological, physiological and reproductive conditions of rafting bull kelp (*Durvillaea antarctica*) in northern-central Chile (30°S). Mar Biol. 2013;160:1339–51.
Tatarenkov A, Bergstrom L, Jonsson RB, Serrão EA, Kautsky L, Johannesson K. Intriguing asexual life in marginal populations of the brown seaweed *Fucus vesiculosus*. Mol Ecol. 2005;14:647–51.
Tegner MJ, Dayton PK, Edwards PB, Riser KL. Sea urchin cavitation of giant kelp *Macrocystis pyrifera* (C. Agardh) holdfasts and its effects on kelp mortality across a large California forest. J Exp Mar Biol Ecol. 1995;191:83–99.
Thiel M. Rafting of benthic macrofauna: important factors determining the temporal succession of the assemblage on detached macroalgae. Hydrobiologia. 2003;503:49–57.
Thiel M, Gutow L. The ecology of rafting in the marine environment. I. The floating substrata. Oceanogr Mar Biol Ann Rev. 2005a;42:181–264.
Thiel M, Gutow L. The ecology of rafting in the marine environment. II. The rafting organisms and community. Oceanogr Mar Biol Ann Rev. 2005b;43:279–418.
Thiel M, Haye P. The ecology of rafting in the marine environment. III. Biogeographical and evolutionary consequences. Oceanogr Mar Biol Ann Rev. 2006;44:323–429.
Vairappan CS. Seasonal occurrences of epiphytic algae on the commercially cultivated red alga *Kappaphycus alvarezii* (Solieriaceae, Gigartinales, Rhodophyta). J Appl Phycol. 2006;18:611–7.
Van Alstyne KL, McCarthy JJ III, Hustead CL, Duggins DO. Geographic variation in polyphenolic levels of Northeastern Pacific kelps and rockweeds. Mar Biol. 1999;133:371–9.
van den Hoek C. The possible significance of long-range dispersal for the biogeography of seaweeds. Helgol Meeresunters. 1987;41:261–72.
Vandendriessche S, Vincx M, Degraer S. Floating seaweed and the influences of temperature, grazing and clump size on raft longevity—a microcosm study. J Exp Mar Biol Ecol. 2007;343:64–73.
Waters JM. Driven by the West Wind Drift? A synthesis of southern temperate marine biogeography, with new directions for dispersalism. J Biogeogr. 2008;35:417–27.

Waters JM, Fraser CI, Hewitt GM. Founder takes all: density-dependent processes structure biodiversity. Trends Ecol Evol. 2013;28:78–85.

West JA, Zuccarello GC, Karsten U. Reproductive biology of *Stictosiphonia hookeri* (Rhodomelaceae, Rhodophyta) from Argentina, Chile, South Africa and Australia in laboratory culture. Hydrobiologia. 1996;326(327):277–82.

Wheeler WN. Effect of boundary layer transport on the fixation of carbon by the giant kelp *Macrocystis pyrifera*. Mar Biol. 1980;56:103–10.

Whittick A. Spatial and temporal distributions of dominant epiphytes on the stipes of *Laminaria hyperborea* (Gunn.) Fosl. (Phaeophyta: Laminariales) in S.E. Scotland. J Exp Mar Biol Ecol. 1983;73:1–10.

Wichmann CS, Hinojosa IA, Thiel M. Floating kelps in Patagonian Fjords: an important vehicle for rafting invertebrates and its relevance for biogeography. Mar Biol. 2012;159:2035–49.

Wiencke C, Bartsch I, Bischoff B, Peters AF, Breeman AM. Temperature requirements and biogeography of Antarctic, Arctic and amphiequatorial seaweeds. Bot Mar. 1994;37:247–59.

Wiencke C, Clayton MN, Gómez I, Iken K, Lüder UH, Amsler CD, Karsten U, Hanelt D, Bischof K, Dunton K. Life strategy, ecophysiology and ecology of seaweeds in polar waters. Rev Environ Sci Bio. 2007;6:95–126.

Wieters EA, Medrano A, Quiroga G. Spatial variation in photosynthetic recovery of intertidal turf algae from acute UVB and temperature stress associated with low tides along the central coast of Chile. J Exp Mar Biol Ecol. 2013;449:340–8.

Wing S, Patterson M. Effects of wave-induced light flecks in the intertidal zone on photosynthesis in the macroalgae *Postelsia palmaeformis* and *Hedophyllum sessile* (Phaeophyceae). Mar Biol. 1993;116:519–25.

Winston JE, Gregory MR, Stevens LM. Encrusters, epibionts, and other biota associated with pelagic plastics: a review of biogeographical, environmental, and conservation issues. In: Coe JM, Rogers DB, editors. Marine debris. New York: Springer; 1997. p. 81–97.

Woelkerling WJ. On the epibiotic and pelagic Chlorophyceae, Phaeophyceae, and Rhodophyceae of the Western Sargasso Sea. Rhodora. 1975;77:1–40.

Won BY, Cho TO, Yates KK, Fredericq S. Characterization of macroalgal epiphytes on *Thalassia testudinum* and *Syringodium filiforme* seagrass in Tampa Bay, Florida. Algae. 2010;25:141–53.

Chapter 5
Change in Southern Hemisphere Intertidal Communities Through Climate Cycles: The Role of Dispersing Algae

Ceridwen I. Fraser

Abstract Macroalgae are fundamental components of most marine ecosystems, creating habitat and food sources for a wide range of other organisms. Macroalgal distributions are strongly linked to water temperature, and global environmental change is therefore likely to drive major shifts both in the distributions and compositions of marine communities. Phylogeographic research on macroalgae and associated organisms can reveal ecological changes that have occurred with global warming since the Last Glacial Maximum (LGM), which can help us to predict what might happen under future climate change scenarios. Such research shows that many macroalgae have changed their distributions, broadly shifting poleward or into deeper waters. Importantly, for organisms to change their distributions in response to climate change, they must be able to disperse, sometimes long distances. Some buoyant, robust macroalgae are extremely good long-distance travellers, and others have apparently been able to disperse across oceans indirectly, such as via rafting. However, not all macroalgal species are capable of long-distance dispersal, and with global warming, ecosystems thus do not simply slide poleward in their entireties, but both move and change. Studies are already showing that contemporary climate change is affecting the distributions of macroalgal-dominated ecosystems. This chapter summarizes some of the ways in which Southern Hemisphere macroalgal distributions are inferred to have shifted with past climate change, and speculates on how they might change in the future. Processes underpinning these changes, such as climate drivers and dispersal capacity, are also discussed.

Keywords Drifting · Last Glacial Maximum (LGM) · Rafting · Sea surface temperature · Seaweed

---

C.I. Fraser (✉)
Fenner School of Environment and Society, Australian National University, Acton, ACT 2601, Australia
e-mail: ceridwen.fraser@gmail.com

Z.-M. Hu and C. Fraser (eds.), *Seaweed Phylogeography*,
DOI 10.1007/978-94-017-7534-2_5

5.1 Introduction

Macroalgae are key components of almost all marine ecosystems; indeed, many macroalgal species create critical habitat for other organisms, and thus act as fundamental ecosystem engineers (Jones et al. 1997). The large brown (phaeophycean) algae, or kelps (Fraser 2012a), often play particularly important roles in the physical and biological structure of the local environment. For example, *Macrocystis* species form kelp 'forests' that can support extraordinarily diverse ecosystems (Graham 2004), pelagic clumps of *Sargassum* act as floating islands inhabited by a wide range of vertebrates and invertebrates (Fine 1970), and stands of bull kelp—*Durvillaea* (in the Southern Hemisphere) and *Nereocystis* (in the Northern Hemisphere)—support diverse assemblages of flora and fauna (Edgar and Burton 2000; Taylor and Schiel 2005; Springer et al. 2010). For many of the large kelps, particularly diverse communities of invertebrates inhabit the structurally complex holdfasts, which are discs or domes at the base of the stipe that attach the plant to the substrate (Smith 2000) (Fig. 5.1). Green (Chlorophyta) and red (Rhodophyta) macroalgae, even in apparently simple structural forms such as crusts, can also create important habitat for many other marine organisms (Marx and Herrnkind 1985; Sánchez-Moyano et al. 2001; Chenelot et al. 2011), and all divisions of macroalgae are primary producers, providing food sources for fauna.

The composition of most marine ecosystems—notably excluding deep-sea ecosystems that are inimical to photosynthesizing organisms—therefore hinges on which algal species are present, which in turn is largely determined by environmental conditions. The large kelps are rare in shallow tropical waters, due to high water temperatures and low nutrient availability (Bolton 2010), but are more common in tropical waters of 30–200 m depth (Graham et al. 2007), and are abundant in cool-temperate and sub-polar zones (Mann 1973). In Antarctic waters, large brown algae do occur but are mostly restricted to members of the order Desmarestiales (Moe and Silva 1977). In coral reef ecosystems, a complex competitive relationship exists between macroalgae and corals, with macroalgal growth generally suppressed on healthy reefs (McCook et al. 2001). The distributional ranges of marine macroalgae are largely related to water temperature (Adey and Steneck 2001), although they may also be influenced by available sunlight (e.g. Schwarz et al. 2003), nutrients (Burkepile and Hay 2006), and other factors such as wave exposure (Underwood and Jernakoff 1984) and the presence of herbivores (Underwood and Jernakoff 1984; McCook et al. 2001; Burkepile and Hay 2006). With environmental conditions driving biogeographic patterns in these important—and often keystone—macroalgae, large-scale environmental change is likely to drive major shifts both in the distributions and compositions of entire marine communities. Indeed, changes in macroalgal distributions, and hence those of associated organisms, have been inferred to have occurred with past climate change (Adey and Steneck 2001; Fraser et al. 2009; Nikula et al. 2010; Fraser 2012b). In the face of ongoing and accelerating (Cox et al. 2000) global warming, inferring

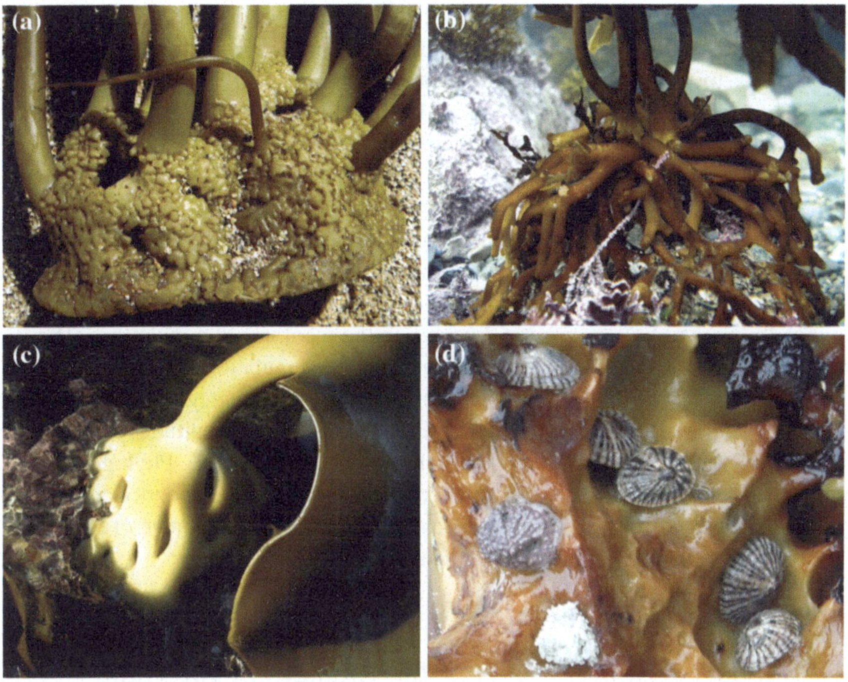

Fig. 5.1 Kelp holdfasts are often structurally complex and can form habitat for diverse organisms such as small invertebrates and algae, including during oceanic dispersal of the kelp (rafting). Pictured are holdfasts of **a** *Lessonia spicata* in Chile (*E. Macaya*, 2007); **b** *Macrocystis pyrifera* in New Zealand (*E. Macaya*, 2009); **c** and **d** *Durvillaea antarctica*, New Zealand (*C. Fraser*, 2006) (underside of detached holdfast, with resident gastropods, pictured in **d**)

past environmentally driven changes to marine ecosystems could help us to predict what might happen under future scenarios.

5.2 Pleistocene Glacial—Interglacial Cycles

During the Pleistocene (from approximately 2.5 Ma until the Holocene, which started 11.7 ka), the Earth underwent several glacial–interglacial cycles. The most recent glacial period peaked between 0.026 and 0.019 Ma, and is referred to as the Last Glacial Maximum (LGM) (Clark et al. 2009). Biological studies indicate that many species shifted their distributions as the planet warmed during interglacial periods, for example, with terrestrial trees generally moving to higher latitudes or higher elevations (Davis and Shaw 2001). Studies of modern taxa indicate that these trends hold true for a wide range of species under contemporary global warming (Parmesan and Yohe 2003; Chen et al. 2011), including marine species (Perry et al. 2005). For example, some Antarctic benthic taxa are inferred to have retreated into deeper waters during glacial periods, and recolonized the continental shelf postglacially (Brey et al. 1996), and some marine species have shifted their distributions towards the poles with recent warming (Perry et al. 2005). When organisms are unable to move to suitable habitat as the environment changes, they face local or total extinction; indeed, a large number of species are predicted to face extinction under future climate change scenarios (Thomas et al. 2004). Some species can survive unfavourable conditions, for example, within glaciated regions, by sheltering in microrefugia—small areas that remain inhabitable, such as localized warm patches formed by geothermal heat (Fraser et al. 2014) or valleys (Stewart and Lister 2001). Recolonization of surrounding regions from such refugia is typically marked by low genetic diversity in the recolonized populations (Provan and Bennett 2008), as only a subset of the original population moves out of each refugium. Likewise, poleward distributional shifts following glacial periods are often characterized by high genetic diversity in the lower latitudes versus low diversity in the higher, recolonized latitudes (Hewitt 2000). Phylogeographic approaches can therefore help us to understand how species responded to past ice ages and subsequent global warming, and which species were capable of dispersing to new areas to track suitable environmental conditions.

5.3 Phylogeographic Evidence for Changes in Intertidal Communities Through Climate Cycles

One of the major environmental changes that occurs as the planet warms is a rise in sea levels, effected through the melting of glaciers and sea ice, and through thermal expansion of ocean water (Rahmstorf 2007). Since the LGM, sea levels have risen about 135 m (Yokoyama et al. 2000), although with some variation according to geographic location (Lambeck and Chappell 2001). This large sea level rise has obscured most LGM shores, impeding direct (e.g. fossil) analysis of LGM shallow-water communities. Molecular research can, however, shed light on ecological changes that have occurred since the LGM, through phylogeographic studies that assess geographic patterns of genetic diversity in representative marine organisms.

5.3.1 *Polar and Sub-polar Latitudes*

Among the subantarctic islands, south-western South America, and New Zealand, intertidal ecosystems are largely dominated by *Durvillaea antarctica* communities (Klemm and Hallam 1988), with other large brown algae, such as *Lessonia*, co-dominant at some locations (Huovinen and Gomez 2012). This large kelp has a strong influence on community structure, for example, by affecting which other macroalgal species are able to settle and grow (Taylor and Schiel 2005), and by supporting a wide range of invertebrate species (Edgar and Burton 2000; Smith and Simpson 2002). In the shallow subtidal of this region, the giant kelp *Macrocystis pyrifera*, which is also an important ecosystem engineer, is more dominant. Both *D. antarctica* and *M. pyrifera* are removed by ice scour, and thus do not occur in regions with much glacial or sea ice (see Fraser 2012b). Both species have also been inferred, based on phylogeographic research that demonstrates higher northern versus lower southern molecular diversity, to have recolonized the higher latitudes postglacially (Fraser et al. 2009, 2010; Macaya and Zuccarello 2010) (Fig. 5.2). Studies of some of the kelp-associated invertebrates support the hypothesis that these communities were absent from higher latitudes at the LGM (Nikula et al. 2010; Cumming et al. 2014). There are few studies of postglacial recolonization patterns for smaller macroalgal species in the high latitudes of the Southern Hemisphere, but recolonization of ice-affected areas of Chile has been inferred for *Mazzaella laminarioides* (Montecinos et al. 2012). Despite the absence of the large kelps and their associated communities from the high latitudes during glacial maxima, these shores are unlikely to have been entirely denuded of life. Some taxa can persist in areas where ice scour is common, for example, by living below the reach of the ice or sheltering in crevices (Barnes 1999). During glacial maxima, subantarctic nearshore ecosystems might therefore have resembled those found today in ice-scoured Antarctic waters, comprising mainly small macroalgae with short or biphasic/seasonal life cycles, and motile invertebrates such as limpets, amphipods, and worms (Fraser 2012b). In the Scotia Arc islands, for example, intertidal algae include *Porphyra*, *Leptosomia*, *Iridaea*, and *Adenocystis*, while genera in the shallow subtidal include *Desmarestia*, *Curdiea*, *Monostroma*, *Plocamium*, *Phyllogigas*, and *Ascosiera* (Mercier and Hamel 2005). Many of these species are also found on subantarctic islands today. A phylogeographic study of *Adenocystis utricularis* and *Bostrychia intricata* indicates that these species may have survived the LGM in situ in the subantarctic, whereas the larger kelps did not (Fraser et al. 2013). Many *Nacella* limpets also show relatively high genetic diversity at subantarctic latitudes, suggesting they are not postglacial recolonists of the region (González-Wevar et al. 2010), although they appear to have been absent from higher latitude, Antarctic shores at the LGM (González-Wevar et al. 2013).

5.3.2 Tropical and Temperate Latitudes

Whereas the marine biogeography of polar and sub-polar regions appears to have been strongly affected by the extent of glacial and sea ice, postglacial climate change impacts in lower-latitude regions have been linked to water temperature, and/or to changes in sea level (and thus changes in shorelines). For example, in south-east Asia, low genetic diversity of the ecologically important brown macroalga *Sargassum polycystum* has been inferred to be a result of the postglacial flooding of the Sunda Shelf (Chan et al. 2013). In New Zealand, postglacial water

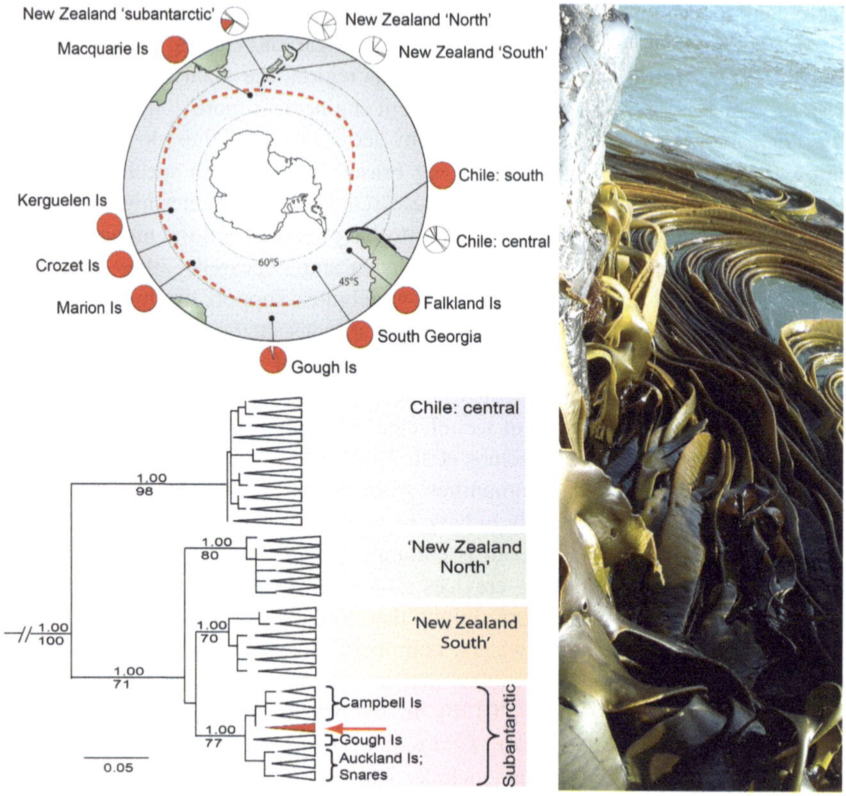

Fig. 5.2 Phylogeography of southern bull kelp, *Durvillaea antarctica* (pictured: *right*), based on mtDNA (COI) data. The phylogenetic tree (*lower left*) shows haplotype relationships, with Bayesian PP values above branches, and ML bootstraps below. The map (*upper left*) shows haplotype proportions at each locality. A single, widespread haplotype (see *arrow*) was detected at many sites throughout the subantarctic and along LGM-glaciated parts of the Chilean coast, whereas more northern regions showed relatively high genetic diversity. Genetically homogeneous regions with the common 'subantarctic' haplotype were generally within the range of the LGM 4 °C isotherm (*dashed line*), which may represent a proxy for the extent of sea ice at the LGM (Fraser et al. 2009). These patterns have been inferred to indicate postglacial recolonization of high latitudes by the buoyant kelp, which would have been removed from these regions by ice scour at the LGM (Fraser et al. 2009, 2010). Similar patterns have been observed for giant kelp, *Macrocystis pyrifera* (Macaya and Zuccarello 2010)

temperature and sea level rise have been invoked to explain patterns of genetic diversity in *Carpophyllum maschalocarpum* (Buchanan and Zuccarello 2012) and in cryptic species within *B. intricata* (Muangmai et al. 2015). In the Southern Californian Bight, in North America, palaeogeographic modelling indicates that a larger area of rocky substrate (a consequence of sea level fall during the glacial period) and increased upwelling supported large, rich, kelp-dominated ecosystems similar to those that are found further north today (Graham et al. 2003).

Fig. 5.3 Change in winter sea surface temperatures in the Southern Hemisphere at the Last Glacial Maximum (adapted from Fig. 5A in CLIMAP 1981). Water temperatures along much of the western coasts of major southern landmasses (Africa, Australia, South America and New Zealand) are estimated to have been considerably (up to 6 °C) cooler than today. Such changes will have had profound impacts on macroalgal communities and associated ecosystems in these regions

Estimates of sea surface temperatures (SST) at the LGM indicate that many of the coastal temperate regions in the Southern Hemisphere experienced less extreme cooling than at comparable latitudes in the Northern Hemisphere (CLIMAP 1981; Trend-Staid and Prell 2002). Some areas may even have had slightly warmer waters than today, such as the sub-tropical latitudes of the eastern coasts of Africa, South America, and Australia (Fig. 5.3). These areas might have been too warm for the large habitat-forming kelps, and may instead have supported more 'tropical' ecosystems of corals and smaller macroalgae. Regions that experienced little change in SST through glacial cycles may have had relatively stable macroalgal populations for up to several millions of years, and may therefore comprise distinctive biogeographic assemblages (Adey and Steneck 2001). The sub-tropical and tropical western coasts of the Southern Hemisphere continents, in contrast, are thought to have been cooler than now during glacial periods, from about 2 °C (Trend-Staid and Prell 2002) to 6 °C (CLIMAP 1981: Fig. 5.3) cooler at the LGM. Under such conditions, large kelps and their associated communities could have occurred considerably farther north than they do today.

5.4 Effects of Dispersal Ability on Community Structure

For organisms to achieve large-scale distributional shifts, they must be able to disperse, and in some cases to disperse hundreds or thousands of kilometres (Gillespie et al. 2012). Long-distance oceanic dispersal capacity is particularly important in the Southern Hemisphere, where landmasses are not as continuous across latitudes as they are in the more continental Northern Hemisphere. For example, coastal macroalgae that are restricted to southern Australia would need to disperse across thousands of kilometres of open ocean to colonise new shores if their distributions continue to shift southward; those that are unable to achieve such long-distance dispersal face extinction (Wernberg et al. 2011).

Some of the key habitat-forming macroalgae are physiologically adapted to be extremely good long-distance travellers, particularly at lower temperatures and when solar radiation is not high (Rothäusler et al. 2011; Graiff et al. 2013; Tala et al. 2013). Both *D. antarctica* and *M. pyrifera*, for example, are robust and buoyant, and are commonly found drifting at sea in the mid- to high latitudes, forming large 'rafts' that are inhabited by a wide range of floral and faunal passengers (Smith 2002; Rothäusler et al. 2012). Other buoyant, raft-forming taxa in the Southern Hemisphere include *Sargassum, Carpophyllum*, *Phyllospora* and *Cystophora* (Rothäusler et al. 2012). Direct observations of kelp rafts, combined with phylogeographic data, have shown that these plants can travel across ocean basins, and can carry diverse epifaunal and epiphyte communities with them. *Durvillaea antarctica* plants have been found washed up on New Zealand shores that molecular data show are hundreds of kilometres from their source populations in the subantarctic (Collins et al. 2010; Fraser et al. 2011; Bussolini and Waters 2015), in some cases with a wide range of living invertebrate passengers (Fraser et al. 2011).

Several non-buoyant and more fragile macroalgae have apparently been able to disperse long, transoceanic distances (such as between New Zealand and South America), presumably via rafting on more robust, buoyant kelps or driftwood. Examples include the red algae *B. intricata* (Fraser et al. 2013) and *Capreolia implexa* (Boo et al. 2014), the green alga *Wittrockiella lyallii* (Boedeker et al. 2010) and others (also see Chapter by Macaya et al. (2016) in this volume). Some otherwise poorly dispersive faunal taxa have also been able to utilise drifting macroalgae to traverse large oceanic distances, particularly kelp-associated marine invertebrates (Donald et al. 2005; Nikula et al. 2010; Haye et al. 2012; Nikula et al. 2013; Cumming et al. 2014). However, not all species are capable of rafting with buoyant kelp; indeed, there is a sharp decline in diversity of associated invertebrates once kelp detaches from the substrate and begins to drift at sea (Gutow et al. 2009), and some species do not attach to the kelp at all. Distributional shifts with climate change must, therefore, involve large-scale movement of some species, and little or no movement of others. With global warming, ecosystems thus do not simply slide poleward in their intireties, but both move and change.

5.5 Summary and Outlook

Phylogeographic evidence has shown that some of the major habitat-forming algae shifted distributions with past climate change, with important consequences for the structure and diversity of marine ecosystems. In general, we are only able to infer what changes might have occurred since the last glacial period, as signals of previous climate cycles are obscured by the most recent events. We are now facing warmer global average temperatures than have occurred for tens of thousands of years. Nonetheless, the insights we can gain by looking at the impacts of the global warming that took place following the LGM can help us to understand the general patterns and processes of climate-induced ecological changes, and thus to predict what might happen in the future.

Studies are already showing some of the impacts of contemporary climate change on macroalgal-dominated ecosystems. A major habitat-forming kelp in Australia, *Ecklonia radiata*, currently faces considerable physiological challenges in the warmer latitudes of its distribution, as well as from lower water clarity, suggesting that its distribution will shift significantly with ongoing global warming (Staehr and Wernberg 2009). Indeed, in South Africa, *Ecklonia maxima* populations have already been documented to have shifted more than 70 km southward with warming waters (Bolton et al. 2012). Observations following extreme weather events (such as heatwaves) also indicate that climate change will lead to major changes to marine ecosystems, as a result of reduced cover and/or altered distributions of habitat-forming seaweeds such as *E. radiata* (Wernberg et al. 2013) and *Scytothalia dorycarpa* (Smale and Wernberg 2013).

Broadly, Southern Hemisphere macroalgal species that can disperse are likely to move southward with future climate change. For those incapable of long-distance dispersal, the extent of southward shifts will be limited by the southern edges of the landmasses, unless they are able to travel with more dispersive taxa (e.g. via rafting). In contrast to terrestrial species, which are predicted to continue moving both poleward and to higher elevations with global warming (Chen et al. 2011), some marine species may shift downhill to deeper, cooler waters. Already, some large kelp species are found in the deep waters of the tropics, where temperatures are cooler and where water clarity allows light to penetrate to tens or hundreds of metres below the surface (Graham et al. 2007). Indeed, water clarity exerts a major influence on macroalgal distributions (e.g. Nielsen et al. 2002; Eriksson and Bergström 2005; De'ath and Fabricius 2010), and changing climates will lead to regional changes in turbidity and water chemistry as well as in temperature (Jickells 1998; Harley et al. 2006; Whitehead et al. 2009). Both forecast temperature and turbidity profiles should therefore be considered when attempting to model future distributions of macroalgae, although there is currently only limited information on predicted changes to water quality.

Importantly, because macroalgae are fundamental parts of many marine ecosystems, shifts in macroalgal distributions with environmental change will drastically alter the structure of many marine ecosystems. Although buoyant macroalgae can carry epifaunal and epifloral passengers when they disperse, not all species are suited to raft with seaweeds, and thus not all community members will shift their distributions in the same ways. With changing algal distributions, therefore, we can expect broad scale changes to marine ecosystems.

Acknowledgements Hu Zi-Min provided feedback on early drafts. Quentin Slade (National Library of Australia) helped with obtaining map resources. The author was supported by an Australian Research Council Discovery Early Career Research Award (DE140101715).

References

Adey WH, Steneck RS. Thermogeography over time creates biogeographic reions: a temperature/space/time-integrated model and an abundance-weighted test for benthic marine algae. J Phycol. 2001;37(5):677–98.

Barnes DKA. The influence of ice on polar nearshore benthos. J Mar Biol Assoc U K. 1999;79 (3):401–7.

Boedeker C, Ramirez ME, Nelson WA. *Cladophoropsis brachyartra* from southern South America is a synonym of *Wittrockiella lyallii* (Cladophorophyceae, Chlorophyta), previously regarded as endemic to New Zealand. Phycologia. 2010;49(6):525–36.

Bolton J. The biogeography of kelps (Laminariales, Phaeophyceae): a global analysis with new insights from recent advances in molecular phylogenetics. Helgol Mar Res. 2010;64(4):263–79.

Bolton JJ, Anderson RJ, Smit AJ, Rothman MD. South African kelp moving eastwards: the discovery of *Ecklonia maxima* (Osbeck) Papenfuss at De Hoop Nature Reserve on the south coast of South Africa. Afr J Mar Sci. 2012;34(1):147–51.

Boo GH, Mansilla A, Nelson W, Bellgrove A, Boo SM. Genetic connectivity between trans-oceanic populations of *Capreolia implexa* (Gelidiales, Rhodophyta) in cool temperate waters of Australasia and Chile. Aquat Bot. 2014;119:73–9.

Brey T, Dahm C, Gorny M, Klages M, Stiller M, Arntz WE. Do Antarctic benthic invertebrates show an extended level of eurybathy? Antarct Sci. 1996;8(1):3–6.

Buchanan J, Zuccarello GC. Decoupling of short- and long-distance dispersal pathways in the endemic New Zealand seaweed *Carpophyllum maschalocarpum* (Phaeophyceae, Fucales). J Phycol. 2012;48(3):518–29.

Burkepile DE, Hay ME. Herbivore vs. nutrient control of marine primary producers: context-dependent effects. Ecology. 2006;87(12):3128–39.

Bussolini LT, Waters JM. Genetic analyses of rafted macroalgae reveal regional oceanographic connectivity patterns. J Biogeogr. 2015;42(7):1319–26.

Chan SW, Cheang CC, Chirapart A, Gerung G, Tharith C, Ang P. Homogeneous population of the brown alga *Sargassum polycystum* in Southeast Asia: possible role of recent expansion and asexual propagation. PLoS ONE. 2013;8(10):e77662.

Chen I-C, Hill JK, Ohlemüller R, Roy DB, Thomas CD. Rapid range shifts of species associated with high levels of climate warming. Science. 2011;333(6045):1024–6.

Chenelot H, Jewett S, Hoberg M. Macrobenthos of the nearshore Aleutian Archipelago, with emphasis on invertebrates associated with *Clathromorphum nereostratum* (Rhodophyta, Corallinaceae). Mar Biodiv. 2011;41(3):413–24.

Clark PU, Dyke AS, Shakun JD, Carlson AE, Clark J, Wohlfarth B, Mitrovica JX, Hostetler SW, McCabe AM. The Last Glacial Maximum. Science. 2009;325(5941):710–4.

CLIMAP. 1981. Seasonal reconstructions of the Earth's surface at the Last Glacial Maximum. Map and Chart Series MC-36. Boulder, Colorado.

Collins CJ, Fraser CI, Ashcroft A, Waters JM. Asymmetric dispersal of southern bull-kelp (*Durvillaea antarctica*) adults in coastal New Zealand: testing an oceanographic hypothesis. Mol Ecol. 2010;19(20):4572–80.

Cox PM, Betts RA, Jones CD, Spall SA, Totterdell IJ. Acceleration of global warming due to carbon-cycle feedbacks in a coupled climate model. Nature. 2000;408(6809):184–7.

Cumming RA, Nikula R, Spencer HG, Waters JM. Transoceanic genetic similarities of kelp-associated sea slug populations: long-distance dispersal via rafting? J Biogeogr. 2014;41(12):2357–70.

Davis MB, Shaw RG. Range shifts and adaptive responses to Quaternary climate change. Science. 2001;292(5517):673–9.

De'ath G, Fabricius K. Water quality as a regional driver of coral biodiversity and macroalgae on the Great Barrier Reef. Ecol Appl. 2010;20(3):840–50.

Donald KM, Kennedy M, Spencer HG. Cladogenesis as the result of long-distance rafting events in South Pacific topshells (Gastropoda, Trochidae). Evolution. 2005;59(8):1701–11.

Edgar GJ, Burton HR. The biogeography of shallow-water macrofauna at Heard Island. Pap Proc R Soc Tasman. 2000;133(2):23–6.

Eriksson BK, Bergström L. Local distribution patterns of macroalgae in relation to environmental variables in the northern Baltic Proper. Estuar Coast Shelf Sci. 2005;62(1–2):109–17.

Fine ML. Faunal variation on pelagic *Sargassum*. Mar Biol. 1970;7(2):112–22.

Fraser C, Thiel M, Spencer H, Waters J. Contemporary habitat discontinuity and historic glacial ice drive genetic divergence in Chilean kelp. BMC Evol Biol. 2010;10(1):203.

Fraser CI, Nikula R, Spencer HG, Waters JM. Kelp genes reveal effects of subantarctic sea ice during the Last Glacial Maximum. Proc Natl Acad Sci USA. 2009;106(9):3249–53.

Fraser CI, Nikula R, Waters JM. Oceanic rafting by a coastal community. Proc R Soc Biol Sci Ser B. 2011;278:649–55.

Fraser CI. Is bull-kelp kelp? The role of common names in science. N Z J Mar Freshw Res. 2012a;46(2):279–84.

Fraser CI. The impacts of past climate change on sub-Antarctic nearshore ecosystems. Pap Proc R Soc Tasman. 2012b;146:89–93.

Fraser CI, Zuccarello GC, Spencer HG, Salvatore LC, Garcia GR, Waters JM. Genetic affinities between trans-oceanic populations of non-buoyant macroalgae in the high latitudes of the Southern Hemisphere. PLoS ONE. 2013;8(7):e69138.

Fraser CI, Terauds A, Smellie J, Convey P, Chown SL. Geothermal activity helps life survive glacial cycles. Proc Natl Acad Sci USA. 2014;111(15):5634–9.

Gillespie RG, Baldwin BG, Waters JM, Fraser CI, Nikula R, Roderick GK. Long-distance dispersal: a framework for hypothesis testing. Trends Ecol Evol. 2012;27(1):47–56.

González-Wevar CA, Nakano T, Cañete JI, Poulin E. Molecular phylogeny and historical biogeography of *Nacella* (Patellogastropoda: Nacellidae) in the Southern Ocean. Mol Phylogenet Evol. 2010;56(1):115–24.

González-Wevar CA, Saucède T, Morley SA, Chown SL, Poulin E. Extinction and recolonization of maritime Antarctica in the limpet *Nacella concinna* (Strebel, 1908) during the last glacial cycle: toward a model of Quaternary biogeography in shallow Antarctic invertebrates. Mol Ecol. 2013;22(20):5221–36.

Graham MH, Dayton PK, Erlandson JM. Ice ages and ecological transitions on temperate coasts. Trends Ecol Evol. 2003;18(1):33–40.

Graham MH. Effects of local deforestation on the diversity and structure of southern California Giant Kelp forest food webs. Ecosystems. 2004;7(4):341–57.

Graham MH, Kinlan BP, Druehl LD, Garske LE, Banks S. Deep-water kelp refugia as potential hotspots of tropical marine diversity and productivity. Proc Natl Acad Sci USA. 2007;104 (42):16576–80.

Graiff A, Karsten U, Meyer S, Pfender D, Tala F, Thiel M. Seasonal variation in floating persistence of detached *Durvillaea antarctica* (Chamisso) Hariot thalli. Bot Mar. 2013;56(1):3.

Gutow L, Gimenez L, Boos K, Saborowski R. Rapid changes in the epifaunal community after detachment of buoyant benthic macroalgae. J Mar Biol Assoc UK. 2009;89(2):323–8.

Harley CDG, Randall Hughes A, Hultgren KM, Miner BG, Sorte CJB, Thornber CS, Rodriguez LF, Tomanek L, Williams SL. The impacts of climate change in coastal marine systems. Ecol Lett. 2006;9(2):228–41.

Haye PA, Varela AI, Thiel M. Genetic signatures of rafting dispersal in algal-dwelling brooders *Limnoria* spp. (Isopoda) along the SE Pacific (Chile). Mar Ecol Prog Ser. 2012;455:111–22.

Hewitt G. The genetic legacy of the Quaternary ice ages. Nature. 2000;405(6789):907–13.

Huovinen P, Gomez I. Cold-temperate seaweed communities of the Southern Hemisphere. In: Wiencke C, Bischof K, editors. Seaweed biology, vol. 219., Ecological studiesBerlin: Springer; 2012. p. 293–313.

Jickells TD. Nutrient biogeochemistry of the coastal zone. Science. 1998;281(5374):217–22.

Jones CG, Lawton JH, Shachak M. Positive and negative effects of organisms as physical ecosystem engineers. Ecology. 1997;78(7):1946–57.

Klemm MF, Hallam ND. Standing crop of *Durvillaea antarctica* (Chamisso) Hariot (Phaeophyta) on the Australian Sub-Antarctic Macquarie and Heard Islands. Phycologia. 1988;27(4):505–9.
Lambeck K, Chappell J. Sea level change through the Last Glacial Cycle. Science. 2001;292 (5517):679–86.
Macaya EC, Zuccarello GC. Genetic structure of the giant kelp *Macrocystis pyrifera* along the southeastern Pacific. Mar Ecol Prog Ser. 2010;420:103–12.
Macaya EC, Lopez B, Tala F, Tellier F, Thiel M. Flot and raft: role of buoyant seaweeds in the phylogeography and genetic structure of non-buoyant associated flora. In: Hu ZM, Fraser CI, editors. Seaweed phylogeography: adaptation and evolution of seaweeds under environmental change. Heidelberg: Springer; 2016.
Mann KH. Seaweeds: their productivity and strategy for growth: the role of large marine algae in coastal productivity is far more important than has been suspected. Science. 1973;182 (4116):975–81.
Marx JM, Herrnkind WF. Macroalgae (Rhodophyta: *Laurencia* spp.) as habitat for young juvenile spiny lobsters, *Panulirus argus*. Bull Mar Sci. 1985;36(3):423–31.
McCook L, Jompa J, Diaz-Pulido G. Competition between corals and algae on coral reefs: a review of evidence and mechanisms. Coral Reefs. 2001;19(4):400–17.
Mercier EC, Hamel J-F. Atlantic Ocean islands, coastal ecology. In: Schwartz ML, editor. Encyclopedia of Coastal Science. Dordrecht: Springer; 2005. p. 84–8.
Moe RL, Silva PC. Antarctic marine flora: uniquely devoid of kelps. Science. 1977;196 (4295):1206–8.
Montecinos A, Broitman BR, Faugeron S, Haye PA, Tellier F, Guillemin ML. Species replacement along a linear coastal habitat: and speciation in the red alga *Mazzaella laminarioides* along the south east Pacific. BMC Evol Biol. 2012;12:97.
Muangmai N, Fraser CI, Zuccarello GC (2015) Contrasting patterns of population structure and demographic history in cryptic species of *Bostrychia intricata* (Rhodomelaceae, Rhodophyta) from New Zealand. J Phycol 51. doi:10.1111/jpy.12305.
Nielsen S, Sand-Jensen K, Borum J, Geertz-Hansen O. Depth colonization of eelgrass (*Zostera marina*) and macroalgae as determined by water transparency in Danish coastal waters. Estuaries. 2002;25(5):1025–32.
Nikula R, Fraser CI, Spencer HG, Waters JM. Circumpolar dispersal by rafting in two subantarctic kelp-dwelling crustaceans. Mar Ecol Prog Ser. 2010;405:221–30.
Nikula R, Spencer HG, Waters JM. Passive rafting is a powerful driver of transoceanic gene flow. Biol Lett. 2013;9:20120821.
Parmesan C, Yohe G. A globally coherent fingerprint of climate change impacts across natural systems. Nature. 2003;421(6918):37–42.
Perry AL, Low PJ, Ellis JR, Reynolds JD. Climate change and distribution shifts in marine fishes. Science. 2005;308(5730):1912–5.
Provan J, Bennett KD. Phylogeographic insights into cryptic glacial refugia. Trends Ecol Evol. 2008;23(10):564–71.
Rahmstorf S. A semi-empirical approach to projecting future sea-level rise. Science. 2007;315 (5810):368–70.
Rothäusler E, Gómez I, Karsten U, Tala F, Thiel M. Physiological acclimation of floating *Macrocystis pyrifera* to temperature and irradiance ensures long-term persistence at the sea surface at mid-latitudes. J Exp Mar Biol Ecol. 2011;405(1–2):33–41.
Rothäusler E, Gutow L, Thiel M. Floating seaweeds and their communities. In: Wiencke C, Bischof K, editors. Seaweed Biology, vol. 219., Ecological studiesHeidelberg: Springer; 2012. p. 359–80.
Sánchez-Moyano EJ, Estacio FJ, García-Adiego EM, Carlos García-Gómez J. Effect of the vegetative cycle of *Caulerpa prolifera* on the spatio-temporal variation of invertebrate macrofauna. Aquat Bot. 2001;70(2):163–74.
Schwarz A-M, Hawes I, Andrew N, Norkko A, Cummings V, Thrush S. Macroalgal photosynthesis near the southern global limit for growth; Cape Evans, Ross Sea, Antarctica. Polar Biol. 2003;26(12):789–99.

Smale DA, Wernberg T. Extreme climatic event drives range contraction of a habitat-forming species. Proc R Soc B: Biol Sci. 2013;280(1754):20122829.

Smith SDA. Evaluating stress in rocky shore and shallow reef habitats using the macrofauna of kelp holdfasts. J Aquat Ecosyst Stress Recovery. 2000;7(4):259–72.

Smith SDA. Kelp rafts in the Southern Ocean. Glob Ecol Biogeogr. 2002;11(1):67–9.

Smith SDA, Simpson RD. Spatial variation in the community structure of intertidal habitats at Macquarie Island (sub-Antarctic). Antarct Sci. 2002;14(4):374–84.

Springer YP, Hays CG, Carr MH, Mackey MR. Toward ecosystem-based management of marine macroalgae—the Bull Kelp, *Nereocystis luetkeana*. Oceanogr Mar Biol. 2010;48:1–41.

Staehr PA, Wernberg T. Physiological responses of *Ecklonia radiata* (Laminariales) to a latitudinal gradient in ocean temperature. J Phycol. 2009;45(1):91–9.

Stewart JR, Lister AM. Cryptic northern refugia and the origins of the modern biota. Trends Ecol Evol. 2001;16(11):608–13.

Tala F, Gómez I, Luna-Jorquera G, Thiel M. Morphological, physiological and reproductive conditions of rafting bull kelp (*Durvillaea antarctica*) in northern-central Chile (30°S). Mar Biol. 2013;160(6):1339–51.

Taylor DI, Schiel DR. Self-replacement and community modification by the southern bull kelp *Durvillaea antarctica*. Mar Ecol Prog Ser. 2005;288:87–102.

Thomas CD, Cameron A, Green RE, Bakkenes M, Beaumont LJ, Collingham YC, Erasmus BFN, de Siqueira MF, Grainger A, Hannah L, Hughes L, Huntley B, van Jaarsveld AS, Midgley GF, Miles L, Ortega-Huerta MA, Townsend Peterson A, Phillips OL, Williams SE. Extinction risk from climate change. Nature. 2004;427(6970):145–8.

Trend-Staid M, Prell WL. Sea surface temperature at the Last Glacial Maximum: a reconstruction using the modern analog technique. Paleoceanography. 2002;17(4):1065.

Underwood AJ, Jernakoff P. The effects of tidal height, wave-exposure, seasonality and rock-pools on grazing and the distribution of intertidal macroalgae in New South Wales. J Exp Mar Biol Ecol. 1984;75(1):71–96.

Wernberg T, Russell BD, Thomsen MS, Gurgel CFD, Bradshaw CJA, Poloczanska ES, Connell SD. Seaweed communities in retreat from ocean warming. Curr Biol. 2011;21(21):1828–32.

Wernberg T, Smale DA, Tuya F, Thomsen MS, Langlois TJ, de Bettignies T, Bennett S, Rousseaux CS. An extreme climatic event alters marine ecosystem structure in a global biodiversity hotspot. Nat Clim Change. 2013;3(1):78–82.

Whitehead PG, Wilby RL, Battarbee RW, Kernan M, Wade AJ. A review of the potential impacts of climate change on surface water quality. Hydrol Sci J. 2009;54(1):101–23.

Yokoyama Y, Lambeck K, De Deckker P, Johnston P, Fifield LK. Timing of the Last Glacial Maximum from observed sea-level minima. Nature. 2000;406(6797):713–6.

Part III
Invasive Seaweeds

Chapter 6
Non-native Seaweeds Drive Changes in Marine Coastal Communities Around the World

Mads Solgaard Thomsen, Thomas Wernberg, Paul M. South and David R. Schiel

Abstract We conducted a bibliographic survey, adding 69 taxa to a published list of 277 seaweeds, thereby updating the total worldwide list of non-native and cryptogenic seaweeds to 346. *Polysiphonia* Greville and *Hypnea* J.V. Lamouroux species were the most common taxa on this list, and the Mediterranean Sea and the NE Atlantic bioregions have received most of the 346 taxa. The most important vectors that carry non-native seaweeds are hull fouling and the transport of aquaculture products including 'blind passengers'. Once a seaweed has arrived in a new location, it can establish a permanent population and spread through natural dispersal or human activity. Non-native seaweeds have negative impacts on native species through competition, habitat destruction and keystone competition, but also positive impacts through habitat formation, food provision and cascading habitat formation. Quantitative meta-analyses have shown that invasive seaweeds typically have a negative effect on local plants, but neutral or positive effects on animal communities. New meta-analyses presented here indicate that impacts increase with the abundance of non-native seaweeds and that non-native seaweeds may increase sample similarity in invaded plant communities, but not in animal communities. The literature on the impact of non-native seaweeds is extensive, but most studies have focused on a few high-profile species. Comprehensive analyses should be

M.S. Thomsen (✉) · P.M. South · D.R. Schiel
Marine Ecology Research Group, School of Biological Sciences,
University of Canterbury, Private Bag 4800, Christchurch 8140, New Zealand
e-mail: mads.solgaard.thomsen@gmail.com

M.S. Thomsen
Centre of Integrative Ecology, School of Biological Sciences,
University of Canterbury, Christchurch, New Zealand

M.S. Thomsen · T. Wernberg
UWA Oceans Institute, School of Plant Biology, The University of Western Australia,
Crawley, WA 6009, Australia
e-mail: thomas.wernberg@uwa.edu.cn

T. Wernberg
Australian Institute of Marine Science, 39 Fairway, Crawley, WA 6009, Australia

Z.-M. Hu and C. Fraser (eds.), *Seaweed Phylogeography*,
DOI 10.1007/978-94-017-7534-2_6

done for more species to allow for better predictions. We conclude that non-native seaweeds have altered shallow coastal communities in most biogeographical regions, and impacts will likely increase along with increases in human populations, transport and associated stressors.

Keywords Invasion impact · Invasion success · Meta-analysis · New invasions · Density-dependent effects · Trophic matching hypothesis

6.1 Introduction—Scope of Seaweed Invasions

Non-native seaweeds are marine macroalgae that have arrived at new locations aided by human activity, either deliberately or accidentally. Today, non-native seaweeds comprise hundreds of species distributed in the photic zone throughout the world's major biogeographical regions. Many non-native seaweeds are 'foundation species' (Dayton 1972) and function as 'ecosystem engineers' (Jones et al. 1994), directly or indirectly affecting the availability of resources to other species by creating, modifying and maintaining habitats (Wallentinus and Nyberg 2007; Thomsen et al. 2010). Consequently, non-native seaweeds play key ecological and biogeochemical roles in their new marine ecosystems, sometimes controlling biodiversity, ecosystem functioning and energy flow (Williams and Smith 2007).

The success of a non-native seaweeds is typically considered a staged process that depends on uptake from a donor-region, transport, arrival and release to a receiver-region, local establishment and proliferation, population growth and regional spread and range expansion (Sakai et al. 2001; Bates et al. 2014). For each of these stages, a non-native seaweed has to overcome a series of physical, physiological, demographic and biological barriers and filters that act as a constant sink for new species and individuals. In this chapter, we first review which seaweed species have invaded various regions and then provide a brief overview of the two main approaches in invasion ecology, 'success' and 'impact' studies and their related frameworks and working hypotheses. We subsequently provide examples of impacts on local populations and evaluate existing quantitative syntheses of invasion impacts and identify current research gaps. We then address key research gaps using meta-analysis to test whether invasive seaweeds have density-dependent effects and decrease variability in community similarity. This 'case study section' is therefore more detailed than the rest of this review chapter. Finally, we provide a summary of the major findings from the chapter. Our focus is on non-native seaweeds, but we also comment on non-native seagrasses (marine or estuarine angiosperms) and terrestrial halophytes (salt marsh and mangrove angiosperms), since they interact with invasive seaweeds in intertidal or brackish habitats (e.g. Thomsen et al. 2012; Williams and Grosholz 2008).

We use the term 'non-native' (=alien, non-indigenous, exotic, introduced) to define species that live outside their native distributional range. We include species that have dispersed by their own means through man-made canals (e.g. Lessepsian migration where non-native species swim or drift through the Suez Canal) but not new species that arrived by their own dispersal across natural barriers, and only survive due to climate changes. We characterise non-native species that are 'highly successful' (e.g. spread rapidly and establish abundant populations) or have a 'strong impact' on resident populations or ecosystems as 'invasive', but acknowledge the ambiguities and lack of a clear definition for when or where non-native species are invasive or not (indeed, it is more appropriate to rank non-natives according to their 'relative invasiveness' within the success and impact criteria). We define cryptogenic species as species that are of uncertain origin: for these species, detailed taxonomic, biogeographical and molecular analyses are required to determine whether they are non-natives. Throughout this chapter, we describe non-native and invasive 'species', but we assert that data are always collected from specific populations (genotypes) that represent a limited subset of a species' gene pool. Researchers should, of course, be cautious when success and impact of a non-native genotype is predicted from a study conducted on a different genotype.

6.2 Research on and Numbers of Non-native Seaweeds

Invasions by marine primary producers have been described by scientists for a long time. For example, the Danish oceanographer Carl Emil Hansen Ostenfeld described the invasion of the planktonic diatom *Biddulphia* (*Odontella*) *sinensis* Greville into the North Sea in 1903, probably transported by ships (Ostenfeld 1908). Over 50 years ago, Charles Elton, in his seminal book on biological invasions, described dramatic ecosystem changes associated with *Spartina* von Schreber invasions, converting marine intertidal mudflats to salt marshes on massive scales (Elton 1958). Elton even applied invasion patterns to support the notion that 'Falkenbergia rufolanosa' was the tetrasporophytic phase of *Asparagopsis armata* Harvey, because these two seaweed morphologies showed simultaneous spread into the north-east Atlantic and Mediterranean Sea in the 1920s–1950s (originating from south-western Australia). However, research on seaweed invasions did not take off until the 1990s. A 'topic' search in the Web of Science using classic invasion terminology [TS = ((invasion* or exotic* or non-native* or alien*) and (seaweed* or macro-alga* or macroalga*))] revealed that the first scientific paper on the topic was published in 1973 (Wassman and Ramus 1973) followed by 2 papers in the 1980s (Russell 1981; Dale 1982), 37 from 1990 to 1999, 210 from 2000 to 2009 and 235 papers from 2010 to 2015. This search does not include all relevant references—and includes a few less relevant studies—but nevertheless documents that few studies predated the 1990s. The growing number of primary research studies on seaweed invasions has been reviewed in some detail (e.g. Thomsen et al. 2009b; Williams and Smith 2007; Schaffelke et al. 2006; Inderjit et al. 2006; Johnson and Chapman 2007). The most comprehensive of these reviews was by Williams and Smith (2007), describing 277 seaweed taxa introduced around the world including a few unidentified taxa within particular genera (e.g. *Bryopsis* J.V. Lamouroux., *Ulva* Linnaeus subspecies (e.g. *Codium fragile* ssp. *scandinavicum* P.C. Silva and *tomentosoides* (van Goor) P.C. Silva) and taxa of cryptogenic origin (including a few that probably are natives, e.g. *Alaria esculenta* (L.) Greville and *Saccharina ochotensis* (Miyabe) C.E. Lane, C. Mayes, Druehl and G.W. Saunders that are likely natives in Denmark).

Williams and Smith's (2007) tally is almost a decade old and we considered it timely to conduct a bibliographic survey to update their list. From this survey, we identified 69 taxa not included on the list (Table 6.1) expanding the number of non-native and cryptogenic seaweed taxa globally to 346. This list includes several cryptogenic taxa that future studies may reclassify as native, but we suggest that the 346 taxa still represent a conservative estimate because (1) our search may have missed studies, (2) seaweeds may have been introduced to new regions prior to modern science, (3) morphologically similar non-native and native species may co-occur and remain taxonomically cryptic, (4) under-studied regions can include non-native taxa that are (5) poorly described, (6) difficult to identify, (7) rare or (8) have invaded inaccessible areas and habitats. The 346 seaweeds are represented by 61 Chlorophycea, 77 Phaeophyceae and 208 Rhodophyceae derived from a global species pool of ca. 2400 Chlorophycea, 1800 Phaeophyceae and 6300 Rhodophyceae. The seaweed genera with most invasive taxa were *Polysiphonia* Greville (15 taxa), *Hypnea* J.V. Lamouroux (13), *Codium* Stackhouse and *Caulerpa* J.V. Lamouroux (11 each) and *Gracilaria* Greville (10) and the most invaded bioregions were the Mediterranean sea (160 taxa), followed by NE Atlantic (93), oceanic islands (73) Australasia (66), NW Atlantic (34) and the NE Pacific (33) (Fig. 6.1).

Finally, we note that in addition to the 346 seaweeds, at least four seagrasses (*Zostera japonica* J.V. Lamouroux, *Zostera tasmanica* Martens ex Ascherson, *Halophila stipulacea* (Forsskål) Ascherson, *H. Halophila decipiens* Ostenfeld), 6 mangroves (*Lumnitzera racemosa* Wild, *Sonneratia caseolaris* (Linnaeus), *Sonneratia apetala* Buch-Ham, *Bruguiera gymnorrhiza* (L.) Lam., *Bruguiera sexangula* (Lour.) Poiret, *Rhizophora mangle* Linnaeus) and 10 salt marsh taxa (*Spartina alterniflora* Loiseleur-Deslongchamps., *S. anglica* C.E. Hubb, *S. patens* (Ait.) Muhl, *S. densiflora* Brögelmann., *S alterniflora* × *foliosa* von Trinius., *S. densiflora* × *foliosa*, *Phragmites australis* (Cav.) von Trinius ex von Steudel., *Elymus athericus* (Link) Kerguélen, *Cotula coronopifolia* Linnaeus, *Ipomoea cairica* (L.) Sweet) have invaded coastlines around the world (Williams 2007; Fourqurean et al. 2010; Chen et al. 2009; Ren et al. 2009; Krauss and Allen 2003; Adam 1990).

Table 6.1 Bibliographic survey of non-native and cryptogenic seaweeds not included in Williams and Smith (2007)

Taxa (Phyla)	Invaded bioregion	Reference
Aglaothamnion cordatum (Børgesen) Feldmann-Mazoyer (R)	Islands	Micael et al. (2014)
Anotrichium okamurae Baldock (R)	Med.	Zenetos et al. (2010)
Antithamnion hubbsii E.Y. Dawson (R)	Med.	Zenetos et al. (2010)
Antithamnionella boergesenii (Cormaci and G. Furnari) Athanasiadis (R)	Islands	Micael et al. (2014)
Asparagopsis taxiformis Delile Trevisan de Siant-Léon Lineage 3 (R)	Multiple, unknown	Dijoux et al. (2014)
Asparagopsis taxiformis Delile Trevisan de Siant-Léon Lineage 4 (R)	NEP	Dijoux et al. (2014)
Botrytella cf. parva (Takamatsu) H.S. Kim (O)	Med.	Zenetos et al. (2010)
Caulerpa racemosa var. [*lamourouxii*] *f. requinii* (Monagne) Weber-van Bosse (C)	Med.	Zenetos et al. (2010)
Caulerpa racemosa turbinate (J. Agardh) Eubank/*uvifera* (J. Agardh) C. Agardh (C)	Med.	Zenetos et al. (2010)
Caulerpa taxifolia distichophylla (Sonder) Verlaque, Huisman and Procaccini (C)	Med.	Jongma et al. (2013)
Caulerpa webbiana Montagne (C)	Islands	Micael et al. (2014)
Ceramium cingulatum Weber-van Bosse (R)	Islands	Micael et al. (2014)
Ceramium gaditanum (Clemente) Cremades (R)	Islands	Micael et al. (2014)
Chondracanthus Kützing sp. (R)	NE Atlantic	Mineur et al. (2012)
Chondria dasyphylla (Woodward) C. Agardh (R)	Islands	Micael et al. (2014)
Cladophora hutchisioides C. Hoek and Wormersley (C)	Med.	Zenetos et al. (2010)
Cladophora ruchingeri (C. Agardh) Kützing (C)	Australasia	Pochon et al. (2015)
Cladophoropsis membranacea (Hofman Bang ex C. Agardh) Børgesen (C)	Islands	Micael et al. (2014)
Cladostephus spongiosus F. *hedwigioides* (Bory de Saint-Vincen) Prud'homme van Reine (O)	NE Pacific	Mazariegos-Villareal et al. (2010)
Codium arabicum Kützing (C)	Med.	Hoffman et al. (2011)
Codium effusum (Rafinesque) Delle Chiaje (C)	Islands	Micael et al. (2014)
Codium parvulum (Bory de Saint Vincent ex Audoin) P.C. Silva (C)	Med.	Zenetos et al. (2010)
Ctenosiphonia hypnoides (Welwitsch ex Agardh) Falkenberg (R)	Islands	Micael et al. (2014)

(continued)

Table 6.1 (continued)

Taxa (Phyla)	Invaded bioregion	Reference
Erythrodermis traillii (Holmes ex Batters) Guiry and Garbary (R)	Islands	Micael et al. (2014)
Erythrotrichia carnea (Dillwyn) J. Agardh (R)	Islands	Micael et al. (2014)
Fredericqia deveauniensis Mags, Le Gall, Mineur, Provan and G.W. Saunders (R)	NW Atlantic, NE Atlantic	Maggs et al. (2013)
Gracilariopsis chorda (Holmes) Ohmi (R)	NE Atlantic	Mineur et al. (2012)
Grateloupia asiatica S. Kawaguchi and H. W. Wang (R)	Med.	Zenetos et al. (2010)
Grateloupia minima P.L. Crouan and H.M. Crouan (R)	Med.	Zenetos et al. (2010)
Grateloupia subpectinata Holmes (R)	Med.	Zenetos et al. (2010)
Grateloupia yinggehaiensis H.W. Wang and R.X. Luan (R)	Med.	Wolf et al. (2014)
Halimeda sp. J.V. Lamouroux (C)	Islands	Sissini et al. (2014)
Haliptilon virgatum (Zanardini) Garbary and H.W. Johansen (R)	Islands	Micael et al. (2014)
Hydropuntia perplexa (K. Byrne and Zuccarello) Conkin (R)	Islands, Australasia	Conklin et al. (2014)
Hypnea anastomosons Papenfuss, Lipkin and P.C. Silva (R)	NE Atlantic	Tsiamis and Verlaque (2011)
Hypnea flagelliformis Greville ex J. Agardh (R)	Islands, Med.	Micael et al. (2014)
Hypnea flexicaulis Y. Yamagishi and M. Masuda (R)	Med.	Wolf et al. (2011)
Hypnea stellulifera (C. Agardh) Kützing (R)	SW Atlantic	de Jesus et al. (2014)
Jania longifurca Zanardini (R)	Islands	(Micael et al. 2014)
Laurencia chondrioides Børgesen (R)	Med.	Hoffman et al. (2014)
Laurencia dendroidea J. Agardh (R)	Islands, Med.	Micael et al. (2014)
Leathesia marina (Lyngbe) Decaisne (O)	Islands	Micael et al. (2014)
Lomentaria flaccida Tanaka (R)	Med.	Zenetos et al. (2010)
Lophosiphonia reptabunda (Suhr) Kylin (R)	Islands	Micael et al. (2014)
Microspongium tenuissimum (Hauck) A.F. Peters (O)	Med.	Zenetos et al. (2010)
Nemalion vermuculare Suringar (R)	Med.	Zenetos et al. (2010)
Neomeris annulata Dickie (C)	Med.	Zenetos et al. (2010)
Nitophyllum stellatocorticatum Okamura (R)	Med.	Zenetos et al. (2010)
Padina antillarum Kützing Piccone (O)	Med.	Zenetos et al. (2010)
Palisada maris-rubri (K.W. Nam and Saito) K.W. Nam (R)	Med.	Zenetos et al. (2012)

(continued)

Table 6.1 (continued)

Taxa (Phyla)	Invaded bioregion	Reference
Papenfussiella kuromo (Yendo) Inagaki (O)	Islands	Micael et al. (2014)
Petalonia binghamiae (J. Agardh) K.L. Vinogradova (O)	Islands	Micael et al. (2014)
Polyopes lancifolius (Harvey) Kawaguchi and Wang (R)	NE Atlantic	(Mineur et al. 2009)
Polysiphonia 'tepida' Holenberg (R)	Islands	Carlton and Eldredge (2015)
Polysiphonia schneideri B. Stuercke and D.W. Freshwater (R)	NE Atlantic	Díaz-Tapia et al. (2013)
Ptilothamnion pluma (Dillwyn) Thuret (R)	Islands	Micael et al. (2014)
Pylaiella littoralis (L.) Kjellman (O)	Med.	Zenetos et al. (2010)
Pyropia suborbiculata (Kjellman) J.E. Sutherland, H.G. Choi, M.S. Hwang and W.A. Nelson (R)	Med., Australasia, Island, NE, SW, NW Atlantic	Verges et al. (2013)
Rugulopteryx okamurae ((E.Y. Dawson) I. K. Hwang, W.J. Lee and H.S. Kim O)	Med.	Verlaque et al. (2009)
Sphacelaria fusca (Hudson) S.F. Gray (O)	Islands	Micael et al. (2014)
Sphacelaria tribuloides Tribuloides (O)	Islands	Micael et al. (2014)
Sphaerotrichia firma (Gepp) A.D. Zinova (O)	Med.	Zenetos et al. (2010)
Spongoclonium caribaeum Børgesen) M. J. Wynne (R)	Islands, Med.	Micael et al. (2014)
Spyridia 'filamentosa' Clade A (Wulfen) Harvey (R)	Islands	Carlton and Eldredge (2015)
Spyridia 'filamentosa' Clade B (R)	Islands	Carlton and Eldredge (2015)
Udotea argentea Zanardini (C)	Islands	Carlton and Eldredge (2015)
Ulva obscura Kützing (C)	Med.	(Zenetos et al. 2010)
Ulva ohnoi M. Hiraoka and S. Shimada (C)	Islands	Carlton and Eldredge (2015)
Uronema marinum Wormersley (C)	Med.	Zenetos et al. (2012)

Phyla are shown in brackets, i.e. Rhodophyta (R), Ocrophyta (O) and Chlorophyta (C). Med. = Mediterranean Sea. Islands = a variety of isolated oceanic islands, e.g. the Azores. We followed Dijoux (2014) on the *Asparagopsis* Montagne complex, assuming that Williams and Smith (2007) included Lineage 2 in their review

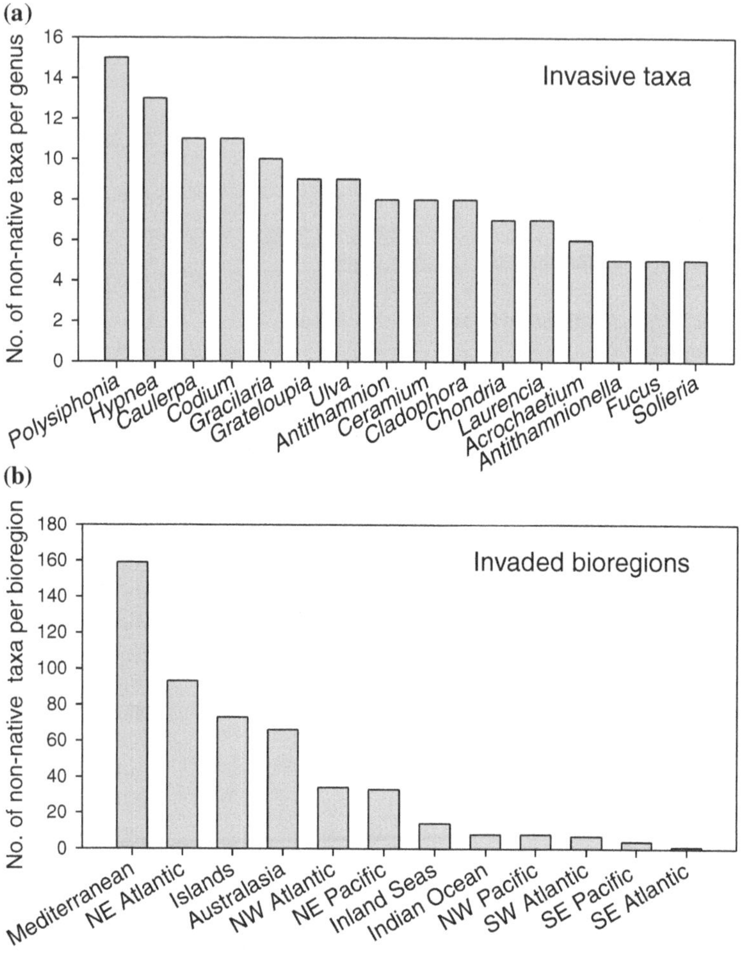

Fig. 6.1 **a** The most invasive genera, compiled from our list of 346 non-native and cryptogenic seaweeds. **b** Regions invaded by most different seaweed taxa [combining our bibliography with that of Williams and Smith (2007)]. The same taxa can invade multiple bioregions and the total number of cases in plot **b** therefore exceeds 346

6.3 How Non-native Seaweeds Get Around

Non-native seaweeds arrive into new regions through intentional or unintentional vectors. Intentional releases are few, perhaps <3 % of reported new incursions (Hewitt et al. 2007), and are mainly associated with aquaculture or, in rare instances, with scientific experimentation (Pickering et al. 2007). New intentional introductions could become even rarer in future because the public, managers, politicians and scientists have all become increasingly aware of the potential impacts that might occur from intentional introductions. However, increased demand for food and seaweed-related products in a growing global population could supersede environmental concerns and lead to more introductions. Furthermore, some scientists advocate 'assisted translocation' (='intentional introduction' in invasion research terminology) of species to new regions, to counter range contractions and extinctions in a warmer future (for discussions for and against, see Seddon 2010; Ricciardi and Simberloff 2009; Sandler 2010). Other researchers may focus on preserving and building 'ecosystem services' (Barbier et al. 2011) and therefore also suggest introductions to add or replace functional traits that are particularly valuable to humans, such as saltmarsh plants, mangroves and oysters that have been introduced to reduce erosion, build land and protect hinterland coastlines.

Most non-native seaweeds are, however, introduced accidentally rather than deliberately. The most important accidental vectors that have facilitated seaweed spread are the aquarium trade as ornamental plants, as 'blind passengers' associated with the intentional introductions, in ballast water and sea chests, attached to ship hulls, and infrastructure breaking down natural barriers such as canal building. Introductions associated with aquarium trade are particularly infamous due the release of *Caulerpa taxifolia* (M. Vahl) C. Agardh into the Mediterranean Sea (Meinesz 1999). However, the aquarium trade has become increasingly regulated with black-listed species, quarantine arrangements and stricter importation rules in many places (Hewitt et al. 2007; Padilla and Williams 2004). Still, non-native species are continually found in both commercial and personal aquaria and consequently pose a future invasion risk (Odom and Walters 2014; Williams et al. 2013). More seaweeds species (>25) have travelled through the Suez Canal, mainly from the Red Sea into the Mediterranean Sea (Williams and Smith 2007). However, there are few analogous marine examples around the world, suggesting that canal building, so far, has been a geographically isolated vector. Still, the future expansion of the Panama Canal could cause a similar spread of non-native species between Atlantic and Pacific bioregions (Ruiz et al. 2009).

Another vector is associated with ballast waters and sea chests (i.e. in waters and sediments held inside ships). Williams and Smith estimated about 25 known seaweed introductions through this vector, and although this vector is considered one of the most important for marine organisms generally, it is of comparatively little importance for seaweeds because of the adverse conditions for photosynthetic macrophytes (dark, nutrient poor, low water movements, high sediment loads) and because most seaweed propagules have relatively short lifespans (but see Flagella et al. (2007, 2010) for contrasting examples). Instead, ballast water is a much more important vector for invertebrates with pelagic larval stages, for sediment-associated fauna, for filter-feeding animals and for species with dormant resistant life stages such as many species of phytoplankton (Davidson and Simkanin 2012; Ibrahim and El-naggar 2012).

By far the most important vectors for seaweed introductions are associated with aquaculture, in particular shellfish transplants, and as hull-fouling organisms. Shellfish transplants have been particularly important in the past, with more than 70 species (mostly corticated rhodophyta) reported to be introduced by this vector (Mineur et al. 2007; Williams and Smith 2007). Oyster transplantations are likely responsible for several high-profile invasions, including those of *Sargassum muticum* (Yendo) Fensholt (Rueness 1989), *Gracilaria vermiculophylla* (Ohmi) Papenfuss (Thomsen et al. 2006a) and *C. fragile* (Suringar) Hariot (Trowbridge 1998). However, methods now exist to kill epibiotic organisms, including seaweeds, while ensuring that the shellfish survive, for example by using short-term bleaching before seeding (Fitridge et al. 2012). Furthermore, increasing use of shellfish cultured in local, land-based facilities, could also lower the importance of this vector in future.

The final major vector is hull fouling, which is responsible for more than 70 seaweed introductions (Hewitt et al. 2007; Williams and Smith 2007; Mineur et al. 2008a). Hull fouling is a particularly important vector for filamentous species (Williams and Smith 2007), although larger species, including high-profile invaders such as *C. fragile* and *Undaria pinnatifida* (Harvey) Suringar likely also have been introduced to many places through hull fouling. Virtually, all seaweeds are capable of hull fouling but survival on ship hulls can be constrained by vessel speed, changes to environmental conditions throughout the voyage (such as changing salinity and temperature), low nutrient levels in offshore waters and the smooth surface of many hulls combined with toxic antifouling paints and coatings (Mineur et al. 2007). Increasing boat activity, both of commercial and recreational vessels, travelled over long or short distances, will likely continue to make this vector important in transporting seaweeds around the globe. Finally, we note that vectors for most non-native seaweeds are poorly described in the scientific literature and even for the better known case studies, described vectors are more like plausible guesses than based on rigorous tests and data (Williams and Smith 2007).

6.4 Success and Impacts of Non-native Seaweeds

After uptake, transport to and release into a new location, the success of a non-native seaweed is measured by how well it survives, grows and reproduces and whether it disperses and expands to nearby habitats, sites and regions. In these 'invasion success' studies, invader attributes are considered dependent variables, thereby contrasting with 'invasion impact' studies where invader attributes are independent variables. This distinction between success and impact is straightforward when interpreting data from manipulative studies but can be blurred for mensurative experiments where key attributes are beyond scientific control. For example, if mensurative data show a negative relationship between *Caulerpa* and seagrass abundances, this can occur because seagrass has a negative effect on *Caulerpa* (a success interpretation), because *Caulerpa* has a negative effect on seagrass (an impact interpretation), or because the relationship may be a spurious effect caused by a third unmeasured factor (Bulleri et al. 2010; Glasby 2012; Klein and Verlaque 2008; Ceccherelli and Campo 2002; Jaubert et al. 1999).

Success studies Success of an invasive seaweed depends on attributes related to the non-native seaweed and the invaded system (Catford et al. 2009). System attributes are typically subdivided into attributes related to abiotic conditions (e.g. salinity, temperature, hydrodynamic forces and desiccation), resources (e.g. space, light, nitrogen and phosphorous) and resident community structures (e.g. the identity and abundance of competitors, facilitators, grazers and other organisms). Many hypotheses have been proposed to explain the invasion success (or lack of) non-native species including seaweeds (Mitchell et al. 2006; Catford et al. 2009; Valentine et al. 2007; Alpert 2006). At least 19 'core' hypotheses are commonly cited, which can be grouped into three clusters that highlight different aspects of invasion success. We use Catford et al.'s hypothesis nomenclature (cf. Table 3 in Catford et al. 2009) but excluded hypotheses that combined multiple core hypotheses ('Darwin's naturalisation', 'Global competition', 'resource-enemy release', 'indirect effects', 'opportunity windows'), that just rebranded other core hypotheses ('adaptation', 'enemy reduction') or that predicted success and failure at the same time and therefore are difficult to test ('enemy inversion', 'dynamic equilibrium model', 'new associations'). We also added 'facilitation- and habitat-cascades' to the list (not included in Catford et al.'s review), because non-native seaweeds often are key players in these ecological interactions (Thomsen et al. 2010).

The first hypothesis cluster proposes that invasion success is determined *by attributes associated with the non-native seaweed.* This cluster includes the 'ideal weed' (Baker and Stebbens 1965), 'propagule pressure' (Lockwood et al. 2009) and 'sampling' (Crawley et al. 1999) hypotheses. The ideal weed hypothesis suggests that certain traits are universally important to successful invaders, particularly *r*-selected traits such as fast growth, high reproductive output and high dispersal capacity. However, there are few consistent patterns for the most successful invasive seaweeds, except perhaps for *U. pinnatifida* which has many weedy traits (Valentine et al. 2007). By comparison, propagule pressure and sampling hypotheses are 'numbers game' hypotheses that suggest that the more propagules (propagule pressure hypothesis, species traits are less important) or the more different species (sampling hypothesis, species traits are more important) that are introduced to a new region, the greater the chance that one of these propagules will establish a self-sustaining population (Britton-Simmons and Abbott 2008; Vaz-Pinto et al. 2012).

The second cluster emphasises that *attributes associated with the invaded system*, such as resource levels and local abiotic conditions, can explain invasion success. These core hypotheses highlight that high 'fluctuating resource availability' (Davis et al. 2000), high 'environmental heterogeneity' (Melbourne et al. 2007), medium-to-high 'disturbances' (Sher and Hyatt 1999), availability of 'empty niches' (Hierro et al. 2005) and low 'habitat filtering' (Procheş et al. 2008) all increase the likelihood that non-native species can establish populations following arrival. These hypotheses suggest that almost any non-native seaweed can be successful in systems that have a plethora of microhabitats, are heterogeneous, are frequently disturbed and have unused resources in space and time. Several of these processes have been suggested to be important for high-profile seaweed invaders including *U. pinnatifida, S. muticum* and *Caulerpa racemosa* (Forsskål) J. Agardh (Valentine et al. 2007; Incera et al. 2010; Olabarria and Arenas 2014).

The final cluster emphasises that *biological interactions, in particular lack of co-evolution with local species,* can explain success of non-native species. In this cluster, 'novel weapons' (Callaway and Ridenour 2004), 'enemy release' (Keane and Crawley 2002), 'evolution of increased competitive ability' (Blossey and Notzold 1995), 'enemy of my enemy' (Eppinga et al. 2006), 'limited similarity' (Emery 2007), 'specialist generalist' (Sax et al. 2007), 'invasional meltdown' (Simberloff and Von Holle 1999) and 'cascading habitat formation' (Thomsen et al. 2010) highlight how non-native species become invasive, whereas 'biotic resistance' (Elton 1958), 'missed mutualisms' (Mitchell et al. 2006) and 'increased susceptibility' (Colautti et al. 2004) highlight why non-native species can fail to become invasive.

We are not aware of any research that has compared the relative merits of all these competing and overlapping core hypotheses for non-native seaweeds. However, we speculate that hypotheses assuming that seaweeds are limited by tight co-evolutionary processes at their place of origin (and lack of co-evolution at the new place) are *less* important for seaweeds, because seaweed–animal interactions are, compared to terrestrial systems, dominated by generalist-type interactions (Bell 1991; Enge et al. 2013; Hay and Steinberg 1992). Importantly, we hope future studies will test for interactions between multiple core hypotheses because we expect that many mechanisms exert selection pressure simultaneously and because mechanisms likely change during the life cycle of non-native seaweeds (Britton-Simmons 2006). For example, *U. pinnatifida* is well adapted to attach to ship hulls and spread in large numbers (~propagule pressure hypothesis) (Hay 1990) but has also many weedy traits, like fast growth and high reproductive output (~ideal weed hypothesis) (Dean and Hurd 2007; Schiel and Thompson 2012). *U. pinnatifida* is also successful in settling onto unoccupied microhabitat space and on biogenic substrata (~environmental heterogeneity hypothesis) (Schiel and Thompson 2012; Thompson and Schiel 2012), often becomes dominant following disturbances (~disturbance hypothesis) (Valentine and Johnson 2003, 2004) and is efficient in converting available resources into biomass (~fluctuating resource hypothesis) (Tait et al. 2015). Finally, *U. pinnatifida* is, of course, also constrained by physiological tolerance levels (~habitat-filtering hypothesis) (Morita et al. 2003; Peteiro and Sánchez 2012) even if *Undaria* colonises sites outside its native environmental range (James et al. 2014). In short, it is important that future studies test multiple core hypotheses simultaneously, rather than a single hypothesis at a time (Britton-Simmons 2006; Olabarria and Arenas 2014; Valentine et al. 2007; Williams and Smith 2007).

Impact studies In invasion impact studies, the non-native species is considered the causal agent of ecological change. Impact is therefore a synonym for 'effect', 'consequence' or 'cause'. Impact studies focus on how invaders affect a particular property of a resident system, and the impact on this property can then be larger or smaller than a reference value (often the non-impacted reference value is defined as zero). Impact can therefore be positive (>0), neutral (0) or negative (<0), which is a statistical measure that differs from the anthropocentric value judgement of whether an impact is 'good' or 'bad'. Impacts can be reported on cultural (economics, health, societal) or natural (biotic, abiotic) properties. Biotic properties can be divided into impacts reported on or above the species level; importantly, impacts reported above the species level can hide opposing effects if some species benefit and others are harmed. Invasion impacts, like successes, also depend on attributes of the non-native seaweeds and attributes associated with the biotic community, resource levels and abiotic conditions of the invaded system (Thomsen et al. 2011a). A highly cited impact formula suggests that the impact from a non-native seaweed is proportional to its abundance, its range (distribution) and its per capita effect (Parker et al. 1999), i.e. what it does to targeted response variables (in the context of its abundance and distribution). This framework highlights that processes which determine invasion success (cf. core hypotheses outlined in the previous section) also modify impact, partly by controlling the distribution and abundance of the non-native seaweed.

6.5 Common Types of Ecological Impacts

The impacts of invasive seaweeds have been reviewed in some detail (Williams 2007; Schaffelke and Hewitt 2007; Thomsen et al. 2009b; Williams and Smith 2007), concluding that impacts have been quantified from a fraction (<10 %) of known taxa, that impacts on native species can be strong and that underlying impact mechanisms are poorly known. Each of the 346 non-native and cryptogenic seaweeds have some level of impact on the invaded communities, contributing their biomass and adding new genotypes to the local species pools, modifying geochemical cycles through metabolic activities and by affecting other species in the community through ecological processes. However, impacts can be subtle or difficult to quantify, particularly if the non-native species is cryptic, if interactions with native species are few and weak, or, if strong interactions only occur in small areas, in short time windows, or in inaccessible habitats. Below we provide examples of ecological impacts grouped as 'negative' or 'positive', occurring through direct or indirect interactions (in indirect effects, an intermediate species is required for the impact to manifest).

Negative effects Many negative effects by non-native seaweeds occur through competition and habitat destruction (direct effects) or keystone competition and keystone habitat destruction (indirect effects). For example, non-native canopy-forming seaweeds, like *S. muticum*, *U. pinnatifida* and *C. fragile* sometimes have strong competitive effects on native seaweeds, competing for limiting resources such as nutrients, light and space (Staehr et al. 2000; Ambrose and Nelson 1982; Britton-Simmons 2004; Schmidt and Scheibling 2006; Levin et al. 2002; Casas et al. 2004). However, these negative effects appear to be weaker in intertidal areas (Sánchez and Fernández 2005; Forrest and Taylor 2003; South et al. 2015; Olabarria et al. 2009), perhaps because desiccation and super-saturation of light are strong stressors here. Furthermore, some of these canopy-forming non-native seaweeds shed much of their thalli (*S. muticum* and *C. fragile*) or the entire thallus (*U. pinnatifida*) (Schiel and Thompson 2012; Wernberg et al. 2001), and effects associated with a canopy cover therefore could only occur seasonally (South et al. 2015). Similar mechanisms have been observed in soft-bottom systems where invasive *Caulerpa* species can compete with native seagrasses (Ceccherelli and Cinelli 1997; de Villele and Verlaque 1995), although other studies suggest that competition from *Caulerpa* is relatively small (Thomsen et al. 2012; Jaubert et al. 1999, 2003; Ceccherelli and Sechi 2002). Non-native seaweeds can also have negative effects on native organisms through modification or destruction of habitats, for example if they convert non-vegetated mud or sand habitats to vegetated meadows. Well-documented examples include *G. vermiculophylla* (Byers et al. 2012; Thomsen et al. 2007, 2010), *Caulerpa* species (Gribben and Wright 2006; Wright and Gribben 2008; Gribben et al. 2009b; McKinnon et al. 2009; Byers et al. 2010; Pacciardi et al. 2011), *S. muticum* (Strong et al. 2006), and invasive seagrasses (Willette and Ambrose 2009, 2012; Baldwin and Lovvorn 1994; Posey 1988; Berkenbusch et al. 2007; Ruesink et al. 2010) and salt marsh plants and mangroves (Demopoulos and Smith 2010; Thomsen et al. 2009a; Wu et al. 2009; Neira et al. 2007). In these examples, local species that depend on sand and mud, such as many infaunal species and burrowing fish, may be negatively affected by the non-native macrophytes (Wright et al. 2007; Gribben et al. 2009b; Tsai et al. 2010), although in some cases, these organisms can also survive under, around or on the macrophytes (Gribben and Wright 2006; Wright and Gribben 2008; Gribben et al. 2009b; McKinnon et al. 2009; Byers et al. 2010; Klein and Verlaque 2011).

Indirect negative effects from invasive seaweeds can occur through keystone competition where an invader reduces a resource that is important for other species within that community. For example, syngnathid and monacanthid fish were more abundant in native seagrass compared to invasive *Caulerpa* meadows (York et al. 2006), juvenile fish were more abundant in native seagrass beds than in the invasive seagrass *H. stipulacea* (Willette and Ambrose 2012) and gastropods and echinoderms were more abundant on native kelp compared to invasive *C. fragile* (Schmidt and Scheibling 2006). These fish, gastropods and echinoderms were probably less abundant around the non-native macrophytes (than around the native macrophytes), because these non-native species provide low-quality foraging grounds and poor protection against predators. In short, if these non-native macrophytes do indeed reduce the abundance of native seagrass and kelps, then they will have indirect negative effects on the same invertebrates and fish (=keystone competition).

Positive effects Invasive seaweeds can also have positive effects on local populations, directly through habitat formation and modification, and by being consumed by grazers, or indirectly through cascading habitat formation, consumption, competition or keystone consumption.

For example, seaweeds invading un-vegetated sediments provide habitat for epibenthic fauna (Thomsen and Wernberg 2015) that live among or on seaweeds, including *C. racemosa* (Klein and Verlaque 2011), *C. Taxifolia* (M. Vahl) C. Agardh (McKinnon et al. 2009) and *G. vermiculophylla* (Byers et al. 2012; Thomsen et al. 2010; Johnston and Lipcius 2012). Similar positive effects on fauna have also been shown for invasive seagrasses such as *H. decipiens* (Willette and Ambrose 2012) and *Z. japonica* Ascherson and Graebner in Engler (Berkenbusch and Rowden 2007; Berkenbusch et al. 2007; Posey 1988) when and where they colonise un-vegetated sediments. Many invasive seaweeds also add structure, biomass and productivity to vegetated rocky coasts (Thomsen et al. 2015). For example, the canopy-forming *S. muticum* and *C. fragile* provide additional habitat for epiphytic plants and invertebrates (Thomsen et al. 2006b; Jones and Thornber 2010; Schmidt and Scheibling 2006, 2007; Wernberg et al. 2004). Invasive seaweeds not only provide habitat, but can also be a food source, and indirectly fuel higher order consumers through trophic cascades. For example, juvenile invasive seaweeds (and seagrass) can provide a seasonal food supply (Sjotun et al. 2007; Thornber et al. 2004; Reynolds et al. 2012): *Littorina* Férussac snails consume *C. fragile* in tide pools (Scheibling et al. 2008), waterfowl graze invasive *Z. japonica* (Baldwin and Lovvorn 1994), and siphonalian seaweeds (*Codium* and *Caulerpa* spp.) are likely to have increased food availability for specialist saccoglossan grazers (Harris and Jones 2005; Trowbridge 2002; Trowbridge and Todd 2001). Still, most invasive seaweeds are not considered preferred food for native larger generalist grazers and smaller meso-grazers and many produce grazer-deterring secondary metabolites, as shown for *S. muticum* (Britton-Simmons 2004; Engelen et al. 2011; Monteiro et al. 2009; Pedersen et al. 2005), *C. fragile* (Scheibling and Anthony 2001), *Caulerpa* species (Gollan and Wright 2006; Boudouresque et al. 1996), *G. vermiculophylla* (Nejrup and Pedersen 2010; Thomsen and McGlathery 2007; Nejrup et al. 2012; Rempt et al. 2012; Nylund et al. 2011) and *Lophocladia lallemandii* (Montagne) F. Schmitz and *Womersleyella setacea* (Hollenberg) R.E. Norris (Cebrian et al. 2011; Tomas et al. 2011a, b).

Invasive marine plants can also have positive indirect effects, for example through cascading habitat formation (Thomsen et al. 2010): invasive *S. muticum, C. fragile* and *G. vermiculophylla* provide primary habitat for sessile plants and animals, such as filamentous seaweeds and ascidians (Nyberg et al. 2009; Thomsen et al. 2006b, 2010; Engelen et al. 2013; Gestoso et al. 2012; Jones and Thornber 2010; Schmidt and Scheibling 2006, 2007; Wernberg et al. 2004). These epiphytic macro-organisms are likely to subsequently provide a secondary habitat for many other smaller organisms, such as hydrozoa, bryozoa and small mobile invertebrates. Invasive plants can also have indirect positive effects through cascading habitat modification. For example, *C. taxifolia* can reduce sediment redox potential, altering abiotic conditions and forcing infaunal bivalves to live at the sediment surface where they are exposed to sessile fouling organisms (Gribben et al. 2009a). Finally, positive effects of seaweed invasions can also be mediated through cascading consumption (trophic cascades), such as when invasive *C. fragile* is consumed by (invasive) *Littorina* snails (Scheibling et al. 2008), which are consumed by (invasive) crabs (Eastwood et al. 2007; Trussell et al. 2002, 2004), which are consumed by native crabs, seabirds and fish (de Rivera et al. 2005).

6.6 Impacts Reviewed Across Studies, Seaweeds and Habitats

The many case studies about impacts from invasive seaweeds (cf. last section) have stimulated quantitative meta-analyses that aim to identify impact generalities across different non-native seaweed species, invaded communities and environmental conditions. In these studies, documented effects are standardised, typically with the Hedges *d* effect size or a log-response ratio (Borenstein et al. 2009), so that effects can be compared between studies. Here, we summarise the main finding from meta-analyses that have included non-native seaweeds as a test factor.

The first meta-analysis of seaweed invasion impacts reviewed field-based impact experiments and showed that invasive seaweeds generally had significant negative effects on local plant abundances, richness and diversity (Thomsen et al. 2009b). By contrast, there were no significant effects on seaweed 'processes' (e.g. growth, photosynthesis, respiration) or the abundance, richness and diversity of animal communities. However, the latter analysis was based on a small sample size and had large confidence limits. A follow-up analysis confirmed that non-native seaweeds, across the reviewed studies, have negative effects on diversity of native seaweeds (Thomsen et al. 2014). This study also tested whether trophic and functional 'matching' (i.e. pairing the trophic level or function of the invader *and* the impacted organism) could provide first-order predictions about impacts on diversity metrics. It was hypothesised that the net effect from non-native seaweeds on diversity would be negative within a trophic level but zero (or even positive) across trophic levels, based on the assumption that competition processes would dominate in the community within trophic levels (in plant–plant interactions) and habitat formation, habitat-modification and food-provisioning processes would dominate across trophic levels (in plant–animal interactions). The hypothesis was confirmed, showing negative effects of non-native seaweeds on the diversity of local seaweed communities, but with positive effects on local animal communities. Furthermore, a recent analysis reconfirmed that non-native seaweeds generally have negative effects on species within the same trophic levels (Maggi et al. 2015), although this study found no effects on species at higher trophic levels (effect size on animals was not statistically different from zero). The discrepancies between these meta-analyses (positive vs. no effect on animal communities) were likely due to there being different study-inclusion criteria. Maggi et al. (2015) included, in contrast to Thomsen et al. (2014), mensurative experiments, laboratory experiments and litter-bag experiments where the invader was dead and decomposing (Rodil et al. 2008; Taylor et al. 2010). These two analyses therefore highlight that readers should be careful when extrapolating meta-analytical results beyond the domain of reviewed primary studies.

Three other meta-analyses have examined aspects of seaweed invasions. An analysis of seaweed impacts on seagrasses, measured in field and laboratory experiments, showed that non-native seaweeds have a less negative impact than native seaweeds (Thomsen et al. 2012). However, this analysis was confounded by the taxonomic status and attachment type of the non-native seaweeds, which were dominated by studies on *Caulerpa* species, a special group of clonal seaweeds attached with rhizomes to sediments. These types of non-native seaweeds have few documented cases of negative effects on seagrasses compared to many native seaweed taxa that are found in seagrass beds entangled around stems or epiphytic on leaves.

More recently, Tamburello et al. (2015) used a subset of the data analysed in Maggi et al. (2015), to test whether impacts co-vary with Halpern's (2008) 'global cumulative human impact index'. In other words, they tested whether seaweed invasion impacts increase or decrease with impacts from human activities (Halpern's index merges a range of human activities including pollution, fisheries, climate changes and eutrophication). This analysis showed that impacts from non-native seaweeds on community biomass and abundance become less negative or even neutral when moving from relatively pristine to heavily impacted environments (but the opposite trend was found on community 'evenness'). Thus, invasion impact appears to co-vary with other human stressors, being greater in stressed systems (presumably with lower resilience). Finally, Thomsen et al. (2015) conducted a meta-analysis to test whether invasion impacts on local plant communities differ when the invader itself was included as part of the total community. This analysis showed again that seaweed invaders have significant negative impacts on plant abundance and plant community richness, but also that these negative effects were cancelled out when the taxonomic status and abundance of the invader were included in the calculations of total community biomass and richness (i.e. the effect size was no longer significantly different from zero). Thus, non-native seaweeds appear, across species and invaded systems, to substitute, rather than increase or decrease, standing biomass and richness.

These meta-analyses all suggested that more impact studies are needed, especially on functional community responses, such as total system productivity, respiration, decomposition rates and nutrient uptake (Altieri et al. 2009; Green et al. 2012, 2013; Tait et al. 2015; Cacabelos et al. 2012; South et al. 2015). Furthermore, although the impact literature is extensive, the vast majority of case studies have tested for impacts of six high-profile invaders (*S. muticum*, *U. pinnatifida*, *C. taxifolia*, *C. racemosa*, *C. fragile* and *G. vermiculophylla*) likely biasing tests towards detecting strong impacts. It is critical that comprehensive impact analyses are conducted on more of the non-native seaweeds that have colonised coastlines around the world, to allow for better generalisations about seaweed invasions across habitats, bioregions and species assemblages.

6.7 New Meta-analysis; Impact Is Density-dependent and Non-native Seaweeds Affect Community Similarity

Two key aspects related to seaweed invasion impacts have not yet been tested with meta-analyses: density dependency and impacts on community structure measured by multivariate metrics. Below we address these research gaps.

6.7.1 New Meta-analysis 1: Density-dependent Effects

We tested whether impacts from non-native seaweeds are density dependent (density is defined loosely to include any abundance metric such as counts, length, coverage, biomass or volume); i.e. whether effects vary with the amount of an invasive seaweed found in a plot, site and region. We identified peer-reviewed papers describing manipulative field experiments in which invader abundance was controlled using addition or removal techniques with replicated treatments and controls. We included only experiments reporting impacts from at least three invasion densities (one of which could be a non-invaded control) so that paired effects within a study could be contrasted between low and high density treatments (Table 6.2).

We used Hedges effect size *d*, corrected for small sample sizes, to standardise effects between treatments, allowing us to include zero-value responses (Borenstein et al. 2009). An all-inclusive unbiased data selection criterion (Englund et al. 1999) was used to extract effect sizes for each paper, calculating *d*-values for *all* reported impacted resident organisms, test combinations (e.g. different depth levels Thomsen 2010) and quantified responses also on the same resident organism, such as seagrass leaf length, above ground biomass and below ground biomass (Drouin et al. 2012). One study presented data as *x–y* points on a graph of abundance of the seaweed versus fauna (because applied densities changed over time) (Byers et al. 2012). We extracted these *x–y* data and reclassified data into low (control), medium and high seaweed densities. To avoid problems associated with temporal autocorrelation, we included only the last data points from repeated measure experiments (Parker et al. 2006). Following the calculations of Hedges *d* between controls and invaded treatments, we calculated a 'Δd' for each paired response (responses are paired because the same un-invaded control data are used to calculate *d* for both the low and high density treatments) (Thomsen et al. 2011b). If more than three invasion densities were tested, we contrasted only the effects between the lowest and highest densities. Nested and orthogonal experiments within research papers were treated as independent studies. Two tests examined whether effect sizes depended on the density of the non-native seaweeds. In the first, we examined whether paired *d*-values were different, using the formula $\Delta d = |d_{\text{high}} - d_{\text{low}}|$. In the second, more conservative, test, we examined whether the high density *d* was numerically larger

Table 6.2 Studies used in our meta-analyses

Taxa	Data from figure	Reference
Caulerpa racemosa (Forsskål) J. Agardh, *C. taxifolia* (M. Vahl) C. Agardh	4	Balata et al. (2004)
Lophocladia lallemandii (Montagne) F. Scmitz	3	Bedini et al. (2014)
Zostera japonica Aschershon and Graebner	2c–d	Berkenbusch and Rowden (2007)
Zostera japonica	5b	Berkenbusch et al. (2007)
Caulerpa racemosa	4	Box et al. (2010)
Sargassum muticum (Yendo) Fensholt	2	Buschbaum et al. (2006)
Gracilaria vermiculophylla (Ohmi) Papenfuss	4c, 5c, 6c	Byers et al. (2012)*
Sargassum muticum	3a	Cacabelos et al. (2010)
Codium fragile (Suringar) Hariot	3a–b, 4a–c	Drouin et al. (2012)*
Multiple (*Eucheuma denticulatum* (N.L. Burman F.S. Collins and Hervey + *Kappaphycus alvarezii*)	4a–b	Eklöf et al. (2005)
Caulerpa scalpeliformis	9	Falcao and de Szechy (2005)
Undaria pinnatifida (Harvey) Suringar	5	Forrest and Taylor (2003)
Multiple (*Gracilaria salicornia* (C. Agardh) E.Y. Dawson + *Acanthophora spicifera* (M. Vahl) Börgesen)	4	Fukunaga et al. (2014)
Caulerpa taxifolia	6a–b	Gallucci et al. (2012)
Caulerpa racemosa	2	Gennaro and Piazzi (2011)
Sargassum muticum	2a, 4	Gestoso et al. (2010)
Caulerpa taxifolia	4	Gribben et al. (2013)
Asparagopsis armata (Harvey)	3	Guerra-García et al. (2012)
Gracilaria vermiculophylla	1–2	Hammann et al. (2013)*
Sargassum muticum	4a–c, 9a–c	Harries et al. (2007)
Undaria pinnatifida	Table 6.1	Irigoyen et al. (2011)

(continued)

Table 6.2 (continued)

Taxa	Data from figure	Reference
Caulerpa racemosa	4	Klein and Verlaque (2011)
Sargassum muticum	3a, 3c	Lang and Buschbaum (2010)
Sargassum muticum	4a–b, 5a–c	Lang and Buschbaum (2010)*
Codium fragile	2a–c	Lutz et al. (2010)
Caulerpa taxifolia	6a	Mateu-Vicens et al. (2010)
Caulerpa taxifolia	2a–b	McKinnon et al. (2009)
Multiple (large group)	3	Mineur et al. (2008b)
Sargassum muticum	3j	Olabarria et al. (2009)
Caulerpa racemosa	2, 5	Pacciardi et al. (2011)
Caulerpa racemosa, Womersleyella setacea (Hollenberg) R.E. Norris	3	Piazzi and Balata (2009)
Caulerpa racemosa	2	(Piazzi and Balata 2008)
Caulerpa racemosa, C. taxifolia, Multiple (*Acrothamnion preissii* (Sonder) E.M. Wollaston + *Womersleyella setacea*)	4a–c	Piazzi and Cinelli (2003)
Acrothamnion preissii, Womersleyella setacea, Multiple (*Acrothamnion preissii* + *Womersleyella setacea*)	5a–b	Piazzi et al. (2002)
Caulerpa racemosa, C. taxifolia, Multiple (*C. racemosa* + *C. taxifolia*)	3	Piazzi et al. (2003)
Caulerpa racemosa	2	Piazzi et al. (2005)
Caulerpa racemosa	2	Piazzi and Ceccherelli (2006)
Sargassum muticum	2	Sánchez et al. (2005)
Sargassum muticum	6, 3a4, 3b4	Sánchez and Fernández (2005)
Codium fragile	5a–b	Schmidt and Scheibling (2006)
Caulacanthus ustulatus (Mertens ex Turner) Kützing	1a, 1c, 1e, 2a–f	Smith et al. (2014)
Undaria pinnatifida	Raw data	South et al. (2015)
Sargassum muticum	5b	Staehr et al. (2000)
Sargassum muticum	2a–b	Strong et al. (2006)

(continued)

Table 6.2 (continued)

Taxa	Data from figure	Reference
Gracilaria vermiculophylla	Raw data	Thomsen and McGlathery (2006)
Gracilaria vermiculophylla	Raw data	Thomsen (2010)
Gracilaria vermiculophylla	1a–h	Thomsen (2010)*
Gracilaria vermiculophylla	Raw data	Thomsen et al. (2010)
Caulerpa racemosa	4	Vázquez-Luis et al. (2012)
Sargassum muticum	2a–d	Wernberg et al. (2004)
Sargassum muticum	4	White and Shurin (2011)*
Caulerpa taxifolia	2a–d	York et al. (2006)

References with asterisks were used for meta-analysis of density effects; all other references were used for meta-analysis of multivariate dispersion effects (i.e. showing MDS plots or where we have access to raw data). Note that multiple effect sizes could often be calculated from a single figure (e.g. as different MDI values for different sites)

than its paired low density *d*, using the formula $\Delta d = |d_{high}| - |d_{low}|$. Analysing numerical Δd-values is necessary to ensure that density-dependent facilitation is not cancelled out by density-dependent inhibition. Non-independent Δd-values from a study were averaged to produce independent Δd-values, using equal weight for each reported type of impact. Finally, un-weighted fixed effects analyses were made on the independent Δd-values in Metawin 2.1, to calculate $\Delta d_{cumulative}$ and 95 % bias-corrected confidence limits (CL) with 999 permutations (Rosenberg et al. 2000). $\Delta d_{cumulative}$ was interpreted to be significantly different from zero if the 95 % CL did not overlap zero.

In the first test, overall heterogeneity of effect sizes was small ($Q_{total} = 1.49$, df = 9, $p = 0.99$), indicating that effect sizes share a common value. The overall $\Delta d_{cumulative}$ was 0.61 and the 95 % bias-corrected confidence limits did not bracket zero (0.40–0.84) indicating that d_{high} was indeed *different* from d_{low}. In the second test, overall heterogeneity of effect sizes was again small ($Q_{total} = 1.29$, df = 9, $p = 0.99$). This $\Delta d_{cumulative}$ was slightly lower (0.52) and the 95 % bias-corrected confidence limits again did not bracket zero (0.31–0.74) highlighting that d_{high} had a *larger magnitude* than d_{low}. We conclude, therefore, based on the few studies that have tested for density-dependent effects, that seaweed invasion impacts, as a general rule, are density dependent. However, more studies are needed to enable specific analyses, for example, to identify non-linear density dependency, changes from positive to negative effect sizes, to locate thresholds and to make regression models that can predict which non-native seaweeds have particular strong density-dependent impacts. This analysis also highlights that *all* seaweed invasion impact studies should report the abundance of the studied invader.

6.7.2 New Meta-analysis 2: Effects on Community Structures

Non-native seaweeds have strong effects on population abundances, growth, survival and community diversity and richness (all univariate metrics, see previous sections). However, non-native seaweeds can also affect multivariate community structure, often assessed through multivariate analysis of variance (Manova, Permanova) and multivariate regressions, and visualised on dimensional reduction methods such as principal coordinates analysis (PCA, PCO) or multidimensional scaling (MDS) (Viejo 1999; Staehr et al. 2000; Piazzi and Balata 2008). These methods typically test whether multivariate 'centroids' or 'dispersion' differ between invaded and non-invaded communities. The 'centroid-analysis' is non-directional in that it tests whether the 'mean' invaded community is different from the 'mean' non-invaded community. However, the 'dispersion-analysis' is directional, with several well-known examples suggesting that invaded communities are *less dispersed* in multivariate space (i.e. are more 'homogenous') than non-invaded communities (Staehr et al. 2000; Piazzi and Balata 2008). Unfortunately, invasion studies do not usually publish the entire species-sample data matrix or the associated sample similarity matrix from which multivariate dispersions can be calculated. Instead, multivariate invasion impact is typically visualised with 2D plots, where each x–y point represents a sample that can contain many species. The spatial distances between these sample points correlate with how similar the communities are, and invaded samples will be clustered compared to non-invaded samples if the invader increases community similarity in space (e.g. Fig. 2, Piazzi and Balata 2008) or over time (e.g. Fig. 5b, Staehr et al. 2000).

Here, we develop a method to test whether non-native seaweeds increase sample similarity across studies, by extracting *x–y* coordinates from published MDS plots. From these coordinates, sample matrices can be reconstructed, composed of two universal 'pseudo-species' from which the (imperfect) sample similarity matrices can be calculated. Reconstructed similarity matrices were then used to calculate Multivariate Dispersion Index values to test whether invasive seaweeds, across studies, increased between-sample similarities. These 'MDI' values have a minimum of −1 (dissimilarities among invaded samples are lower than any dissimilarities among non-invaded samples = non-native seaweeds 'homogenise' local communities) and a maximum of 1 (the opposite case = non-native seaweeds make local communities more heterogeneous (Warwick and Clarke 1993).

We located peer-reviewed studies that compared samples from invaded and non-invaded communities in 2D MDS plots (Table 6.2) by searching Google Scholar and by back-tracking references in past reviews (Williams 2007; Thomsen et al. 2009b, 2011b, 2014; Schaffelke and Hewitt 2007; Maggi et al. 2015; Williams and Smith 2007). We included a few studies that did not show MDS plots for which we had access to raw data and therefore could calculate the sample similarity matrices (using common methods; Bray–Curtis similarity coefficient, square root transformed data and excluding the abundance of the invasive species) (Thomsen et al. 2010; Thomsen and McGlathery 2006; South et al. 2015). MDS plots were imported into TechDig and rescaled from 0 to 1 for the longest axis. Each sample was classified as invaded or non-invaded and its *x*–*y* coordinates extracted. A few samples overlapped on the plots, making it difficult to extract all samples (but >90 % of all samples were extracted from each plot). Extracted *x*–*y* coordinates represent the relative abundance of two 'universal pseudo-species' in the invaded versus non-invaded samples. Paired similarity matrices between the invaded and non-invaded samples were reconstructed from the two-species communities using Euclidean distances for untransformed data. This method provides a perfect spatial fit between the abundance of the pseudo-species and the published MDS plots they were derived from (see Fig. 6.2 for examples). For each plot, we also extracted data to test whether MDI values depended on whether (1) the impacted community was

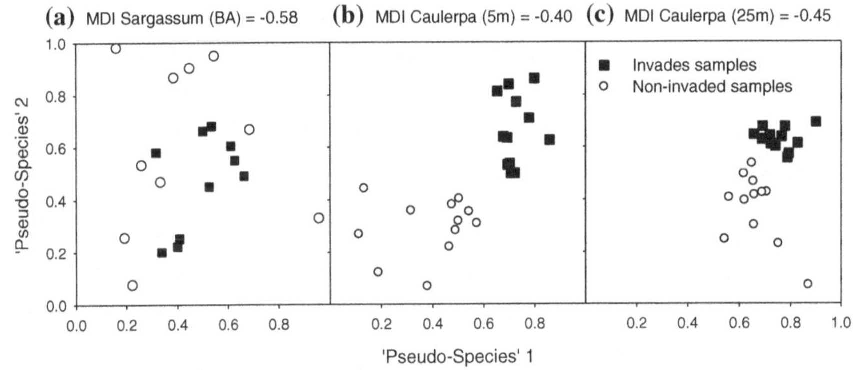

Fig. 6.2 Relative abundance of universal 'pseudo-species' 1 and 2 in invaded (*black squares*) and non-invaded (*white circles*) samples, reconstructed from two MDS plots; Fig. 5 in Staehr et al. (2000) and Fig. 2 in Piazzi and Balata (2008). The MDS plots were imported into TechDig and rescaled from 0 to 1 for the longest axis. Each sample was classified as invaded or non-invaded and its coordinates extracted (representing the relative abundance of universal 'pseudo-species 1 and 2'). Similarity matrices were calculated for invaded and non-invaded samples using Euclidean distances based on the relative abundances of the two pseudo-species. Finally, MDI values were calculated for each set of paired similarity matrices (Warwick and Clarke 1993). The negative MDI values shown above each plot indicate invasion-driven 'homogenisation', i.e. invaded samples are spatially clustered both in time (**a** before vs. after invasion) and space (**b** at 5 m depth, **c** at 25 m depth)

composed of plants, animals or both (Thomsen et al. 2014), (2) the impacted community was composed of sessile, mobile organisms or both (Thomsen et al. 2014), (3) the field method was based on a manipulative or mensurative experiment (Maggi et al. 2015) and (4) analysis included (Staehr et al. 2000; Klein and Verlaque 2011) or excluded (Balata et al. 2004) the abundance of the invader in the similarity matrix. We finally calculated a MDI value for each reconstructed paired invaded versus non-invaded similarity matrix (Warwick and Clarke 1993).

Meta-analytical methods followed the previous density analysis; factorial and nested experiments were treated as independent data, as were effects from the same invasive species reported in different studies. Only the last data point was used from repeated measure data and only effects that compared the highest invader density to non-invaded samples were used from multidensity experiments. Invasion effects reported on different communities within a single study (e.g. on endobenthic and epibenthic communities, Lang and Buschbaum 2010) were also treated as independent effects. Cumulative effect sizes ($MDI_{cummulative}$) with 95 % bias-corrected CL were calculated in Metawin 2.1 with 999 permutations (Rosenberg et al. 2000). $MDI_{cummulative}$ was interpreted to be significantly different from zero or another $MDI_{cummulative}$ if the 95 % CL did not overlap zero or each other, respectively.

We found no significant effects, i.e. treatments were not significantly different from zero or each other (Fig. 6.3a–d). Still, there were several interesting (non-significant) trends, indicating that effect sizes were larger and more negative *within* rather than *between* trophic or functional groups (Fig. 6.3a, b). Perhaps with more studies, and smaller confidence limits, it will become clearer whether non-native seaweeds, as a general rule, homogenise 'similar' invaded plant communities but not 'different' invaded animal communities. Our results also highlight that conclusions based on a few well-cited studies (Staehr et al. 2000; Piazzi and Balata 2008) should only cautiously be interpreted out of the context of the methods, invasive species, invaded community and surrounding abiotic environment. We expect analogous multivariate directional tests in future will have stronger predictive power, when (1) more studies report impacts on different types of communities, (2) more test factors and levels within factors are included in tests (e.g. on form-groups such as epiphytes, canopies and understory communities) and (3) tests also include interactions between factors (e.g. between form-groups and field and data collection methods). Finally, future studies should also test how robust our 'dispersion meta-analysis' is by modelling how much dispersion values can vary in direction and magnitude depending on MDS stress values, chosen dispersion index (e.g. Permdisp vs. MDI), plot type (e.g. PCO vs. MDS), distance metric (e.g. Bray–Curtis vs. Gower), measurements variables (e.g. cover vs. biomass data) and transformation methods (e.g. untransformed vs. log-transformed).

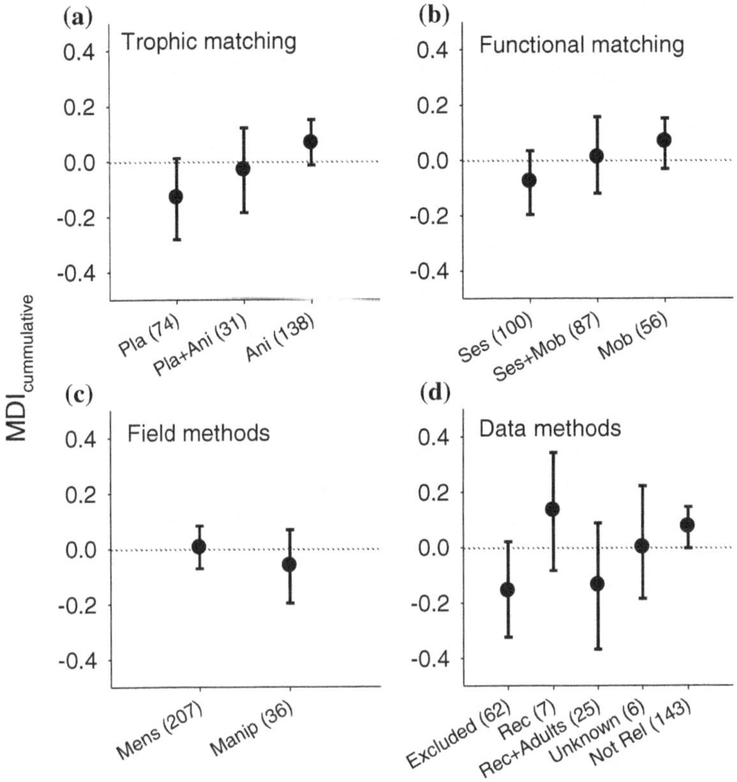

Fig. 6.3 Meta-analysis of 243 MDI values extracted from MDS plots that contrast invaded and non-invaded community samples. Negative values correspond to samples being more similar ('homogenisation') and numbers in brackets correspond to replication levels. Impacts of invasive seaweeds were evaluated on (**a**) plant (*Pla*), animal (*Ani*) or combined (Pla + Ani) communities, (**b**) sessile (*Ses*), mobile (*Mob*) or combined (Ses + Mob) communities, (**c**) communities measured in mensurative (*Mens*) or manipulative (*Manip*) field experiments and (**d**) communities where the invader itself was included or excluded. For plot **d**, MDI values were calculated for 5 types of reported similarity matrices; (1) excluding the invader (Excluded), (2) including the invader's new recruits (Rec), (3) including both the new recruits and the abundance of the invader itself (Rec + Adult), (4) studies where it could not be determined (Unknown) and (5) irrelevant studies (Not Rel), i.e. where the invaded community is different from the invasive species (here animal communities)

6.8 Summary

We conducted a bibliographic survey, adding 69 taxa to a published list of 277 seaweeds, thereby updating the total worldwide list of non-native and cryptogenic seaweeds to 346. None of these new non-native seaweeds have received much scientific scrutiny, but several may over time become abundant in invaded regions.

The seaweed genera with most non-native taxa are *Polysiphonia*, *Hypnea*, *Codium*, *Gracilaria* and *Caulerpa*. Bioregions that have received most non-native seaweeds are the Mediterranean Sea, followed by NE Atlantic, Australasia, NE Pacific and the NW Atlantic. Our tally is most likely an underestimate because (1) seaweeds may have been introduced to new regions prior to modern science, (2) morphologically similar non-native and native species may co-occur and remain taxonomically cryptic, (3) under-studied regions can include non-native taxa that are (4) poorly described, (5) difficult to identify, (6) rare or (7) have invaded inaccessible areas and habitats. The most important vectors that carry seaweeds around the world are hull fouling and as 'blind passengers' associated with aquaculture (in particular seaweeds attached to shell transplants), together accounting for >75 % of known seaweed introductions. These vectors are likely to continue to spread seaweeds around the world.

Once a non-native seaweed has arrived in a new location, it can establish a permanent population, spread (secondarily) through natural dispersal or associated with human vectors, increase in abundance and affect local species and ecosystem properties. Establishment, increase in abundances and secondary spread is quantified in 'success studies', whereas the effects on the invaded communities are quantified in 'impact studies'. The success of non-native seaweeds depends on attributes of the non-native seaweeds and attributes associated with the invaded system, including the biotic community, resource levels and abiotic conditions. Many 'core hypotheses' have been suggested to explain why some non-native species (seaweeds included) become invasive or fail, of which the 'ideal weed', 'propagule pressure', 'sampling', 'fluctuating resource availability', 'environmental heterogeneity', 'disturbances', 'empty niches', 'habitat filtering', 'limited similarity' and 'specialist-generalist matching' are important to understand. Impact framework also suggests the direction and magnitude of impacts depends on attributes of the non-native seaweeds and attributes associated with the biotic community, resource levels and abiotic conditions of the invaded system. Thus, any process that increases success (cf. core hypotheses outlined above) should modify impact, by controlling the distribution and abundance of the non-native seaweed. Impacts can, like success, be difficult to quantify, if the non-native seaweed is inconspicuous or difficult to identify or find, if interactions with native species are few and weak, or if strong interactions occur only in small areas, short time windows and in inaccessible habitats. Case studies have shown that many negative effects by non-native seaweeds occur though competition and habitat destruction (direct effects) or keystone competition and keystone habitat destruction (indirect effects). However, invasive seaweeds can also have positive effects on local populations, directly through habitat formation and modification and by being consumed, or indirectly through cascading habitat formation, consumption, competition or keystone consumption.

Case studies documenting seaweed invasion impacts have stimulated meta-analytical synthesis that aims to identify impact generalities across different invasive seaweeds, invaded communities and environmental conditions. Meta-analyses have shown that invasive seaweeds typically have a negative effect on local plant abundances, richness and diversity but positive or neutral effects on animal communities. Furthermore, negative effects reported on local plants can become less negative or neutral when moving from pristine to heavily human-impacted environments. Finally, our new meta-analyses presented in this chapter indicate that impacts increase with the abundance of non-native seaweeds and that non-native seaweeds may increase sample similarity in invaded plant communities, but not in animal communities (but these findings were based on a non-significant trend). Although the impact literature is extensive, most studies have tested for impacts of only 6 high-profile invaders (*S. muticum*, *U. pinnatifida*, *C. taxifolia*, *C. racemosa*, *C. fragile* and *G. vermiculophylla*). It is therefore important that similar comprehensive analyses are conducted on non-native seaweeds to allow for better generalisations and predictions about seaweed invasion impacts.

In conclusion, at least 346 non-native and cryptogenic seaweeds have altered local shallow water coastal communities in most major biogeographical regions around the world, and this pattern is likely to increase as human populations, transport and other co-occurring stressors continue to increase.

Acknowledgements MST was supported by the Marsden Fund Council from Government funding, administered by the Royal Society of New Zealand. TW was supported by a Future Fellows grant from the Australian Research Council. DRS gratefully acknowledges the continued support by the New Zealand Ministry of Science and Innovation and the National Institute of Water and Atmospheric Research (contract C01X0501).

References

Adam P. Saltmarsh ecology. Cambridge: Cambridge University Press; 1990. p. 461.

Alpert P. The advantages and disadvantages of being introduced. Biol Invasions. 2006;8(7): 1523–34. doi:10.1007/s10530-005-5844-z.

Altieri A, Trussell G, Ewanchuck P, Bernatchez G. Consumers control diversity and functioning of a natural marine ecosystem. PLoS Biol. 2009;4:e5291.

Ambrose WG, Nelson BW. Inhibition of giant kelp recruitment by an introduced brown alga. Bot Mar. 1982;25:265–7.

Baker HG, Stebbens GL. The genetics of colonizing species. New York: Academic Press; 1965.

Balata D, Piazzi L, Cinelli F. A comparison among assemblages in areas invaded by *Caulerpa taxifolia* and *C. racemosa* on a subtidal Mediterranean rocky bottom. Marine Ecology-Pubblicazioni Della Stazione Zoologica Di Napoli I. 2004;25(1):1–13. doi:10.1111/j.1439-0485.2004.00013.x

Baldwin JR, Lovvorn JR. Expansion of seagrass habitat by the exotic *Zostera japonica*, and its use by dabbling ducks and brant in Boundary Bay, British Colombia. Mar Ecol Prog Ser. 1994;103:119–27.

Barbier EB, Hacker SD, Kennedy CJ, Koch EW, Stier AC, Silliman BR. The value of estuarine and coastal ecosystem services. Ecol Monogr. 2011;81:169–93.

Bates AE, Pecl GT, Frusher S, Hobday AJ, Wernberg T, Smale DA, Sunday JM, Hill NA, Dulvy NK, Colwell RK. Defining and observing stages of climate-mediated range shifts in marine systems. Glob Environ Change. 2014;26:27–38.

Bedini R, Bonechi L, Piazzi L. Mobile epifaunal assemblages associated with *Cystoseira* beds: comparison between areas invaded and not invaded by *Lophocladia lallemandii*. Scientia Marina. 2014;78(3):425–32. doi:10.3989/scimar.03995.28B.

Bell SS. Amphipods as insect equivalents? An alternative view. Ecology. 1991;72:350–4.

Berkenbusch K, Rowden AA. An examination of the spatial and temporal generality of the influence of ecosystem engineers on the composition of associated assemblages. Aquat Ecol. 2007;41:129–47.

Berkenbusch K, Rowden AA, Myers TE. Interactions between seagrasses and burrowing ghost shrimps and their influence on infaunal assemblages. J Exp Mar Biol Ecol. 2007;341:70–84.

Blossey B, Notzold R. Evolution of increased competitive ability in invasive nonindigenous plants: a hypothesis. J Ecol. 1995;83(5):887–9. doi:10.2307/2261425.

Borenstein M, Hedges LV, Higgins JPT, Rothstein HR. Introduction to meta-analysis. West Sussex, United Kingdom: Wiley; 2009. p. 421.

Boudouresque CF, Lemée R, Mari X, Meinesz A. The invasive alga *Caulerpa taxifolia* is not a suitable diet for the sea urchin *Paracentrotus lividus*. Aquat Bot. 1996;53:245–50.

Box A, Martin D, Deudero S. Changes in seagrass polychaete assemblages after invasion by *Caulerpa racemosa* var. *cylindracea* (Chlorophyta: Caulerpales): community structure, trophic guilds and taxonomic distinctness. Scientia Marina. 2010;74(2):317–29.

Britton-Simmons KH. Direct and indirect effects of the introduced alga *Sargassum muticum* on benthic, subtidal communities of Washinton State, USA. Mar Ecol Prog Ser. 2004;277:61–78.

Britton-Simmons KH. Functional group diversity, resource preemption and the genesis of invasion resistance in a community of marine algae. Oikos. 2006;113(3):395–401.

Britton-Simmons KH, Abbott KC. Short-and long-term effects of disturbance and propagule pressure on a biological invasion. J Ecol. 2008;96(1):68–77.

Bulleri F, Balata D, Bertocci I, Tamburello L, Benedetti-Cecchi L. The seaweed *Caulerpa racemosa* on Mediterranean rocky reefs: from passenger to driver of ecological change. Ecology. 2010;91(8):2205–12. doi:10.1890/09-1857.1.

Buschbaum C, Chapman AS, Saier B. How an introduced seaweed can affect epibiota diversity in different coastal systems. Mar Biol. 2006;148(4):743–54. doi:10.1007/s00227-005-0128-9.

Byers J, Wright JT, Gribben PE. Variable direct and indirect effects of a habitat-modifying invasive species on mortality of native fauna. Ecology. 2010;91:1787–98.

Byers JE, Gribben PE, Yeager C, Sotka EE. Impacts of an abundant introduced ecosystem engineer within mudflats of the southeastern US coast. Biol Invasions. 2012;14(12):2587–600. doi:10.1007/s10530-012-0254-5.

Cacabelos E, Olabarria C, Incera M, Troncoso JS. Effects of habitat structure and tidal height on epifaunal assemblages associated with macroalgae. Estuar Coast Shelf Sci. 2010;89(1):43–52. doi:10.1016/j.ecss.2010.05.012.

Cacabelos E, Engelen AH, Mejia A, Arenas F. Comparison of the assemblage functioning of estuary systems dominated by the seagrass *Nanozostera noltii* versus the invasive drift seaweed *Gracilaria vermiculophylla*. J Sea Res. 2012;72:99–105. doi:10.1016/j.seares.2012.02.003.

Callaway RM, Ridenour WM. Novel weapons: invasive success and the evolution of increased competitive ability. Front Ecol Environ. 2004;2:436–43.

Carlton JT, Eldredge LG. Update and revisions of the marine bioinvasions of Hawai'i: the introduced and cryptogenic marine and estuarine animals and plants of the Hawaiian Archipelago. In: Lucius G Eldredge III Memorial Volume: Tribute to a Polymath. 2015. p. 25.

Casas G, Scrosati R, Piriz ML. The invasive kelp *Undaria pinnatifida* (Phaeophyceae, Laminariales) reduces native seaweed diversity in Nuevo Gulf (Patagonia, Argentina). Biol Invasions. 2004;6(4):411–6. doi:10.1023/b:binv.0000041555.29305.41.

Catford JA, Jansson R, Nilsson C. Reducing redundancy in invasion ecology by integrating hypothesis into a single theoretical framework. Divers Distrib. 2009;15:22–40.

Cebrian E, Ballesteros E, Linares C, Tomas F. Do native herbivores provide resistance to Mediterranean marine bioinvasions? A seaweed example. Biol Invasions. 2011;13(6): 1397–408. doi:10.1007/s10530-010-9898-1.

Ceccherelli G, Campo D. Different effects of *Caulerpa racemosa* on two co-occuring seagrasses in the Mediterranean. Bot Mar. 2002;45:71–6.

Ceccherelli G, Cinelli F. Short-term effects of nutrient enrichment of the sediment and interactions between the seagrass *Cymodocea nodosa* and the introduced green alga *Caulerpa taxifolia* in a Mediterranean bay. J Exp Mar Biol Ecol. 1997;217:165–77.

Ceccherelli G, Sechi N. Nutrient availability in the sediment and the reciprocal effects between the native seagrass *Cymodocea nodosa* and the introduced green alga *Caulerpa taxifolia* in a Mediterranean bay. Hydrobiologia. 2002;474:57–66.

Chen L, Zan Q, Li M, Shen J, Liao W. Litter dynamics and forest structure of the introduced *Sonneratia caseolaris* mangrove forest in Shenzhen, China. Estuar Coast Shelf Sci. 2009;85 (2):241–6.

Colautti RI, Ricciardi A, Grigorovich LA, MacIsaac HJ. Is invasion success explained by the enemy release hypothesis? Ecol Lett. 2004;7:721–33.

Conklin KY, O'Doherty DC, Sherwood AR. *Hydropuntia perplexa*, n. comb. (Gracilariaceae, Rhodophyta), first record of the genus in Hawai'i. Pac Sci. 2014;68(3):421–34. doi:10.2984/68.3.9.

Crawley MJ, Brown SL, Heard MS, Edwards GR. Invasion-resistance in experimental grassland communities: species richness or species identity? Ecol Lett. 1999;2(3):140–8.

Dale M. Phytosociological structure of seaweed communities and the invasion of *Fucus serratus* in Nova Scotia. Can J Bot. 1982;60(12):2652–8.

Davidson IC, Simkanin C. The biology of ballast water 25 years later. Biol Invasions. 2012;14: 9–13.

Davis MA, Grime JP, Thompsen K. Fluctuating resources in plant communities: a general theory of invasibility. J Ecol. 2000;88:528–34.

Dayton PK. Towards an understanding of community resilience and the potential effects of enrichment to the benthos of McMurdo Sound, Antarctica. In: Proceedings of the colloquium on conservation problems in Antartica. 1972. pp. 81–96.

Dean PR, Hurd CL. Seasonal growth, erosion rates, and nitrogen and photosynthetic ecophysiology of *Undaria pinnatifida* (Heterokontophyta) in southern New Zealand1. J Phycol. 2007;43(6):1138–48.

Demopoulos AWJ, Smith CR. Invasive mangroves alter macrofaunal community structure and facilitate opportunistic exotics. Mar Ecol Prog Ser. 2010;404:51–67. doi:10.3354/meps08483.

de Jesus PB, Silva MS, de Mattos Lyra G, de Castro Nunes JM, Schnadelbach AS. Extension of the distribution range of *Hypnea stellulifera* (Cystocloniaceae, Rhodophyta) to the South Atlantic: morphological and molecular evidence. Aquat Bot. 2014;123:26–36.

de Rivera CE, Ruiz GM, Hines AH, Jivoff P. Biotic resistance to invasion: native predator limits abundance and distribution of an introduced crab. Ecology. 2005;86(12):3364–76. doi:10.1890/05-0479.

de Villele X, Verlaque M. Changes and degradation in a *Posidonia oceanica* bed invaded by the introduced tropical alga *Caulerpa taxifolia* in the North Western Mediterranean. Bot Mar. 1995;38(1):79–87.

Díaz-Tapia P, Sook Kim M, Secilla A, Bárbara I, Cremades J. Taxonomic reassessment of *Polysiphonia foetidissima* (Rhodomelaceae, Rhodophyta) and similar species, including *P. schneideri*, a newly introduced species in Europe. Eur J Phycol. 2013;48(4):345–62. doi:10.1080/09670262.2013.842655

Dijoux L, Viard F, Payri C. The more we search, the more we find: discovery of a new lineage and a new species complex in the Genus *Asparagopsis*. PLoS ONE. 2014;9(7):e103826.

Drouin A, McKindsey CW, Johnson LE. Detecting the impacts of notorious invaders: experiments versus observations in the invasion of eelgrass meadows by the green seaweed *Codium fragile*. Oecologia. 2012;168(2):491–502. doi:10.1007/s00442-011-2086-x.

Eastwood MM, Donahue MJ, Fowler AE. Reconstructing past biological invasions: niche shifts in response to invasive predators and competitors. Biol Invasions. 2007;9:397–407.

Eklöf JS, de la Torre Castro M, Adelsköld L, Jiddawi NS, Kautsky N. Differences in macrofaunal and seagrass assemblages in seagrass beds with and without seaweed farms. Estuar Coast Shelf Sci. 2005;63(3):385–96.

Elton CS (1958) The ecology of invasions by animals and plants. London: Mathuess; 1996.

Emery SM. Limiting similarity between invaders and dominant species in herbaceous plant communities? J Ecol. 2007;95(5):1027–35. doi:10.1111/j.1365-2745.2007.01274.x.

Enge S, Nylund GM, Pavia H. Native generalist herbivores promote invasion of a chemically defended seaweed via refuge-mediated apparent competition. Ecol Lett. 2013;16(4):487–92.

Engelen A, Henriques N, Monteiro C, Santos R. Mesograzers prefer mostly native seaweeds over the invasive brown seaweed *Sargassum muticum*. Hydrobiologia. 2011;669(1):157–65. doi:10.1007/s10750-011-0680-x.

Engelen A, Primo A, Cruz T, Santos R. Faunal differences between the invasive brown macroalga *Sargassum muticum* and competing native macroalgae. Biol Invasions. 2013;15(1):171–83. doi:10.1007/s10530-012-0276-z.

Englund G, Sarnelle O, Cooper SD. The importance of data-selection criteria: meta-analysis of stream predation experiments. Ecology. 1999;80:1132–41.

Eppinga MB, Rietkerk M, Dekker SC, De Ruiter PC, Van der Putten WH. Accumulation of local pathogens: a new hypothesis to explain exotic plant invasions. Oikos. 2006;114(1):168–76. doi:10.1111/j.2006.0030-1299.14625.x.

Falcao C, de Szechy MTM. Changes in shallow phytobenthic assemblages in southeastern Brazil, following the replacement of *Sargassum vulgare* (Phaeophyta) by *Caulerpa scalpelliformis* (Chlorophyta). Bot Mar. 2005;48:208–17.

Fitridge I, Dempster T, Guenther J, de Nys R. The impact and control of biofouling in marine aquaculture: a review. Biofouling. 2012;28(7):649–69.

Flagella MM, Verlaque M, Soria A, Buia MC. Macroalgal survival in ballast water tanks. Mar Pollut Bull. 2007;54(9):1395–401. doi:10.1016/j.marpolbul.2007.05.015.

Flagella MM, Andreakis N, Hiraoka M, Verlaque M, Buia MC. Identification of cryptic *Ulva* species (Chlorophyta, Ulvales) transported by ballast water. J Biol Res Thessaloniki. 2010;13:47–57.

Forrest BM, Taylor MD. Assessing invasion impact: survey design considerations and implications for management of an invasive marine plant. Biol Invasions. 2003;4:375–86.

Fourqurean JW, Smith TJ III, Possley J, Collins TM, Lee D, Namoff S. Are mangroves in the tropical Atlantic ripe for invasion? Exotic mangrove trees in the forests of South Florida. Biol Invasions. 2010;12(8):2509–22.

Fukunaga A, Peyton KA, Thomas FIM. Epifaunal community structure and ammonium uptake compared for the invasive algae, *Gracilaria salicornia* and *Acanthophora specifera*, and the native alga, Padina thivyi. J Exp Mar Biol Ecol. 2014;456:78–86. doi:10.1016/j.jembe.2014.03.013.

Gallucci F, Hutchings P, Gribben P, Fonseca G. Habitat alteration and community-level effects of an invasive ecosystem engineer: a case study along the coast of NSW, Australia. Mar Ecol Prog Ser. 2012;449:95–108.

Gennaro P, Piazzi L. Synergism between two anthropic impacts: *Caulerpa racemosa* var. *cylindracea* invasion and seawater nutrient enrichment. Mar Ecol Prog Ser. 2011;427:59–70.

Gestoso I, Olabarria C, Troncoso JS. Variability of epifaunal assemblages associated with native and invasive macroalgae. Mar Freshw Res. 2010;61(6):724–31. doi:10.1071/mf09251.

Gestoso I, Olabarria C, Troncoso J. Effects of macroalgal identity on epifaunal assemblages: native species vs the invasive species *Sargassum muticum*. Helgol Mar Res. 2012;66(2):159–66. doi:10.1007/s10152-011-0257-0.

Glasby T. *Caulerpa taxifolia* in seagrass meadows: killer or opportunistic weed? Biol Invasions. 2012;1–19. doi:10.1007/s10530-012-0347-1

Gollan JR, Wright JT. Limited grazing pressure by native herbivores on the invasive seaweed *Caulerpa taxifolia* in a temperate Australian estuary. Mar Freshw Res. 2006;57:685–94.

Green DS, Boots B, Crowe TP. Effects of non-indigenous oysters on microbial diversity and ecosystem functioning. PLoS ONE. 2012;7(10):e48410.

Green D, Rocha C, Crowe T. Effects of non-indigenous oysters on ecosystem processes vary with abundance and context. Ecosystems. 2013;16:881–93. doi:10.1007/s10021-013-9659-y.

Gribben PE, Wright JT. Sublethal effects on reproduction in native fauna: are females more vulnerable to biological invasion? Oecologia. 2006;149:352–61.

Gribben PE, Byers J, Clements M, McKenzie LA, Steinberg PD, Wright JT. Behavioural interactions between ecosystem engineers control community species richness. Ecol Lett. 2009a;12:1127–36.

Gribben PE, Wright JT, O'Connor WA, Doblin MA, Eyre B, Steinberg PD. Reduced performance of native infauna following recruitment to a habitat-forming invasive marine alga. Oecologia. 2009b;158:733–45.

Gribben PE, Byers JE, Wright JT, Glasby TM. Positive versus negative effects of an invasive ecosystem engineer on different components of a marine ecosystem. Oikos. 2013;122(6):816–24. doi:10.1111/j.1600-0706.2012.20868.x.

Guerra-García JM, Ros M, Izquierdo D, Soler-Hurtado MM. The invasive *Asparagopsis armata* versus the native *Corallina elongata*: differences in associated peracarid assemblages. J Exp Mar Biol Ecol. 2012;416(417):121–8.

Halpern BS, Walbridge S, Selkoe KA, Kappel CV, Micheli F, Agrosa C. A global map of human impact on marine ecosystems. Science. 2008;319:948–52.

Hammann M, Buchholz B, Karez R, Weinberger F. Direct and indirect effects of *Gracilaria vermiculophylla* on native *Fucus vesiculosus*. Aquat Invasions. 2013;8(2):121–32. doi:10.3391/ai.2013.8.2.01.

Harries DB, Harrow S, Wilson JR, Mair JM, Donnan DW. The establishment of the invasive alga *Sargassum muticum* on the west coast of Scotland: a preliminary assessment of community effects. J Mar Biol Ass UK. 2007;87:1057–67.

Harris LG, Jones AC. Temperature, herbivory and epibiont acquisition as factors controlling the distribution and ecological role of an invasive seaweed. Biol Invasions. 2005;7:913–24.

Hay CH. The dispersal of sporophytes of *Undaria pinnatifida* by coastal shipping in New Zealand, and implications for further dispersal of *Undaria* in France. Brit Phycol J. 1990;25(4):301–13.

Hay M, Steinberg P. The chemical ecology of plant-herbivore interactions in marine versus terrestrial communities. Herbivores Interact Secondary Plant Metabolites. 1992;2:371–413.

Hewitt CL, Campbell ML, Schaffelke B. Introductions of seaweeds: accidental transfer pathways and mechanisms. Bot Mar. 2007;50(5–6):326–37. doi:10.1515/bot.2007.038.

Hierro JL, Maron JL, Callaway RM. A biogeographical approach to plant invasions: the importance of studying exotics in their introduced and native range. J Ecol. 2005;93:5–15.

Hoffman R, Shemesh E, Ramot M, Dubinsky Z, Pinchasov-Grinblat Y, Iluz D. First record of the Indo-Pacific seaweed *Codium arabicum* Kutz. (Bryopsidales, Chlorophyta) in the Mediterranean Sea. Bot Mar. 2011;54(5):487–495. doi:10.1515/bot.2011.056.

Hoffman R, Sternberg M, Serio D. First report of *Laurencia chondrioides* (Ceramiales, Rhodophyta) and its potential to be an invasive in the eastern Mediterranean Sea. Bot Mar. 2014;57(6):449–57. doi:10.1515/bot-2014-0053.

Ibrahim AM, El-naggar MM. Ballast water review: impacts, treatments and management. Middle-East J Sci Res. 2012;12(7):976–84.

Incera M, Bertocci I, Benedetti-Cecchi L. Effects of mean intensity and temporal variability of disturbance on the invasion of *Caulerpa racemosa* var. *cylindracea* (Caulerpales) in rock pools. Biol Invasions. 2010;12(3):501–14.

Inderjit Chapman D, Ranelletti M, Kaushik S. Invasive marine algae: an ecological perspective. Bot Rev. 2006;72(2):153–78. doi:10.1663/0006-8101(2006)72[153:imaaep]2.0.co;2.

Irigoyen AJ, Trobbiani G, Sgarlatta MP, Raffo MP. Effects of the alien algae *Undaria pinnatifida* (Phaeophyceae, Laminariales) on the diversity and abundance of benthic macrofauna in Golfo Nuevo (Patagonia, Argentina): potential implications for local food webs. Biol Invasions. 2011;13(7):1521–32. doi:10.1007/s10530-010-9910-9.

James K, Middleton I, Middleton C, Shears NT. Discovery of *Undaria pinnatifida* (Harvey) Suringar, 1873 in northern New Zealand indicates increased invasion threat in subtropical regions. BioInvasions Rec. 2014;3(1):21–4.

Jaubert JM, Chisholm JRM, Ducrot D, Ripley HT, Roy L, Passeron GS. No deleterious alterations in *Posidonia* beds in the Bay of Menton (France) eight years after *Caulerpa taxifolia* colonization. J Phycol. 1999;35:1113–9.

Jaubert JM, Chisholm JRM, Minghelli-Roman A, Marchioretti M, Morrow JH, Ripley HT. Re-evaluation of the extent of *Caulerpa taxifolia* development in the northern Mediterranean using airborne spectrographic sensing. Mar Ecol Prog Ser. 2003;263:75–82.

Johnson CR, Chapman ARO. Seaweed invasions: introduction and scope. Bot Mar. 2007;50(5–6): 321–5. doi:10.1515/bot.2007.037.

Johnston CA, Lipcius RN. Exotic macroalga *Gracilaria vermiculophylla* provides superior nursery habitat for native blue crab in Chesapeake Bay. Mar Ecol Prog Ser. 2012;467:137–46. doi:10. 3354/meps09935.

Jones E, Thornber CS. Effects of habitat-modifying invasive macroalgae on epiphytic algal communities. Mar Ecol Prog Ser. 2010;400:87–100.

Jones CG, Lawton JH, Shachak M. Organisms as ecosystem engineers. Oikos. 1994;69:373–86.

Jongma DN, Campo D, Dattolo E, D'Esposito D, Duchi A, Grewe P, Huisman J, Verlaque M, Yokes MB, Procaccini G. Identity and origin of a slender *Caulerpa taxifolia* strain introduced into the Mediterranean Sea. Bot Mar. 2013;56(1):27–39. doi:10.1515/bot-2012-0175.

Keane RM, Crawley MJ. Exotic plant invasions and the enemy release hypothesis. TREE. 2002;17:164–70.

Klein J, Verlaque M. The *Caulerpa racemosa* invasion: a critical review. Mar Pollut Bull. 2008;56 (2):205–25. doi:10.1016/j.marpolbul.2007.09.043.

Klein JC, Verlaque M. Experimental removal of the invasive *Caulerpa racemosa* triggers partial assemblage recovery. J Mar Biol Assoc UK. 2011;91(1):117–25. doi:10.1017/ s0025315410000792.

Krauss KW, Allen JA. Influences of salinity and shade on seedling photosynthesis and growth of two mangrove species, *Rhizophora mangle* and *Bruguiera sexangula*, introduced to Hawaii. Aquat Bot. 2003;77(4):311–24.

Lang AC, Buschbaum C. Facilitative effects of introduced Pacific oysters on native macroalgae are limited by a secondary invader, the seaweed *Sargassum muticum*. J Sea Res. 2010;63(2): 119–28. doi:10.1016/j.seares.2009.11.002.

Levin PS, Coyer JA, Petrik R, Good TP. Community-wide effects of nonindigenous species on temperate rocky reefs. Ecology. 2002;83(11):3182–93. doi:10.2307/3071852.

Lockwood JL, Cassey P, Blackburn T. The more you introduce the more you get: the role of colonization pressure and propagule pressure in invasion ecology. Divers Distrib. 2009;15:904–10.

Lutz M, Davis A, Minchinton T. Non-indigenous macroalga hosts different epiphytic assemblages to conspecific natives in southeast Australia. Mar Biol. 2010;157(5):1095–03. doi:10.1007/ s00227-010-1391-y.

Maggi E, Benedetti-Cecchi L, Castelli A, Chatzinikolaou E, Crowe TP, Ghedini G, Kotta J, Lyons DA, Ravaglioli C, Rilov G, Rindi L, Bulleri F. Ecological impacts of invading seaweeds: a meta-analysis of their effects at different trophic levels. Divers Distrib. 2015;21 (1):1–12. doi:10.1111/ddi.12264.

Maggs CA, Gall LL, Mineur F, Provan J, Saunders GW. *Fredericqia deveauniensis*, gen. et sp. nov. (Phyllophoraceae, Rhodophyta), a new cryptogenic species. Cryptogam Algologie. 2013;34(3):273–296.

Mateu-Vicens G, Box A, Deudero S, Rodriguez B. Comparative analysis of epiphytic foraminifera in sediments colonized by seagrass *Posidonia oceanica* and invasive macroalgae *Caulerpa* spp. J Foramin Res. 2010;40(2):134–47.

Mazariegos-Villareal A, Riosmena-Rodríguez R, Rosa Rivera-Camacho A, Serviere-Zaragoza E. First report of *Cladostephus spongiosus* (Sphacelariales: Phaeophyta) from the Pacific coast of Mexico. Bot Mar. 2010;53(2):153–7.

McKinnon JG, Gribben PE, Davis AR, Jolley DF, Wright JT. Differences in soft-sediment macrobenthic assemblages invaded by *Caulerpa taxifolia* compared to uninvaded habitats. Mar Ecol Prog Ser. 2009;380:59–71. doi:10.3354/meps07926.

Meinesz A. Killer algae—the true tale of a biological invasion. Chicago: The University of Chicago Press; 1999. p. 360.

Melbourne BA, Cornell HV, Davies KF, Dugaw CJ, Elmendorf S, Freestone AL, Hall RJ, Harrison S, Hastings A, Holland M, Holyoak M, Lambrinos L, Moore K, Yokomizo H. Invasion in a heterogeneous world: resistance, coexistence or hostile takeover? Ecol Lett. 2007;10:77–97.

Micael J, Parente MI, Costa AC. Tracking macroalgae introductions in North Atlantic oceanic Islands. Helgol Mar Res. 2014;68(2):209–19. doi:10.1007/s10152-014-0382-7.

Mineur F, Johnson MP, Maggs CA, Stegenga H. Hull fouling on commercial ships as a vector of macroalgal introduction. Mar Biol. 2007;151(4):1299–307.

Mineur F, Johnson MP, Maggs CA. Macroalgal introductions by hull fouling on recreational vessels: seaweeds and sailors. Environ Manage. 2008a;42(4):667–76. doi:10.1007/s00267-008-9185-4.

Mineur F, Johnson MP, Maggs CA. Non-indigenous marine macroalgae in native communities: a case study in the British Isles. J Mar Biol Assoc UK. 2008b;88(4):693–8. doi:10.1017/s0025315408001409.

Mineur F, De Clerck O, Le Roux A, Maggs CA, Verlaque M. *Polyopes lancifolius* (Halymeniales, Rhodophyta), a new component of the Japanese marine flora introduced to Europe. Phycologia. 2009;49(1):86–96. doi:10.2216/09-45.1.

Mineur F, Le Roux A, Stegenga H, Verlaque M, Maggs CA. Four new exotic red seaweeds on European shores. Biol Invasions. 2012;14(8):1635–41. doi:10.1007/s10530-012-0186-0.

Mitchell CE, Agrawal AA, Bever JD, Gilbert GS, Hufbauer RA, Klironomos JN, Maron JL, Morris WF, Parker IM, Power AG, Seabloom EW, Torchin ME, Vazquez DP. Biotic interactions and plant invasions. Ecol Lett. 2006;9:726–40.

Monteiro C, Engelen AH, Santos RO. Macro- and mesoherbivores prefer native seaweeds over the invasive brown seaweed *Sargassum muticum*: a potential regulating role on invasions. Mar Biol. 2009;156:2505–15.

Morita T, Kurashima A, Maegawa M. Temperature requirements for the growth and maturation of the gametophytes of *Undaria pinnatifida* and *U. undarioides* (Laminariales, Phaeophyceae). Phycol Res. 2003;51(3):154–60.

Neira C, Levin LA, Grosholz ED, Mendoza G. Influence of invasive *Spartina* growth stages on associated macrofaunal communities. Biol Invasions. 2007;9(8):975–93. doi:10.1007/s10530-007-9097-x.

Nejrup LB, Pedersen MF. Growth and biomass development of the introduced red alga *Gracilaria vermiculophylla* is unaffected by nutrient limitation and grazing. Aquat Biol. 2010;10(3):249–59. doi:10.3354/ab00281.

Nejrup L, Pedersen M, Vinzent J. Grazer avoidance may explain the invasiveness of the red alga *Gracilaria vermiculophylla* in Scandinavian waters. Mar Biol. 2012;159(8):1703–12. doi:10.1007/s00227-012-1959-9.

Nyberg CD, Thomsen MS, Wallentinus I. Flora and fauna associated with the introduced red alga *Gracilaria vermiculophylla*. Eur J Phycol. 2009;44(3):395–403. doi:10.1080/09670260802592808.

Nylund GM, Weinberger F, Rempt M, Pohnert G. Metabolomic assessment of induced and activated chemical defence in the invasive red alga *Gracilaria vermiculophylla*. PLoS ONE. 2011;6(12):e29359.

Odom RL, Walters LJ. A safe alternative to invasive *Caulerpa taxifolia* (Chlorophtya)? Assessing aquarium-release invasion potential of aquarium strains of the macroalgal genus *Chaetomorpha* (Chlorophyta). Biol Invasions. 2014;16(8):1589–97.

Olabarria C, Arenas F. Understanding biological invasions by seaweeds. Mar Algae Biodiver Taxonomy Environ Assess Biotechnol. 2014;140.

Olabarria C, Rodil IF, Incera M, Troncoso JS. Limited impact of *Sargassum muticum* on native algal assemblages from rocky intertidal shores. Mar Environ Res. 2009;67(3):153–8. doi:10.1016/j.marenvres.2008.12.007.

Ostenfeld CH. On the immigration of *Biddulphia sinensis* Grev. and its occurrence in the North Sea during 1903–1907 and on its use for the study of the direction and rate of flow of the currents. Meddelelser fra Kommissionen for Danmarks Fiskeri- og Havundersøgelser: Serie Plankton. 1908;6:1–44.

Pacciardi L, De Biasi AM, Piazzi L. Effects of *Caulerpa racemosa* invasion on soft-bottom assemblages in the Western Mediterranean Sea. Biol Invasions. 2011;13(12):2677–90. doi:10.1007/s10530-011-9938-5.

Padilla DK, Williams SL. Beyond ballast water: aquarium and ornamental trades as sources of invasive species in aquatic ecosystems. Front Ecol Environ. 2004;2(3):131–8.

Parker IM, Simberloff D, Lonsdale WM, Goodell K, Wonham MJ, Kareiva PM, Williamson MH, Von Holle B, Moyle PB, Byers JL, Goldwasser L. Impact: toward a framework for understanding the ecological effects of invaders. Biol Invasions. 1999;1:3–19.

Parker DJ, Burkepile DE, Hay ME. Opposing effects of native and exotic herbivores on plant invasion. Science. 2006;311:1459–61.

Pedersen MF, Stæhr PA, Wernberg T, Thomsen M. Biomass dynamics of exotic *Sargassum muticum* and native *Halidrys siliquosa* in Limfjorden, Denmark—implications of species replacements on turnover rates. Aquat Bot. 2005;83:31–47.

Peteiro C, Sánchez N. Comparing salinity tolerance in early stages of the sporophytes of a non-indigenous kelp (*Undaria pinnatifida*) and a native kelp (*Saccharina latissima*). Russ J Mar Biol. 2012;38(2):197–200.

Piazzi L, Balata D. The spread of *Caulerpa racemosa* var. *cylindracea* in the Mediterranean Sea: an example of how biological invasions can influence beta diversity. Mar Environ Res. 2008;65(1):50–61.

Piazzi L, Balata D. Invasion of alien macroalgae in different Mediterranean habitats. Biol Invasions. 2009;11(2):193–204. doi:10.1007/s10530-008-9224-3.

Piazzi L, Ceccherelli G. Persistence of biological invasion effects: recovery of macroalgal assemblages after removal of *Caulerpa racemosa* var. *cylindracea*. Estuar Coast Shelf Sci. 2006;68(3–4):455–61. doi:10.1016/j.ecss.2006.02.011.

Piazzi L, Cinelli F. Evaluation of benthic macroalgal invasion in a harbour area of the western Mediterranean Sea. Eur J Phycol. 2003;38(3):223–31. doi:10.1080/1364253031000136358.

Piazzi L, Balata D, Cinelli F. Epiphytic macroalgal assemblages of *Posidonia oceanica* rhizomes in the western Mediterranean. Eur J Phycol. 2002;37(1):69–76. doi:10.1017/s0967026201003432.

Piazzi L, Balata D, Cecchi E, Cinelli F. Co-occurrence of *Caulerpa taxifolia* and *C. racemosa* in the Mediterranean Sea: interspecific interactions and influence on native macroalgal assemblages. Cryptogamie Algologie. 2003;24(3):233–43.

Piazzi L, Balata D, Ceccherelli G, Cinellia F. Interactive effect of sedimentation and *Caulerpa racemosa* var. *cylindracea* invasion on macroalgal assemblages in the Mediterranean Sea. Estuar Coast Shelf Sci. 2005;64:467–74.

Pickering TD, Skelton P, Sulu RJ. Intentional introductions of commercially harvested alien seaweeds. Bot Mar. 2007;50(5–6):338–50. doi:10.1515/bot.2007.039.

Pochon X, Atalah J, Wood SA, Hopkins GA, Watts A, Boedeker C. *Cladophora ruchingeri* (C. Agardh) Kützing, 1845 (Cladophorales, Chlorophyta): a new biofouling pest of green-lipped mussel *Perna canaliculus* (Gmelin, 1791) farms in New Zealand; 2015.

Posey MH. Community changes associated with the spread of an introduced seagrass, *Zostera Japonica*. Ecology. 1988;69:974–83.

Procheş Ş, Wilson JRU, Richardson DM, Rejmánek M. Searching for phylogenetic pattern in biological invasions. Glob Ecol Biogeogr. 2008;17(1):5–10. doi:10.1111/j.1466-8238.2007.00333.x.

Rempt M, Weinberger F, Grosser K, Pohnert G. Conserved and species-specific oxylipin pathways in the wound-activated chemical defense of the noninvasive red alga *Gracilaria chilensis* and the invasive *Gracilaria vermiculophylla*. Beilstein J Org Chem. 2012;8(1):283–9.

Ren H, Lu H, Shen W, Huang C, Guo Q, Za Li, Jian S. *Sonneratia apetala* Buch. Ham in the mangrove ecosystems of China: an invasive species or restoration species? Ecol Eng. 2009;35(8):1243–8.

Reynolds LK, Carr LA, Boyer KE. A non-native amphipod consumes eelgrass inflorescences in San Francisco Bay. Mar Ecol Prog Ser. 2012;451:107–18. doi:10.3354/meps09569.

Ricciardi A, Simberloff D. Assisted colonization is not a viable conservation strategy. Trends Ecol Evol. 2009;24(5):248–53.

Rodil IF, Olabarria C, Lastra M, Lopez J. Differential effects of native and invasive algal wrack on macrofaunal assemblages inhabiting exposed sandy beaches. J Exp Mar Biol Ecol. 2008;358 (1):1–13. doi:10.1016/j.jembe.2007.12.030.

Rosenberg MS, Adams DC, Gurevitch J. Metawin: statistical software for meta-analysis, vol Version 2. Massachusetts: Sinauer Associates; 2000.

Rueness J. *Sargassum muticum* and other introduced Japanese macroalgae: biological pollution of European coasts. Mar Pollut Bull. 1989;20:173–6.

Ruesink JL, Hong JS, Wisehart L, Hacker SD, Dumbauld BR, Hessing-Lewis M, Trimble AC. Congener comparison of native (*Zostera marina*) and introduced (*Z. japonica*) eelgrass at multiple scales within a Pacific Northwest estuary. Biol Invasions. 2010;12(6):1773–89. doi:10.1007/s10530-009-9588-z.

Ruiz GM, Torchin ME, Grant K. Using the Panama Canal to test predictions about tropical marine invasions. In: Proceedings of the Smithsonian Marine Science Symposium; 2009. pp. 291–299.

Russell DJ. Introduction of alien seaweeds to Hawaii. Phycologia. 1981;20(2):112.

Sakai AK, Allendorf FW, Holt JS, Lodge DM, Molofsky J, With KA, Baughman S, Cabin RJ, Cohen JE, Ellstrand N, McCauley DE, O'Neil P, Parker IM, Thompson JN, Weller SG. The population biology of invasive species. Annu Rev Ecol Syst. 2001;32:305–32.

Sánchez I, Fernández C. Impact of the invasive seaweed *Sargassum muticum* (Phaeophyta) on an intertidal macroalgal assemblage. J Phycol. 2005;41:923–30.

Sánchez I, Fernández C, Arrontes J. Long-term changes in the structure of intertidal assemblages after invasion by *Sargassum muticum* (Phaeophyta). J Phycol. 2005;41:942–9.

Sandler R. The value of species and the ethical foundations of assisted colonization. Conserv Biol. 2010;24(2):424–31. doi:10.1111/j.1523-1739.2009.01351.x.

Sax DF, Stachowicz JJ, Brown JH, Bruno JF, Dawson MN, Gaines SD, Grosberg RK, Hastings H, Holt RD, Mayfield MM, O'Connor MI, Rice WR. Ecological and evolutionary insights from species invasions. Trends Ecol Evol. 2007;22:465–71.

Schaffelke B, Hewitt CL. Impacts of introduced seaweeds. Bot Mar. 2007;50:397–417.

Schaffelke B, Smith JE, Hewitt CL. Introduced macroalgae—a growing concern. J Appl Phycol. 2006;18(3–5):529–41. doi:10.1007/s10811-006-9074-2.

Scheibling RE, Anthony SX. Feeding, growth and reproduction of sea urchins (*Strongylocentrotus droebachiensis*) on single and mixed diets of kelp (*Laminaria* spp.) and the invasive alga *Codium fragile* spp. *tometosoides*. Mar Biol. 2001;139:139–46.

Scheibling RE, Lyons DA, Sumi CBT. Grazing of the invasive alga *Codium fragile* ssp. *tomentosoides* by the common periwinkle *Littorina littorea*: effects of thallus size, age and condition. J Exp Mar Biol Ecol. 2008;355(2):103–13. doi:10.1016/j.jembe.2007.12.002.

Schiel DR, Thompson GA. Demography and population biology of the invasive kelp *Undaria pinnatifida* on shallow reefs in southern New Zealand. J Exp Mar Biol Ecol. 2012;434:25–33.

Schmidt AL, Scheibling RE. A comparison of epifauna and epiphytes on native kelps (*Laminaria* species) and an invasive alga (*Codium fragile* ssp. *tomentosoides*) in Nova Scotia, Canada. Bot Mar. 2006;49:315–330.

Schmidt AL, Scheibling RE. Effects of native and invasive macroalgal canopies on composition and abundance of mobile benthic macrofauna and turf-forming algae. J Exp Mar Biol Ecol. 2007;341:110–30.

Seddon PJ. From reintroduction to assisted colonization: moving along the conservation translocation spectrum. Restor Ecol. 2010;18(6):796–802. doi:10.1111/j.1526-100X.2010.00724.x.

Sher AA, Hyatt LA. The disturbed resource-flux invasion matrix: a new framework for patterns of plant invasions. Biol Invasions. 1999;1:107–14.

Simberloff D, Von Holle B. Positive interactions of nonindigenous species: invasional meltdown? Biol Invasions. 1999;1:21–32.

Sissini MN, Longo GO, Martins CDL, Floeter SR, Pereira SB, Horta PA. First record of the green alga *Halimeda* (Bryopsidales: Chlorophyta) at Rocas Atoll—natural dispersion or anthropogenic causes? Mar Biodivers Rec. 2014;7:e104.

Sjotun K, Eggereide SF, Hoisaeter T. Grazer-controlled recruitment of the introduced *Sargassum muticum* (Phaeophycae, Fucales) in Northern Europe. Mar Ecol Prog Ser. 2007;342:127–38.

Smith JR, Vogt SC, Creedon F, Lucas BJ, Eernisse DJ. The non-native turf-forming alga *Caulacanthus ustulatus* displaces space-occupants but increases diversity. Biol Invasions. 2014;16(10):2195–208. doi:10.1007/s10530-014-0658-5.

South PM, Lilley S, Tait LW, Alestra T, Hickford MJH, Thomsen MS, Schiel DR. Transient effects of an invasive kelp on the community structure and primary productivity of an intertidal assemblage. Marine and Freshwater Research MF14152. 2015. Accepted 14 Aug 2014.

Staehr PA, Pedersen MF, Thomsen MS, Wernberg T, Krause-Jensen D. Invasion of *Sargassum muticum* in Limfjorden (Denmark) and its possible impact on the indigenous macroalgal community. Mar Ecol Prog Ser. 2000;207:79–88.

Strong JA, Dring MJ, Maggs CA. Colonisation and modification of soft substratum habitats by the invasive macroalga *Sargassum muticum*. Mar Ecol Prog Ser. 2006;321:87–97.

Tait LW, South PM, Lilley SA, Thomsen MS, Schiel DR. Assemblage and understory carbon production of native and invasive canopy-forming macroalgae. J Exp Mar Biol Ecol. 2015;469:10–7. doi:10.1016/j.jembe.2015.04.007.

Tamburello L, Maggi E, Benedetti-Cecchi L, Bellistri G, Rattray AJ, Ravaglioli C, Rindi L, Roberts J, Bulleri F. Variation in the impact of non-native seaweeds along gradients of habitat degradation: a meta-analysis and an experimental test. 2015. Oikos:n/a-n/a. doi:10.1111/oik.02197

Taylor SL, Bishop MJ, Kelaher BP, Glasby TM. Impacts of detritus from the invasive alga *Caulerpa taxifolia* on a soft sediment community. Mar Ecol Prog Ser. 2010;420:73–81.

Thomsen MS. Experimental evidence for positive effects of invasive seaweed on native invertebrates via habitat-formation in a seagrass bed. Aquat Invasions. 2010;5(4):341–6. doi:10.3391/ai.2010.5.4.02.

Thomsen MS, McGlathery K. Effects of accumulations of sediments and drift algae on recruitment of sessile organisms associated with oyster reefs. J Exp Mar Biol Ecol. 2006;328(1):22–34. doi:10.1016/j.jembe.2005.06.016.

Thomsen MS, McGlathery KJ. Stress tolerance of the invasive macroalgae *Codium fragile* and *Gracilaria vermiculophylla* in a soft-bottom turbid lagoon. Biol Invasions. 2007;9(5):499–513. doi:10.1007/s10530-006-9043-3.

Thompson GA, Schiel DR. Resistance and facilitation by native algal communities in the invasion success of *Undaria pinnatifida*. Mar Ecol Prog Ser. 2012;468:95.

Thomsen MS, Wernberg T. The devil in the detail: harmful seaweeds are not harmful to everyone. Global Change Biology. 2015;21(4):1381–2.

Thomsen MS, Gurgel CFD, Fredericq S, McGlathery KJ. *Gracilaria vermiculophylla* (Rhodophyta, Gracilariales) in Hog Island Bay, Virginia: a cryptic alien and invasive macroalga and taxonomic correction. J Phycol. 2006a;42(1):139–41. doi:10.1111/j.1529-8817.2005.00160.x.

Thomsen MS, Wernberg T, Stæhr PA, Pedersen MF. Spatio-temporal distribution patterns of the invasive macroalga *Sargassum muticum* within a Danish *Sargassum*-bed. Helgol Mar Res. 2006b;60:50–8.

Thomsen MS, Stæhr P, Nyberg CD, Krause-Jensen D, Schwærter S, Silliman B. *Gracilaria vermiculophylla* in Northern Europe, with focus on Denmark, and what to expect in the future. Aquat Invasions. 2007;2:83–94.

Thomsen MS, Adam P, Silliman B. Anthropogenic threats to Australasian coastal salt marshes. In: Silliman BR, Bertness MD, Strong D, editors. Anthropogenic modification of North American salt marshes. California: University of California Press; 2009a. pp. 361–390.

Thomsen MS, Wernberg T, Tuya F, Silliman BR. Evidence for impacts of non-indigenous macroalgae: a meta-analysis of experimental field studies. J Phycol. 2009b;45:812–9.

Thomsen MS, Wernberg T, Altieri AH, Tuya F, Gulbransen D, McGlathery KJ, Holmer M, Silliman BR. Habitat cascades: the conceptual context and global relevance of facilitation cascades via habitat formation and modification. Integr Comp Biol. 2010;50(2):158–75. doi:10.1093/icb/icq042.

Thomsen MS, Olden JD, Wernberg T, Griffin JN, Silliman BR. A broad framework to organize and compare ecological invasion impacts. Environ Res. 2011a;111:899–908.

Thomsen MS, Wernberg T, Olden JD, Griffin JN, Silliman BR. A framework to study the context-dependent impacts of marine invasions. J Exp Mar Biol Ecol. 2011b;400(1–2):322–7. doi:10.1016/j.jembe.2011.02.033.

Thomsen MS, Wernberg T, Engelen AH, Tuya F, Vanderklift MA, Holmer M, McGlathery KJ, Arenas F, Kotta J, Silliman BR. A meta-analysis of seaweed impacts on seagrasses: generalities and knowledge gaps. PLoS ONE. 2012;7(1):e28595.

Thomsen MS, Byers JE, Schiel DR, Bruno JF, Olden JD, Wernberg T, Silliman BR. Impacts of marine invaders on biodiversity depend on trophic position and functional similarity. Mar Ecol Prog Ser. 2014;495:39–47. doi:10.3354/meps10566.

Thomsen MS, Wernberg T, Schiel DR. Invasions by non-indigenous species. In: Crowe TP, Frid CLJ, editors. Marine ecosystems: human impacts on biodiversity, functioning and services. Cambridge: Cambridge University Press; 2015. p. 274–332.

Thornber CS, Kinlan BP, Graham MH, Stachowicz JJ. Population ecology of the invasive kelp *Undaria pinnatifida* in California: environmental and biological controls on demography. Mar Ecol Prog Ser. 2004;268:69–80.

Tomas F, Box A, Terrados J. Effects of invasive seaweeds on feeding preference and performance of a keystone Mediterranean herbivore. Biol Invasions. 2011a;13(7):1559–70. doi:10.1007/s10530-010-9913-6.

Tomas F, Cebrian E, Ballesteros E. Differential herbivory of invasive algae by native fish in the Mediterranean Sea. Estuar Coast Shelf Sci. 2011b;92(1):27–34.

Trowbridge CD. Ecology of the green macroalga *Codium fragile* (Suringar) Hariot 1889: invasive and non-invasive subspecies. Oceanogr Mar Biol. 1998;36:1–64.

Trowbridge CD. Local elimination of *Codium fragile* ssp. *tomentosoides*: indirect evidence of sacoglossan herbivory. J Mar Biol Ass UK. 2002;82:1029–30.

Trowbridge CD, Todd CD. Host-plant change in marine specialist herbivores: ascoglossan sea slugs on introduced Macroalgae. Ecol Monogr. 2001;71:219–43.

Trussell G, Ewanchuck P, Bertness MD. Field evidence of trait-mediated indirect interactions in a rocky intertidal food web. Ecol Lett. 2002;5:241–5.

Trussell G, Ewanchuk P, Bertness MD, Silliman BR. Trophic cascades in rocky shore tide pools: distinguishing lethal and non-lethal effects. Oecologia. 2004;139:427–32.

Tsai C, Yang S, Trimble AC, Ruesink JL. Interactions betweem two introduced species: *Zostera japonica* (dwarf eelgrass) facilitates itself and reduces condition of *Ruditapes philippinarum* (Manila clam) on intertidal mudflats. Mar Biol. 2010;157:1929–36.

Tsiamis K, Verlaque M. A new contribution to the alien red macroalgal flora of Greece (Eastern Mediterranean) with emphasis on *Hypnea* species. Cryptogamie Algologie. 2011;32(4):393–410.

Valentine JP, Johnson CR. Establishment of the introduced kelp *Undaria pinnatifida* in Tasmania depends on disturbance to native algal assemblages. J Exp Mar Biol Ecol. 2003;295(1):63–90. doi:10.1016/s0022-0981(03)00272-7.

Valentine JP, Johnson CR. Establishment of the introduced kelp *Undaria pinnatifida* following dieback of the native macroalga *Phyllospora comosa* in Tasmania, Australia. Mar Freshw Res. 2004;55(3):223–30. doi:10.1071/mf03048.

Valentine JP, Magierowski RH, Johnson CR. Mechanisms of invasion: establishment, spread and persistence of introduced seaweed populations. Bot Mar. 2007;50(5–6):351–60. doi:10.1515/bot.2007.040.

Vaz-Pinto F, Olabarria C, Arenas F. Propagule pressure and functional diversity: interactive effects on a macroalgal invasion process. Mar Ecol Prog Ser. 2012;471:51–60. doi:10.3354/meps10024.

Vázquez-Luis M, Borg JA, Sanchez-Jerez P, Bayle-Sempere JT. Habitat colonisation by amphipods: comparison between native and alien algae. J Exp Mar Biol Ecol. 2012; 432–433:162–70.

Verges A, Sanchez N, Peteiro C, Polo L, Brodie J. *Pyropia suborbiculata* (Bangiales, Rhodophyta): first records from the northeastern Atlantic and Mediterranean of this North Pacific species. Phycologia. 2013;52(2):121–9. doi:10.2216/12-003.1.

Verlaque M, Steen F, De Clerck O. *Rugulopteryx* (Dictyotales, Phaeophyceae), a genus recently introduced to the Mediterranean. Phycologia. 2009;48(6):536–42.

Viejo RM. Mobile epifauna inhabiting the invasive *Sargassum muticum* and two local seaweeds in northern Spain. Aquat Bot. 1999;64:131–49.

Wallentinus I, Nyberg CD. Introduced marine organisms as habitat modifiers. Mar Pollut Bull. 2007;55:323–32.

Warwick RM, Clarke KR. Increased variability as a symptom of stress in marine communities. J Exp Mar Biol Ecol. 1993;172:215–26.

Wassman R, Ramus J. Seaweed invasion. Nat Hist. 1973;82(10):25.

Wernberg T, Thomsen MS, Stæhr PA, Pedersen MF. Comparative phenology of *Sargassum muticum* and *Halidrys siliquosa* (Phaeophyceae: Fucales) in Limfjorden, Denmark. Bot Mar. 2001;44:31–9. doi:10.1515/BOT.2001.005.

Wernberg T, Thomsen MS, Staerh PA, Pedersen MF. Epibiota communities of the introduced and indigenous macroalgal relatives *Sargassum muticum* and *Halidrys siliquosa* in Limfjorden (Denmark). Helgol Mar Res. 2004;58:154–61.

White LF, Shurin JB. Density dependent effects of an exotic marine macroalga on native community structure. J Exp Mar Biol Ecol. 2011;405:111–9.

Willette DA, Ambrose RF. The distribution and expansion of the invasive seagrass *Halophila stipulacea* in Dominica, West Indies, with a preliminary report from St.Lucia. Aquat Bot. 2009;91(3):137–42.

Willette DA, Ambrose RF. Effects of the invasive seagrass *Halophila stipulacea* on the native seagrass, *Syringodium filiforme*, and associated fish and epibiota communities in the Eastern Caribbean. Aquat Bot. 2012;103:74–82.

Williams SL. Introduced species in seagrass ecosystems: status and concerns. J Exp Mar Biol Ecol. 2007;350(1–2):89–110. doi:10.1016/j.jembe.2007.05.032.

Williams SL, Grosholz ED. The invasive species challenge in estuarine and coastal environments: marrying management and science. Estuaries Coasts. 2008;31(1):3–20.

Williams SL, Smith JE. A global review of the distribution, taxonomy, and impacts of introduced seaweeds. Annu Rev Ecol Evol Syst. 2007;38:327–59. doi:10.1146/annurev.ecolsys.38.091206.095543

Williams SL, Davidson IC, Pasari JR, Ashton GV, Carlton JT, Crafton RE, Fontana RE, Grosholz ED, Miller AW, Ruiz GM. Managing multiple vectors for marine invasions in an increasingly connected world. Bioscience. 2013;63(12):952–66.

Wolf MA, Sfriso A, Andreoli C, Moro I. The presence of exotic *Hypnea flexicaulis* (Rhodophyta) in the Mediterranean Sea as indicated by morphology, rbcL and cox1 analyses. Aquat Bot. 2011;95(1):55–8. doi:10.1016/j.aquabot.2011.02.009.

Wolf MA, Sfriso A, Moro I. Thermal pollution and settlement of new tropical alien species: the case of *Grateloupia yinggehaiensis* (Rhodophyta) in the Venice Lagoon. Estuar Coast Shelf Sci. 2014;147:11–6. doi:10.1016/j.ecss.2014.05.020.

Wright JT, Gribben PE. Predicting the impact of an invasive seaweed on the fitness of native fauna. J Appl Ecol. 2008;45:1540–9.

Wright JT, McKenzie LA, Gribben PE. A decline in the abundance and condition of a native bivalve associated with *Caulerpa taxifolia* invasion. Mar Freshw Res. 2007;58:263–72.

Wu YT, Wang CH, Zhang XD, Zhao B, Jiang LF, Chen JK, Li B. Effects of saltmarsh invasion by *Spartina alterniflora* on arthropod community structure and diets. Biol Invasions. 2009;11 (3):635–49. doi:10.1007/s10530-008-9279-1.

York PH, Booth DJ, Glasby TM, Pease BC. Fish assemblages in habitats dominated by *Caulerpa taxifolia* and native seagrasses in south-eastern Australia. Mar Ecol Prog Ser. 2006;312:223–34.

Zenetos A, Gofas S, Verlaque M, Çinar ME, Garcia Raso J, Bianchi C, Morri C, Azzurro E, Bilecenoglu M, Froglia C. Alien species in the Mediterranean Sea by 2010. A contribution to the application of European Union's Marine Strategy Framework Directive (MSFD). Part I. Spatial distribution. Mediterr Mar Sci. 2010;11:381–493.

Zenetos A, Gofas S, Morri C, Rosso A, Violanti D, Raso JEG, Cinar ME, Almogi-Labin A, Ates AS, Azzurro E, Ballesteros E, Bianchi CN, Bilecenoglu M, Gambi MC, Giangrande A, Gravili C, Hyams-Kaphzan O, Karachle PK, Katsanevakis S, Lipej L, Mastrototaro F, Mineur F, Pancucci-Papadopoulou MA, Espla AR, Salas C, San Martin G, Sfriso A, Streftaris N, Verlaque M. Alien species in the Mediterranean Sea by 2012. A contribution to the application of European Union's Marine Strategy Framework Directive (MSFD). Part 2. Introduction trends and pathways. Mediterr Mar Sci. 2012;13(2):328–52.

Chapter 7
Towards an Integrative Phylogeography of Invasive Marine Seaweeds, Based on Multiple Lines of Evidence

Marianela Zanolla and Nikos Andreakis

Abstract Molecular phylogeography has for decades been a frequently used approach to delineate novel evolutionarily significant units (ESUs) and to study the dynamics of invasive species. Next-generation sequencing technology (NGS) and the use of environmental DNA (eDNA) have the potential to revolutionize our way of understanding biodiversity and to establish rapid protocols for early-stage detection of invasive species. In seaweeds, however, several years of research on iconic invasive taxa of ambiguous taxonomic status (e.g. *Caulerpa*, *Codium*, *Asparagopsis*) have suggested that an integrative approach, namely the combination of multiple lines of evidence (e.g. phylogeographic, ecological, physiological and predictive modelling), is necessary to accurately resolve the taxonomy and their invasive potential. At present, integrative approaches in these fields are often weak because of incongruences among species delineation, newly discovered ESUs which remain undescribed taxonomically, and because databases containing vouchers of barcoded specimens are incomplete. As relocations of marine biota accelerate and climatic changes offer new potential niches for invasive seaweeds, new, transferable and internationally adopted protocols are necessary for exploring, monitoring and managing marine biodiversity. This is particularly urgent in areas of intense maritime traffic, such as the Mediterranean Sea and the Hawaiian archipelago, in order to achieve sustainable socio-economic development without compromising the local marine resources.

M. Zanolla (✉)
Universidad de Málaga, Departamento de Biología Vegetal (Botánica), Facultad de Ciencias, Campus de Teatinos s/n, 29071 Málaga, Spain
e-mail: marianela@uma.es

N. Andreakis
Australian Institute of Marine Science, PMB no. 3, Townsville, QLD 4810, Australia
e-mail: n.andreakis@aims.gov.au

N. Andreakis
College of Marine and Environmental Sciences, James Cook University, Townsville, QLD, Australia

Z.-M. Hu and C. Fraser (eds.), *Seaweed Phylogeography*,
DOI 10.1007/978-94-017-7534-2_7

Keywords Cosmopolitan species · Cryptic · Distribution · Introductions · Invasion · Life stages · Molecular · Species complex

7.1 Introduction

Humans have relied on macroalgae for food since the very early stages of civilization. Today, seaweeds also represent a promising resource for biocompounds, alternative energy, bioindicators for environmental health monitoring and bioextractors for recovering eutrophic areas (Gupta and Abu-Ghannam 2011). From an ecological perspective, seaweeds hold a fundamental role in regulating the nutrient composition of the water column, the hydrodynamic forces and sedimentation. Further, as ecosystem engineers, seaweeds provide shelter for the development and maintenance of benthic communities, occupy empty space and are placed at the base of trophic nets (Lüning et al. 1990). However, in the era of global climate change and accelerated trade, seaweeds are of major concern as invasive species (in the sense of Boudouresque and Verlaque 2010), posing severe social and economic threats in the coastal economy of many countries (Andreakis and Schaffelke 2012; Schaffelke and Hewitt 2007; Schaffelke et al. 2006). Estimates indicate that seaweeds represent up to 40 % of the non-native introduced species in the world's oceans (Schaffelke et al. 2006). The outstanding success of some invasive aliens is mostly attributed to remarkable levels of propagule persistence during transport across the globe, together with a suite of demographic traits that support adaptation and elevated growth rates in the recipient environments (Anderson 2007; Engelen and Santos 2009; Flagella et al. 2006, 2007, 2010; Zanolla et al. 2015).

A combination of environmental variables such as temperature, light, salinity and nutrients affect seaweed survival and distributions locally, regionally and globally (Breeman et al. 1988, 2002; Eggert 2012). These variables are subject to change either slowly through geological time at large geographical scales, or rapidly within decades locally or globally, in response to climate change and ocean acidification (O'Hara et al. 2011; Harley et al. 2012). Historical climatic fluctuations and large scale geological events have often altered the ecophysiological optima for survival in many marine groups including seaweeds, causing extensive shifts in species distributions, and may explain some present-day biogeographic patterns (Maggs et al. 2008; Payo et al. 2013). Signatures of diversification and distributional shifts driven by the reduction of global sea levels by more than a hundred metre, during Pleistocene glacial maxima, can be detected in many marine communities as a periodic, slow, yet naturally occurring process (Ludt and Rocha 2015). The same, however, cannot be assumed for the present-day changes observed on the distribution patterns of many marine groups. In the last decades, we have witnessed unprecedented rates of species range shifts (Chen et al. 2011) and relocation among bioregions both intentionally or accidentally via human-mediated transport (Sorte et al. 2010). The rates and the distances within which species are moving across oceans cannot be compared to the macroecological changes observed in historical times and they have been interpreted as a consequence of climate change (Boudouresque and Verlaque 2002, 2010; Andreakis and Schaffelke 2012).

Phylogeography is concerned with identifying the processes responsible for the geographic distribution of genealogical lineages in space and time. A gene genealogy can be inferred from genetic data extracted at individual, population or species level to test how historical, geological, climatic or ecological events have influenced their distribution patterns (Avise 2000). Methodological approaches based mostly on phylogeographic inference and species spatial distribution modelling have recently become the main tools for identifying introduced species, deciphering sources of introduction and assessing the success and invasive potential of new colonists at multiple stages across the invasion process (Peterson 2003; Booth et al. 2007; Bolton et al. 2011). Given that human-mediated transport and global change facilitate diffusion of biota that would otherwise have limited dispersal potential, it becomes obvious that surveys aiming to identify endemisms and detect introduced species will have profound consequences in conservation biogeography and ecosystem management (Bickford et al. 2007; Andreakis and Schaffelke 2012).

In this chapter, we discuss the importance of properly delineating taxonomic units in phylogeographic research of invasive seaweeds. The case studies discussed below are not necessarily considered pests with demonstrated economic impact,

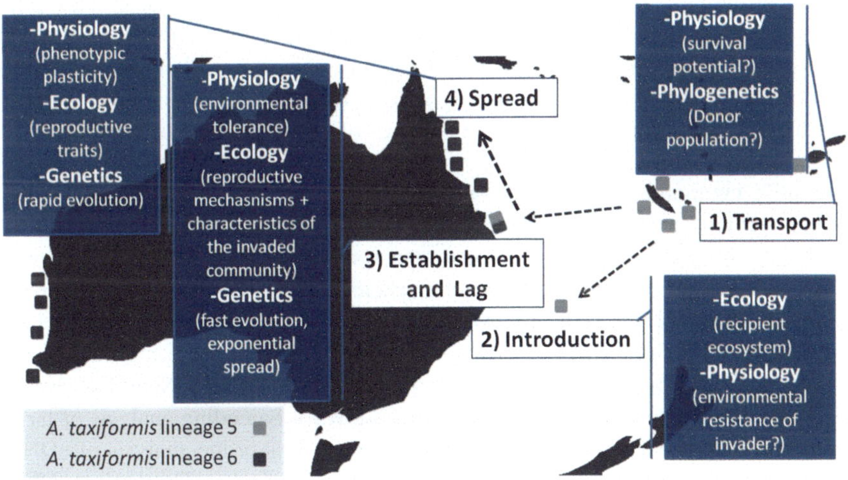

Fig. 7.1 The integrative phylogeographic approach for studying biological invasions. Progression of the invasive process (*1–4*) and the disciplines necessary to study the main variables involved in each stage are reported in brackets. The representation is based on the introduction of multiple *Asparagopsis* cryptic lineages in Australia

since this has been established only in a few cases (Andreakis and Schaffelke 2012). We further argue that current knowledge and methodologies are insufficient for accurate predictions of whether introduced seaweeds will become invasive, pests or neither of the above. However, in cases of already-established invaders or species of recognized invasive potential, it is possible to accrue evidence for forecasting source populations, direction of range expansion and to predict hypothetical distribution ranges based on environmental suitability. In invasion biology, the adoption of integrative approaches based on species distribution modelling, phylogeographic inference and ecophysiological data is necessary for successful predictions and conservation planning (Fig. 7.1).

7.2 The Advantages of Molecular Tools in Delineating Species

Accurate taxonomic delineation is essential for identifying the organisms being transported, understanding the dynamics of the invasion processes, and tracking the species' historical root or cryptogenic status in situ from the analysis of historical collections (Sherwood 2008). Despite several species definitions proposed previously (Mayden 1997) and approaches for testing them (Leliaert et al. 2009), an unifying concept of species still represents a hot debate within the scientific and environmental community (Wattier and Maggs 2001; Carstens et al. 2013). Modern biology in the post-genomic era calls for a convincing and universal species definition to be used as the basic unit in biodiversity research. This is crucial for corroborating ecological, biological and evolutionary interpretations, executing practical applications such as estimating species richness and bio-invasion control programmes and comparing invasive processes of the same organism from distant areas (Pante et al. 2015). Morphology has traditionally been the dominant criterion to identify taxonomical units in seaweeds (Wattier and Maggs 2001). Molecular markers provide a solid alternative when morphological data are insufficient or inexistent. Indeed, a few cases of new species descriptions are today accepted without molecular corroboration. Genetic species delineation offers several undeniable advantages. First, neutrally evolving DNA regions do not suffer homoplasy compared to morphological and ecophysiological traits which are responsible for the species' functional reaction to its surrounding environment (Saunders 2005). Second, modern molecular biology platforms allow for barcoding analysis of large numbers of samples for the validation of a specific ESUs or the analysis of eDNA (environmental DNA) for the identification of cryptic species. Third, since DNA composition at the sequence level remains the same across the ontogenetic development of any organism, molecular tools have the power to accurately identify heteromorphic life stages or microscopic forms from within the species life cycle or even fragments of propagules. Disadvantages of molecular taxonomy are mostly related to the adoption of inconsistent laboratory methodologies amongst laboratories and the production of heterogeneous data from multiple DNA regions instead of consistently targeting the same markers across the group in question (Verbruggen et al. 2010).

Molecular taxonomy involves (a) the collection of molecular data (e.g. DNA sequences from one or more genomic regions) from multiple specimens of the group under examination and (b) their phylogenetic analysis, in order to identify well-supported clusters of closely related specimens corresponding to ESUs (Moritz 1994). ESUs are here intended as taxa which are reciprocally monophyletic for mtDNA alleles and show significant divergence of allele frequencies at nuclear loci (Moritz 1994). Different molecular markers are characterized by different evolutionary rates. The choice is based on their ability to achieve unambiguous identifications given the taxonomical level questioned (Hebert et al. 2003). Examples of different markers developed for systematic and evolutionary studies in seaweeds are shown in Table 7.1. In phylogeography of invasive seaweeds, the choice of the marker is a fundamental decision for identifying introduced species. The selected molecular marker should have a suitable resolution for detecting the target taxa either in their vectors of transport or within their suspected introduced range (Geller et al. 2010).

Molecular taxonomy and phylogenetics have revolutionized our perception of seaweed biodiversity by revealing hitherto unknown levels of diversity, and have provided a statistically robust framework for testing evolutionary hypotheses (Bickford et al. 2007). Cryptic speciation is common in many marine organisms (Knowlton 1993). Leaving morphological delineation aside, a preliminary genetic screening is always necessary to draw the bottom line when it comes to investigations within widely distributed taxa (Geller et al. 2010). In light of combined molecular and morphological evidence, a choice must be made on which taxonomical level to focus on and bypass potential incongruences that may exist between molecular data, morphological descriptions and previous reports. A classic example of an extremely morphologically plastic group, in which genetic surveys have been essential tools for species delineation, is the green algal genus *Caulerpa* J.V. Lamouroux. *Caulerpa* represents an iconic case in which the phylogenetic approach has led to controversy because of differences in the chosen molecular markers (and their resolution) among studies. The original results of the first studies based mostly on morphology led to misidentifications and constant taxonomic restructuring. Previous morphological identifications of *Caulerpa cylindracea* Sonder led to erroneous records of the invasive strain (Verlaque et al. 2003; Yeh and Chen 2004; Nuber et al. 2007), later emended by molecular studies based on the ITS marker (Klein and Verlaque 2008). Further, *C. cylindracea* started as a "variety" (*C. racemosa* var. *cylindracea* (Sonder) Verlaque, Huisman & Boudouresque), went through a "form" (*C. racemosa* f. *complanata* (J. Agardh) Weber-van Bosse), and ended as a "species" into which a new cryptic variety has been included (Belton et al. 2014).

Table 7.1 Molecular markers developed for systematic and evolutionary studies in seaweeds

Phyllum	Genus	Marker	Resolution power	Reference
Chlorophyta	*Caulerpa*	ITS1 & ITS2	Varieties (strains)	Olsen et al. (1998)
				Meusnier et al. (2004)
			Haplotypes	Jousson et al. (1998)
				Schaffelke et al. (2002)
	Codium	*psb*J-*psb*L, *rpl*16	Subspecies	Provan et al. (2005)
		*rbc*L, *rps*3–*rpl*16	Species	Verbruggen et al. (2007)
Phaeophyta	*Fucus*	Rubisco spacer	Species	Coyer et al. (2004)
	Sargassum	A198, B113, B128 & F4	Hybrids	Johnson et al. (2012)
		TrnW_I spacer	Haplotypes	Johnson et al. (2012)
		RUBISCO, TrnW_I, ITS2	Haplotypes	Cheang et al. (2010)
		COI, ITS	Species	McDevit and Saunders (2009)
	Undaria	atp8-S, W-I	Haplotypes	Voisin et al. (2005)
		*cox*3, ITS1	Haplotypes	Uwai et al. (2006)
		20 loci	Microsatellites	Daguin et al. (2005)
Rhodophyta	*Polysiphonia*	*rbc*L	Haplotypes	McIvor et al. (2001)
				Geoffroy et al. (2012)
	Gracilaria	*cox*2-3	Haplotypes	Thomsen (2005)
		COI	Species	Saunders (2009)
		*cox*1	Haplotypes	Kim et al. (2010)
	Asparagopsis	*cox*2-3, LSU, RuBisCo	Species	Andreakis et al. (2004)
		*cox*2-3, LSU, RuBisCo	Lineages	Andreakis et al. (2007a, b)
		Microsatellites		Andreakis et al. (2009)
		COI, *cox*2-3	Haplotypes, lineages	Sherwood (2008)
		*cox*2-3	Haplotypes	Bolton et al. (2011)
	Eucheuma	LSU, UPA, *cox*2-3		Conklin et al. (2009)
	Kappaphycus	*cox*2-3	Haplotypes	Halling et al. (2013)
	Acanthophora	LSU, *cox*2-3	Fail to uncover genetic structure	O'Doherty and Sherwood (2007)
	Grateloupia	*rbc*L, *cox*2-3	Species	D'Archino et al. (2007)
		RAPDs haplotypes	Genetic variability	Marston and Villalard-Bohnsack (2002)

ITS: internal transcribed spacer; UPA: Universal Plastid Amplicon; COX: cytochrome c oxidase; COI cytochrome c oxidase 1; LSU: large-subunit of ribosomal RNA; rbcL: ribulose bisphosphate carboxylase/oxygenase

7.3 Multiple Cryptic Endemisms or Introduced Lineages Within Cosmopolitan Species?

Sorting out endemic lineages from cryptic introductions may be difficult without information gathered from historical collections, since the discovery of a new species may be the result of unrecognized parapatric or sympatric speciation (Voisin et al. 2005; Schaffelke et al. 2006; Andreakis and Schaffelke 2012). Unnoticed propagules of cryptic lineages can successfully establish founder populations following human-mediated long-range dispersal in areas of environmental suitability for that lineage (Breeman 1988). Megadiverse areas such as the coral triangle and the Great Barrier Reef or bioregions affected by intense maritime traffic such as the Mediterranean Sea and the Hawaiian archipelago are considered highly vulnerable to biological invasions. In these areas, the identification of cryptic endemisms, versus cryptic introductions, represents both a major endeavour for conservation biogeography and a knowledge gap capable of compromising future biodiversity management strategies (Bickford et al. 2007; Trontelj and Fišer 2009). Misidentifications of one or the other will result in a complete overlook of either the invasive organism, leading to a cryptic invasion, or unrecognized endemic lineage (McIvor et al. 2001; Geller et al. 2010). Given the importance of biodiversity fluctuations for ecological marine conservation, control plans and assessments of the long-term ecological and evolutionary consequences of cryptic species must be a priority for management agencies.

7.4 The Impact of Multiple Invasive Life Stages

Life-history strategies characterized by multiple phases of distinct morphology and ploidy level are common in seaweeds (Drew 1955). Each of the life stages can therefore contribute differently in the expansion potential of the species across the course of an invasion, assuring its success acting either as dispersal units (Hewitt et al. 2007; Zanolla et al. 2015) or seed banks (Hewitt et al. 2005). For instance, the red seaweed genus *Asparagopsis* has a triphasic diplohaplontic heteromorphic life cycle (Fig. 7.2), in which gametophytes, microscopic carposporophytes and filamentous tetrasporophytes (Falkenbergia) of unknown ploidy level alternate (Feldmann and Feldmann 1942; Rojas et al. 1982).

Depending on how pronounced the heteromorphy is, each of the life-history stages may eventually belong to a different functional group and is thus expected to present different thermal ranges of reproduction, growth and survival (Breeman 1988; Eggert 2012), as well as to be subject to distinct biotic and abiotic pressures (Littler and Littler 1980). Microscopic life stages are believed to be more resistant (Breeman 1988) and are thus considered good candidates for long distance

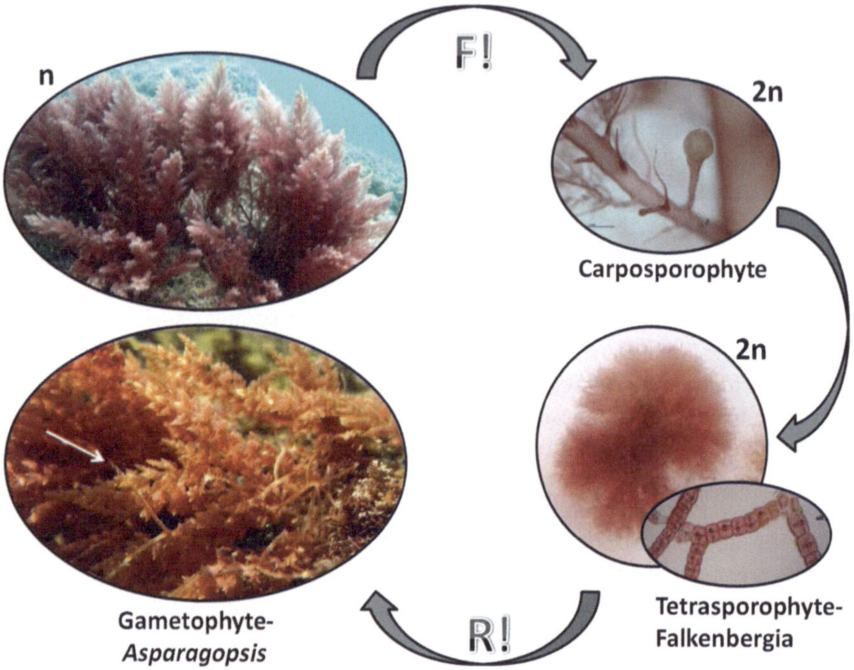

Fig. 7.2 Life cycle of the invasive red genus *Asparagopsis*. White arrow indicates the harpoon-like structures characteristic of *A. armata* which are absent in *A. taxiformis*. F!: fecundation; R!: meiosis, n: ploidy level

dispersal. Macroscopic phases on the other hand would be responsible for increased biomass production and will guarantee population persistence and short distance dispersal. The contribution to the population persistence of each of the life-history stages, ploidy levels, clones or sexually reproduced individuals may vary. Their impact will depend on the functional group affected and may reflect shifts in the dominance of one versus the other according to their adaptation potentials (Van der Strate et al. 2002) and the environmental characteristics of the invaded location.

7.5 Vectors of Introduction Promote Relocations of Seaweeds: Range Shifts Versus Niche Shifts

A major cause of increased invasion rates is the reinforcement of multiple introduction vectors rather than global climate change itself (Boudouresque and Verlaque 2010). Aquaculture, maritime transport and the opening of interoceanic canals, to mention a few, have been widely accepted as major causes for seaweed introductions (Williams and Smith 2007). In Europe, globalization has promoted an increased number of maritime commercial shipping routes, considered responsible for more than half of the introduced species in the marine environment, followed by the opening of artificial water corridors between basins. Intentional or unintentional introductions of invasive species through aquaculture and aquarium trade are considered less relevant pathways for the spread of invasive species in Europe (<1.2 %) (Katsanevakis et al. 2013). Invasive species provide unique model systems for ecology and evolutionary biology since the species range shift following introduction can potentially be accompanied by niche shifts because of the species' environmental tolerance in the introduced region (Rödder and Lötters 2009; Tingley et al. 2014). A 'niche shift' refers to "any change in the position of either the fundamental or realized (Hutchinsonian) niche of a species" (Pearman et al. 2008). As the exploitation of novel habitats and niches may lead to adaptive radiations, the study of such organisms can provide important insights in understanding species diversification. Further, ecological niche models for predicting potential distribution of invasive species may allow us to anticipate invasions (Guisan et al. 2014). Niche conservatism, i.e. the extent to which niches are conserved over space and time, is a useful approach for extrapolating invasive species' distribution ranges and predicting their invasive risk. Niche shifts among seaweeds have only been rarely documented. For instance, in *Halimeda* J.V. Lamouroux (Verbruggen et al. 2009), despite the dispersal limitations and the conservatism of the genus to tropical habitats, successful colonisation of colder environments has occurred than once in the past. However, range shifts of tropical seaweeds are expected to occur more frequently as a result of ongoing global warming (Thuiller et al. 2005; Boudouresque and Verlaque 2010).

7.6 Multiple Introductions from a Single Source

Biological invasions may involve the introduction of multiple contingents of individuals from the same source or multiple introduced individuals from more than one source. *Asparagopsis armata* Harvey is common in temperate seas and is believed to be native in Southern Australia and New Zealand (Dixon 1964). Two cryptic lineages have been recently reported within this species (lineages 1 and 2) (Dijoux et al. 2014). Lineage 1 has been introduced to northern Mediterranean Sea, Western Europe, Japan and the US west coast. Analysis of nuclear, mitochondrial, and plastid molecular markers revealed a unique southern Australian origin of invasive Mediterranean populations of lineage 1 of *A. armata* (Andreakis et al. 2007b). This conclusion is supported by the lack of genetic structuring among invasive populations, the shared haplotypes between recipient and donor regions and the invasive history of this species (Feldmann and Feldmann 1942; Andreakis et al. 2007b).

Codium fragile subsp. *tomentosoides* (van Goor) P.C. Silva is a well-recognized invasive green seaweed. This taxon is characterized by low genetic variation both within its introduced (Mediterranean Sea, Northwest Atlantic, Northern European and South Pacific) and native (Japan) range although parthenogenesis, prevalent in this genus, may also contribute to this (Prince and Trowbridge 2004). Introduced *Codium* populations shared unique haplotypes with the subspecies donor region. Eight different haplotypes were recovered in Japan and only one of them could be found exclusively in the Mediterranean Sea. The latter was clearly different from the haplotypes present in other introduced locations (Provan et al. 2005, 2008).

The "aquarium-Mediterranean strain" of *Caulerpa taxifolia* (M. Vahl) C. Agardh has invaded the Mediterranean Sea through aquarium trade (Meusnier et al. 2002). Phylogenetic studies of the aforementioned strain which was released from aquarium facilities revealed the origin of these invasive populations to be the tropical and subtropical region of Australia (Meusnier et al. 2002). However, invasive Australian populations of this same taxon were suggested to be derived from different source regions (Schaffelke et al. 2002).

7.7 Introductions from Multiple Sources

In many cases, invasive populations are the result of introductions from multiple sources. This type of introduction is often overlooked when the so-called cosmopolitan species are believed to be native in many regions or when successfully established invasive populations act as propagule donors. *Asparagopsis taxiformis* (Delile) Trevisan de Saint León represents a clear example (Fig. 7.3). Multiple lineages are known from the Mediterranean Sea, the Atlantic and the Indo-Pacific Oceans. Interestingly, invasive behaviour has been reported for solely lineages 2 and 3 in the Mediterranean region (Andreakis et al. 2007b; Boudouresque and Verlaque 2010) and South Africa (Bolton et al. 2011), respectively. Further, lineage 2 recently expanded its distributional range in southern Portugal and the Mediterranean Sea, and it is considered an invasive taxon characterized by high genotypic diversity (Andreakis et al. 2009). Two more lineages have been recently described, both confined to South Pacific and Western Australia (Dijoux et al. 2014; Andreakis et al. 2016). To what extent these *Asparagopsis* lineages represent biologically different species, ESUs or groups of distinct genotypes, still requires further assessments (Dijoux et al. 2014; Zanolla et al. 2014, 2015). However, Mediterranean strains of lineage 2 might represent a distinct ecotype for that lineage found in Australia (Andreakis et al. 2007b; Dijoux et al. 2014) and the Hawaiian Islands (Sherwood 2008) and not a distinct genetic variant. This is confirmed by its distinct morphology, photosynthetic performance and the identical mitochondrial haplotypes shared among Mediterranean Australian, Hawaiian and African isolates (Zanolla et al. 2014, 2015).

Fig. 7.3 Updated distribution occurrence map in the Mediterranean Sea (**a**), the Alboran Sea (**b**) and the Hawaiian Islands (**c**) for each of the *Asparagopsis* lineages based on genetically delineated specimens (Andreakis et al. 2004, 2007b; Sherwood 2008; Dijoux et al. 2014). *A. armata* (*filled triangle*); *A. taxiformis* L1 (X); *A. taxiformis* L2 (*filled circle*); *A. taxiformis* L3 (*filled square*); *A. taxiformis* L4 (*filled diamond*), modified from Zanolla et al. (Submitted)

7.8 Differences Between Donor and Introduced Populations

Genetic variation of introduced seaweed populations may differ significantly from the donor populations because (a) individuals transported under suboptimal conditions are subjected to negative selection, (b) introduction events may be multiple (propagule pressure) and (c) introduced species undergo strong bottleneck events following introduction. Consequently, as a general trend, the genetic profile of a recently introduced population may appear less variable due to the perpetuation of few successful genotypes (Voisin et al. 2005). In addition, genetic variability among invasive populations may differ because of dissimilar population propagation mechanisms (i.e. sexual vs. vegetative propagation) and/or introduction events from multiple sources (Andreakis et al. 2009).

Invasive species characterized by identical genetic profiles between their native and introduced strains may develop adaptive phenotypic plasticity and ecophysiological tolerance in response to the novel environmental conditions (Eggert 2012). Rapid adaptation will promote a superior fitness to the introduced individuals which will be characterized by distinct ecophysiological and/or morphological traits, compared to populations found in the species native range (Andreakis and Schaffelke 2012). Adaptive plasticity can therefore confer evolutionary advantages to invasive species by optimizing their acclimation mechanisms (Davidson et al. 2011).

7.9 Integrative Taxonomy and Phylogeography: Combining Multiple Lines of Evidence

In several cases, a molecular phylogeographic approach has been decisive for the identification of cryptic invaders, the detection of organisms imported via vectors of transport such as ballast tanks and to infer the colonization route and the donor population of introduced taxa (Deagle et al. 2003; Bolton et al. 2011). However, for robust species delineation and prediction making, the combination of data from multiple lines of evidence will give more realistic results (Figs. 7.1 and 7.4). Molecular, morphological, physiological and geographic distribution data tested against multiple species concepts (e.g. phenetic, biological and phylogenetic) render species delineation unambiguous. A species integrative profile can thereafter be tested against observed phenetic, physiological or behavioural variants for that species and assess whether these changes are (a) part of the species' adaptation potential within its plasticity range, (b) associated with genome level variations or specific gene expression profiles or (c) are the result of transgenerational adaptation

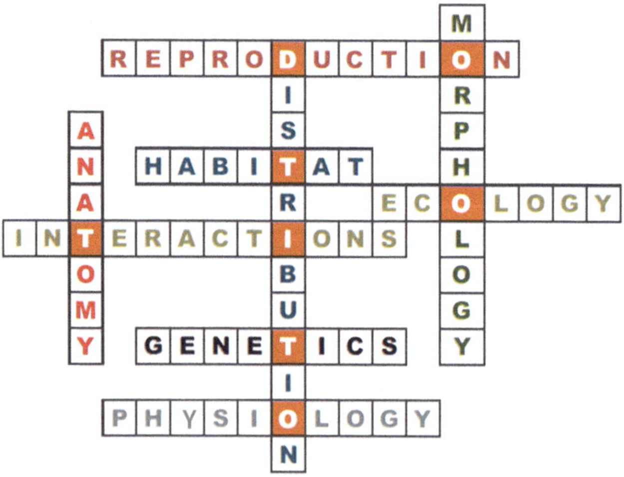

Fig. 7.4 Integrating baseline disciplines from which to construct a species working hypothesis in invasion biology

mechanisms based on epigenetic modifications. The latter is often induced by diversifying selection among populations invading different habitats (Weinig 2000; Lee 2002).

The invasive red seaweed genus *Asparagopsis* (Montagne) provides a good example of combining molecular, morphological and ecophysiological data in resolving taxonomic issues. Previous reports on the presence of *Asparagopsis* species worldwide were limited to reporting the species *sensu lato* rather than the corresponding cryptic lineage itself. This misidentification resulted in erroneous distribution maps and a general failure to identify species where molecular studies had not been employed. Immediate *Asparagopsis* lineage delineation is now possible, even without molecular screening, by means of a set of vegetative and reproductive diagnostic morphological characters. These characters have been identified from morphological studies performed on genetically delineated specimens of *Asparagopsis* lineages collected from sites where these lineages are considered either introduced or native (Zanolla et al. 2014).

7.10 The Utility of Combining Multiple Lines of Evidence in the Study of Invasive Seaweeds

Species distribution modelling (SDM) calculates the similarities of environmental affinities between locations where a species occurs and locations where the same species has never been reported. As a consequence, SDM can be used to forecast distribution ranges using environmental variables as predictors (Austin 2002). In invasion biology, SDMs are routinely used to identify potentially suitable settlement areas of invasive species. However, since SDMs require the compilation of georeferenced data (species presence or absence), the precise taxonomic identification and deep knowledge of the target species' ecophysiological optima have become of paramount importance for the reliable performance of the models. Accurate information on the taxonomy, distribution and ecophysiological limits of species or lineages can be combined in SDM to provide crucial knowledge in biodiversity research and invasion biology. For instance, in the case of a cosmopolitan species, the occurrence of that species in locations not suggested by the models may indicate a novel cryptic ESU, which can represent either an endemism or a cryptic introduction that requires further attention for management and conservation.

A combination of field and laboratory work, aiming to identify the taxonomic position and understand the physiology, demography, phenology, population dynamics and impact on the local community of an invasive organism, is necessary to detect biological invasions but also designates them as pests (Meinesz 2007). The astonishing success of invasive species relies on their extraordinary adaptation potential. As a process, adaptation relies on the improvement of multiple functional traits in the life cycle of an organism capable of enhancing its fitness, survival and resilience against novel environmental conditions. Consequently, adaptation potential cannot be understood by means of one approach. The extreme adaptive responses in invasive seaweeds can be clearly visible (e.g. noticeable ecophysiological and morphological differences between native and introduced populations) or invisible. The latter category includes adaptive responses associated with positively selected genetic variants in response to novel conditions or responses associated with transgenerationally inherited epigenetic polymorphisms. The latter in their turn are induced by environmental cues, and might not be detectable by traditional sequence analysis approaches (i.e. no differences at the DNA sequence level between native and introduced populations). Thus, the invasive behaviour observed in an organism cannot be explained on the basis of one line of evidence. The importance of the integrative approach becomes further obvious when the communication of scientific knowledge goes through stakeholders responsible for the implementation of management plans and decision making.

An integrative approach has been implemented in the genus *Asparagopsis*, one of the few cases where haplotype analysis, historical demography and SDM were combined to assess lineage-specific invasive risk (Zanolla et al., submitted), lineage-specific photosynthetic plasticity in response to a range of temperatures (Zanolla et al. 2015) and morphological differentiation among cryptic lineages of genetically delineated tetrasporophytes and gametophytes locally and globally (Zanolla et al. 2014). Further, vegetative and reproductive traits were examined taking as a model an established population of this invasive taxon (Zanolla et al., submitted). This comprehensive multifaceted approach allowed characterization of the genetic composition, colonization strategy and lineage-specific potential for short and long distance dispersals as well as invasion risk.

7.11 Modern Technology and Metabarcoding in the Study of Invasive Seaweeds

Global climate change, human-mediated transport and extensive urban development of coastal zones are responsible for increasing rates of marine species introductions but also conspicuous behavioural and/or phenetic changes in many endemic, bloom-forming taxa. Because of eutrophication or accidental release, both native and introduced organisms can possibly and unexpectedly turn into pests. Therefore, the complete prevention of seaweeds invasions or native algal blooms appears nearly impossible in the long term. However, focusing on taxonomic groups associated with high rates of invasibility risk and the establishment of early warning protocols, especially in vulnerable areas, can help reduce the speed and ecological impact of invasions. Modern high throughput molecular approaches possess the resolution power and the affordability to unravel many aspects of the biology of invasive organisms at functional and molecular levels. Novel technology has the potential to address questions and examine behavioural changes associated with the metabolomic, proteomic, genomic and transcriptomic profiles of an invasive organism but also the influences of bacterial symbionts in the species response against environmental cues. A major outcome is expected in genomic and/or proteomic identification of biomarkers to be used for assessing invasive potential in high-risk groups. The same marker profiles can be used to monitor the possibilities of invasiveness in endemic taxa. In addition, genomic approaches are expected to resolve the influence of polyploidy in inducing invasiveness in seaweeds. Polyploidy has already been proven to be relevant in the invasiveness of higher plants (Pandit et al. 2011) and has been proposed as a mechanism capable of supporting invasiveness in green (*Caulerpa*) and red (*Asparagopsis*) algal genera (Andreakis et al. 2007a; Varela-Álvarez et al. 2012). The correlation between invasive potential and increased ploidy levels in these species is likely explained by increased levels of heterozygosity associated with polyploidy (Brochmann et al. 2004). Increased heterozygosity may support the ecological success of an introduced alien by balancing the loss of diversity due to population bottleneck and low sexual recombination that take place at early stages of introduction (reviewed in Varela-Alvarez et al. 2012).

Next-generation sequencing-based eDNA/RNA metabarcoding from environmental samples is expected to play a crucial role in the monitoring, detection and identification of introduced, transported or invasive species (Chown et al. 2015). Although not extensively applied in assessing biological invasions, eDNA/RNA metabarcoding has so far showed promising results (Armstrong and Ball 2005; Ficetola et al. 2008; Saunders 2009). The technology relies on the same general principle of the regular DNA barcoding for species identification. Short DNA sequences (barcodes) originated upon a previously agreed, high-resolution DNA marker are compared to barcodes produced from well-identified voucher specimens deposited in reference databases. This approach can be designed to capture more than one species and can be applied in all steps of the invasion history (Fig. 7.1), or the species life stages (Fig. 7.2), to provide constant monitoring support and information for early detection and identification of dispersal vectors, therefore allowing estimates on the demography, population dynamics and dispersal of the invasive organism (Metzker 2010). Although initially the short DNA reads (ca. 100 bp) limited the use of NGS for DNA metabarcoding purposes, this downside has been now attenuated by the sequencing of longer reads produced by many platforms (e.g. >500 bp; Illumina, MiSeq, http://www.illumina.com/systems/miseq/performance_specifications.ilmn). At present, however, a real drawback of metabarcoding remains the current limited availability of reference databases (Cristescu 2014).

7.12 Conclusion

The combination of novel DNA-based analytical methodologies with traditional approaches has the potential to alleviate the methodological and conceptual critiques charged at invasion science (Richardson and Ricciardi 2013).

Several European countries have officially recognized the need for surveillance and monitoring of invasive species (No 1143/2014 of 22 October 2014; http://ec.europa.eu/environment/nature/invasivealien/index_en.htm). With the rapid identification and detection of non-native species being the top priority, DNA metabarcoding, in particular, represents an efficient and affordable method to identify transported species or detect the presence of invasive pests from minute quantities of DNA from environmental samples (Darling and Mahon 2011). To be successful, this approach requires the rapid description of the cryptic species and the parallel development of accurate reference databases, consisting of type voucher specimens associated with specific environmental profiles and lineage-specific barcodes (e.g. BoLD) (Ratnasingham and Hebert 2007).

An integrative approach to biological invasions can provide realistic solutions by documenting global patterns of the invasion process and by identifying areas into which direct management actions are immediately required (i.e. introduction vectors, marine protected zones, areas of intense maritime traffic). As the vertiginous socio-economic development of many countries of the world struggles to stay ecologically sustainable, the consequences of biological invasions still remain global and irreversible. This interface requires the use of common language and sense from both scientists and politicians towards rapid and effective balancing of management actions at national and international levels.

Acknowledgments Marianela Zanolla is supported by the project P09-RNM-5187 from the Consejería de Innovación, Ciencia y Empresa, Junta de Andalucía, Spain; NA by the Commonwealth Research and Environment Facilities (CERF) Marine Biodiversity Hub. The CERF programme is an Australian Government initiative supporting world class, public good research and is a collaborative partnership between the University of Tasmania, CSIRO Wealth from Oceans Flagship, Geoscience Australia, Australian Institute of Marine Science and Museum Victoria.

References

Anderson LW. Control of invasive seaweeds. Bot Mar. 2007;50:418–37.

Andreakis N, Procaccini G, Kooistra WHCF. *Asparagopsis taxiformis* and *Asparagopsis armata* (Bonnemaisoniales, Rhodophyta): genetic and morphological identification of Mediterranean populations. Eur J Phycol. 2004;39:273–83.

Andreakis N, Koolstra WHCF, Procaccini G. Microsatellite markers in an invasive strain of *Asparagopsis taxiformis* (Bonnemaisoniales, Rhodophyta): insights in ploidy level and sexual reproduction. Gene. 2007a;406:144–51.

Andreakis N, Procaccini G, Maggs C, Kooistra WHCF. Phylogeography of the invasive seaweed *Asparagopsis* (Bonnemaisoniales, Rhodophyta) reveals cryptic diversity. Mol Ecol. 2007b;16:2285–99.

Andreakis N, Kooistra WHCF, Procaccini G. High genetic diversity and connectivity in the polyploid invasive seaweed *Asparagopsis taxiformis* (Bonnemaisoniales) in the Mediterranean, explored with microsatellite alleles and multilocus genotypes. Mol Ecol. 2009;18:212–26.

Andreakis N, Schaffelke B. Invasive marine seaweeds: pest or prize? In: Wiencke C, Bischof K, editors. Seaweed biology. Berlin: Springer; 2012. p. 235–62.

Andreakis N, Costello P, Zanolla M, Saunders GW, Mata L. Endemic or introduced? Phylogeography of Asparagopsis (Florideophyceae) in Australia reveals multiple introductions and a new mitochondrial lineage. J Phycol. 52:2016. doi:10.1111/jpy.12373-15-166.

Armstrong K, Ball S. DNA barcodes for biosecurity: invasive species identification. Phil Trans R Soc Lond B Biol Sci. 2005;360:1813–23.

Austin M. Spatial prediction of species distribution: an interface between ecological theory and statistical modelling. Ecol Model. 2002;157:101–18.

Avise JC. Phylogeography: the history and formation of species. Massachusetts: Harvard University Press; 2000.

Belton GS, Reine WF, Huisman JM, Draisma SG, Gurgel D, Frederico C. Resolving phenotypic plasticity and species designation in the morphologically challenging *Caulerpa racemosa–peltata* complex (Chlorophyta, Caulerpaceae). J Phycol. 2014;50:32–54.

Bickford D, Lohman DJ, Sodhi NS, Ng PKL, Meier R, Winker K, Ingram KK, Das I. Cryptic species as a window on diversity and conservation. Rend Ecol Evol. 2007;22:148–55.

Bolton JJ, Andreakis N, Anderson RJ. Molecular evidence for three separate cryptic introductions of the red seaweed *Asparagopsis* (Bonnemaisoniales, Rhodophyta) in South Africa. Afr J Mar Sci. 2011;33:263–71.

Booth D, Provan J, Maggs CA. Molecular approaches to the study of invasive seaweeds. Bot Mar. 2007;50:385–96.

Boudouresque C, Verlaque M. Assessing scale and impact of ship-transported alien macrophytes in the Mediterranean Sea CIESM Workshop Monographs. 2002;53–61.

Boudouresque CF, Verlaque M. Is global warming involved in the success of seaweed introductions in the Mediterranean Sea? In seaweeds and their role in globally changing environments. Berlin: Springer; 2010. p. 31–50.

Breeman AM. Relative importance of temperature and other factors in determining geographic boundaries of seaweeds—experimental and phenological evidence. Helgol Meeresunters. 1988;42:199–241.

Breeman A, Oh YS, Hwang MS, Van Den Hoek C. Evolution of temperature responses in the *Cladophora vagabunda* complex and the *C. albida/sericea* complex (Chlorophyta). Eur J Phycol. 2002;37:45–58.

Brochmann C, Brysting A, Alsos I, Borgen L, Grundt H, Scheen AC, Elven R. Polyploidy in arctic plants. Biol J Linnean Soc. 2004;82:521–36.

Carstens BC, Pelletier TA, Reid NM, Satler JD. How to fail at species delimitation. Mol Ecol. 2013;22:4369.

Cheang CC, Chu KH, Ang PO. Phylogeography of the marine macroalga *Sargassum hemiphyllum* (Phaeophyceae, Heterokontophyta) in northwestern Pacific. Mol Ecol. 2010;19(14):2933–48.

Chen I-C, Hill JK, Ohlemüller R, Roy DB, Thomas CD. Rapid range shifts of species associated with high levels of climate warming. Science. 2011;333(6045):1024–6.

Chown SL, Hodgins KA, Griffin PC, Oakeshott JG, Byrne M, Hoffmann AA. Biological invasions, climate change and genomics. Evol Appl. 2015;8:23–46.

Conklin KY, Kurihara A, Sherwood AR. A molecular method for identification of the morphologically plastic invasive algal genera *Eucheuma* and *Kappaphycus* (Rhodophyta, Gigartinales) in Hawaii. J Appl Phycol. 2009;21(6):691–9.

Coyer JA, Hoarau G, Stam WT, Olsen JL. Geographically specific heteroplasmy of mitochondrial DNA in the seaweed, *Fucus serratus* (Heterokontophyta: Phaeophyceae, Fucales). Mol Ecol. 2004;13(5):1323–6.

Cristescu ME. From barcoding single individuals to metabarcoding biological communities: towards an integrative approach to the study of global biodiversity. Trends Ecol Evol. 2014;29(10):566–71.

Daguin C, Voisin M, Engel C, Viard F. Microsatellites isolation and polymorphism in introduced populations of the cultivated seaweed *Undaria pinnatifida* (Phaeophyceae, Laminariales). Conserv Genet. 2005;6(4):647–50.

D'Archino R, Nelson WA, Zuccarello GC. Invasive marine red alga introduced to New Zealand waters: first record of *Grateloupia turuturu* (Halymeniaceae, Rhodophyta). New Zeal J Mar Fresh Res. 2007;41(1):35–42.
Darling JA, Mahon AR. From molecules to management: adopting DNA-based methods for monitoring biological invasions in aquatic environments. Env Res. 2011;111:978–88.
Davidson AM, Jennions M, Nicotra AB. Do invasive species show higher phenotypic plasticity than native species and if so, is it adaptive? A meta-analysis. Ecol Lett. 2011;14:419–31.
Deagle B, Bax N, Hewitt C, Patil J. Development and evaluation of a PCR-based test for detection of Asterias (Echinodermata: Asteroidea) larvae in Australian plankton samples from ballast water. Mar Fresh Res. 2003;54:709–19.
Dijoux L, Viard F, Payri C. The more we search, the more we find: discovery of a new lineage and a new species complex in the genus *Asparagopsis*. PLoS ONE. 2014;9:e103826.
Dixon PS. *Asparagopsis* in Europe. Nature. 1964;204:902.
Drew KM. Life histories in the algae with special reference to the Chlorophyta, Phaeophyta and Rhodophyta. Biol Rev. 1955;30:343–90.
Eggert A. Seaweed responses to temperature seaweed biology. Berlin: Springer; 2012. p. 47–66.
Engelen A, Santos R. Which demographic traits determine population growth in the invasive brown seaweed *Sargassum muticum*? J Ecol. 2009;97(4):675–84.
Feldmann J, Feldmann G. Recherches sur les Bonnemaisoniacées et leur alternance de générations Annales des Sciences Naturelles. Bot Sér. 1942;11:75–175.
Ficetola GF, Miaud C, Pompanon F, Taberlet P. Species detection using environmental DNA from water samples. Biol Lett. 2008;4:423–5.
Flagella MM, Soria A, Buia MC. Shipping traffic and introduction of non-indigenous organisms: study case in two Italian harbours. Ocean Coast Manag. 2006;49:947–60.
Flagella MM, Verlaque M, Soria A, Buia MC. Macroalgal survival in ballast water tanks. Mar Poll Bull. 2007;54(9):1395–401.
Flagella MM, Andreakis N, Hiraoka M, Verlaque M, Buia MC. Identification of cryptic *Ulva* species (Chlorophyta, Ulvales) transported by ballast water. J Biol Res. 2010;13:47–57.
Geller JB, Darling JA, Carlton JT. Genetic perspectives on marine biological invasions. Annu Rev Mar Sci. 2010;2:367–93.
Geoffroy A, Le Gall L, Destombe C. Cryptic introduction of the red alga *Polysiphonia morrowii* Harvey (Rhodomelaceae, Rhodophyta) in the North Atlantic Ocean highlighted by a DNA barcoding approach. Aq Bot. 2012;100:67–71.
Guisan A, Petitpierre B, Broennimann O, Daehler C, Kueffer C. Unifying niche shift studies: insights from biological invasions. Trends Ecol Evol. 2014;29(5):260–69.
Gupta S, Abu-Ghannam N. Recent developments in the application of seaweeds or seaweed extracts as a means for enhancing the safety and quality attributes of foods. Innov Food Sci Emerg. 2011;12:600–9.
Halling C, Wikström SA, Lilliesköld-Sjöö G, Mörk E, Lundsør E, Zuccarello GC. Introduction of Asian strains and low genetic variation in farmed seaweeds: indications for new management practices. J Appl Phycol. 2013;25(1):89–95.
Harley CD, Anderson KM, Demes KW, Jorve JP, Kordas RL, Coyle TA, Graham MH. Effects of climate change on global seaweed communities. J Phycol. 2012;48(5):1064–78.
Hebert PDN, Cywinska A, Ball SL, DeWaard JR. Biological identifications through DNA barcodes. P R Soc London B Bio. 2003;270:313–21.
Hewitt CL, Campbell ML, McEnnulty F, Moore KM, Murfet NB, Robertson B, Schaffelke B. Efficacy of physical removal of a marine pest: the introduced kelp *Undaria pinnatifida* in a Tasmanian Marine Reserve. Biol Inv. 2005;7:251–63.
Hewitt CL, Campbell ML, Schaffelke B. Introductions of seaweeds: accidental transfer pathways and mechanisms. Bot Mar. 2007;50:326–37.
Johnson MT, Carpenter EJ, Zhijian Tian, et al. Evaluating methods for isolating total RNA and predicting the success of sequencing phylogenetically diverse plant transcriptomes. PLoS ONE. 2012;7:e50226.

Jousson O, Pawlowski J, Zaninetti L, Meinesz A, Boudouresque CF. Molecular evidence for the aquarium origin of the green alga *Caulerpa taxifolia* introduced to the Mediterranean Sea. Mar Ecol Progr Ser. 1998;172:275–80.

Katsanevakis S, Zenetos A, Belchior C, Cardoso AC. Invading European seas: assessing pathways of introduction of marine aliens. Ocean Coast Manag. 2013;76:64–74.

Kim SY, Weinberger F, Boo SM. Genetic data hint at a common donor region for the invasive Atlantic and Pacific populations of *Gracilaria vermiculophylla* (Gracilariales, Rhodophyta). J Phycol. 2010;46(6):1346–9.

Klein J, Verlaque M. The *Caulerpa racemosa* invasion: a critical review. Mar Poll Bull. 2008;56:205–25.

Knowlton N. Sibling Species in the Sea. Ann Rev Ecol Systemat. 1993;24:189–216.

Lee CE. Evolutionary genetics of invasive species. Trends Ecol Evol. 2002;17:386–91.

Leliaert F, Verbruggen H, Wysor B, De Clerck O. DNA taxonomy in morphologically plastic taxa: algorithmic species delimitation in the *Boodlea* complex (Chlorophyta: Cladophorales). Mol Phylogenet Evol. 2009;53:122–33.

Littler MM, Littler DS. The evolution of thallus form and survival strategies in benthic marine macroalgae: field and laboratory tests of a functional form model. Am Nat. 1980;116:25–44.

Ludt WB, Rocha LA. Shifting seas: the impacts of Pleistocene sea-level fluctuations on the evolution of tropical marine taxa. J Biogeogr. 2015;42:25–38.

Lüning K, Yarish C, Kirkman H. Seaweeds: their environment, biogeography, and ecophysiology. New York: Wiley; 1990.

Maggs CA, Castilho R, Foltz D, Henzler C, Jolly MT, Kelly J, Olsen J, Perez KE, Stam W, Vainola R, Viard F, Wares J. Evaluating signatures of glacial refugia for North Atlantic Benthic Marine Taxa. Ecology. 2008;89:S108–22.

Marston M, Villalard-Bohnsack M. Genetic variability and potential sources of *Grateloupia doryphora* (Halymeniaceae, Rhodophyta), an invasive species in Rhode Island waters (USA). J Phycol. 2002;38(4):649–58.

Mayden RL. A hierarchy of species concepts: the denouement in the saga of the species problem. In: Claridge MF, Dawah HA, Wilson MR, editors. Species: the units of biodiversity. New York: Chapman & Hall; 1997. p. 381–424.

McDevit DC, Saunders GW. On the utility of DNA barcoding for species differentiation among brown macroalgae (Phaeophyceae) including a novel extraction protocol. Phycol Res. 2009; 57(2):131–41.

McIvor L, Maggs CA, Provan J, Stanhope MJ. rbcL sequences reveal multiple cryptic introductions of the Japanese red alga *Polysiphonia harveyi*. Mol Ecol. 2001;10:911–9.

Meinesz A. Methods for identifying and tracking seaweed invasions. Bot Mar. 2007;50:373–84.

Metzker ML. Sequencing technologies—the next generation. Nat Rev Gen. 2010;11:31–46.

Meusnier I, Valero M, Destombe C, Gode C, Desmarais E, Bonhomme F, Stam W, Olsen J. Polymerase chain reaction–single strand conformation polymorphism analyses of nuclear and chloroplast DNA provide evidence for recombination, multiple introductions and nascent speciation in the *Caulerpa taxifolia* complex. Mol Ecol. 2002;11:2317–25.

Meusnier I, Valero M, Olsen JL, Stam WT. Analysis of rDNA ITS1 indels in *Caulerpa taxifolia* (Chlorophyta) supports a derived, incipient species status for the invasive strain. Eur J Phycol. 2004;39(1):83–92.

Moritz C. Defining 'evolutionarily significant units' for conservation. Trends Ecol Evol. 1994;9:373–5.

Nuber N, Gornik O, Lauc G, Bauer N, Zuljevic A, Papes D, Zoldos V. Genetic evidence for the identity of *Caulerpa racemosa* (Forsskal) J. Agardh (Caulerpales, Chlorophyta) in the Adriatic sea. Eur J Phycol. 2007;42:113–20.

O'Hara T, Gledhill D, Butler A, Bax N, Wilson R, Poore G, McCallum A, Last P, England P, Andreakis N. Report on application of evolutionary history to inform how species and/or communities might respond to changes in climate. CERF Marine Biodiversity Hub. 2011.

O'Doherty DC, Sherwood AR. Genetic population structure of the Hawaiian Alien invasive seaweed *Acanthophora spicifera* (Rhodophyta) as revealed by DNA sequencing and ISSR analyses. Pac Sci. 2007;61(2):223–33.

Olsen JL, Valero M, Meusnier I, Boele-Bos S, Stam WT. Mediterranean *Caulerpa taxifolia* and *C. mexicana* (Chlorophyta) are not conspecific. J Phycol. 1998;34(5):850–6.

Pandit MK, Pocock MJ, Kunin WE. Ploidy influences rarity and invasiveness in plants. J Ecol. 2011;99:1108–15.

Pante E, Puillandre N, Viricel A, Arnaud-Haond S, Aurelle D, Castelin M, Chenuil A, Destombe C, Forcioli D, Valero M. Species are hypotheses: avoid connectivity assessments based on pillars of sand. Mol Ecol. 2015;24:525–44.

Payo DA, Leliaert F, Verbruggen H, D'hondt S, Calumpong HP, De Clerck O. Extensive cryptic species diversity and fine-scale endemism in the marine red alga *Portieria* in the Philippines. Proc R Soc B. 2013;280:20122660.

Pearman PB, Guisan A, Broennimann O, Randin CF. Niche dynamics in space and time. Trends Ecol Evol. 2008;23(3):149–58.

Peterson AT. Predicting the geography of species' invasions via ecological niche modeling. Quart Rev Biol. 2003;78:419–33.

Prince JS, Trowbridge CD. Reproduction in the green macroalga *Codium* (Chlorophyta): characterization of gametes. Bot Mar. 2004;47:461–70.

Provan J, Murphy S, Maggs CA. Tracking the invasive history of the green alga *Codium fragile* ssp *tomentosoides*. Mol Ecol. 2005;14:189–94.

Provan J, Booth D, Todd NP, Beatty GE, Maggs CA. Tracking biological invasions in space and time: elucidating the invasive history of the green alga *Codium fragile* using old DNA. Divers Distrib. 2008;14:343–54.

Ratnasingham S, Hebert PD. BOLD: the barcode of life data system. Mol Ecol Notes. 2007;7:355–64. http://www.barcodinglife.org.

Richardson DM, Ricciardi A. Misleading criticisms of invasion science: a field guide. Divers Distrib. 2013;19:1461–7.

Rödder D, Lötters S. Niche shift versus niche conservatism? Climatic characteristics of the native and invasive ranges of the Mediterranean house gecko (*Hemidactylus turcicus*). Glob Ecol Biogeogr. 2009;18:674–87.

Rojas J, Lemus A, Ganesan E. El ciclo vital "in vitro" del alga marina roja *Asparagopsis taxiformis* (Delile) Collins & Hervey (Bonnemaisoniales, Rhodophyta) del Mar Caribe. Bol Inst Oceanogr (Cumaná). 1982;21:101–12.

Saunders GW. Applying DNA barcoding to red macroalgae: a preliminary appraisal holds promise for future applications. Phil Tran R Soc B. 2005;360:1879–88.

Saunders GW. Routine DNA barcoding of Canadian Gracilariales (Rhodophyta) reveals the invasive species *Gracilaria vermiculophylla* in British Columbia. Mol Ecol Res. 2009;9:140–50.

Schaffelke B, Murphy N, Uthicke S. Using genetic techniques to investigate the sources of the invasive alga *Caulerpa taxifolia* in three new locations in Australia. Mar Poll Bull. 2002;44:204–10.

Schaffelke B, Smith JE, Hewitt CL. Introduced macroalgae—a growing concern. J Appl Phycol. 2006;18:529–41.

Schaffelke B, Hewitt CL. Impacts of introduced seaweeds. Bot Mar. 2007;50:397–17.

Sherwood AR. Phylogeography of *Asparagopsis taxiformis* (Bonnemaisoniales, Rhodophyta) in the Hawaiian Islands: two mtDNA markers support three separate introductions. Phycologia. 2008;47:79–88.

Sorte CJ, Williams SL, Carlton JT. Marine range shifts and species introductions: comparative spread rates and community impacts. Global Ecol Biogeogr. 2010;19:303–16.

Thomsen MS, Gurgel CFD, Fredericq S, McGlathery KJ. *Gracilaria vermiculophylla* (Rhodophyta, Gracilariales) in Hog Island Bay, Virginia: a cryptic alien and invasive macroalga and taxonomic correction. J Phycol. 2005;42:139–41.

Thuiller W, Richardson DM, Pysek P, Midgley GF, Hughes GO, Rouget M. Niche-based modelling as a tool for predicting the risk of alien plant invasions at a global scale. Glob Change Biol. 2005;11:2234–50.

Tingley R, Vallinoto M, Sequeira F, Kearney MR. Realized niche shift during a global biological invasion. Proc Natl Acad Sci USA. 2014;111:10233–8.

Trontelj P, Fišer C. Perspectives: cryptic species diversity should not be trivialised. Syst Biodivers. 2009;7:1–3.

Uwai S, Nelson W, Neill K, Wang WD, Aguilar-Rosas LE, Boo SM, Kawai H. Genetic diversity in *Undaria pinnatifida* (Laminariales, Phaeophyceae) deduced from mitochondria genes-origins and succession of introduced populations. Phycologia. 2006;45(6):687–95.

Van der Strate HJ, Van de Zande L, Stam WT, Olsen JL. The contribution of haploids, diploids and clones to fine-scale population structure in the seaweed *Cladophoropsis membranacea* (Chlorophyta). Mol Ecol. 2002;11:329–45.

Varela-Álvarez E, Gómez Garreta A, Rull Lluch J, Salvador Soler N, Serrao EA, Siguán MAR. Mediterranean species of *Caulerpa* are polyploid with smaller genomes in the invasive ones. PLoS ONE. 2012;7:e47728.

Verbruggen H, Leliaert F, Maggs CA, Shimada S, Schils T, Provan J, Coppejans E. Species boundaries and phylogenetic relationships within the green algal genus *Codium* (Bryopsidales) based on plastid DNA sequences. Mol Phyl Evol. 2007;44(1):240–54.

Verbruggen H, Tyberghein L, Pauly K, Vlaeminck C, Nieuwenhuyze KV, Kooistra WH, Leliaert F, Clerck OD. Macroecology meets macroevolution: evolutionary niche dynamics in the seaweed Halimeda. Glob Ecol Biogeogr. 2009;18:393–405.

Verbruggen H, Maggs CA, Saunders GW, Le Gall L, Yoon HS, De Clerck O. Data mining approach identifies research priorities and data requirements for resolving the red algal tree of life. BMC Evol Biol. 2010;10(1):16.

Verlaque M, Durand C, Huisman JM, Boudouresque CF, Le Parco Y. On the identity and origin of the Mediterranean invasive *Caulerpla racemosa* (Caulerpales, Chlorophyta). Eur J Phycol. 2003;38:325–39.

Voisin M, Engel CR, Viard F. Differential shuffling of native genetic diversity across introduced regions in a brown alga: aquaculture vs. maritime traffic effects. Proc Natl Acad Sci USA. 2005;102:5432–7.

Wattier R, Maggs CA. Intraspecific variation in seaweeds: the application of new tools and approaches. Adv Bot Res. 2001;35:171–212.

Weinig C. Plasticity versus canalization: population differences in the timing of shade-avoidance responses. Evolution. 2000;54(2):441–51.

Williams SL, Smith JE. A global review of the distribution, taxonomy, and impacts of introduced seaweeds. Ann Rev Ecol Evol Syst. 2007;38:327–59.

Yeh WJ, Chen GY. Nuclear rDNA and internal transcribed spacer sequences clarify *Caulerpa racemosa* vars. from other *Caulerpa* species. Aq Bot. 2004;80:193–207.

Zanolla M, Carmona R, De La Rosa J, Salvador N, Sherwood AR, Andreakis N, Altamirano M. Morphological differentiation of cryptic lineages within the invasive genus *Asparagopsis* (Bonnemaisoniales, Rhodophyta). Phycologia. 2014;53(3):233.

Zanolla M, Altamirano Jeschke M, Carmona R, de la Rosa Álamos JC, Sherwood AR, Andreakis N. Photosynthetic plasticity of the genus *Asparagopsis* (Bonnemaisoniales, Rhodophyta) in response to temperature: implications for invasiveness. Biol Inv. 2015;17(5): 1–13.

Part IV
Comparative Phylogeography of Seaweeds

Chapter 8
Phylogeography of Tropical Pacific Marine Algae

Alison R. Sherwood and Giuseppe C. Zuccarello

Abstract The tropical Pacific is a unique region to study marine algal phylogeographic patterns. The ancient age of the ocean basin, combined with the presence of numerous islands and archipelagos derived from a variety of geological and biological processes, has yielded several "cosmopolitan" algal species that likely achieved a broad distribution during the times of the Tethys Ocean. These cosmopolitan species consist, in almost all cases, of a series of lineages that can be interpreted as cryptic or pseudo-cryptic species. We review several example studies from the literature that examine phylogeographic patterns of marine algae from the tropical Pacific, and conclude (1) that in all cases the number of species discovered by molecular methods is large, (2) that the increase in diversity is correlated with sampling effort, and (3) that while morphological species are widespread in the tropics, cryptic or pseudo-cryptic species are often more localized, and even appear to have neighboring distribution patterns. These conclusions lead to a call for more large-scale collaborative studies to examine the phylogeographic trends of purportedly cosmopolitan species across the tropical Pacific.

Keywords Biodiversity · Cosmopolitan species · Endemism · Marine phylogeography · Tropical seaweeds

A.R. Sherwood (✉)
Department of Botany, University of Hawaii, 3190 Maile Way, Hawaii,
Honolulu 96822, USA
e-mail: asherwoo@hawaii.edu

G.C. Zuccarello
School of Biological Sciences, Victoria University of Wellington,
Wellington 6140, New Zealand
e-mail: joe.zuccarello@vuw.ac.nz

Z.-M. Hu and C. Fraser (eds.), *Seaweed Phylogeography*,
DOI 10.1007/978-94-017-7534-2_8

8.1 Introduction

Phylogeography examines the spatial arrangement of genetic lineages, with a focus on intraspecific patterns and those of closely related species (Avise 2009). Phylogeographic studies have greatly increased our understanding of the processes that have produced the great biological diversity found in the world. Understanding the factors that have influenced species and population diversity and distribution can help us predict how species and populations will change under scenarios of future environmental and climate change. The study of marine algal phylogeography has emerged over the last 10–15 years as the use of molecular tools for the investigation of genetic variation have become commonplace, and advances have been made in both phylogeographic methodologies and knowledge of patterns of other marine lineages (Knowles 2009). Marine algal phylogeographic patterns in the tropical Pacific are a result of many influences, including historical geography of the region, dispersal capabilities of individual species, tolerance to environmental conditions, and, more recently, the influence of anthropogenic transport. Macroalgae are starting to be studied more extensively, including in the tropics, and both the number of species examined and the analytical methodologies and sampling methods used will be summarized in this chapter.

8.2 Geology of the Tropical Oceans

The tropical oceans of the world have unique histories that have influenced and continue to influence species living in these environments. Tropical oceans span a latitudinal range from approximately the Tropic of Cancer in the north (23.5°N) to the Tropic of Capricorn (23.5°S) in the south, or the 20 °C winter isotherm. These boundaries also coincide with the distribution of hermatypic corals (Lüning 1990). The tropics, at present, span three ocean basins (Atlantic, Indian, and Pacific), which have had connections with each other in historical times and have had barriers, of various effectiveness, separating and joining them (Lomolino et al. 2009).

A historical geological feature linking all tropical oceans was the Tethys Ocean, which was a long-lived ocean basin (ca. 250–60 Ma) that separated the supercontinents of Laurasia and Gondwana (Lomolino et al. 2009). By the late Cretaceous (100–60 Ma) this ocean began to break apart with the splitting up of the supercontinents, followed by movement of the continents. The circum-equatorial Tethys Ocean (the Tethys Seaway) likely led to pan-tropical distributions of tropical marine species, which subsequently became isolated as individual ocean basins formed. Thus, the historical joining of oceanic basins in the Tethys Ocean was likely responsible for the evolution of widespread tropical algal species, or groups of related species (Lüning 1990). In the Eocene, the Tethys Seaway continued to close with the movement of Africa and India northward. This closure of the circum-equatorial tropical Tethys led to the semi-isolated ocean basins of today (Lomolino et al. 2009).

The Pacific Ocean originated following the breakup of Pangea. Its tropical regions were influenced by movements in the Tethys Seaway (Neall and Trewick 2008). While the Pacific is the largest ocean, and dispersal distances between land masses are enormous, it is also covered with a variety of islands. These islands break up distances between shallow marine habitats (strongly influencing the ecology of many marine species) and also give rise to the possibility of partial isolation of populations. The current islands of the tropical Pacific originated in a variety of ways, and include volcanically formed oceanic islands and seamounts (e.g., the Hawaiian Archipelago, the Line Islands), atolls that developed with coral growth resulting in reefs (e.g., Rarotonga, Aitutaki, and Palmerston Atoll of the Cook Islands), uplifted coral reefs and atolls (e.g., Rennell Island of the Solomon Islands), fragments of continental crust (e.g., New Zealand and New Caledonia), and the formation of island arcs on the Pacific margins (e.g., the Tonga-Kermadec Islands) (Neall and Trewick 2008). This extraordinarily rich range of mechanisms for the development of the islands of the tropical Pacific has enabled the evolution of a phenomenal level of biodiversity in both the marine and terrestrial realms.

8.3 Speciation in the Marine Tropical Pacific

Most marine invertebrates and many fishes disperse widely via planktonic larval stages (e.g., Grosberg and Levitan 1992), and this potential for dispersal for many marine animals has shaped assumptions about the degree of connectivity in the marine environment. Most marine phylogeographic research has been conducted under the premise that as population connectivity decreases through decreased planktonic-stage dispersal (e.g., at larger geographic distances from the larval source), genetic differentiation increases and populations become more isolated, which leads to speciation in the marine realm (Mayr 1963; Rocha and Bowen 2008).

The most commonly accepted mechanism of speciation is allopatry (isolation of populations due to a barrier), and it has been largely accepted that speciation in the sea occurs via this mechanism (Rocha and Bowen 2008). However, allopatric speciation does not adequately explain all observed patterns of biodiversity in the oceans. Given the lack of obvious breaks in continuity in the marine realm, under a model of pure allopatric speciation one would expect to find relatively few unique species in the sea; it is clear, given levels of marine biodiversity, that this is not the case.

Other mechanisms of speciation have, therefore, been proposed for marine environments (Bowen et al. 2013). Ecological speciation can occur under parapatry and even sympatry in the oceans (Bowen et al. 2013) and has been proposed for a variety of coral reef taxa including fishes (Rocha et al. 2008), sponges (Rützler et al. 2007), corals (Bongaerts et al. 2010), limpets (Bird et al. 2011), and nudibranchs

(Faucci et al. 2007). The rich diversity of available habitats in coral reefs ecosystems (i.e., high physical and ecological heterogeneity), combined with an elevated intensity of competition among their occupants, is believed to drive these speciation processes. How the process of ecological speciation would function in macroalgae is less clear, however, given their lack of mobility.

Evidence is accumulating that speciation is high in macroalgae that mostly have poorly dispersing propagules. For example, up to 21 genetic species are recognized in the red alga *Portieria* in the Philippine archipelago alone (Payo et al. 2013). Whether these speciation events occurred allopatrically is unknown; perhaps more importantly, the scale at which allopatric speciation operates is not fully understood. Full or partial allopatric speciation may be possible even at small geographical distances for taxa such as macroalgae, which, on the whole, disperse poorly.

8.4 Dispersal of Seaweeds

Macroalgae lack "long-lived" propagules, which distinguishes them from many marine animals. Spores and gametes, as single cells, have very limited survival times (on the order of hours or days), and this is especially true for gametes that need to find companions before dilution increases distances from con-specifics (Coyer et al. 2003). Moreover, the locomotory powers of even the strongest algal swimming cells are simply ineffective against wave action and general oceanic water circulation, and algal gametes may be best suited to finding a nearby mate or settling in an appropriate location within the minute boundary layer, rather than influencing dispersal at larger geographical scales (Norton 1992). This problem is perhaps most profound for the red algae, which lack flagellated cells altogether, and instead rely on water motion for dispersal of gametes and spores. Local population genetic studies indicate that macroalgal populations are strongly structured (i.e., partially isolated) even at small scales (e.g., Valero et al. 2011; Kruger-Hadfield et al. 2013; Provan et al. 2013) and isolation by distance can explain some of this structure (e.g., Coyer et al. 2003; Provan et al. 2013). Of course, not all studies show isolation by distance, which indicates that stochastic processes (i.e., rare long distance dispersal events) can be significant (Zuccarello et al. 2011). One known method of frequent long distance dispersal is through floating, or rafting, of algal fronds, especially of large brown algae and their associated flora and fauna (Buchanan and Zuccarello 2012; Fraser et al. 2009).

Thus, both isolation, which drives population differentiation, ultimately leading to speciation if continued, and infrequent long distance dispersal, lead to the phylogeographic patterns of tropical Pacific macroalgae. The wide expanse of the tropical oceans and the many islands situated at various distances from each other result in the processes of speciation and differentiation being especially important in these marine environments.

8.5 General Patterns of Pacific Marine Tropical Biodiversity

The Coral Triangle (the region between Indonesia, New Guinea, and the Philippines) of the Indo-Pacific has documented high species richness for corals, reef fishes, and some gastropod and crustacean groups relative to other Pacific marine regions (Bellwood and Meyer 2009), and thus is widely recognized as the tropical marine biodiversity hotspot of the Pacific Ocean (e.g., Briggs 2005). The Coral Triangle has a long history of species accumulation, dating back 20–12 Ma (the Miocene), and with exportation of biodiversity in the Pliocene, Pleistocene and Holocene (7 Ma until present) (Cowman and Bellwood 2013; Briggs and Bowen 2013). Three biogeographic explanations have been put forth for the observed pattern of increased biodiversity in the Coral Triangle marine biodiversity hotspot, termed the Center of Speciation, Center of Accumulation and Center of Overlap models (Bowen et al. 2013), which differ in interpretation of where rates of speciation are highest. Individual studies have provided evidence for all three models. The Center of Speciation model posits that the speciation process is most intense at biodiversity hotspots themselves (with evidence in support of this model from studies of sea turtles; Bowen et al. 1998). The Center of Accumulation model differs in that it assumes that new species arise peripherally to biodiversity hotspots and that these species are concentrated in the center of the range by prevailing oceanographic currents, and studies of fish have yielded support for this model (e.g., Bernardi et al. 2004). Finally, the Center of Overlap hypothesis promotes the idea of isolated distributions of taxa overlapping at their distributional edges, establishing a region of elevated biodiversity (with evidence provided, for example, from studies of fish; Barber and Bellwood 2005). More recently, some researchers have argued in favor of a Biodiversity Feedback model, which integrates elements of several or all of the above models to describe a more dynamic interchange of biodiversity among peripheral and central regions of tropical marine biodiversity (Bowen et al. 2013). Which, if any, of these models applies to the tropical marine algal diversity of the Pacific, has yet to be investigated in detail.

8.6 Marine Algal Phylogeographic Patterns in the Tropical Pacific

Phylogeographic studies of seemingly cosmopolitan tropical marine algal species most often reveal what was formerly considered to be one species to actually be a species complex (e.g., Zuccarello et al. 2002; Andreakis et al. 2009; Sherwood et al. 2011; Payo et al. 2013). Some of these genetic species have restricted geographical distributions, while others are broader in distribution. These patterns clearly demonstrate that there are two differing processes affecting algal phylogeography: (1) differentiation due to drift, or selection, leading to isolated populations with the potential to produce locally adapted populations that could be considered species. These results contradict the assumption that marine species are cosmopolitan because of high dispersal capability in connected habitats (i.e., a lack of dispersal barriers in the sea); (2) long distance dispersal, which can lead to widely distributed species, and can be anthropogenic or natural. Observed phylogeographic distributions of macroalgae may in fact be a confluence of patterns from the above two forces.

Below we compare and contrast several studies of marine algal phylogeography in the tropical Pacific. Most marine algal phylogeographic research thus far has focused on the red algae, which perhaps reflects their species richness in these environments (e.g., the marine algal flora of the Hawaiian Islands comprises approximately 70 % red algal species; Abbott 1999; Abbott and Huisman 2004). The number of phylogeographic studies of tropical marine algae is still low, but some general patterns are beginning to emerge. For example, in one recent study, Kerswell (2006) suggested that richness of marine algae is actually lower in the tropics than at higher latitudes, which contrasts with the pattern found for many other marine organisms. We attempt to use the examples below to summarize phylogeographic patterns for marine algae in the tropical Pacific.

***Amansia glomerata* C. Agardh (Rhodophyta)**—The tropical marine red alga *Amansia glomerata* was described from an unknown location on the Hawaiian Island of Oahu (Agardh 1822), but is recognized as being very broadly distributed throughout the tropical and subtropical Pacific and Indian Ocean (Fig. 8.1a, Guiry and Guiry 2015). Extreme morphological variability has been noted for this taxon (Phillips 2009). Sherwood et al. (2011) compared 61 specimens representing all eight Main Hawaiian Islands plus French Frigate Shoals of the Northwestern Hawaiian Islands using several molecular markers. Three distinct mitochondrial lineages were recognized, and the possibility of the specimens representing a species complex (either cryptic or incipient species) was discussed. Sampling of areas outside the Hawaiian Islands is needed to determine the full extent of molecular variation in this widespread "species" and to elucidate phylogeographical patterns beyond its eastern limit of distribution in the Hawaiian Islands.

***Asparagopsis taxiformis* (Delile) Trevisan de Saint-Léon (Rhodophyta)**—This red algal morphospecies was considered to be widely distributed and found in all tropical ocean basins (Fig. 8.1b). Initial phylogeographic analyses revealed it consisted of four cryptic lineages (Andreakis et al. 2009). Continued sampling has discovered a fifth cryptic lineage (Dijoux et al. 2014), indicating that cryptic diversity discovery is sensitive to sampling effort. These results support a presumed cosmopolitan species consisting of several genetic entities (designated as lineages) with each lineage often having a much more limited distribution. For example, *A. taxiformis* lineage 5 is found in the southern Pacific, indicating isolation in this area (area of origin) with subsequent limited dispersal. Lineage 4, on the other hand, is distributed widely (the Indian Ocean, both sides of the Pacific, and Hawaii). This lineage has only moderate levels of sequence divergence (3.59 % for the mitochondrial *cox*2-3 spacer), indicating that dispersal is a relatively recent historical event. Some haplotype distributions of *A. taxiformis* (e.g., a single *cox*2-3 spacer

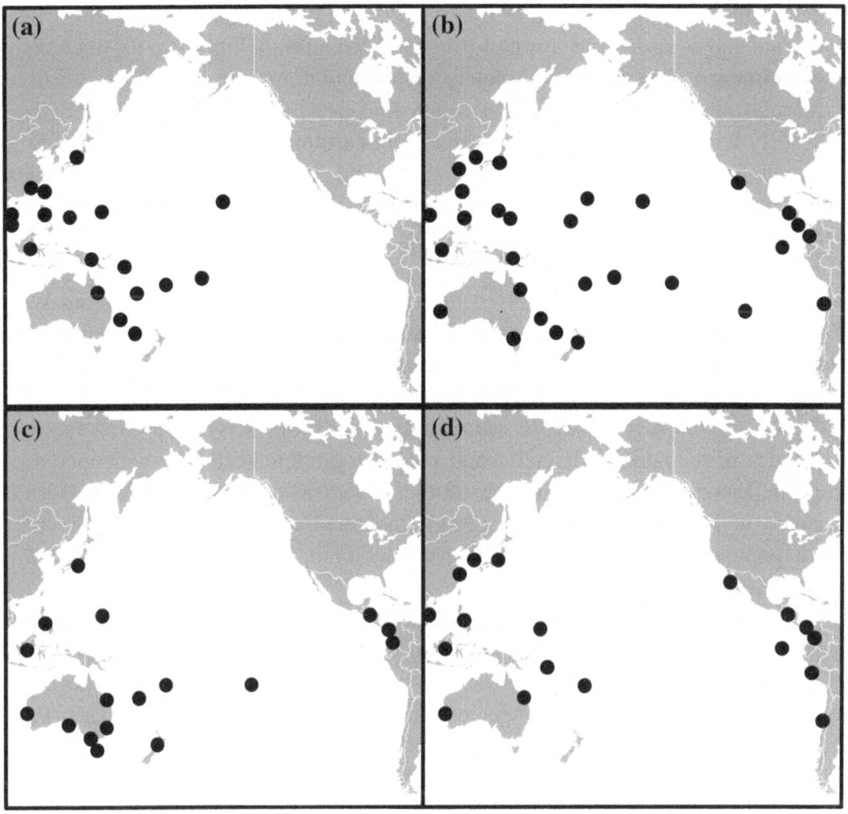

Fig. 8.1 Pacific basin distributions of case study taxa discussed in the current chapter (indicated with *black circles*). **a** *Amansia glomerata* **b** *Asparagopsis taxiformis* **c** *Bostrychia moritziana* **d** *Bostrychia radicans*. Distributional data from AlgaeBase (Guiry and Guiry 2015)

haplotype found in Hawaii, Taiwan and Costa Rica) also indicate that some of this dispersal is recent and possibly human-mediated (Sherwood 2008; Dijoux et al. 2014). The ability of *A. taxiformis* to disperse is potentially due to its heteromorphic alternation of generations in which the sporophytic stage (*Falkenbergia*-stage) is cryptic, epiphytic on other algae (acting as possible dispersal agents) and physiologically flexible (i.e., eurythermal; Chualáin et al. 2004). Recent research has established that the *Falkenbergia*-stages of the *A. taxiformis* lineages can in fact be distinguished morphologically, although the commonly encountered gametophytes remain cryptic (Zanolla et al. 2014). Another characteristic of *Asparagopsis* that may facilitate establishment in new environments is its chemical defense against herbivory and biofouling (e.g., in *A. armata*, see Vergés et al. 2008).

Bostrychia moritziana **(Sonder ex Kützing) J. Agardh/B.** ***radicans*** **(Montagne) (Rhodophyta)**—Another genus in which molecular evidence has shown that the morphological diversity underestimates that determined by molecular methods is the red alga *Bostrychia*. The two morphospecies *Bostrychia moritziana* and *B. radicans* have pan-tropical distributions (Fig. 8.1c, d), but consist of seven lineages that do not completely correspond to morphospecies designations (Zuccarello and West 2003). These seven lineages have distributions that range from fairly wide ranging to more localized. For example, lineage 1 is confined to the southern Pacific (Australia, Indonesia) and Indian Ocean (South Africa, Madagascar), while lineage 3 is found only in Atlantic South America, and other lineages have much wider distributions (e.g., lineage 6 in the Atlantic, Pacific and Indian Oceans) (Zuccarello and West 2003). Furthermore, these genetic lineages are known to be mostly reproductively isolated from each other (Zuccarello and West 1997; Zuccarello et al. 1999), demonstrating that the genetic species meet the criteria for recognition as distinct species under other species concepts (i.e., reproductive isolation under the Biological Species Concept).

The *Bostrychia moritziana* /*B. radicans* species complex is also a good example of changes in recognized diversity and phylogeographic scenarios with increased sampling. Increased sampling of the southern USA as an expansion of the results presented in Zuccarello and West (2003) indicated a different level of diversity within populations in addition to altered phylogeographic interpretation of this diversity (Zuccarello et al. 2006).

***Portieria hornemannii* (Lyngbye) P.C. Silva (Rhodophyta)**—Phylogeographic patterns within this red algal species (Fig. 8.2a) were studied from samples collected in the Philippines in one of the first algal studies to use formal species delimitation methodologies to determine the "species status" of distinct molecular lineages (Payo et al. 2013). These species delimitation methods (Leliaert et al. 2014) are becoming widely used to remove some of the subjective nature of species determination from phylogenetic trees (e.g., Vieira et al. 2014; Muangmai et al. 2014). Payo et al. (2013) discovered that collections of *Portieria* throughout the Philippines constituted up to 21 species based on these species delimitation methods. Most species were found to be highly restricted in distribution to single islands within the archipelago. They concluded that speciation in the marine realm can occur at spatial scales of less than 100 km; thus, this was a critical study clearly highlighting the potential of tropical marine algae with limited dispersal to speciate while maintaining close proximity. If other tropical marine algal species show similar patterns (speciation within tropical archipelagos), the tropics will be supported as one of the major regions of marine algal species generation.

***Spyridia filamentosa* (Wulfen) Harvey (Rhodophyta)**—The red alga *Spyridia filamentosa* is well known due to its wide distribution (Fig. 8.2b), ease of identification and potential for being a nuisance species (Guiry and Guiry 2015). This "species" has received a considerable amount of attention throughout its distributional range (Zuccarello et al. 2002; Conklin and Sherwood 2012). Although this taxon was described from the Adriatic Sea, it is currently recognized as one of the most widely distributed marine algae in the world (although its distribution in the temperate regions of the North Atlantic may be due to misidentifications; Zuccarello et al. 2004).

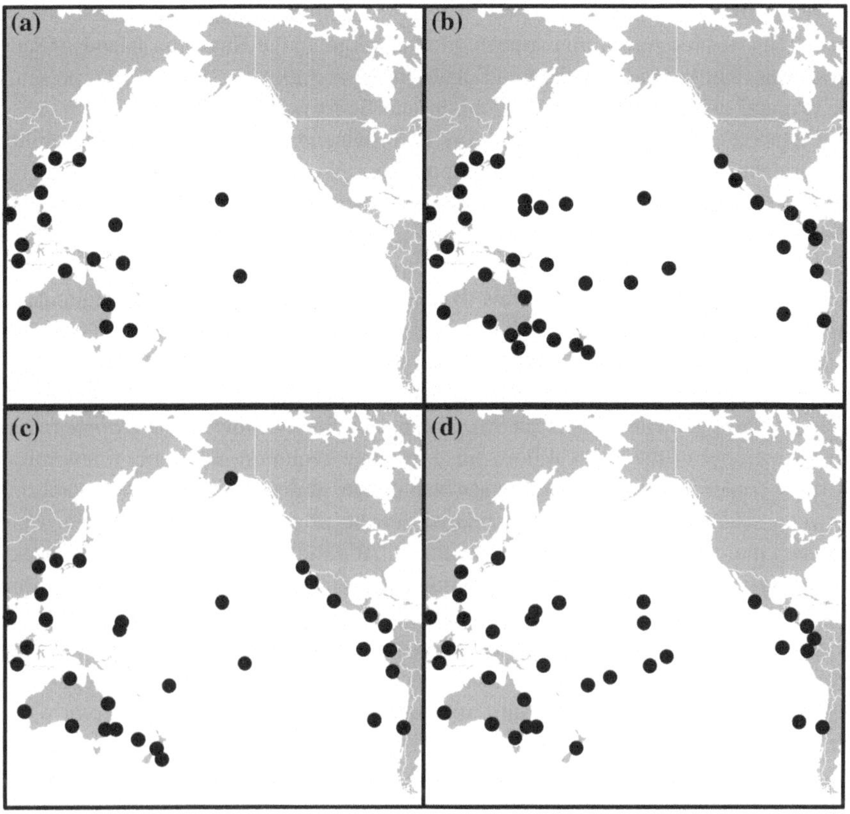

Fig. 8.2 Pacific basin distributions of case study taxa discussed in the current chapter (indicated with *black circles*). **a** *Portieria hornemannii* **b** *Spyridia filamentosa* **c** *Colpomenia sinuosa* **d** *Lobophora variegata. Distributional data from AlgaeBase* (Guiry and Guiry 2015)

Zuccarello et al. (2002) sequenced markers from the nuclear, plastid, and mitochondrial genomes of specimens of *S. filamentosa* from around the world and demonstrated that samples from different oceanic basins are usually well separated, and mostly reproductively isolated (with more complex patterns observed in the Mediterranean specimens explained as back-colonizations to that ocean after the Messinian salinity crisis ca. 5 Ma). Two main lineages were proposed—an Atlantic/Indian Ocean lineage and a Pacific lineage, consisting of multiple, cryptic species.

Subsequently, Conklin and Sherwood (2012) analyzed 124 specimens of *Spyridia filamentosa* from islands and atolls of the Hawaiian Islands in the subtropical Pacific, and demonstrated the presence of 5–6 mitochondrial lineages, again highlighting that increased sampling uncovers new diversity and can alter phylogeographic scenarios. Comparisons of these Hawaiian sequences with others for this "species" from other parts of its range illustrated that *Spyridia filamentosa* arrived in the Hawaiian Islands on at least six different occasions. One clade of specimens was recovered that appeared to be unique to the Hawaiian Islands (Clade D). It was also the most widespread clade in the archipelago, and likely represents the original clade in the islands (Conklin and Sherwood 2012).

The apparent ease of dispersal of *Spyridia filamentosa* could be due to its ability to fragment easily and produce large populations (West and Calumpong 1989). This dispersal ability could lead to the observed widespread distributions, but also obscure native lineages (or species) due to masking by more recent arrivals. These kinds of complicated phylogeographic patterns can be eventually unraveled, as they were for Hawaiian *Spyridia*, but this depends on intensive sampling throughout the "species" range.

***Halimeda* spp. (Chlorophyta)**—Broad-scale phylogeographic patterns of species of *Halimeda* have been elucidated using molecular analyses and niche modeling (Verbruggen et al. 2005, 2009a). These data clearly showed that *Halimeda* has been a member of the tropical flora since its early evolution, with only a few range shifts to cooler waters. These analyses also demonstrated that the main separations between sister species, or closely related cryptic or pseudo-cryptic species, were between major ocean basins (Indo-Pacific vs. Atlantic). It is likely that these ocean basin phylogeographic separations are due to the limited dispersal since the closure of the Tethys Ocean as the diversification of genus commenced 150 Ma (Verbruggen et al. 2009b) and proceeded for several millions of years, with the sections of *Halimeda* (groups of related species that show an ocean basin separation within the genus) diversifying until approximately 90 Ma, the time interval when the Tethys Seaway was closing. Verbruggen et al. (2009a) also proposed that species have not dispersed between ocean basins since this time because of lack of suitable or available habitat (Waters et al. 2013).

***Colpomenia sinuosa* (Mertens ex Roth) Derbès & Solier**—In one of the most geographically broad phylogeographic studies of a marine alga (Fig. 8.2c), mitochondrial and plastid DNA from specimens of the brown alga *Colpomenia sinuosa* from 19 countries were compared. Cryptic diversity was revealed within the taxon as three distinct genetic groups, with multiple lineages within two of the three groups (Lee et al. 2013); Group I was the largest, with pan-tropical representation, Group II was found in the Red Sea and the western Mediterranean, and Group III was found in the central and western Pacific, as well as the Indian Ocean and the western Atlantic. *Colpomenia sinuosa* was demonstrated to be a wide ranging yet morphologically variable taxon, and pairwise divergences among the recognized groups were high, suggesting a long evolutionary history. Additionally, the complex biogeographical patterns observed, especially for Group I, suggested that anthropogenic dispersal events have played a major role in shaping the distribution of this taxon. The authors noted that most lineages within Group I were distributed at relatively temperate latitudes, which contrasts with the distributions for the other Groups, and also with the idea of peripheral populations having reduced diversity (Lee et al. 2013).

***Lobophora variegata* (J.V. Lamouroux) Womersley ex E.C. Oliveira**—*Lobophora* is a common tropical brown algal genus, with one cosmopolitan species reported from most tropical and temperate areas (*L. variegata*) (Fig. 8.2d). Early molecular diversity studies in the western Pacific indicated that several species existed within this taxon, and some of these morphologically distinguishable lineages (pseudo-cryptic species) were named (Sun et al. 2012), while other lineages were left as undescribed species. Recently, diversity studies of *Lobophora* focused on the southwestern Pacific islands of New Caledonia, and employed various species delimitation methods (see Leliaert et al. 2014) resulting in the recognition of 29 MOTUs (Molecular Operational Taxonomic Units) from this one, geographically limited, location (Vieira et al. 2014). Some of these MOTUs could not be delineated morphologically, but 10 new species of *Lobophora* were described from the specimens and had distinctive morphological features associated with them; thus, this study demonstrated that many genetic species segregate based on both morphological and ecological characters.

8.7 Summary of Tropical Marine Algal Phylogeographic Patterns

It is clear that morphology-based measures of biodiversity consistently underestimate that determined using molecular methods. Although this finding is hardly new, its implications for the study of algal phylogeography are profound; speciation in the marine tropical Pacific can occur at geographical scales much smaller than typically assumed (e.g., less than 100 km for *Portieria* in the Philippines; Payo et al. 2013), yielding presumed radiations of marine algal species. Additionally, it has become evident that while "morphological" algal species appear to be widespread in the tropics, many molecularly defined species have much more limited distributions.

Our understanding of seaweed diversity in the tropics has progressed a long way since the introduction of molecular data for examining phylogeographic trends. However, only a few species have been studied extensively (for further examples see: *Padina*, Win et al. 2011; Silberfeld et al. 2013 and *Caloglossa*, Kamiya and West 2014). Briefly, we see three main trends from the available data: (1) that in all studies examined the number of species discovered by molecular methods is large, (2) that the increase in estimates of diversity is correlated with sampling effort, and (3) that while morphological species are widespread in the tropics, cryptic or pseudo-cryptic species are often more localized, and even appear to have neighboring distribution patterns.

The tropics are well known as an area of high species diversity and richness (e.g., under the Latitudinal Diversity Gradient, e.g., Hillebrand 2004). It could be hypothesized that this pattern should also be found in the tropical Pacific for marine algae. Certainly, the Coral Triangle is recognized as a biodiversity hotspot for many marine species, with almost all supporting data stemming from studies of reef fishes and some invertebrate groups (Bowen et al. 2013). Some of the phylogeographic examples discussed in this chapter from or near the Coral Triangle (e.g., *Portieria* from the Philippines, Payo et al. 2013; *Lobophora* from New Caledonia, Vieira et al. 2014) support the idea of elevated speciation levels in this region. However, not all biogeographical studies based on species numbers support this idea. Kerswell (2006) suggested that richness of marine algae is actually lower in the tropics than at higher latitudes, but some questions remain as to the conclusiveness of this pattern given that the analysis was restricted to the genus level and was based on a limited number of literature records of mostly morphological identifications. Later, studies following up on this idea suggested that environmental tolerance influences species richness in the high latitudes, whereas biotic interactions are more important for species richness in the tropics (Keith et al. 2014). Schils et al. (2013) analyzed substantially larger numbers of algal records and included representation from many islands in the western and central Pacific Ocean, and concluded that marine algal richness was defined by local habitat diversity and availability; they did not uncover a latitudinal diversity gradient of marine algae in the region of study. However, all of these biogeographical studies are based on morphologically derived species lists that are almost certainly inaccurate, perhaps even more so in the tropics. The degree to which the phylogeographic studies discussed in previous sections of this chapter support the presence of numerous cryptic or pseudo-cryptic species within "cosmopolitan" marine taxa suggests that much work remains to be done defining and characterizing the marine algal species of the tropical Pacific before biogeographical patterns can be investigated with confidence.

Why would the discoveries of cryptic species be higher in the tropics than other areas? We believe that there are several possibilities. First, the historically connected nature of the tropics in ancient oceans (i.e., Tethys Sea) may have produced widely distributed taxa in a tropical climate lacking the abiotic stressors that dominate in temperate and polar environments, and that these taxa became increasingly isolated as these regions became disconnected with the movements of land masses. Second, the combination of the wide expanse of the tropical oceans, the large number of isolated islands, and the poor dispersal ability of many macroalgae likely produced many opportunities for speciation under allopatry, sympatry or peripatry. Third, it is possible that biotic interactions have little influence on species morphology.

8.8 Perspectives for Future Research

Phylogeographical studies of marine algae have progressed substantially in the last several years, yet it is clear that many more example taxa need to be investigated in a high level of detail, and from a much broader geographical area than has been typically sampled in the past, in order to elucidate large-scale phylogeographical patterns for the tropical marine Pacific. As such, it would be beneficial to see a series of collaborative projects undertaken (in the spirit of the "*Portieria* Evolution

Consortium" spearheaded by O. De Clerck, F. Leliaert and colleagues, which aims to assess lineage diversity patterns within that genus throughout its global range) to investigate phylogeographic patterns of some of the most widespread tropical marine algae, including representatives from the red, green, and brown algae. The tropical Pacific is particularly difficult for any one research group to sample effectively because the area is dauntingly large and it is geopolitically divided into many nations, yet many "species" are widespread, making adequate sampling for phylogeographic studies a challenge. Combining strong sampling effort at fine geographical scales with large sample sizes and novel molecular analysis techniques (e.g., recent species delimitation methods), or even novel kinds of molecular data (e.g., analyses of single nucleotide polymorphisms with RAD sequencing) have the potential to take the field of tropical marine algal phylogeography forward at a fast pace in the near future.

References

Abbott IA. Marine red algae of the Hawaiian Islands. Honolulu: Bishop Museum Press; 1999. p. 477.

Abbott IA, Huisman JM. Marine green and brown algae of the Hawaiian Islands. Honolulu: Bishop Museum Press; 2004. p. 259.

Agardh CA. Species algarum. 1, part 2. Lund; 1822. pp viii + 169–531.

Andreakis N, Kooistra WHCF, Procaccini G. High genetic diversity and connectivity in the polyploid invasive seaweed *Asparagopsis taxiformis* (Bonnemaisoniales) in the Mediterranean, explored with microsatellite alleles and multilocus genotypes. Mol Ecol. 2009;18:212–26.

Avise JC. Phylogeography: retrospect and prospect. J Biogeogr. 2009;36:3–15.

Barber PH, Bellwood DR. Biodiversity hotspots: evolutionary origins of biodiversity in wrasses (*Halichoeres*: Labridae) in the Indo-Pacific and New World tropics. Mol Phylogenet Evol. 2005;35:235–53.

Bellwood DR, Meyer CP. Searching for heat in a marine biodiversity hotspot. J Biogeogr. 2009;36:569–76.

Bernardi G, Bucciarelli G, Costagliola D, Robertson DR, Heiser JB. Evolution of coral reef fish *Thalassoma* spp. (Labridae). 1. Molecular phylogeny and biogeography. Mar Biol. 2004;144:369–75.

Bird CE, Holland BS, Bowen BW, Toonen RJ. Diversification of endemic sympatric limpets (*Cellana* spp.) in the Hawaiian Archipelago. Mol Ecol. 2011;20:2128–41.

Bongaerts P, Riginos C, Ridgway T, Sampayo EM, van Oppen MJH, Englebert N, Vermeulen F, Hoegh-Guldberg O. Genetic divergence across habitats in the widespread coral *Seriatopora hystrix* and its associated *Symbiodinium*. PLoS ONE. 2010;5:e10871.

Bowen BW, Rocha LA, Toonen RJ, Karl SA, the ToBo Laboratory. The origins of tropical marine biodiversity. Trends Ecol Evol. 2013;28:359–366.

Bowen BW, Clark AM, Abreu-Grobois AF, Chaves A, Reichart HA, Ferl RJ. Global phylogeography of the ridley sea turtles (*Lepidochelys* spp.) as inferred from mitochondrial DNA sequences. Genetica. 1998;101:179–89.

Briggs JC. The marine East Indies: diversity and speciation. J Biogeogr. 2005;32:1517–22.

Briggs JC, Bowen BW. Evolutionary patterns: marine shelf habitat. J Biogeogr. 2013;40:1023–35.

Buchanan J, Zuccarello GC. Decoupling of short and long distance dispersal pathways in the endemic New Zealand seaweed *Carpophyllum maschalocarpum* (Phaeophyceae, Fucales). J Phycol. 2012;48:518–29.

Chualáin FN, Maggs CA, Saunders GW, Guiry MD. The invasive genus *Asparagopsis* (Bonnemaisoniaceae, Rhodophyta): molecular systematics, morphology, and ecophysiology of *Falkenbergia* isolates. J Phycol. 2004;40:1112–26.

Conklin KY, Sherwood AR. Molecular and morphological variation of the red alga *Spyridia filamentosa* (Ceramiales, Rhodophyta) in the Hawaiian Archipelago. Phycologia. 2012;51:347–57.

Cowman PF, Bellwood DR. The historical biogeography of coral reef fishes: global patterns of origination and dispersal. J Biogeogr. 2013;40:209–24.

Coyer JA, Peters AF, Stam WT, Olsen JL. Post-ice age recolonization and differentiation of *Fucus serratus* L. (Phaeophyceae; Fucaceae) populations in Northern Europe. Mol Ecol. 2003;12:1817–29.

Dijoux L, Viard F, Payri C. The more we search, the more we find: discovery of a new lineage and a new species complex in the genus *Asparagopsis*. PLoS ONE. 2014;9:e103826.

Faucci A, Toonen RJ, Hadfield MG. Host shift and speciation in a coral-feeding nudibranch. Proc R Soc B. 2007;274:111–9.

Fraser C, Nikula R, Spencer H, Waters J. Kelp genes reveal effects of subantarctic sea ice during the last glacial maximum. Proc Nat Acad Sci USA. 2009;106:3249–53.

Grosberg RK, Levitan DR. For adults only? Supply-side ecology and the history of larval biology. Trends Ecol Evol. 1992;7:130–3.

Guiry MD, Guiry GM AlgaeBase. World-wide electronic publication, National University of Ireland, Galway. 2015. http://www.algaebase.org; searched on 25 Mar 2015.

Hillebrand H. On the generality of the latitudinal diversity gradient. Am Nat. 2004;163:192–211.

Kamiya M, West JA. Cryptic diversity in the euryhaline red alga *Caloglossa ogasawaraensis* (Delesseriaceae, Ceramiales). Phycologia. 2014;53:374–82.

Keith SA, Kerswell AP, Connolly SR. Global diversity of marine macroalgae: environmental conditions explain less variation in the tropics. Glob Ecol Biogeogr. 2014;23:517–29.

Kerswell AP. Global biodiversity patterns of benthic marine algae. Ecology. 2006;87:2479–88.

Knowles L. Statistical phylogeography. Ann Rev Ecol Evol Syst. 2009;40:593–612.

Krueger-Hadfield SA, Roze D, Mauger S, Valero M. Intergametophytic selfing and microgeographic genetic structure shape populations of the intertidal red seaweed *Chondrus crispus*. Mol Ecol. 2013;22:3242–60.

Lee KM, Boo SM, Kain (Jones) JM, Sherwood AR. Cryptic diversity and biogeography of the widespread brown alga *Colpomenia sinuosa* (Ectocarpales, Phaeophyceae). Bot Mar. 2013;56:15–25.

Leliaert F, Verbruggen H, Vanormelingen P, Steen F, Lopez-Bautista JM, Zuccarello GC, De Clerck O. DNA-based species delimitation in algae. Eur J Phycol. 2014;49:179–96.

Lomolino MV, Riddle BR, Brown JH. Biogeography. 3rd ed. Sunderland: Sinauer Associates Inc; 2009. pp. 845.

Lüning K. Seaweeds: their environment, biogeography, and ecophysiology. New York: Wiley; 1990. pp 527.

Mayr E. Animal species and evolution. Cambridge: Harvard University Press; 1963. p. 797.

Muangmai N, West JA, Zuccarello GC. Evolution of four Southern Hemisphere *Bostrychia* (Rhodomelaceae, Rhodophyta) species: phylogeny, species delimitation and divergence times. Phycologia. 2014;53:593–601.

Neall VE, Trewick SA. The age and origin of the Pacific Islands: a geological overview. Philos Trans R Soc B. 2008;363:3293–308.

Norton TA. Dispersal by macroalgae. Br Phycol J. 1992;27:293–301.

Payo DA, Leliaert F, Verbruggen H, D'hondt S, Calumpong HP, De Clerck O. Extensive cryptic species diversity and fine-scale endemism in the marine red alga *Portieria* in the Philippines. Proc R Soc B. 2013;280:20122660.

Phillips L. The taxonomy of selected tribes and genera of the Rhodomelaceae: a molecular and anatomical appraisal of selected members of the red algal family Rhodomelacae (Ceramiales, Rhodophyta). Köln: Lambert Academic Publishing; 2009. p. 486.

Provan J, Glendinning K, Kelly R, Maggs CA. Levels and patterns of population genetic diversity in the red seaweed *Chondrus crispus* (Florideophyceae): a direct comparison of single nucleotide polymorphisms and microsatellites. Biol J Linn Soc. 2013;108:251–62.

Rocha LA, Bowen BW. Speciation in coral-reef fishes. J Fish Biol. 2008;72:1101–21.

Rocha LA, Lindeman KC, Rocha CR, Lessios HA. Historical biogeography and speciation in the reef fish genus *Haemulon* (Teleostei: Haemulidae). Mol Phylogenet Evol. 2008;48:918–28.

Rützler K, Duran S, Piantoni C. Adaptation of reef and mangrove sponges to stress: evidence for ecological speciation exemplified by *Chondrilla caribensis* new species (Demospongiae, Chondrosida). Mar Ecol. 2007;28:95–111.

Schils T, Vroom PS, Tribollet AD. Geographical partitioning of marine macrophyte assemblages in the tropical Pacific: a result of local and regional diversity processes. J Biogeogr. 2013;40:1266–77.

Sherwood AR. Phylogeography of *Asparagopsis taxiformis* (Bonnemaisoniales, Rhodophyta) in the Hawaiian Islands: two mtDNA markers support three separate introductions. Phycologia. 2008;47:79–88.

Sherwood AR, Kurihara A, Conklin KY. Molecular diversity of Amansieae (Ceramiales, Rhodophyta) from the Hawaiian Islands: a multi-marker assessment reveals unexpected diversity within *Amansia glomerata*. Phycol Res. 2011;59:16–23.

Silberfeld T, Bittner L, Fernández-García C, Cruaud C, Rousseau F, de Reviers B, Leliaert F, Payri CE, De Clerck O. Species diversity, phylogeny and large scale biogeographic patterns of the genus *Padina* (Phaeophyceae, Dictyotales). J Phycol. 2013;49:130–42.

Sun Z, Hanyuda T, Lim PE, Tanaka J, Gurgel CFD, Kawai H. Taxonomic revision of the genus *Lobophora* (Dictyotales, Phaeophyceae) based on morphological evidence and analyses *rbcL* and *cox*3 gene sequences. Phycologia. 2012;51:500–12.

Valero M, Destombe C, Mauger S, Ribout C, Engel C R, Daguin-Thiebaut C, Tellier F. Using genetic tools for sustainable management of kelps: a literature review and the example of *Laminaria digitata*. Cah Biol Mar. 2011;52:467–483.

Verbruggen H, DeClerck O, Schils T, Kooistra WHCF, Coppejans E. Evolution and phylogeography of *Halimeda* section *Halimeda* (Bryopsidales, Chlorophyta). Mol Phylogenet Evol. 2005;37:789–803.

Verbruggen H, Tyberghein L, Pauly K, Vlaeminck C, Van Nieuwenhuyze K, Kooistra WHCF, Leliaert F, De Clerck O. Macroecology meets macroevolution: evolutionary niche dynamics in the seaweed *Halimeda*. Glob Ecol Biogeogr. 2009a;18:393–405.

Verbruggen H, Ashworth M, LoDuca ST, Vlaeminck C, Cocquyt E, Sauvage T, Zechman FW, Littler DS, Littler MM, Leliaert F, De Clerck O. A multi-locus time-calibrated phylogeny of the siphonous green algae. Mol Phylogenet Evol. 2009b;50:642–53.

Vergés A, Paul NA, Steinberg PD. Sex and life-history stage alter herbivore responses to a chemically defended red alga. Ecology. 2008;89:1334–43.

Vieira C, D'hondt S, De Clerck O, Payri CE. Toward an inordinate fondness for stars, beetles and *Lobophora*? Species diversity of the genus *Lobophora* (Dictyotales, Phaeophyceae) in New Caledonia. J Phycol. 2014;50:1101–19.

Waters JM, Fraser CI, Hewitt GM. Founder takes all: density-dependent processes structure biodiversity. Trends Ecol Evol. 2013;28:78–85.

West JA, Calumpong HP. Reproductive biology of *Spyridia filamentosa* (Wulfen) Harvey (Rhodophyta) in culture. Bot Mar. 1989;32:379–387.

Win NN, Hanyuda T, Arai S, Uchimura M, Prathep A, Draisma SGA, Phang SM, Abbott IA, Millar AJK, Kawai H. A taxonomic study of the genus *Padina* (Dictyotales, Phaeophyceae) including the descriptions of four new species from Japan, Hawaii, and the Andaman Sea. J Phycol. 2011;47:193–191.

Zanolla M, Carmona R, De la Rosa J, Salvador N, Sherwood AR, Andreakis N, Altamirano M. Morphological differentiation of cryptic lineages within the invasive genus *Asparagopsis* (Bonnemaisoniales, Rhodophyta). Phycologia. 2014;53:233–42.

Zuccarello GC, Sandercock B, West JA. Diversity within red algal species: variation in world-wide samples of *Spyridia filamentosa* (Ceramiales) and *Murrayella periclados* (Rhodomelaceae) using DNA markers and breeding studies. Eur J Phycol. 2002;37:403–17.
Zuccarello GC, West JA. Multiple cryptic species: molecular diversity and reproductive isolation in the *Bostrychia radicans / B. moritziana* complex (Rhodomelaceae, Rhodophyta) with a focus on North American isolates. J Phycol. 2003;39:948–59.
Zuccarello GC, Prud'homme van Reine WF, Stegenga H. Recognition of *Spyridia griffithsiana* comb. nov. (Ceramiales, Rhodophyta): a taxon previously misidentified as *Spyridia filamentosa* from Europe. Bot Mar. 2004;47:481–489.
Zuccarello GC, Buchanan J, West JA, Pedroche FF. Genetic diversity of the mangrove-associated alga *Bostrychia radicans/Bostrychia moritziana* (Ceramiales, Rhodophyta) from southern Central America. Phycol Res. 2011;59:98–104.
Zuccarello G, West JA. Hybridization studies in *Bostrychia*: 2. Correlation of crossing data and plastid DNA sequence data within *B. radicans* and *B. moritziana* (Ceramiales, Rhodophyta). Phycologia. 1997;36:293–304.
Zuccarello G, West JA, King R. Evolutionary divergence in the *Bostrychia moritziana/B. radicans* complex (Rhodomelaceae, Rhodophyta): molecular and hybridization data. Phycologia. 1999;38:234–44.
Zuccarello GC, Buchanan J, West JA. Increased sampling for inferring phylogeographic patterns in *Bostrychia radicans/Bostrychia moritziana* in the eastern USA. J Phycol. 2006;42:1349–52.

Chapter 9
Evolution and Biogeography of Laminarialean Kelps

Hiroshi Kawai, Takeaki Hanyuda and Shinya Uwai

Abstract This review covers the evolution of Laminariales and recent phylogeographic studies focusing on *Chorda* and *Undaria*. In Laminariales, the phylogenetic relationships between basal families (i.e., Akkesiphycaceae, Pseudochordaceae, and Chordaceae) with simple thallus structures, and derived families (i.e., Alariaceae, Laminariaceae, and Lessoniaceae) with large elaborate sporophytes, have remained unclear. Derived Laminariales have been suggested to consist of three major clades roughly corresponding to Alariaceae, Agaraceae (=Costariaceae), and Laminariaceae/Lessoniaceae. Recently, a novel species *Aureophycus aleuticus,* basal to all derived Laminariales, was found in the Bering Sea, and was shown to be basal to all derived Laminariales. Geographically, the majority of derived families in the Laminariales only occur in the Northern Hemisphere, and the Laminariales are therefore considered to have originated in the Northwestern Pacific and spread to the other regions including the Atlantic and the Southern Hemisphere. The limited distributional range of the basal families Akkesiphycaceae and Pseudochordaceae in northeastern Asia, and that of Aureophycaceae in the Bering Sea, supports the notion that the Laminariales originated in the Northwestern Pacific and evolved to giant taxa such as *Macrocystis* in the course of dispersal to the Northeastern Pacific, perhaps through the Aleutian Archipelago. Chordaceae has wide distributional ranges both in the Atlantic and Pacific oceans, but of the four clearly recognized species of *Chorda*, *C. asiatica*, *C. rigida,* and *C. kikonaiensis* are found only in the Northwestern Pacific, whereas *C. filum* is also found in the North Atlantic. In addition, the genetic diversity within *C. asiatica* is greater than that of *C. filum*, and it is suggested that *Chorda* also originated and diverged in the Pacific, then spread into the Atlantic. In the genus *Undaria* (Alariaceae), three species, *U. pinnatifida*, *U. undarioides,* and *U. peterseniana* have been traditionally recognized based on morphological characters, and *U. crenata* was recently described. Based on the genetic studies, *U. crenata* was

---

H. Kawai (✉) · T. Hanyuda
Kobe University Research Center for Inland Seas, Rokkodai, Nadaku, Kobe 657-8501, Japan
e-mail: kawai@kobe-u.ac.jp

S. Uwai
Faculty of Science, Niigata University, Ikarashi-2, Nishiku, Niigata 950-2181, Japan

Z.-M. Hu and C. Fraser (eds.), *Seaweed Phylogeography*,
DOI 10.1007/978-94-017-7534-2_9

considered to be conspecific with *U. pinnatifida* or *U. peterseniana*, although the three species were shown to be genetically rather close, at present we suggest retaining the species-level taxonomy. Based on genetic analyses, we discuss the likely geographic origin and dispersal pathways for nonindigenous populations of *U. pinnatifida.*

Keywords *Aureophycus* · Biogeography · *Chorda* · Laminariales · Phylogeography · *Undaria*

9.1 Background

Members of the order Laminariales (often referred to as the kelps) are the largest marine macroalgae in terms of physical size, constituting a significant ecological element of coastal ecosystems in temperate and colder seas, for example, by providing habitat for diverse organisms (Bold and Wynne 1985; Dayton 1985; Graham and Wilcox 2000). They are the most structurally complex macroalgae with two distinctive generations, large sporophytes and microscopic gametophytes. Sporophytes of the largest laminarialean species such as *Macrocystis* and *Nereocystis* show distinctive differentiation between stipe and blade, as well as buoyancy structures to maintain the distal portion at the water's surface for photosynthesis. Their remarkably heteromorphic life history is explained as an adaptation to seasonality, enabling tolerance of warm and low water transparency summer conditions as small gametophytes and rapid winter growth as the large sporophytes. Although Laminariales has been considered a distinctive order characterized by the above-mentioned elaborate anatomy and life history pattern, there are considerable gaps in morphological complexities and thallus sizes among the major basal lineages such as *Akkesiphycus*, *Pseudochorda*, and *Chorda*, and other derived families with complex thallus morphology such as Alariaceae, Laminariaceae, and Lessoniaceae (Setchell and Gardner 1925; Tilden 1935; Kawai and Kurogi 1985; Henry and South 1987).

Most members of the order are distributed only in the Northern Hemisphere, with only four genera (*Ecklonia*, *Eisenia*, *Lessonia,* and *Macrocystis*) reported from the Southern Hemisphere as indigenous populations. Furthermore, because of the higher biodiversity in the Pacific Ocean, the Laminariales are considered to have originated in the North Pacific Ocean and spread to other regions including the Atlantic Ocean and the Southern Hemisphere (Lüning and tom Dieck 1990; Bolton 2010). Occurrence of basal taxa such as *Akkesiphycus* and *Pseudochorda* only in the Northwestern Pacific Ocean supports this notion. The distributional ranges of the basal taxa Akkesiphycaceae and Pseudochordaceae are remarkably narrow, and to date, the monotypic species *Akkesiphycus lubricus* has been reported only from the eastern coast of Hokkaido. *Pseudochorda nagaii* has a somewhat broader distribution, but is restricted to the Sea of Okhotsk coasts in Japan and Russia, and *P. gracilis* has been recorded only from the Sea of Japan coast of Hokkaido. In contrast, the other basal taxon, Chordaceae, is distributed both in the Northern Pacific and Northern Atlantic Ocean. Historically, the traditional inclusion of Phyllariaceae Tilden and *Halosiphon tomentosum* (=*Chorda tomentosa*; since moved to Stschapoviales) hindered elucidation of the biogeographical origin of the Laminariales (for details see below), because their occurrence only in the Atlantic Ocean appeared to contradict a Pacific origin of the order (Lüning and tom Dieck 1990) (Fig. 9.1).

Because of the lack of any significant fossil record and incomplete molecular phylogeny, the evolution of laminarialean specialization and history of dispersal have remained unclear. The divergence time of Laminariales from its sister taxon Ectocarpales has been suggested to be relatively recent (<100 Ma), but the phylogeny is still poorly understood (Silberfeld et al. 2010; Kawai et al. 2015a). However, detailed analyses of the genetic diversities in local populations of selected laminarialean taxa have revealed the geographical origin of these taxa, and their dispersal history.

In order to discuss the biogeography and dispersal history of each taxon, it is essential that the phylogeny and the species boundaries are clearly understood. This review aims to: (1) summarize the current status of higher rank taxonomy of Laminariales, focusing on molecular phylogenies, and (2) to review phylogeographical studies of two representative genera, with a focus on the work carried out by the present authors.

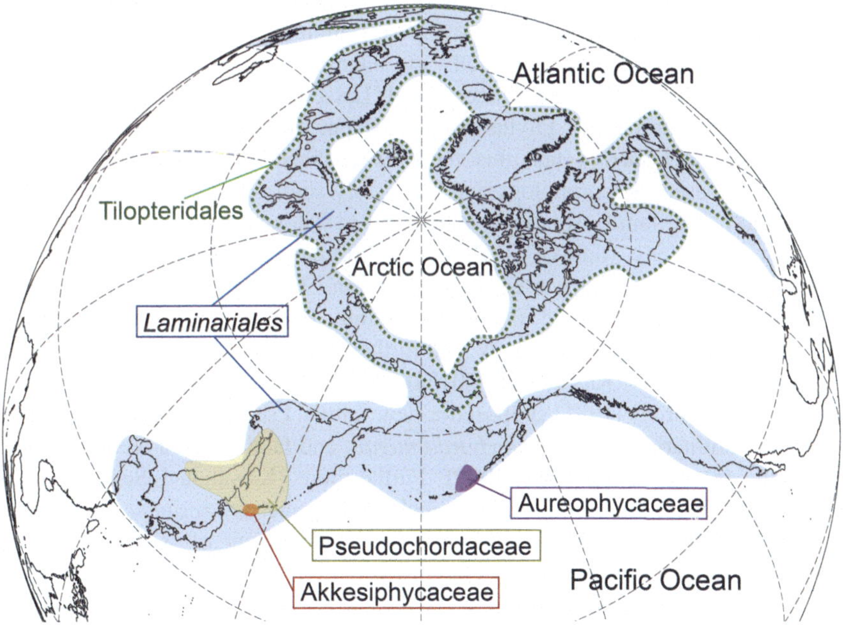

Fig. 9.1 Geographical distributions of Laminariales and Tilopteridales in the Northern Hemisphere. Distributional ranges of the basal laminarialean taxa Akkesiphycaceae, Pseudochordaceae, and Aureophycaceae are separately indicated. The other basal family Chordaceae is shown in Fig. 9.4

9.2 Historical Review of the Taxonomy of Laminariales

Fossil evidence of the evolution of the Phaeophyceae is scant, because of their generally soft tissues composed of polysaccharides such as alginates, fucoidans and cellulose, very limited occurrence of calcified taxa, preference for exposed habitats where sedimentation is not common, and relatively recent evolution compared with red and green algae. Therefore, it is very difficult to assess the divergence times of phaeophycean taxa. However, Lim et al. (1986) estimated the divergence of Phaeophyceae from Bacillariophyceae to be around 200 Ma based on molecular phylogenetic analyses using rRNA sequences. Later Silberfeld et al. (2010) published a time-calibrated molecular phylogenetic tree based on combined DNA sequence data, and suggested the branching time of Laminariales from Ectocarpales to be around 100 Ma, and that of Chordaceae and derived families to be around 85 Ma. Kawai et al. (2015a) reexamined the time-calibrated molecular phylogenetic tree adding the sister and basal taxa of Phaeophyceae (i.e., Schizocladiophyceae and Discosporangiales, respectively) and suggested that the branching times were more recent: 90 Ma (Laminariales/Ectocarpales) and 75 Ma (Chordaceae/derived Laminariales).

Among the Phaeophycean orders, the order Laminariales Migula (1909) has long been regarded as a well-defined order characterized by a heteromorphic life history, alternating between a large parenchymatous sporophyte and filamentous, oogamous gametophytes (Kylin 1916; Sauvageau 1916; Oltmanns 1922; Setchell and Gardner 1925; Fritsch 1945; Bold and Wynne 1985). Laminariales traditionally included four families, Alariaceae Setchell et Gardner, Chordaceae Dumortier, Laminariaceae Bory de Saint-Vincent, and Lessoniaceae Setchell et Gardner. Tilden (1935) proposed Phyllariaceae Tilden, including *Saccorhiza* and *Phyllariopsis* (=*Phyllaria*), and this family was considered a basal member of Laminariales based on morphology and life history studies (Henry and South 1987; Henry 1987). The family was later transferred to Tilopteridales Bessey based on molecular phylogeny (Sasaki et al. 2001). Similarly, *Halosiphon tomentosus* (=*Chorda tomentosa*), which used to be placed in Chordaceae, Laminariales, was transferred to Tilopteridales in its own new family Halosiphonaceae (Sasaki et al. 2001) based on molecular phylogeny, and further moved to a newly proposed order Stschapoviales Kawai (Kawai et al. 2015a).

Kawai and Kurogi (1985) proposed a new family Pseudochordaceae Kawai et Kurogi in Laminariales for *Pseudochorda nagaii*, and later added a second species *P. gracilis* (Kawai and Nabata 1990). Furthermore, Kawai and Sasaki (2000) extended the definition of the order to include a species (*Akkesiphycus lubricus*) with plano-anisogamy (Kawai 1986) by the establishment of Akkesiphycaceae Kawai et Sasaki and its inclusion in the order.

Lane et al. (2006) examined the molecular phylogeny of derived Laminariales, and suggested that they consist of three major clades roughly corresponding to Alariaceae, Agaraceae Postels et Ruprecht (=Costariaceae C.E. Lane, C. Mayes, Druehl et G.W. Saunders), and Laminariaceae/Lessoniaceae. Later, Kawai et al.

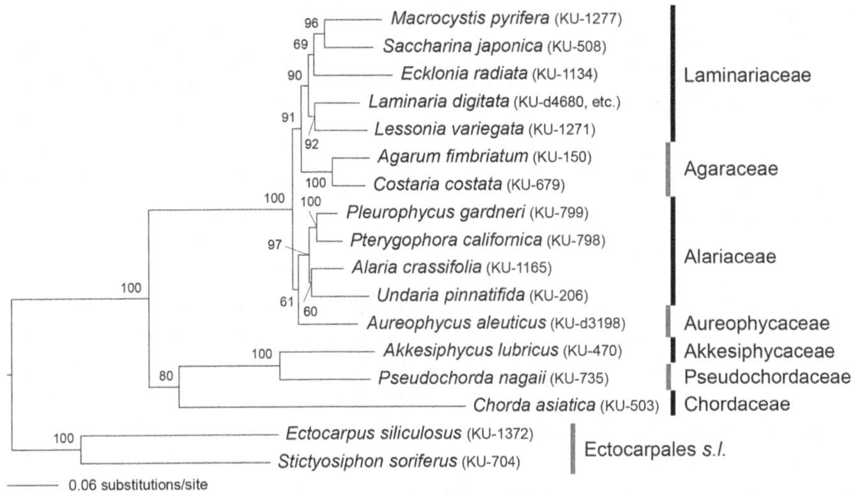

Fig. 9.2 Maximum likelihood phylogenetic tree of Laminariales based on the concatenated DNA sequences of chloroplast *atp*B, *psa*A, *psa*B, *psb*A, *psb*C, *rbc*L, and mitochondrial *cox*1 and *cox*3 genes. Node numbers indicate bootstrap values (only values >50 % are shown). KU-### means culture strains housed in Kobe University Macroalgal Culture Collections (KU-MACC), and KU-d### means silica gel dried field-collected specimens housed in Kobe University Research Center for Inland Seas

(2008) described a novel genus/species *Aureophycus aleuticus*, which is distinct from any known laminarialean species, from islands in the Bering Sea, and later proposed a new family Aureophycaceae H. Kawai, T. Hanyuda, Lindeberg et S.C.Lindstrom to accommodate this species (Kawai et al. 2013). Molecular phylogeny suggested that *Aureophycus* is basal to all derived Laminariales, and this notion was supported by its remarkably simple sporophyte morphology, lacking a rhizoidal meristematic holdfast hapteron and any mucilaginous organs. The limited distributional range of *Aureophycus* in the Bering Sea, together with the limited distribution of the basalmost taxa of the order in the Northwestern Pacific, supports the hypothesis that Laminariales originated in the Northwestern Pacific Ocean, and evolved to giant taxa such as *Macrocystis* and *Nereocystis* in the course of their dispersal to the Northeastern Pacific Ocean, perhaps along the Aleutian Archipelago (Fig. 9.2).

9.3 Systematics and Phylogeography of Selected Genera

9.3.1 Phylogeny and Congeneric Phylogeography of Chorda

In spite of its remarkably different thallus morphology, *Chorda* has been regarded as a member of Laminariales because of the anatomical similarity of its sori, composed of unicellular paraphyses and unilocular zoidangia (Reinke 1892).

However, in addition to the simpler sporophyte anatomy lacking differentiations between blade and stipe, *Chorda* has been regarded as a basal ('primitive') member of the order because of the differences in the following features: lack of a meristematic rhizoidal holdfast haptera; annual nature of sporophytes; lack of mucilage organs such as mucilage gland cells and ducts; lack of mucilage caps on paraphyses; presence of eyespots in zoospores; and occurrence of monoecious gametophytes in *C. tomentosa* = *H. tomentosus*.

The placement of *H. tomentosus* in the family Chordaceae has been controversial because the species differs from the generitype *C. filum* in various basic features: occurrence of long assimilatory filaments instead of unicellular paraphyses; absence of an intercalary meristem; absence of trumpet-shaped hyphae or an obvious differentiation between cortical layer and peripheral (meristodermal) layer; occurrence of monoecious gametophytes; and presence of different sexual pheromones (Maier 1984, 1995; Kogame and Kawai 1996). More recently Peters (1998) established that there is a relatively substantial genetic distance between *C. filum* and *H. tomentosus* based on molecular phylogenetic data. Consequently, Peters (1998) reinstated the generic name *Halosiphon*, suggesting the possibility that *Halosiphon* was incorrectly placed in the Chordaceae, but did not present a formal taxonomic treatment. Only one *Chorda* species, *C. filum* (except for some doubtful species such as *C. munuta*), was then recognized after the transfer of *C. tomentosa* to *Halosiphon* (as *H. tomentosus*; Peters 1998). Later, *Halosiphon* was placed in an independent family Halosiphonaceae Jaasund ex Kawai et Sasaki (Sasaki et al. 2001) in Tilopteridales.

Later, within *Chorda*, an additional species *C. rigida* was described based on morphological and molecular phylogenetic data (Kawai et al. 2001). This species is similar to *C. filum* in habit, but it differs by its more robust thallus with an intercalary meristem even at maturity. Later, largely based on molecular phylogeny, further taxonomic divergence within the genus became evident: A third species, *C. kikonaiensis* was described, and Asian *C. filum* was shown to be independent from Atlantic *C. filum* and was therefore designated as *C. asiatica* (Sasaki and Kawai 2007). *C. kikonaiensis* resembles *C. filum* and *C. asiatica*, but is distinguished by the shorter (0.4–1.3 m) and softer sporophyte, and the thinner cortex composed of fewer (2–4) cells. The independence of this species is further supported by molecular phylogenetic analyses using *rbc*L gene and ITS rDNA sequences (Fig. 9.3). *C. asiatica* is more variable in morphology (length of erect thallus and number of cell layers composing the cortex) than *C. kikoniensis* and *C. rigida*, and is difficult to distinguish from *C. filum* based on morphology, but is clearly separated from the other species (*C. filum, C. kikoniensis* and *C. rigida*) based on ITS rDNA data. *C. filum* is distributed in the Atlantic, whereas *C. asiatica*, *C. kikonaiensis,* and *C. rigida* are distributed in the Pacific Ocean. It is interesting that Atlantic *C. filum*, which is considered based on molecular phylogenetic data to have diverged more recently compared to Asian taxa (Kawai et al. 2015b), demonstrates considerably greater morphological diversity in thallus anatomy (South and Burrows 1967) than is observed within each Asian species, and this hindered recognition of the species divergences in the Pacific Ocean (Sasaki and Kawai 2007).

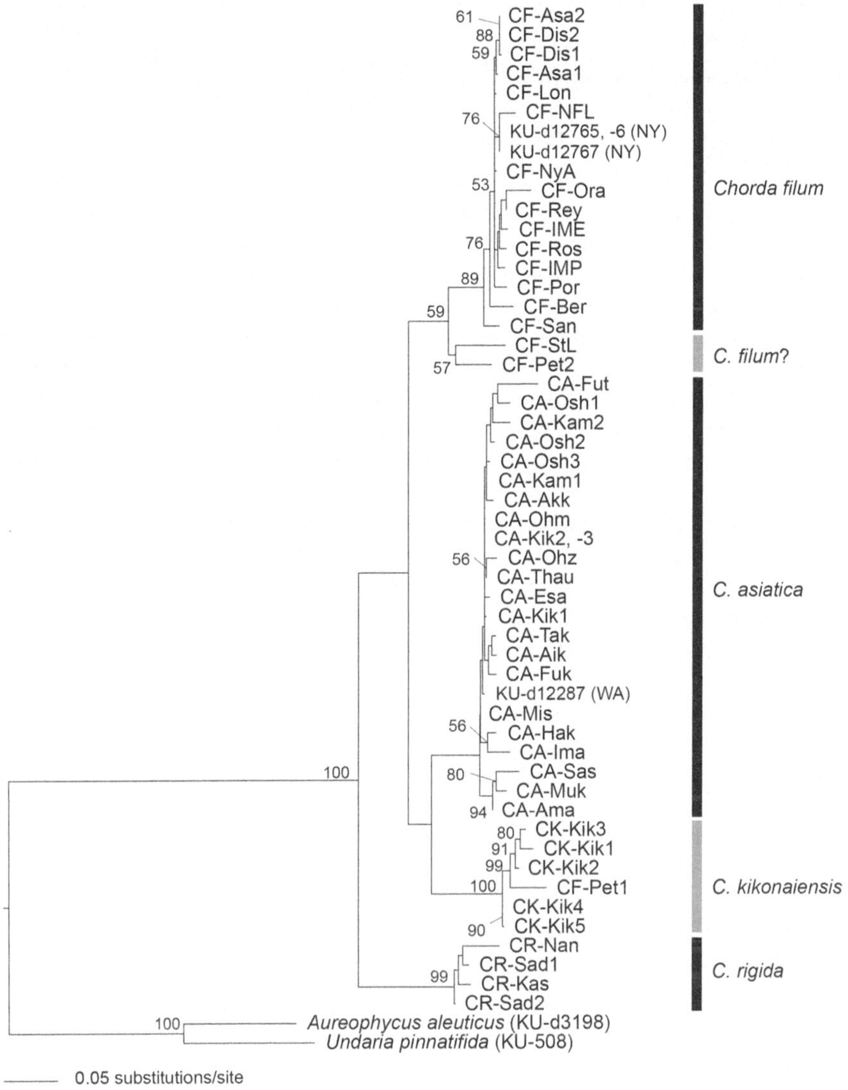

Fig. 9.3 Maximum Likelihood phylogenetic tree of Chordaceae based on ITS1-5.8S-ITS2 region of rDNA sequences. Bootstrap values (>50 %) based on 10,000 replicates are shown. For specimen codes, see Kawai et al. (2015b). KU-###, KU-MACC culture strains; KU-d####, silica gel dried field-collected specimens housed in Kobe University Research Center for Inland Seas

Sasaki and Kawai (2007) demonstrated the occurrence of genetically distinct populations of *Chorda* species on the northwestern American coast based on the sequence data of specimens collected in Puget Sound, WA, USA. In contrast, examination of multiple newly collected specimens from the same locality (Hood Canal, Puget Sound) gave the same ITS rDNA sequence, which was placed in the clade of *C. asiatica*. Considering the historical introductions of oysters from Japan, Kawai et al. (2015b) concluded that they are most likely anthropogenically introduced populations, associated with the widespread practice in the shellfish industry of Puget Sound from 1903 until the 1970s of bringing in oyster spat (*Crassostrea gigas*) on adult shells from Japan (Bonnott 1935; Steele 1964). Similar (and possibly simultaneous) introductions of *C. asiatica* (reported as *C. filum*) associated with young oysters from Japan have been reported in the Mediterranean Sea (Riouall 1985; Boudresque and Verlaque 2002; Kawai et al. 2015b).

Furthermore, the presence of one or two additional cryptic species is suggested in the northern Pacific based on the molecular data. Therefore, it is shown that the genus *Chorda* has considerably greater taxonomic and genetic diversity in the Pacific than in the Atlantic. Although no molecular data are available for other eastern Asian (southeastern Russian coast, Korea and China) *Chorda* species, they are likely referable to *C. asiatica,* based on morphology and geographical distributions. *C. kikonaiensis* grows on upper intertidal rocks of relatively sheltered, somewhat muddy, wide flat rock beds, whereas *C. asiatica* typically grows on upper subtidal rocks. Young sporophytes of *C. asiatica*, retaining the intact distal portion of the intercalary meristem, are found in early March, and they retain the intercalary meristem throughout the spring. The sporophytes reach their maximum size (normally up to 0.8 m, but sometimes attaining 1.3 m in length, and 2.5 mm in diameter) during April and May, become fertile in April, and disappear in July–August.

Water temperature is the primary environmental factor delimiting the distributional ranges of most marine macroalgae. The sporophytes of *C. rigida* grew well in culture under 10 to 25 °C conditions, but did not grow well below 5 °C. They became fertile at 15 °C forming unilocular sporangia among unicellular paraphyses. In contrast, the strains of *C. filum* (Bergen, Norway) showed relatively good growth at 5 to 10 °C, and they did not grow in higher temperatures of 15 and 20 °C respectively. *C. kikonaiensis* grew well at 2 to 15 °C but did not grow at 20 and 25 °C. In contrast, the sporophytes of *C. asiatica* grew at 5, 10 and 15 °C, but did not grow at 2 °C (Sasaki and Kawai 2007).

In the cooler parts of the distributional range of *C. filum*, (e.g., northern Europe where monthly average surface water temperature in August is 11–13 °C; South and Burrows 1967), the growth of *C. filum* may be slow and the sporophytes often persist over summer (Lund 1959; South and Burrows 1967). However, even in such localities, the sporophytes do not remain vegetative after attaining their maximum length in summer (South and Burrows 1967; Russell 1985).

Of the four clearly recognized species of *Chorda*, *C. asiatica*, *C. rigida*, and *C. kikonaiensis* are distributed only in the Northwest Pacific (Kawai et al. 2001, 2015b). In addition, genetic diversity within *C. asiatica* is greater than that of Atlantic *C. filum*, although the data for *C. asiatica* are based only on Japanese specimens. This suggests that *Chorda* originated and diverged in the Pacific Ocean, and spread into the Atlantic (Fig. 9.4). These findings are consistent with the notion that the Laminariales originated in the Pacific Ocean, based on the rich diversity of laminarialean taxa in the Pacific, the occurrence of basal taxa and the limited

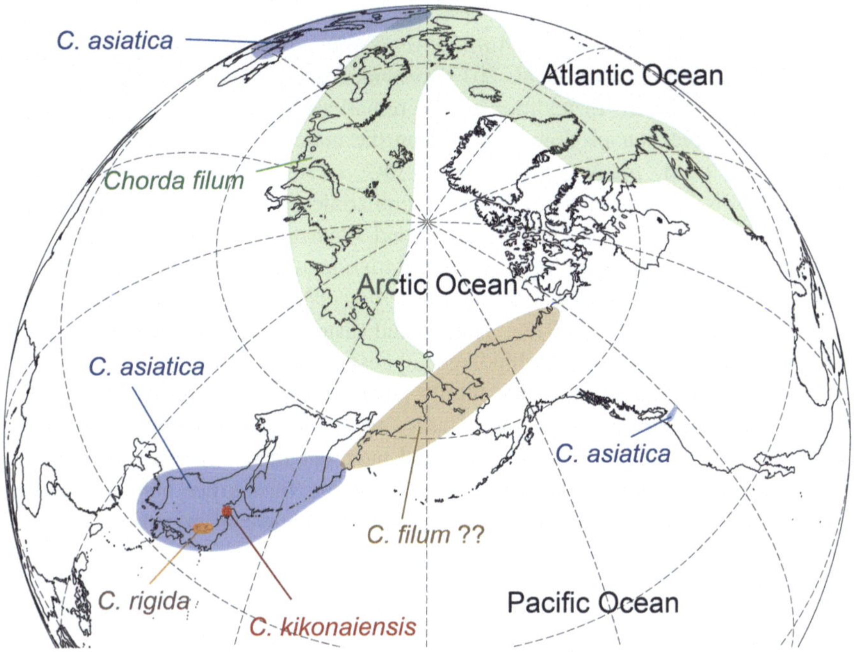

Fig. 9.4 Geographical distribution of *Chorda* species including introduced populations of *C. asiatica* in the Mediterranean Sea and Northwestern America

distribution of the Pseudochordaceae (Kawai and Kurogi 1985; Kawai and Nabata 1990; Lüning and tom Dieck 1990) and Akkesiphycaceae (Kawai 1986; Kawai and Sasaki 2000) in this area.

9.4 Phylogeny of Laminarialean Crown Taxa

Family-level taxonomy of derived Laminariales based on morphology has been reexamined by many authors using different genetic markers, raising questions about the monophyly of each family (Fain et al. 1988; Saunders and Druehl 1992; Boo et al. 1999).

Yoon et al. (2001), using *rbc* spacer and rDNA ITS sequences, obtained eight clades (corresponding to the genera *Hedophyllum*, *Macrocystis*, *Alaria*, *Agarum*, *Ecklonia*, *Lessonia*, *Laminaria*, and *Egregia*) and showed that some of the clades corresponded to the Tribes suggested by Kützing, Bory de Saint Vincent, Setchell, and Setchell and Gardner (Setchell and Gardner 1925). The authors discussed subdivision of the order into eight families, although they did not provide a taxonomic treatment. Lane et al. (2006) suggested that Laminariales consist of three major clades roughly corresponding to Alariaceae, newly proposed Costariaceae, and Laminariaceae/Lessoniaceae.

As mentioned above, there has been a large gap between basal and derived Laminariales but the discovery of *Aureophycus aleuticus* (Kawai et al. 2008) and its familial assignment (Kawai et al. 2013) provided a clue for elucidating the relationship between them. In addition to the remarkably simple morphology of the sporophytes, molecular phylogeny based on the concatenated sequences of eight genes (*rbc*L, *atp*B, *psa*A, *psa*B, *psb*A, *psb*C, *cox*1, and *cox*3) suggested that *Aureophycus* is either basal to all derived Laminariales, or to the clade of Alariaceae including *Alaria, Undaria, Pterygophora*, and *Pleurophycus*. However, based on the results of character mapping of representative taxonomic features onto the two topology options, the authors concluded that *Aureophycus* was basal to all derived Laminariales (Kawai et al. 2013). The limited distribution of *Aureophycus* in the Bering Sea and the Aleutian Archipelago suggest the importance of this area as the center for the differentiation of the family-level lineages of derived Laminariales: evolution of the genera with very large and elaborate sporophytes such as *Macrocystis*, *Egregia* and *Nereocystis* could have occurred in the process of eastward dispersal, whereas evolution of taxa adapting to warmer habitats such as *Undaria* and *Ecklonia* could have occurred in the course of westward dispersal from the region.

9.5 Phylogeny and Phylogeography of *Undaria*

9.5.1 Taxonomy of Undaria Species

The genus *Undaria* Suringar (1873) is distinctive from other laminarialean genera, and is characterized by the soft annual sporophytes forming reproductive structures at the basal portion of the membranous thalli (blades) and/or sporophylls at the basal part of the stipe. Three to four species are currently recognized in the genus and they have relatively limited geographical distributions in the temperate regions in northeastern Asia. Because of the high morphological plasticity and close genetic relationships among the reported taxa, their species-level and intraspecific taxonomy has been somewhat confused (Yendo 1911; Okamura 1915; Castric-Fey et al. 1999; Cecere et al. 2000).

Three species, *U. pinnatifida*, *U. undarioides* and *U. peterseniana*, have been traditionally recognized (Okamura 1915, 1926) based on morphological characters. These species have been distinguished by combinations of characters such as presence/absence of pinnae on the blade, presence/absence of sporophyll, and presence/absence of a midrib. *U. pinnatifida* typically has a pinnately lobed blade (membranous thallus) with a midrib throughout the entire length, and a wrinkled sporophyll on the flattened stipe. *U. peterseniana* has a long foliose blade without a midrib, forming sori on the basal and middle portion of the blade. *U. undarioides* has an intermediate morphology between *U. pinnatifida* and *U. peterseniana*, having an elliptical to ovate blade without pinnae, and the presence of midrib is variable. *U. undarioides* sometimes has sporophylls on the basal stipe in addition to the sorus on the blade (Yendo 1903, 1911; Okamura 1915). Furthermore, the occurrence of sori on the basal part of the blade and midrib in *U. pinnatifida* has also been found in the field (H. Kawai, unpublished data).

U. pinnatifida has the broadest distribution, along the Sea of Japan and the Pacific coast of Japan, whereas *U. peterseniana* and *U. undarioides* are distributed in narrower ranges within that of *U. pinnatifida*. Later a new species *U. crenata* Y. P. Lee was described from Korea, having a blade with laciniate margin, considered to be an intermediate between *U. pinnatifida* and *U. peterseniana* (Lee and Yoon 1998). Additionally, Lee (1998) established *Undariella* to accommodate *Undariella peterseniana*, although *Undariella* was later reduced to synonymy with *Undariopsis* Miyabe et Okamura (Okamura 1902) by the author (Lee 1999).

Muraoka and Saito (2005) reported genetic distinctness between *U. pinnatifida* and *U. undarioides* based on the mitochondrial 23S rDNA sequence. Later, Uwai et al. (2007) examined morphology and genetic diversity using mitochondrial *cox*3 sequences among *Undaria* species collected from throughout their distribution around Japan (Fig. 9.5). In the statistical parsimony network, *U. peterseniana* haplotypes were divided into two lineages (type-c17, -c18 and -c19), and they were closely related to haplotypes found in *U. pinnatifida* in the Sea of Japan and type-c15, -c16 found in *U. undarioides*, respectively. Similarly, type-c12, found in *U. undarioides*, was closely related to type-c13 and -c14 found in *U. pinnatifida*. In the network, direct connection among *U. peterseniana* haplotypes (between type-c17, -c18 and type-c19) and among *U. undarioides* haplotypes (between type-c15, -c16, and -c12) were not observed. In the ML molecular phylogenetic tree based on the *cox*3 sequences, none of the three species were monophyletic, and the statistical support for the clades was moderate (<80 %, data not shown).

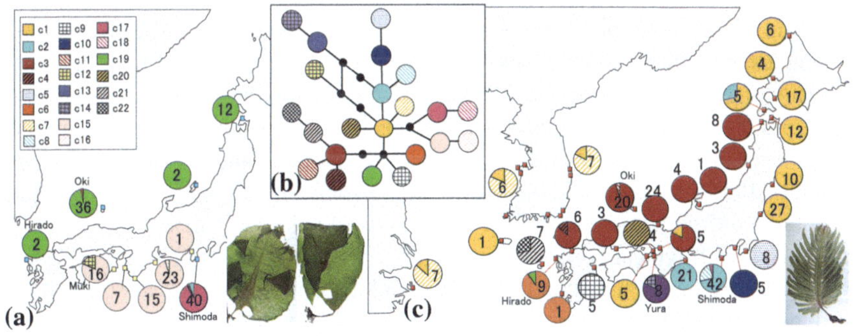

Fig. 9.5 Network and geographic distribution of the mitochondrial *cox*3 haplotypes of *Undaria* spp. **b** haplotypes are illustrated by different combinations of color and pattern and identified with a haplotype code (*inset* in **a**). In the maps, pie charts represent local haplotype frequencies, with numerals indicating the number of samples in each site for (**a**) *U. undarioides* (*yellow*) and *U. peterseniana* (*blue*), and (**c**) *U. pinnatifida*. Plants with the typical morphology of *U. undarioides* (**a**, *left*), *U. peterseniana* (**a**, *right*), and *U. pinnatifida* (**c**), are also shown

Geographical distributions of each haplotype elucidated in Uwai et al. (2007) are shown in Fig. 9.5. Each of three species had unique *cox*3 haplotypes, with a few exceptions, and showed apparent geographical structure within species. Each population of *U. undarioides* and *U. peterseniana*, as well as *U. pinnatifida* populations, had only one or two haplotypes. A *U. peterseniana* population on the Pacific coast (Shimoda, Shizuoka, Japan) had type-c17 and -c18, both of which were unique in this population, in addition to type-c2 which was found from *U. pinnatifida* populations in the same region. *U. peterseniana* populations along the Sea of Japan had unique type-c19. Almost all *U. undarioides* had species-specific type-c15 in addition to type-c16 and type-c12 as minor haplotypes.

Specimens of each species with atypical morphology did not show unique *cox3* haplotypes. For example, the specimens with well-developed sporophyll had type-c17 (*U. peterseniana* collected in the Pacific coast) and type-c15 (*U. undarioides*), identical to the sporophytes having blade sori in *U. peterseniana* and *U. undarioides*, respectively (Uwai et al. 2007). Similarly, the *U. pinnatifida* specimens morphologically similar to *U. crenata*, collected at Oki, central region of the Sea of Japan, had type-c3, which was commonly distributed along the *U. pinnatifida* populations in the Sea of Japan. On the other hand, the sporophytes of *U. pinnatifida* collected in Hirado, Northern Kyusyu, had type-c19, which was identical to that of *U. peterseniana* in the Sea of Japan. Exceptionally, a Pacific sample with typical *U. peterseniana* morphology showed type-c2, found in a *U. pinnatifida* population in that region (Shimoda). Three sporophytes of six *U. undarioides* with dentate margin collected in Muki, the westernmost sampling site for the species, had a unique haplotype, type-c12, which was closely related to the haplotypes found in nearby *U. pinnatifida* populations (Yura, Awaji Island). However, three other sporophytes with dentate margin had type-c15, which was identical to *U. undarioides* with typical morphology.

These results based on mitochondrial *cox*3 haplotypes strongly suggest that there is, in principle, genetic distinctness among the Japanese *Undaria* species. Almost all samples examined were easily identified based on morphological features, especially based on presence or absence of pinnae on the blade, and the *cox*3 haplotypes were different depending on the species. This result supports the present species-level taxonomy of *Undaria*, especially the significance of the blade shape (Okamura 1915); *U. peterseniana* and *U. undarioides* plants with well-developed sporophylls did not show any differences in *cox*3 haplotype when compared with sporophytes with typical sori.

On the other hand, an exceptional *cox*3 haplotype was observed in some of the sporophytes with intermediate morphologies. An *U. pinnatifida* specimen from Oki that had only short pinnae and a rudimentary midrib had a haplotype that was identical to those of *U. pinnatifida* in the Sea of Japan, however, an *U. pinnatifida* specimen from Hirado, morphologically similar to the specimen from Oki, had the same haplotype as *U. peterseniana*. Interestingly, those *U. pinnatifida* were similar to *U. crenata* (Lee and Yoon 1998) as well as the experimental hybrids that resulted from crosses between *U. pinnatifida* and *U. peterseniana* (Migita 1967; Saito 1972). Therefore, Uwai et al. (2007) considered that *U. crenata* is conspecific with

U. pinnatifida or *U. peterseniana*; those plants could be considered as one extreme example of the morphological variations of *U. pinnatifida* or *U. peterseniana*. Alternatively, it is also possible that *U. crenata* and the *U. pinnatifida* samples from Oki and Hirado were hybrids originating from crosses between *U. pinnatifida* and *U. peterseniana* (Kikuchi et al. 1996). Similarly, three *U. undarioides* plants with dentate margin and type-c12 also could be considered a result of interspecies hybridization, rather than resulting from a polyphyletic origin of species or lineage sorting (discussed below), although type-c12 has not been found from *U. pinnatifida*. More significantly, occurrence of interspecies hybridization was suggested based on a sporophyte collected in Shimoda, Japan, with typical *U. peterseniana* morphology but having the *cox*3 haplotype that was common in *U. pinnatifida* in that region. These results suggest frequent occurrence of interspecies hybridizations among *Undaria* spp.

In addition to the exceptional haplotypes, the phylogeny of the *cox*3 haplotypes suggests that taxonomic problems still remain in the genus. Conspecific populations had different haplotypes depending on their geographical origins, and such haplotypes, in some cases, had no direct connection in the network, both of which suggest that each species includes two to several sibling species. An alternative explanation for the observed polyphyly is incomplete lineage sorting (Avise 2000); through speciation, each daughter species could take a subset of haplotypes, which are not necessarily monophyletic, from the ancestral species, and therefore the daughter species is polyphyletic in the gene tree for a certain period after speciation. The reported small genetic distance (Muraoka and Saito 2005; Uwai et al. 2007) suggests this could be the case for *Undaria* species. Besides the polyphyly, the haplotypic differentiation between conspecific populations also suggests the presence of sibling species, or an ongoing speciation event. Only little has been known about the isolation mechanisms between such local groups, as well as species; based on laboratory culture experiments with strict temperature control, Morita et al. (2003a, b) reported differences in temperature requirements between parapatric *U. pinnatifida* and *U. undarioides* populations, which possibly functions as the isolation mechanism between them. Gene flow between conspecific populations, as well as between species, and studies on the isolation mechanism are warranted to evaluate the taxonomic status of local groups.

Within *U. pinnatifida*, several intraspecific taxa have been recognized as varieties (Suringar 1873) and forms (Miyabe 1902; Yendo 1911; Okamura 1915). The two forms *U. pinnatifida* f. *typica* and *U. pinnatifida* f. *distans* (Yendo 1911; Okamura 1915) are distinguished by the distance between the sporophyll and blade, and the depth of the incisions of the blade. Besides these forms, Yendo (1911) recognized *U. pinnatifida* f. *narutensis*, characterized by a short stipe with less-folded sporophylls, although Okamura (1915) regarded this as a synonym of f. *typica*. Later, Lee and Yoon (1998) suggested the reinstatement of *U. pinntifida* var. *elongata* and var. *vulgaris* for those forms because of nomenclatural priority. Relationships between these morphological varieties/forms and the local groups characterized by the haplotypes remain unknown.

All molecular analyses for interspecies-level phylogeny (Muraoka and Saito 2005; Uwai et al. 2007) have shown close genetic relationships among members of *Undaria*. The number of substitutions among *Undaria* spp. is up to 1.9 % (nine of 470 bp) in the mitochondrial *cox*3 gene, which is remarkably smaller than those often observed within single species; up to 10.3 % in *Scytosiphon lomentaria* (Kogame et al. 2005), 5.9 % in *Ishige okamurae* (Lee et al. 2012), and 3.7 % in *Colpomenia peregrina* (Lee et al. 2014). Several genera have been established to accommodate *Undaria* species; Lee (1998) established *Undariella* (=*Undariopsis*) based on *U. peterseniana*; *Hirome* for *U. undarioides* by Yendo (1903); *Undariopsis* for *U. peterseniana* by Miyabe et Okamura (Okamura 1902). As Okamura (1915) concluded, however, morphological variations of *Undaria* spp. could be considered continuous when three species (*U. pinnatifida, U. undarioides*, and *U. peterseniana*) were compared. Therefore, we suspend any changes in the present taxonomic system and classify three species in the single genus *Undaria*.

In addition to taxonomic research, because of the commercial importance of *Undaria*, there have been a number of studies examining the influences of genetic and environmental traits on the morphological plasticity of *Undaria* species (Saito 1972; Stuart et al. 1999; Choi et al. 2009; Nanba et al. 2011; Park et al. 2012). By the application of genetic analyses, intraspecific taxonomy has been updated, although the species-level taxonomy still appears to need further studies.

9.5.2 Phylogeography of Undaria Pinnatifida

Several studies on the molecular diversity of *U. pinnatifida* have been done for the purpose of deducing the geographic origin of introduced populations (Voisin et al. 2005; Uwai et al. 2006a), to establish a method for authenticating the origin of commercial products (Endo et al. 2009), and to analyze the phylogeographic history of the species (Uwai et al. 2006b), based on mitochondrial DNA sequences and the nuclear ribosomal ITS regions. All authors commonly reported high genetic diversity within indigenous populations as well as introduced ones.

9.5.3 Indigenous Populations

The native distributional range of *U. pinnatifida* has been regarded as temperate waters of northeastern Asia (Japan, Korea and China) (Akiyama and Kurogi 1982; Tseng 1983), although Chinese populations are regarded as introduced by Lutaenko et al. (2013). Techniques for *U. pinnatifida* mariculture were established in the 1950s, and intentional introductions within the indigenous region for mariculture have been a concern in relation to genetic disturbance through escape and hybridization between introduced and native populations. High and geographically structured genetic diversity has been reported for the "wild" populations, especially along Japanese coasts (Voisin et al. 2005; Uwai et al. 2006a, b; Endo et al. 2009), suggesting limited genetic disturbance by anthropogenic movement, although the Bay of Osaka is an exceptional area (discussed below).

Uwai et al. (2006a) reported a strong geographic structure throughout the native range of *U. pinnatifida* based on concatenated sequences (ca. 900 bp) of the mitochondrial *cox*3 and *tat*C-*trn*L region. Haplotypes found on Japanese coasts were divided into three local types based on their geographic distributions and phylogenetic relationships in the statistical parsimony network (Fig. 9.6): (1) Pacific central Japan type from northern Kyushu to central Honshu, (2) northern Japan type from Hokkaido and the Pacific coast of northern Honshu, and (3) the Sea of Japan type found from the Sea of Japan coast of Honshu. Besides these local types, the populations in Korea and China had unique major haplotypes (continental type), in addition to the northern Japan haplotype as a minority. The northern Japan haplotype and the continental type had sequences similar to each other, but differed by a long (20 bp) indel in the intergenic spacer region between *trn*W and *trn*I. Haplotypes comprising each local type differed by only a few base pairs, except for the Pacific central Japan type. Their geographical distributions in northeastern Asia are shown Fig. 9.6a.

Based on the small number of steps and the star-like topology among haplotypes of the Sea of Japan type, Uwai et al. (2006a) concluded that regional populations there have expanded recently and linked this notion with geological events in this region. Ohba et al. (1991) concluded, based on the microfossil assemblages in piston core samples, that a large input of freshwater into the surface layer and stratification of the water column caused severe anoxic conditions in the Sea of Japan during the middle period of the last glacial age (27,000–20,000 years ago). These perturbations led to the extinction of benthic fauna in the Sea of Japan, and are likely to have impacted the seaweed flora of this region as well. Both seaweed flora and benthic fauna of the Sea of Japan might have been established after the beginning of the inflow of seawater after the Last Glacial Maximum, especially after the inflow of the Tsushima warm current through the Tsushima strait from 8000 years ago. A similar colonization pattern into the Sea of Japan was reported for the Japanese turban shell (Kojima et al. 1997) and the ulvophycean genus *Blidingia* (Woolcott et al. 2000). The diverged haplotypes of the central Pacific type, especially those of central Pacific Honshu, show clear contrast with the Sea of Japan type and might indicate differences in population demographics between both sides of Japan.

9.5.4 Nonindigenous Populations

Undaria pinnatifida has been introduced into Europe in the 1970s associated with young oysters introduced for mariculture, initially to the French Mediterranean coast. Within a decade the species had spread to a broader area in Europe (Perez

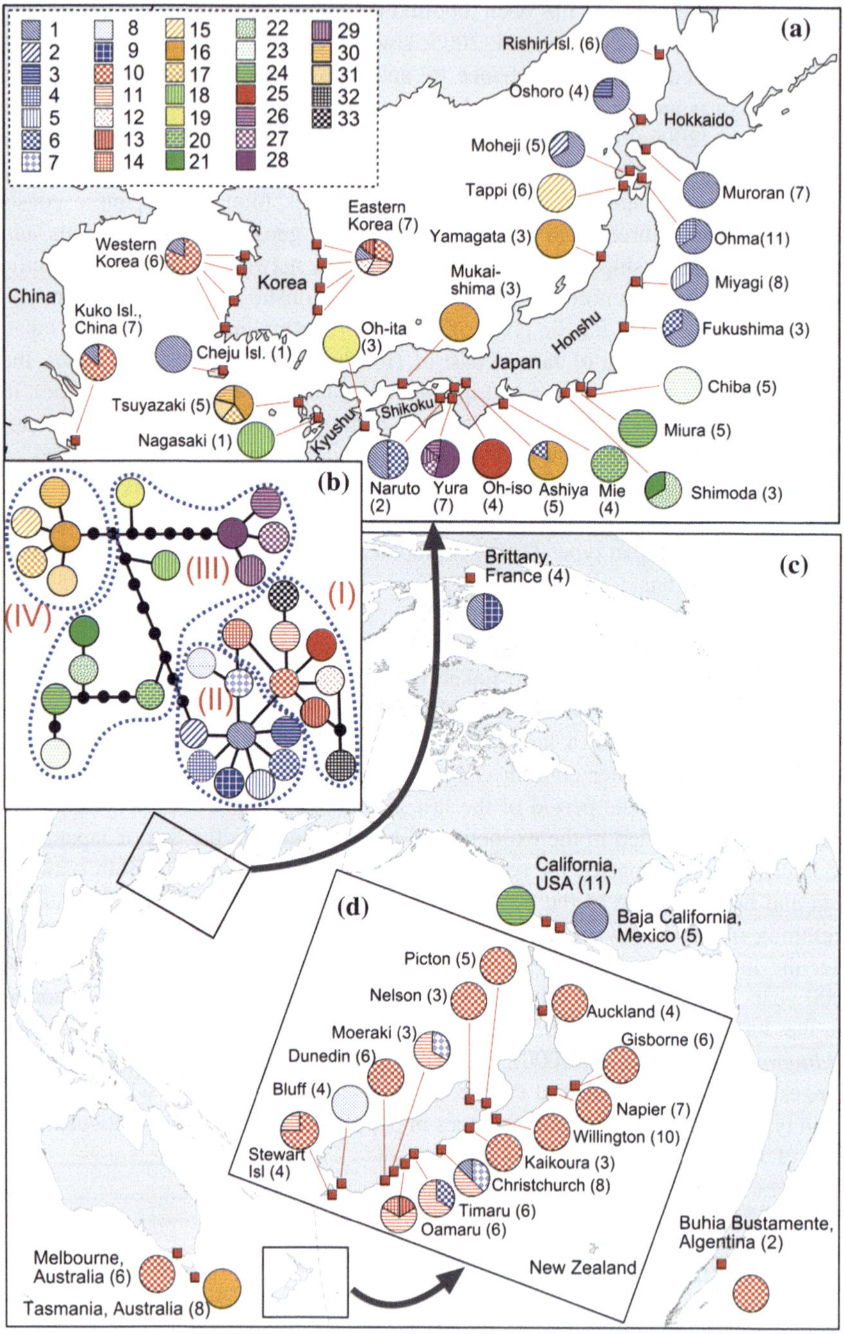
(a)
1
2
3
4
5
6
7
8
9
10
11
12
13
14
15
16
17
18
19
20
21
22
23
24
25
26
27
28
29
30
31
32
33
Rishiri Isl. (6)
Oshoro (4)
Hokkaido
Moheji (5)
Tappi (6)
Muroran (7)
Ohma(11)
Yamagata (3)
Miyagi (8)
Eastern Korea (7)
Western Korea (6)
Korea
China
Mukai-shima (3)
Fukushima (3)
Kuko Isl., China (7)
Oh-ita (3)
Cheju Isl. (1)
Japan
Honshu
Chiba (5)
Tsuyazaki (5)
Shikoku
Miura (5)
Nagasaki (1)
Kyushu
Naruto (2)
Yura (7)
Oh-iso (4)
Ashiya (5)
Mie (4)
Shimoda (3)
(b)
(III)
(I)
(IV)
(II)
(c)
Brittany, France (4)
California, USA (11)
Baja California, Mexico (5)
(d)
Picton (5)
Nelson (3)
Auckland (4)
Moeraki (3)
Gisborne (6)
Dunedin (6)
Bluff (4)
Napier (7)
Willington (10)
Stewart Isl (4)
Kaikoura (3)
Christchurch (8)
Timaru (6)
Oamaru (6)
Buhia Bustamente, Algentina (2)
Melbourne, Australia (6)
Tasmania, Australia (8)
New Zealand

◀ **Fig. 9.6** Parsimony network and geographical distributions of mitochondrial DNA haplotypes (*cox*3 and *tat*C-*tLeu* regions) of *Undaria pinnatifida* in the native and introduced populations. **b** Haplotypes are illustrated by different combinations of color and pattern and identified with an haplotype code (*inset* in **a**). Pie charts depict haplotype frequencies in each populations, with numerals in parentheses representing the number of samples in each population in **a** Native range, **c** Worldwide introduced populations except New Zealand, and **d** Introduced populations in New Zealand. Reprint from Uwai et al. (2006a) with permission

et al. 1981; Boudouresque et al. 1985; Castric-Fay et al. 1999; Fletcher and Farrell 1998). In the late 1980s the species was reported worldwide, in New Zealand (Hay and Luckens 1987; Hay 1990), Tasmania, Australia (Sanderson 1990), Argentina (Casas and Piriz 1996), Victoria, Australia (Campbell and Burridge 1998), California, USA (Silva et al. 2002) and in Baja California, Mexico (Aguilar-Rosas et al. 2004).

Apart from the introductions to Europe (non-intentional introductions associated with oyster farming), processes allowing introductions into other regions are not clearly understood. Although ballast water and hull fouling have been considered probable dispersal vectors to other areas, there has been limited scientific information supporting these assumptions. Furthermore, possible secondary introductions within the same ranges have not been investigated.

Voisin et al. (2005) as well as Uwai et al. (2006a) studied the genetic diversity of *U. pinnatifida* populations based on newly collected specimens worldwide using mitochondrial gene sequences. Uwai et al. (2006a) also examined historical voucher specimens collected from various localities in New Zealand since the time of early introduction and discussed the succession of the dominant haplotypes in populations in New Zealand. Based on the analyses, they suggested multiple introductions in *U. pinnatifida*, because all local types were found in the introduced populations; the population in Brittany, France and that in Baja California, Mexico, as well as some of the New Zealand populations, had the northern Japan types; almost all populations in the Southern Hemisphere (most New Zealand populations, the Argentina population and the Melbourne population), had the continental types; one population in California, USA and in Tasmania, Australia, had the central Japan type and the Sea of Japan type, respectively (Fig. 9.6c). Furthermore, an analysis of populations throughout almost all introduced areas in New Zealand, indicated that the populations have origins in continental Asia as well as northern Japan (Uwai et al. 2006a). Historical samples from the times of the early introductions (1987–1990) showed that an introduced population at Wellington had the continental types (type-10 and -11), whereas those in Timaru and Oamaru, both in central region of the South Island, had the northern Japan type (type-1). Nine haplotypes, 7 from 28 historical samples (1987–2003) and seven from contemporary samples (five the same as those in historical samples), were found in the New Zealand samples. This suggests that possibly more than two introductions have occurred in this country, especially on the South Island (Uwai et al. 2006a). In addition to New Zealand, the Sea of Japan type was found in the Tasmania population, suggesting that another equator-crossing introduction occurred. Since

Undaria gametophytes are relatively tolerant to high temperatures up to 30 °C (tom Dieck 1993), they could be transported alive across the tropics on ships' hulls and/or in ballast tanks (Sanderson 1990). *Undaria* sporophytes were collected from the hull of a Korean trawler in New Zealand in 1987, suggesting the introduction across the equator was achieved in this way.

In contrast to high haplotype diversity on the South Island, only a single haplotype (type-10) was found on the North Island, suggests successive introductions, possibly through domestic vessel transport (Stuart 2004), although scientific evidence for such hull fouling as an introduction vector, as well as successive secondary introductions, has been limited. This is possibly true for the populations in Argentina and Melbourne, i.e., they could be carried from New Zealand. The optimal temperature for growth of both gametophytes (Akiyama 1965; Morita et al. 2003a) and sporophytes (Morita et al. 2003b; Gao et al. 2013) are below 23 °C; at 25 °C or above, growth is stunted. The survival of *Undaria* gametophytes and sporophytes is regarded as much more likely in the course of ship operations within the temperate region than those crossing the high temperature tropical regions.

Although the haplotypes found in the Bay of Osaka area showed high haplotype divergences, and included all of the local types reported in Uwai et al. (2006a), this is explained as the result of intentional introductions for breeding of strains used for mariculture in the area.

9.6 Future Work

The familial taxonomy of the Laminariales has been considerably revised in the last two to three decades by the discoveries of new key taxa and the application of molecular phylogenetic approaches. Based on this new information, combined with existing biogeographic information, it has been possible to elucidate the phylogeography of the basal taxa, especially their global dispersal history. Despite these advances, the familial or generic level phylogeographies of the derived Laminariales remain poorly understood. For example, although Laminariales is considered to have originated in cool temperate or cold water regions, adaptation to warm temperate habitat has occurred in some lineages but it is not clear where these evolutionary events occurred (Uwai et al. 2007; Rothman et al. 2015). We consider that detailed phylogenetic analyses of such derived families and genera will provide insights for understanding the processes driving genetic and morphological diversification in Laminariales. For example, development of large and elaborate sporophytes could have resulted from adaptations to drastic changes of sea temperatures and sea levels with plate tectonic movements and global climate change cycles.

As summarized above, phylogeographic patterns at the genus and species levels have suggested greater species diversities within of *Chorda* and *Undaria* species than formerly estimated by morphological taxonomy alone, and the occurrence of cryptic species has been demonstrated in both genera. These findings suggest there may also be unrecognized, cryptic species in other laminarialean genera and species. Species-level taxonomy needs further revisions especially in the genera such as *Alaria* (Kraan and Guiry 2000), *Saccharina* (Yotsukura et al. 2008), and *Ecklonia* (Rothman et al. 2015). However, clarification of species boundaries is difficult if natural hybridization is possible, and both experimental (Migita 1967; Lewis and Neushul 1995; Kraan and Guiry 2000) and observational (Coyer et al. 1992; Kikuchi et al. 1996) studies indicate that hybridization among laminarialean species can occur.

On the other hand, a considerable range of morphological and physiological adaptations could be expected within a single species having a wide distributional range, as suggested in the population structure of *Undaria pinnatifida*. Therefore, in addition to detailed molecular phylogenetic analyses for each genus and species, population genetic analyses and ecophysiological comparisons between local populations are necessary for understanding species-level phylogeographic patterns. Such studies would also provide useful information for discussing the possible impacts of rising seawater temperatures on existing populations of the laminarialean species, many of which are fundamental elements of coastal ecosystems.

Acknowledgments We are grateful to Dr. E.C. Henry for his critical comments on the manuscript. A part of this work was supported by JSPS Grants-in-Aid for Scientific Research (No. 22370034) to H. K.

References

Aguilar-Rosas R, Aguilar-Rosas LE, Avila-Serrano G, Marcos-Ramirez R. First record of *Undaria pinnatifida* (Harvey) Suringar (Laminariales, Phaeophyta) on the Pacific coast of Mexico. Bot Mar. 2004;47:255–8.

Akiyama K. Studies of ecology and culture of *Undaria pinnatifida* (Harv.) Sur. II. Environmental factors affecting the growth and maturation of gametophyte. Bull Tohoku Nat Fish Res Inst. 1965;25:143–70.

Akiyama K, Kurogi M. Cultivation of *Undaria pinnatifida* (Harvey) Suringar, the decrease in crops from natural plants following crop increase from cultivation. Bull Tohoku Reg Fish Res Lab. 1982;44:91–100.

Avise JC. Phylogeography. The history and formation of species. Cambridge: Harvard University Press; 2000.

Bold HC, Wynne MJ. Introduction to the Algae: structure and reproduction. 2nd ed. Englewood Cliffs: Prentice-Hall; 1985. p. 720.

Bolton J. The biogeography of kelps (Laminariales, Phaeophyceae): a global analysis with new insights from recent advances in molecular phylogenies. Helgol Mar Res. 2010;64:263–79.

Bonnott P. A recent introduction of exotic species of mollusks into California waters from Japan. The Nautilus. 1935;49:1–2.

Boo SM, Lee WJ, Yoon HS, Kato A, Kawai H. Molecular phylogeny of Laminariales (Phaeophyceae) inferred from small subunit ribosomal DNA sequences. Phycol Res. 1999;47:109–14.

Boudouresque CF, Verlaque M. Biological pollution in the Mediterranean Sea: invasive versus introduced macrophytes. Mar Poll Bull. 2002;44:32–8.

Boudouresque CF, Gerbal M, Knoepffler-Peguy M. L'algue japonnaise *Undaria pinnatifida* (Phaeophyta, Laminariales) en Mediterranee. Phycologia. 1985;24:364–6.

Campbell SJ, Burridge TR. Occurrence of *Undaria pinnatifida* (Phaeophyta: Laminariales) in Port Philip Bay, Victoria, Australia. Mar Fresh Wat Res. 1998;49:379–81.

Casas GN, Piriz ML. Survey on *Undaria pinnatifida* (Laminariales, Phaeophyta) in Golfo Nuevo, Argentina. Hydrobiologia. 1996;326(327):213–5.

Castric-Fey A, Beaupoil C, Bouchain J, Pradier E, L'Hardy-Halos MT. The introduced alga *Undaria pinnatifida* (Laminariales, Alariaceae) in the rocky shore ecosystem of the St Malo Area: morphology and growth of the sporophyte. Bot Mar. 1999; 42:71–82.

Cecere E, Petrocelli A, Saracino OD. *Undaria pinnatifida* (Fucophyceae, Laminariales) spread in the Central Mediterranean: its occurrence in the Mar Piccolo of Taranto (Ionian Sea, Southern Italy). Cryptogam Algologie. 2000;21:305–9.

Choi CG, Oh SJ, Kang IJ. Growth and morphological characteristics of *Undaria pinnatifida* in the cultivation ground at Busan, Korea. J Fac Agr Kyushu Univ. 2009;54:47–51.

Coyer JA, Engel JM, Zimmerman RC. Discovery of a fertile *Pelagophycus* × *Macrocystis* (Phaeophyta) putative hybrid and subsequent production of F_2 sporophytes in the laboratory. J Phycol. 1992;28:127–30.

Dayton PK. Ecology of kelp communities. Ann Rev Ecol Syst. 1985;16:215–45.

Endo H, Park EJ, Sato Y, Mizuta H, Saga N. Intraspecific diversity of *Undaria pinnatifida* (Harvey) Suringar (Laminariales, Phaeophyta) from Japan, China and Korea based on the cox1 gene and ITS2 sequences. Fish Sci. 2009;75:393–400.

Fain SR, Druehl LD, Baillie DL. Repeat and single copy sequences are differentially conserved in the evolution of kelp chloroplast DNA. J Phycol. 1988;24:292–302.

Fletcher RL, Farrell P. Introduced brown algae in the North Atlantic, with particular respect to *Undaria pinnatifida* (Harvey) Suringar. Helgol Meeresun. 1998;52:259–75.

Fritsch FE. The structure and reproduction of the Algae, vol. 2. Cambridge: Cambridge University Press; 1945.

Gao X, Endo H, Taniguchi K, Agatsuma Y. Combined effects of seawater temperature and nutrient condition on growth and survival of juvenile sporophytes of the kelp *Undaria pinnatifida* (Laminariales; Phaeophyta) cultivated in northern Honshu, Japan. J Appl Phycol. 2013;25:269–75.

Graham LE, Wilcox LW. Algae. London: Prentice-Hall; 2000.

Hay CH. The dispersal of sporophytes of *Undaria pinnatifida* by coastal shipping in New Zealand and implications for further dispersal of Undaria in France. Br Phycol J. 1990;25:301–13.

Hay C, Luckens P. The Asian kelp *Undaria pinnatifida* (Phaeophyta: Laminariales) found in a New Zealand harbour. NZ J Bot. 1987;25:364–6.

Henry EC. Primitive reproductive characters and a photoperiodic response in *Saccorhiza dermatodea* (Laminariales, Phaeophyceae). Br Phycol J. 1987;22:23–31.

Henry EC, South GR. *Phyllariopsis* gen. nov. and a reappraisal of the Phyllariaceae Tilden 1935 (Laminariales, Phaeophyceae). Phycologia. 1987;26:9–16.

Kawai H. Life history and systematic position of *Akkesiphycus lubricus* (Phaeophyceae). J Phycol. 1986;22:286–91.

Kawai H, Kurogi M. On the life history of *Pseudochorda nagaii* (Pseudochordaceae fam. nov.) and its transfer from the Chordariales to the Laminariales (Phaeophyta). Phycologia. 1985;24:289–96.

Kawai H, Nabata S. Life history and systematic position of *Pseudochorda gracilis* sp. nov. (Laminariales, Phaeophyceae). J Phycol. 1990;26:721–7.

Kawai H, Sasaki H. Molecular phylogeny of brown algal genera *Akkesiphycus* and *Halosiphon* (Laminariales), resulting in the circumscription of the new families Akkesiphycaceae and Halosiphonaceae. Phycologia. 2000;39:416–28.

Kawai H, Sasaki H, Maeda Y, Arai S. Morphology, life history, and molecular phylogeny of *Chorda rigida*, sp. nov. (Laminariales, Phaeophyceae) from the Sea of Japan and the genetic diversity of *Chorda filum*. J Phycol. 2001;37:130–42.

Kawai H, Hanyuda T, Lindeberg M, Lindstrom SC. Morphology and molecular phylogeny of *Aureophycus aleuticus* gen. et sp. nov. (Laminariales, Phaeophyceae) from the Aleutian Islands. J Phycol. 2008;44:1013–21.

Kawai H, Hanyuda T, Ridgway LM, Holser K. Ancestral reproductive structure in basal kelp *Aureophycus aleuticus*. Sci Rep. 2013;3:2491.

Kawai H, Hanyuda T, Draisma S, Wilce R, Andersen, R. Molecular phylogeny of two unusual brown algae, *Phaeostrophion irregulare*, *Platysiphon verticillatus*, the proposal of the Stschapoviales ord. nov. and Platysiphonaceae fam. nov. and a re-examination of the divergence time for brown algal orders. J Phycol. 2015a;51:918–28. doi:10.1111/jpy.12332

Kawai H, Hanyuda T, Mumford T, Waaland JR. An introduced population of *Chorda asiatica* (Chordaceae, Laminariales) in Puget Sound, Pacific coast of North America. Phycol Res. 2015b;63:154–8.

Kikuchi N, Miyaki K, Maesako N, Oda K. Occurrence of a possible hybrid between *Undaria pinnatifida* (Harvey) Suringar and *U. peterseniana* (Kjellman) Okamura (Phaeophyceae) in Tsushima Island, Nagasaki Prefecture. Trans Nagasaki Biol Soc. 1996;47:20–2 (in Japanese).

Kogame Y, Kawai H. Development of the intercalary meristem in *Chorda filum* (Laminariales, Phaeophyceae) and other primitive Laminariales. Phycol Res. 1996;44:247–60.

Kogame K, Uwai S, Shimada S, Masuda M. A study of sexual and asexual populations of *Scytosiphon lomentaria* (Scytosiphonaceae, Phaeophyceae) in Hokkaido, Northern Japan, using molecular markers. Eur J Phycol. 2005;40:313–22.

Kojima S, Segawa R, Hayashi I. Genetic differentiation among populations of the Japanese turban shell *Turbo* (*Batillus*) *cornutus* corresponding to warm currents. Mar Ecol Prog Ser. 1997;150:149–55.

Kraan S, Guiry MD. Sexual hybridization experiments and phylogenetic relationships as inferred from *Rubisco* spacer sequences in the genus *Alaria* (Phaeophyceae). J Phycol. 2000;35:190–8.

Kylin H. Über dem Generationswechsel bei *Laminaria digitata*. Svensk Bot Tidsk. 1916;10:551–61.

Lane CE, Mayes C, Druehl LD, Saunders GW. A multi-gene molecular investigation of the kelp (Laminariales, Phaeophyceae) supports substantial taxonomic re-organization. J Phycol. 2006;42:493–512.

Lee YP. *Undariella*, a new genus of the Alariaceae. Algae. 1998;13:419–26.

Lee YP. Nomenclatural correction of *Undariella*, a new genus of the Alariaceae (Laminariales, Phaeophyta). Algae. 1999;14:79.

Lee YP, Yoon JT. Taxonomy and morphology of *Undaria* (Alariaceae, Phaeophyceae). Algae. 1998;13:427–46.

Lee KM, Yang EC, Coyer JA, Zuccarello GC, Wang WL, Choi CG, Boo SM. Phylogeography of the seaweed *Ishige okamurae* (Phaeophyceae): evidence for glacial refugia in the northwest Pacific region. Mar Biol. 2012;159:1021–8.

Lee KM, Boo GH, Coyer JA, Nelson WA, Miller KA, Boo SM. Distribution patterns and introduction pathways of the cosmopolitan brown alga *Colpomenia peregrina* using mt *cox*3 and *atp*6 sequences. J Appl Phycol. 2014;26:491–504.

Lewis RJ, Neushul M. Intergeneric hybridization among five genera of the family Lessoniaceae (Phaeophyceae) and evidence for polyploidy in a fertile *Pelagophycus* × *Macrocystis* hybrid. J Phycol. 1995;31:1012–7.

Lim BL, Kawai H, Hori H, Osawa S. Molecular evolution of 5S ribosomal RNA from red and brown algae. Jpn J Genet. 1986;61:169–76.

Lund S. The marine algae of East Greenland 1. Taxonomical part. Medd Grønland. 1959;156:1–247.

Lüning K, tom Dieck I. The distribution and evolution of the Laminariales: North Pacific-Atlantic relationships. In: Garbary DJ, South GR, editors. NATO ASI Series G22. Evolutionary biogeography of the marine algae of the North Atlantic. Berlin: Springer; 1990. p. 187–204.

Lutaenko K, Furota T, Nakayama S, Shin K, Jing X. Atlas of marine invasive species. Beijing: Nowpap Dinrac; 2013.

Maier I. Culture studies of *Chorda tomentosa* (Phaeophyta, Laminariales). Br Phycol J. 1984;19: 95–106.

Maier I. Brown algal pheromones. In: Round FE, Chapman DJ, editors. Progress in phycological research, vol. 11. Bristol: Biopress; 1995. p. 51–102.

Migita S. Studies on artificial hybrids between *Undaria peterseniana* (Kjellm.) Okam. and *U. pinnatifida* (Harv) Sur. Bull Fac Fish Nagasaki Univ. 1967;24:9–20 (in Japanese with English abstract).

Migula W. Kryptogamen-Flora von Deutschland, Deutsch-Österreich und der Schweiz. Band II. Algen. 2. Teil. Rhodophyceae, Phaeophyceae, Characeae. Friedriech von Zezschwitz, Gera; 1909. pp i–iv, 1–382, 122 (41 col) pls.

Miyabe K Konbu. Hokkaido Suisan Hokoku, vol 3. 1902 (in Japanese).

Morita T, Kurashima A, Maegawa M. Temperature requirements for growth and maturation of the gametophytes of *Undaria pinnatifida* and *Undaria undarioides* (Laminariales, Phaeophyceae). Phycol Res. 2003a;51:154–60.

Morita T, Kurashima A, Maegawa M. Temperature requirements for growth of young sporophytes of *Undaria pinnatifida* and *Undaria undarioides* (Laminariales, Phaeophyceae). Phycol Res. 2003b;51:266–70.

Muraoka D, Saito K. Identification of *Undaria pinnatifida* and *Undaria undarioides* Laminariales, Phaeophyceae using mitochondrial 23S ribosomal DNA sequences. Fish Sci. 2005;71:1365–9.

Nanba N, Fujiwara T, Kuwano K, Ishikawa Y, Ogawa H, Kado R. Effect of water flow velocity on growth and morphology of cultured *Undaria pinnatifida* sporophytes (Laminariales, Phaeophyceae) in Okiray Bay on the Sanriku coast, Northeast Japan. J Appl Phycol. 2011;23:1023–30.

Ohba T, Kato M, Kitazato H, Koizumi I, Omura A, Sakai T, Takayama T. Paleoenvironmental changes in the Japan Sea during the last 85,000 years. Paleoceanography. 1991;6:499–518.

Okamura K. Nippon Sourui Meii. Tokyo: Seibido; 1902.

Okamura K. *Undaria* and its species. Bot Mag Tokyo. 1915;29:265–78.

Okamura K. Icones of Japanese Algae. Vol 5(7) Privately published, Tokyo. 1926. p. 117–130.

Oltmanns F. Morphologie und Biologie der Algen. Vol II 2nd ed. Fischer, Jena; 1922. p. 439.

Park KJ, Kim BY, Park SK, Lee JH, Kim YS, Choi HG, Nam KW. Morphological and biochemical differences in three *Undaria pinnatifida* populations in Korea. Algae. 2012;27:189–96.

Perez P, Lee JY, Juge C. Observations sur la biologie de l'algue japoniase *Undaria pinnatifida* (Harvey) Suringar introduite accidentellement dans l'etang de Thau. Science et Peche. 1981;315:1–12.

Peters A. Ribosomal DNA sequences support taxonomic separation of the two species of *Chorda*: reinstatement of *Halosiphon tomentosus* (Lyngbye) Jaasund (Phaeophyceae, Laminariales). Eur J Phycol. 1998;33:65–71.

Reinke J. Atlas Deutscher Meeresalgen. Berlin: Heft II; 1892.

Riouall R. Sur la présence dans l'étang de Thau (Hérault-France) de *Sphaerotrichia divaricata* (C. Ag.) Kylin et *Chorda filum* (L) Stackhouse. Bot Mar. 1985; 28:83–86.

Rothman MD, Mattio L, Wernberg T, Anderson RJ, Uwai S, Mohring MB, Bolton JJ. A molecular investigation of the genus *Ecklonia* (Phaeophyceae, Laminariales) with special focus on the Southern Hemisphere. J Phycol. 2015;51:236–46.

Russell G. Some anatomical and physiological differences in *Chorda filum* from coastal waters of Finland and Great Britain. J Mar Biol Assoc UK. 1985;65:343–9.

Saito Y. On the effect of environmental factors on morphological characteristics of *Undaria pinnatifida* and the breeding of hybrids in the genus *Undaria*. In: Abbott IA, Kurogi M, editors. Contributions to the systematics of benthic marine algae of the North Pacific. Kobe: Jpn Phycol Soc; 1972. p. 117–32.

Sanderson JC. A preliminary survey of the distribution of the introduced macroalga *Undaria pinnatifida* (Harvey) Suringar on the coast of Tasmania, Australia. Bot Mar. 1990; 33:153–157.

Sasaki H, Kawai H. Taxonomic revision of the genus *Chorda* (Chordaceae, Laminariales) on the basis of sporophyte anatomy and molecular phylogeny. Phycologia. 2007;46:10–21.

Sasaki H, Flores-Moya A, Henry EC, Müller DG, Kawai H. Molecular phylogeny of Phyllariaceae, Halosiphonaceae and Tilopteridales (Phaeophyceae). Phycologia. 2001;40:123–34.

Saunders GW, Druehl LD. Nucleotide sequences of the small subunit ribosomal RNA genes from selected Laminariales (Phaeophyta): implications for kelp evolution. J Phycol. 1992;28:544–9.

Sauvageau C. Sur la sexualité heterogamique d'une Laminaire (*Alaria esculenta*) C R Acad Sci Paris. 1916;161:840–842.

Setchell WA, Gardner NL. The marine algae of the Pacific Coast of North America Part III Melanophyceae. Univ Calif Publ Bot. 1925;8:383–898.

Silberfeld T, Jeigh JW, Verbruggen H, Cruaud C, de Reviers B, Rousseau F. A multi-locus time-calibrated phylogeny of the brown algae (Heterokonta, Ochrophyta, Phaeophyceae): Investigating the evolutionary nature of the brown algal crown radiation. Mol Phylogenet Evol. 2010;56:659–74.

Silva PC, Woodfield RA, Cohen AN, Harris LH, Goddard JHR. First report of the Asian kelp *Undaria pinnatifida* in the northeastern Pacific Ocean. Biol Invasions. 2002;4:333–8.

South GR, Burrows EM. Studies on marine algae of the British Isles 5 *Chorda filum* (L) Stackh. Br Phycol Bull. 1967;3:379–402.

Steele EN. The Immigrant Oyster (*Ostrea gigas*) now known as the Pacific Oyster. Warren's Quick Print, Washington; 1964. p. 179.

Stuart MD. Review of research on *Undaria pinnatifida* in New Zealand and its potential impacts on the eastern coast of the South Island. DOC Science Internal Series 166. Department of Conservation. Wellington, New Zealand; 2004.

Stuart MD, Hurd CL, Brown MT. Effects of seasonal growth rate on morphological variation of *Undaria pinnatifida* (Alariaceae, Phaeophyceae). Hydrobiologia. 1999;398(399):191–9.

Suringar WFR. Illustrationes des algues du Japon. Musée Botanique de Leide. 1873;1:77–90, pls 26–33.

Tilden JE. The algae and their life relations. Minneapolis: University of Minnesota Press; 1935.

tom Dieck I. Temperature tolerance and survival in darkness of kelp gametophytes (Laminariales, Phaeophyta): ecological and biogeographical implications. Mar Ecol Prog Ser. 1993;100:253–64.

Tseng CK. Common seaweeds of China. Beijing: Science Press; 1983. p. 316.

Uwai S, Nelson W, Neill K, Wang WD, Aguilar-Rosas LE, Boo S-M, Kitayama T, Kawai H. Genetic diversity in *Undaria pinnatifida* deduced from mitochondria genes—origins and succession of introduced populations. Phycologia. 2006a;45:687–95.

Uwai S, Yotsukura N, Serisawa Y, Muraoka D, Hiraoka M, Kogame K. Interspecific genetic diversity of *Undaria pinnatifida* in Japan, based on the mitochondrial *cox3* gene and the ITS1 of nrDNA. Hydrobiologia. 2006b;553:345–56.

Uwai S, Arai S, Morita T, Kawai H. Genetic distinctness and phylogenetic relationships among *Undaria* spp. based on mitochondrial *cox3* sequence. Phycol Res. 2007;45:687–95.

Voisin M, Engel CR, Viard F. Differential shuffling of native genetic diversity across introduced regions in a brown alga: Aquaculture vs. maritime traffic effects. Proc Natl Acad Sci USA. 2005;102:5432–7.

Woolcott GW, Iima M, King RJ. Speciation within *Blidingia minima* (Chlorophyta) in Japan: evidence from morphology, ontogeny, and analyses of nuclear rDNA ITS sequence. J Phycol. 2000;36:227–36.

Yendo K. Three new marine algae from Japan. Bot Mag Tokyo. 1903;17:99–104.

Yendo K. Kaisan-Syokubutsugaku. Hakubunkan: Tokyo; 1911. p. 748.

Yoon HS, Lee JY, Boo SM, Bhattacharya D. Phylogeny of Alariaceae, Laminariaceae, and Lessoniaceae (Phaeophyceae) based on plastid-encoded Rubisco spacer and nuclear-encoded ITS sequence comparisons. Mol Phylogenet Evol. 2001;21:231–43.

Yotsukura N, Kawashima S, Kawai T, Abe T, Druehl LD. A systematic re-examination of four *Laminaria* species: *L. japonica*, *L. religiosa*, *L. ochotensis* and *L. diabolica*. J Jpn Bot. 2008;83:165–76.

Chapter 10
Phylogeography of Seaweeds in the South East Pacific: Complex Evolutionary Processes Along a Latitudinal Gradient

Marie-Laure Guillemin, Myriam Valero, Florence Tellier, Erasmo C. Macaya, Christophe Destombe and Sylvain Faugeron

Abstract The coast along the temperate South East Pacific (SEP) presents a simple linear topography with a north-south orientation spanning more than 4600 km. However, environmental heterogeneity associated with two major biogeographic boundaries has been described along the SEP (30–33°S and 42°S). Recent phylogeographic studies of seaweeds revealed the existence of different cryptic species along the SEP coast and that most of the genetic breaks between them are broadly congruent with the biogeographic boundaries. These phylogeographic patterns characterized by genetic discontinuities could be attributed to historical vicariance or to budding speciation. For SEP seaweeds, two major phylogeographic patterns are observed. Endemic species living north of 42°S show complex haplotype networks and an almost complete genetic isolation between populations located only a few kilometres from each other. This extreme genetic patchiness has been related to the combined effects of limited dispersal, reduced population size and high population turnover of these intertidal seaweeds due to stochastic effects of climatic and

M.-L. Guillemin (✉)
Instituto de Ciencias Ambientales y Evolutivas, Facultad de Ciencias, Universidad Austral de Chile, Casilla 567, Valdivia, Chile
e-mail: marielaure.guillemin@gmail.com

M.-L. Guillemin · M. Valero · C. Destombe · S. Faugeron
CNRS, Sorbonne Universités, UPMC University Paris VI, UMI 3614, Evolutionary Biology and Ecology of Algae, Station Biologique de Roscoff, CS 90074, Place G. Teissier, 29680 Roscoff, France
e-mail: myriam.valero@sb-roscoff.fr

C. Destombe
e-mail: destombe@sb-roscoff.fr

S. Faugeron
e-mail: sfaugeron@bio.puc.cl

F. Tellier
Departamento de Ecología, Facultad de Ciencias, Universidad Católica de la Santísima Concepción, Casilla 297, Concepción, Chile
e-mail: ftellier@gmail.com

Z.-M. Hu and C. Fraser (eds.), *Seaweed Phylogeography*,
DOI 10.1007/978-94-017-7534-2_10

tectonic catastrophes. On the other hand, species with a range distribution limited to the south of 42°S and inhabiting the area highly affected by the coastal ice cap during the Last Glacial Maximum (LGM), show typical signatures of post-glacial demographic expansion. Finally, molecular studies reveal that several species are recent immigrants from New Zealand, demonstrating the importance of oceanic dispersal in shaping the diversity of the SEP.

Keywords Biogeographic boundaries · Budding speciation · Chile · Cryptic species · Phylogeographic breaks · Post-glacial recolonization

10.1 Introduction

The temperate South East Pacific (SEP) represents one of the most productive marine ecosystems of the world. Research interest in this region has increased during the recent years and consequently, phylogeographic studies have started to accumulate at a rapidly increasing rate. Areas of research range from simple exploratory analyses of genetic variation focusing on the identification of genetic resources, to specific assessments of the role of biogeographic discontinuities in the evolutionary history of an organism. Most studies are based on a 'single marker-single species' sampling design and focus on different areas of the SEP which have limited the description of general genetic trends in the region. Indeed, some studies reveal strong genetic structure and have stated that highly regionalized historical and contemporary factors are driving these patterns (see, for example Tellier et al. 2009; Fraser et al. 2010a; Brante et al. 2012; Montecinos et al. 2012; Varela and Haye 2012), whereas other studies suggest that species form single genetic units that share a common demographic history (for examples see Cárdenas et al. 2009; Haye et al. 2010, 2012). Recently, the first multispecies comparative study was conducted in central and northern Chile to investigate the role of oceanographic and biogeographic boundaries present around 30°S (Haye et al. 2014). This study suggests that genetic structure among populations in eight marine

E.C. Macaya
Laboratorio de Estudios Algales (ALGALAB), Departamento de Oceanografía, Universidad de Concepción, Casilla 160-C, Concepción, Chile
e-mail: emacaya@oceanografia.udec.cl

E.C. Macaya
Millennium Nucleus Ecology and Sustainable Management of Oceanic Island (ESMOI), Coquimbo, Chile

S. Faugeron
Centro de Conservación Marina and CeBiB, Facultad de Ciencias Biológicas, Pontificia Universidad Católica de Chile, Santiago, Chile

invertebrate species can be largely explained by dispersal potential and historical processes that limit gene flow near 30°S (Haye et al. 2014). Indeed, species with short pelagic larval duration display a clear-cut phylogeographic discontinuity coincident with the biogeographic transition zone at 30°S whereas the species with long-term pelagic stage do not show a genetic discontinuity. Another biogeographic boundary described at 42°S (Camus 2001) was also reported to match the location of genetic discontinuities (i.e. cryptic species were defined on the both sides of the 42°S; Fraser et al. 2009). Along the SEP, the concordance between biogeographical boundaries and genetic breaks has been attributed to historical vicariance caused by oceanographic or climatic features (Tellier et al. 2009; Zakas et al. 2009; Brante et al. 2012; Montecinos et al. 2012; Varela and Haye 2012; Haye and Muñoz-Herrera 2013; Haye et al. 2014).

The use of molecular markers has allowed for the identification of many cryptic species in marine organisms. Molecular identification has been especially useful for studying organisms, such as seaweeds, that display few diagnostic characters and are morphologically simple (Leliaert et al. 2014). The lack of morphological characteristics that differ between evolutionary units (i.e. species as defined by the lineage species concept) has for some time obscured the study of biogeographic boundaries, species distributions, and speciation processes in seaweeds. For example, in a molecular study of the red alga *Portieria hornemannii* in the Philippine archipelago, Payo et al. (2013) demonstrates that this unique morphospecies is in fact composed of 21 cryptic taxa. These remarkable findings led to the complete reinterpretation of *Portieria*'s biogeography in the region. Even if many cryptic species have been reported in marine algae along the SEP, the general processes that contribute to speciation and genetic breaks, and how these relate to biogeographic boundaries and vicariance, have not yet been investigated for these organisms.

In this chapter, we intend to discuss new findings of seaweed phylogeographic structure and cryptic sister species distributions along the SEP to gain some biogeographic insight and to identify some of the major drivers of evolutionary processes that have affected Chilean seaweeds.

10.2 Major Biogeographical Characteristics of the SEP Coast: Linear Gradient or Strong Regional Pattern?

The SEP coast (from ca. 14°S to 56°S) is characterized by a linear topography with a north-south orientation spanning more than 4600 km (Fig. 10.1). From southern Peru (i.e. 18°S) to the Island of Chiloé (i.e. 42°S), the coastline is straight and continuous and presents no major topographic discontinuities except for a few relatively small open bays. South of Chiloé Island, the topography differs greatly, being dominated by insular systems surrounded by numerous fjords and channels (Thiel et al. 2007). The oceanic circulation of the SEP is dominated by the northward flowing Humboldt Current (HC) that extends to 5°S where it meets the

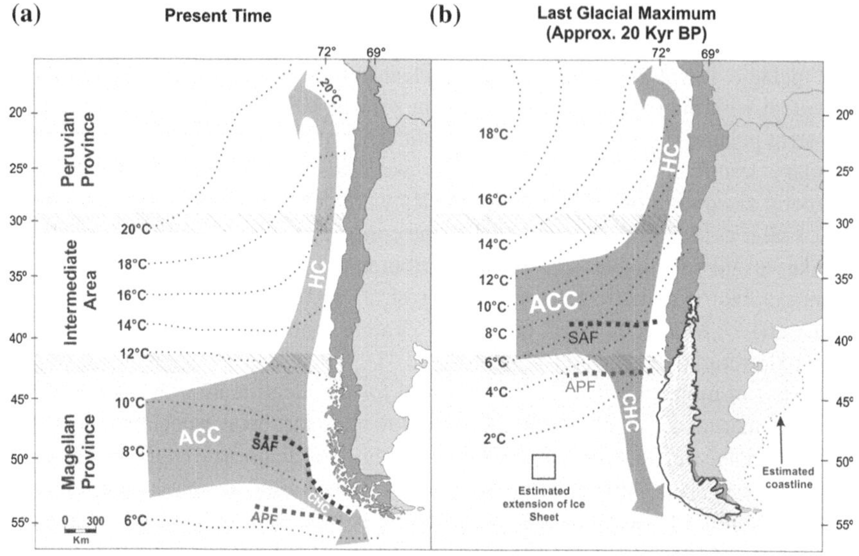

Fig. 10.1 Map of the coastal South East Pacific (SEP) at present time (**a**) and reconstruction of the Quaternary environmental changes in the region (**b**). Present time map shows the three marine biogeographic provinces (after Camus 2001) with grey shaded lines representing the reported biogeographic breaks around 30–33°S and 42–43°S. Mean annual sea surface temperatures (i.e. SST, dashed lines, Locarnini et al. 2010), the directions of the major surface currents (*grey arrows*, Kaiser et al. 2005: Antarctic Circumpolar Current (ACC), Humboldt Current (HC), Cape Horn Current (CHC)) and the position of the Subantarctic Front (SAF) and the Antarctic Polar Front (APF) (after Belkin and Gordon 1996) are indicated. A sketched reconstruction of the Last Glacial Maximum (LGM) environmental changes in the region shows an estimated extension of the ice sheet (McCulloch et al. 2000), the latitudinal shift of the major surface currents and SST (modified after Kaiser et al. 2005), the shift in the SAF and APF position (after Verleye and Louwye 2010) and the changes in the coast line occurring in Patagonia during this period (Ponce et al. 2011)

Equatorial front, and by the southward flowing Cape Horn Current (CHC; Strub et al. 1998; Fig. 10.1a). Both circulation features originate between 40°S and 45°S where the Antarctic Circumpolar Current (ACC) approaches South America. The northern latitude at which the ACC meets the continent shifts seasonally from 35–40°S during austral winter to 45°S during austral summer (Thiel et al. 2007). The direct influence of the Humboldt Current system promotes upwelling of cold and nutrient-rich waters to coastal ecosystems (Hormazabal et al. 2004; Thiel et al. 2007). This peculiarity favours primary productivity (Lachkar and Gruber 2012) even at low latitudes, as far north as 5°S off the coast of Peru, and is thus a major determinant of marine species' latitudinal ranges.

Based partly on these contrasting topological and oceanic characteristics, several studies have proposed the existence of two main biogeographic provinces along the SEP: the Peruvian and Magellan provinces (Camus 2001; Thiel et al. 2007, and references therein; Fig. 10.1a). The Peruvian province is characterized by the presence of a strong tropical component in both its flora and fauna while there is a predominance of species of subantarctic origin in the Magellan province. An Intermediate Area that combines mixed components of the two neighbouring provinces extends from 30–33°S to 42°S (Camus 2001; Thiel et al. 2007). The southern limit of the Peruvian province at 30–33°S is considered to be a major transition zone in the oceanographic features, with strong but seasonal upwelling south of this limit and weak but persistent upwelling conditions in the north (Broitman et al. 2001; Thiel et al. 2007; Tapia et al. 2014). Fenberg et al. (2015) have shown that upwelling-related variables are the best overall predictors of biogeographic structure of northeast Pacific rocky intertidal species, and upwelling seems to be especially influential on assemblages of algae and invertebrates with low-to-medium dispersal capabilities. In this context, it is interesting to note that spatial heterogeneity in the upwelling regime seems to be a major determinant of biogeographic boundaries in the SEP. A second major habitat discontinuity at 42°S, linked with differences in topography, freshwater input, and current dynamics, determines the northern limit for the Magellan province (Brattström and Johanssen 1983; Fernández et al. 2000; Camus 2001; Thiel et al. 2007) (Fig. 10.1a).

The SEP biogeographic boundaries are well correlated with present-day oceanographic conditions but these have not been fully persistent throughout evolutionary history (Fig. 10.1b). However, despite climatic and oceanographic fluctuations, the regions of both 30–33°S and 42°S also represented major habitat discontinuities in the past. First, the northernmost limit of the coastal ice sheet during the Last Glacial Maximum (LGM, 0.026–0.019 Ma) was established near 42°S (McCulloch et al. 2000; Fig. 10.1b). It is well accepted that no coastal ice formed north of that latitude, where potential refugia are proposed for cold-water species. Models of ice sheet extension propose that the southernmost tip of South America, near Cape Horn, was also free of coastal ice and provided the second glacial refugium for marine species of the Magellan province (Hulton et al. 1994, 2002). Indeed, many seaweeds are endemic to the Strait of Magellan, the Beagle Channel, and the Cape Horn archipelago (53–55°S) and are completely absent from the rest of the Magellan province (reviewed by Fernández et al. 2000; Santelices and Meneses 2000). Second, the Intermediate Area experienced historical changes in habitat quality. During glacial periods, the influence of the ACC shifted towards the north leading to a northward shift of the Humboldt/Cape Horn current split (Fig. 10.1b). This change in oceanic circulation likely caused a strong reduction in upwelling in the Intermediate Area, as shown by a reduction of plankton deposits in the region associated with the LGM (Mohtadi and Hebbeln 2004; Kaiser et al. 2005; Verleye and Louwye 2010). In contrast, the area north of 30–33°S is considered to be an area with long-term persistence of upwelling (Mohtadi and Hebbeln 2004).

In addition to the historical geographic and climatic barriers in this region, short-term environmental disturbances have also been taken place repeatedly throughout the SEP's history. Two major factors have been shown to cause such disturbances. First, the El Niño–Southern Oscillation (ENSO) events that consist of a 2–7 years alternation of a cold period (La Niña) and a warm period (El Niño) (Tarazona and Arnzt 2001). ENSO fluctuations began during the Holocene (Moy et al. 2002) and have intensified during the last 5000–3000 years. El Niño events produce strong mortality in seaweeds due to abrupt increases in the temperature of coastal waters and decreases in nutrient concentrations along the northern part of the SEP, from 6°S to ~30°S (Camus 1990; Martínez et al. 2003). Second, tectonic activity also causes strong seaweed mortality along the extensive coastline of the SEP. Specifically, the sudden coastal up-lifts and downward drops of several meters during major earthquakes, accompanied by major tsunamis, are detrimental to coastal seaweeds (Darwin 1839; Castilla 1988; Castilla et al. 2010; Jaramillo et al. 2012). Several of these coastal up-lifting events have been reported in recent decades and their effects on intertidal communities have been well described (Castilla 1988; Castilla et al. 2010; Jaramillo et al. 2012; Fuentes and Brante 2014; Hernández-Miranda et al. 2014; Ortega et al. 2014).

10.3 Phylogeography: The Problem of Cryptic Species and Consequences for the Delineation of Species Range Distributions

The SEP region hosts nearly 600 seaweed morphospecies of which about 27 % are endemic (Meneses and Santelices 2000). While global diversity gradients for a wide range of taxa follow a classic latitudinal pattern, characterized by a decrease in richness from the tropics to the poles, seaweed morphospecies richness tends to increase with latitude along the SEP (Santelices 1980; Santelices and Marquet 1998; Ramírez 2010; Keith et al. 2014). This inverted latitudinal seaweed diversity pattern, which exists worldwide (Kerswell 2006), was recognized recently by Keith et al. (2014) as a potential consequence of niche conservatism (Pyron and Burbrink 2009). Indeed, contrary to the majority of taxa, macroalgae originating in temperate zones and may have been limited in their ability to colonize tropical regions due to competition with established corals and predation by herbivores (Keith et al. 2014). Along the SEP, the high diversity of seaweeds at high latitudes has been linked to the presence of a highly diverse subantarctic flora with a range distribution that is restricted to the tip of South America (Santelices 1980; Santelices and Marquet 1998; Ramírez 2010). Based on clustering of the flora, Santelices (1980) distinguished three distinct areas along the coastline: the tropical area of northern Peru (4–5°S), a broad intermediate area (5°–53°S) and the southern tip of South America (53–55°S). Within the broad intermediate area, both brown and red seaweeds show clear biogeographical breaks at 30–33°S and 42–43°S (Meneses and Santelices 2000). The dominance of endemic species diminishes southward as they are replaced by subantarctic species, which are distributed in the cold waters of the South Pacific and the Southern Ocean (Santelices 1980).

Species richness along the SEP has probably been underestimated in the above cited studies as species determination was based solely on the use of morphological characters. The development of molecular genetic tools over the past two decades has strongly demonstrated our inability to correctly identify species on the basis of morphological characters alone. Using molecular markers, the presence of cryptic species has been revealed in many taxonomic groups and in various habitats (e.g. in marine environments Knowlton 1993, 2000; and in particular, in algae, see the recent review Leliaert et al. 2014). Studies based on molecular markers have accumulated compelling evidence that algal names have been applied incorrectly and morphological identification could lead to erroneous conclusions about trait evolution in seaweeds (see, for example the comparative analysis of chemical defence in Gracilariaceae, Weinberger et al. 2010). These erroneous taxonomic assignments have also revealed that seaweeds could have higher species diversity, and species could be more regionally confined than previously thought (see for example Tronholm et al. 2012; Payo et al. 2013; Pardo et al. 2014; Vieira et al. 2014).

Along the SEP, molecular studies have revealed the presence of cryptic/sister species in both red and brown seaweeds (see Table 10.1), and new species have recently been described in the region. For example, *Lessonia berteroana* and *Lessonia spicata* now stand in for the well studied morpho-species *Lessonia nigrescens* in the central and northern part of the SEP (González et al. 2012). Similarly, the species *Pyropia columbina* reported to be one of the most common and ubiquitous bladed Bangiales in the Southern Pacific is now restricted to a subantarctic range, while new endemic species were described for the temperate coast of New Zealand (*Pyropia plicata*, Nelson 2013) and Chile (*P. orbicularis*, Ramírez et al. 2014). Several new taxa yet unnamed were identified based on phylogenetic and phylogeographic studies of *Ectocarpus* (Peters et al. 2010), *Durvillaea* (Fraser et al. 2009, 2010a), *Adenocystis* (Fraser et al. 2013), *Mazzaella* (Montecinos et al. 2012) and *Nothogenia* (Lindstrom et al. 2015) (Table 10.1).

Interestingly, molecular data seem to strengthen the existence of the biogeographic boundaries proposed by Meneses and Santelices (2000) for seaweeds. Indeed, the genetic breaks found between *Durvillaea antarctica* "central Chile" and *D. antarctica* "subantarctic", between *Mazzaella laminarioides* "north" and *M. laminarioides* "center" and between *L. berteroana* and *L. spicata* are broadly congruent with the 42–43°S and the 30–33°S biogeographic boundaries (Table 10.1; Fig. 10.2). For other taxa studied, the lack of extensive sampling does not allow a clear pinpointing of the location of the phylogeographic breaks. However some clades, such as *Adenocystis utricularis* "Clade 3-COI" (Fraser et al. 2013) or *Nothogenia* "Taxon A" and "Taxon B" (Lindstrom et al. 2015), seem to be restricted to the Intermediate Area (Table 10.1). Finally, molecular studies have revealed that numerous species are shared between the SEP and New Zealand, the Subantarctic Islands and the Falkland Islands (in the genera *Macrocystis*, Macaya and Zuccarello 2010a, b; *Durvillaea*, Fraser et al. 2009, 2010a; *Adenocystis* Fraser et al. 2013; *Gigartina*, Billard et al. 2015; *Gracilaria*, Guillemin et al. 2014; *Nothogenia*, Lindstrom et al. 2015; *Capreolia*, Boo et al. 2014; *Bostrychia*, Fraser et al. 2013). These results demonstrate the importance of oceanic dispersal in shaping the diversity of the SEP, especially in the Magellan province and in the Intermediate Area.

Although the sample sites, sample sizes and level of diversity differ among taxa (Table 10.1), two major phylogeographic structuring patterns can be observed (Fig. 10.2). While endemic species living north of 42°S show complex reticulated

Table 10.1 Summary of phylogeographic studies of seaweed taxa along the South East Pacific coast

	Morpho-species	Cryptic species and taxa	Sampling distribution	Life cycle characteristics	Habitat	Buoyancy	Number of sites Individuals sampled for each marker	References
Phaeophyceae Laminariales	*Lessonia nigrescens*	*Lessonia berteroana*	16°58'S - 30°14'S	Heteromorphic haploid diploid cycle	Intertidal	Non buoyant	18 sites 522[a]: *atp8/trnS*[M]; 63: RuBisCo spacer[C]; 25: ITS1[N]; 31: ITS2[N]	Tellier et al. (2009)
		Lessonia spicata	29°03'S - 41°48'S	Heteromorphic haploid diploid cycle	Intertidal	Non buoyant	19 sites 523[a]: *atp8/trnS*[M]; 61: RuBisCo spacer[C]; 27: ITS1[N]; 35: ITS2[N]	Tellier et al. (2009)
	Macrocystis pyrifera		13°55'S - 53°28'S	Heteromorphic haploid diploid cycle	Subtidal	Buoyant thallus	39 sites 730[a]: *atp8/trnS*[M]; 45: *cox*1[M]	Macaya and Zuccarello (2010a, b)
Fucales	*Durvillaea antarctica*	"central Chile"	32°20'S - 42°59'S	Diploid cycle	Intertidal	Buoyant thallus	19 sites 130: *cox*1[M], 54: *rbc*L[C]	Fraser et al. (2009, 2010a)
		"subantarctic"	49°09'S - 55°59'S	Diploid cycle	Intertidal	Buoyant thallus	5 sites 34: *cox*1[M], 24: *rbc*L[C]	Fraser et al. (2009, 2010a)
Ectocarpales	*Adenocystis utricularis*	"Clade 1-COI"	53°37'S	Heteromorphic haploid diploid cycle	Intertidal	Non buoyant	1 site 2: *cox*1[M], 2: *rbc*L[C]	Fraser et al. (2013)
		"Clade 2-COI"	51°09'S - 53°37'S	Heteromorphic haploid diploid cycle	Intertidal	Non buoyant	4 sites 6: *cox*1[M], 12: *rbc*L[C]	Fraser et al. (2013)
		"Clade 3-COI"	36°35'S	Heteromorphic haploid diploid cycle	Intertidal	Non buoyant	1 site 8: *cox*1[M], 6: *rbc*L[C]	Fraser et al. (2013)
		"Clade 4-COI"	51°28'S - 53°37'S	Heteromorphic haploid diploid cycle	Intertidal	Non buoyant	3 sites 16: *cox*1[M], 7: *rbc*L[C]	Fraser et al. (2013)

Table 10.1 (continued)

	Morpho-species	Cryptic species and taxa	Sampling distribution	Life cycle characteristics cycle	Habitat	Buoyancy	Number of sites Individuals sampled for each marker	References
Rhodophyta Gigartinales	*Mazzaella laminarioides*	"north"	28°55'S - 32°37'S	Isomorphic haploid diploid cycle	Intertidal	Non buoyant	5 sites 96: *cox*1[M], 61: *rbc*L[C]	Montecinos et al. (2012)
		"center"	34°05'S - 37°38'S	Isomorphic haploid diploid cycle	Intertidal	Non buoyant	5 sites 105: *cox*1[M], 71: *rbc*L[C]	Montecinos et al. (2012)
		"south"	39°40'S - 54°03'S	Isomorphic haploid diploid cycle	Intertidal	Non buoyant	8 sites 151: *cox*1[M], 101: *rbc*L[C]	Montecinos et al. (2012)
	Gigartina skottsbergii		41°51'S - 55°06'S	Isomorphic haploid diploid cycle	Subtidal	Non buoyant	10 sites 149: cox2-3[M], 10: *rbc*L[C]	Billard et al. (2015)
Gracilariales	*Gracilaria chilensis*		36°45'S - 43°46'S	Isomorphic haploid diploid cycle	Intertidal	Non buoyant	11 sites 108: ITS2[N], 5: *rbc*L[C]	Guillemin et al. (2014)
Nemaliales	*Nothogenia chilensis*	*Nothogenia chilensis*	32°42'S - 34°22'S	Isomorphic haploid diploid cycle	Intertidal	Non buoyant	3 sites 3: ITS [N], 3: *rbc*L[C], 3: *psb*A[C], 3: *cox*1[M]	Lindstrom et al. (2015)
		"Taxon A"	35°20'S - 36°48'S	Isomorphic haploid diploid cycle	Intertidal	Non buoyant	2 sites 3: ITS [N], 3: *rbc*L[C], 3: *psb*A[C], 1: *cox*1[M]	Lindstrom et al. (2015)
		"Taxon B"	33°11'S	Isomorphic haploid diploid cycle	Intertidal	Non buoyant	1 site 2: ITS [N], 2: *rbc*L[C], 2: *psb*A[C], 1: *cox*1[M]	Lindstrom et al. (2015)
		"Taxon C"	36°08'S - 43°53'S	Isomorphic haploid diploid cycle	Intertidal	Non buoyant	5 sites 6: ITS [N], 8: *rbc*L[C], 2: *psb*A[C]	Lindstrom et al. (2015)
	Nothogenia fastigiata		39°51'S - 53°37'S	Isomorphic haploid diploid cycle	Intertidal	Non buoyant	5 sites 5: ITS [N], 9: *rbc*L[C], 2: *psb*A[C], 1: *cox*1[M]	Lindstrom et al. (2015)
	Nothogenia fragilis		23°28'S - 36°31'S	Isomorphic haploid diploid	Intertidal	Non buoyant	2 sites 2: ITS [N], 3: *rbc*L[C], 3: *psb*A[C], 1: *cox*1[M]	Lindstrom et al. (2015)

Table 10.1 (continued)

	Morpho-species	Cryptic species and taxa	Sampling distribution	Life cycle characteristics cycle	Habitat	Buoyancy	Number of sites Individuals sampled for each marker	References
Gelidiales	*Capreolia implexa*		39°54'S - 41°46'S	Heteromorphic diphasic cycle	Intertidal	Non buoyant	3 sites 4: *rbc*L[C], 27: *cox*1[M]	Boo et al. 2014
Ceramiales	*Bostrychia intricata*	"Clade 1-COI"	41°28'S - 53°37'S	Isomorphic haploid diploid cycle	Intertidal	Non buoyant	6 sites 20: *cox*1[M], 13: *rbc*L[C]	Fraser et al. (2013)
		"Clade 4-COI"	41°28'S - 52°15'S	Isomorphic haploid diploid cycle	Intertidal	Non buoyant	5 sites 16: *cox*1[M], 6: *rbc*L[C]	Fraser et al. (2013)
Corallinales	*Spongites* sp.		28°16'S - 36°48'S	Isomorphic haploid diploid cycle	Intertidal	Non buoyant	9 sites 37: SSU[N], 37: cox2-3[C]	Vidal et al. (2008)

Sampling range, habitat, life-history, molecular data available and reference are given for each taxon. The name of the morpho-species for which the studies were undertaken is given in the second column. The species and taxa including extended geographical coverage and population sampling design are shaded in light grey. *Abbreviations cox*1: cytochrome c oxidase subunit I, *cox*2–3: intergenic sequence between the cytochrome c oxidase subunit II and III, *atp8/trnS*: intergenic sequence between the ATPase subunit 8 gene and the trnS, ITS: internal transcribed spacer, *psb*A: photosystem II thylakoid membrane protein D1, *rbc*L: large subunit of the Rubisco; RuBisCo spacer: intergenic sequence between the large and the small subunits of Rubisco, SSU: Small subunit rRNA gene; [M] mitochondrial, [C] chloroplastic, [N] nuclear

[a]Part of the data set was obtained using single-strand conformation polymorphism (SSCP)

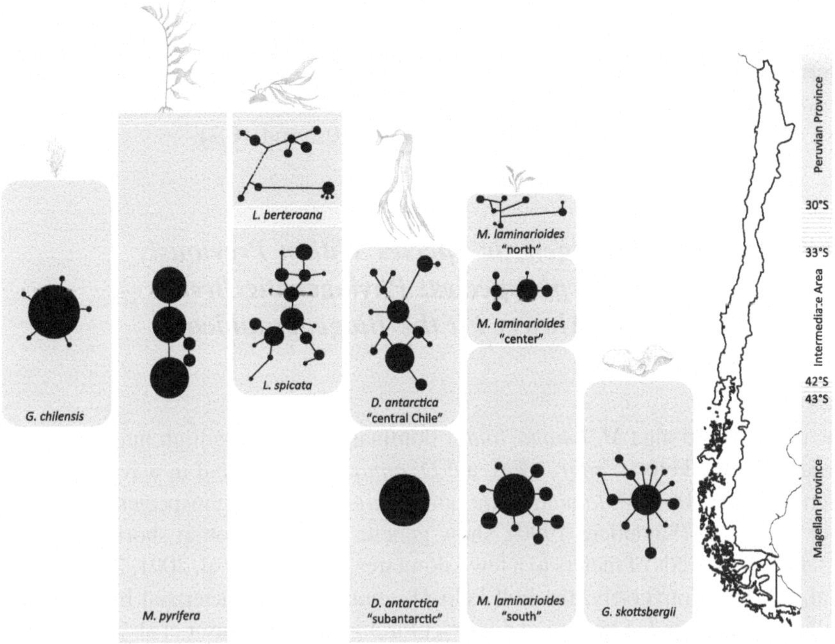

Fig. 10.2 Schematized median joining haplotype networks of 10 seaweed taxa along the South East Pacific coast for which extensive phylogeographic studies have been published (see Table 10.1). Range distribution of each taxon is represented by a *grey box* (for more information about latitudinal limits of each taxon see Table 10.1). In all haplotype networks each *circle* represents a haplotype and, while the sizes of circles are proportional to haplotype frequencies in each taxon, they are not comparable across studies. *Black lines* represent mutational steps between haplotypes and the length of each line is proportional to the number of different base pairs between them. For more information about each taxon see Table 10.1

haplotype networks, species that have recently colonized the SEP or endemic species with a range distribution limited to the south of 42°S generally show simple star-like haplotype networks (Fig. 10.2). Because these differences reflect distinct evolutionary trajectories and demographic histories, in the following sections, we discuss the historical and contemporary scenarios that likely shaped these different patterns of phylogeographic structure.

10.4 Parapatric Distribution and Speciation Processes Along a Linear Coast

As 42°S is considered to be the northern limit of the coastal ice cap (McCulloch et al. 2000; Fig. 10.1b), comparing the phylogeography of seaweeds endemic to the SEP that are distributed north of 42°S allows for the inference of historical processes that occurred along this coast long before the LGM. We will focus on three case studies of previously recognized morphospecies distributed in the Intermediate Area and/or the Peruvian Province (*Lessonia nigrescens*, *M. laminarioides* and *Durvillaea antarctica*) that were in fact proven to include divergent taxa or cryptic sister species (Table 10.1; Figs. 10.2 and 10.3).

10.4.1 Cryptic Phylogenetic Species Within Previously Reported Morphospecies: Phylogenetic Breaks that Do not Always Fit the Biogeographical Boundaries

The endemic red alga *M. laminarioides* dominates the middle-high intertidal rocky shore while the kelps *L. nigrescens* and *D. antarctica* are found in wave-swept low intertidal areas. Both *M. laminarioides* and *L. nigrescens* morphospecies, considered poor dispersers (Santelices 1990), show genetic differentiation at short geographic distances (hundreds of meters to a few kilometres, Faugeron et al. 2001, 2005; Tellier et al. 2011a). Conversely, the bull kelp *D. antarctica*, characterized by a buoyant thallus is considered to be a good disperser through rafting (Fraser et al. 2009, 2010a, b; also see chapters by Macaya et al. 2016 and Fraser 2016 in this volume).

Despite ecological differences among these morphospecies, molecular studies have revealed that each is subdivided into phylogenetic species, i.e. reciprocally monophyletic and highly divergent clades (see Fig. 10.3). Two clades were recovered for *L. nigrescens* (Tellier et al. 2009), whereas three clades were recovered for *M. laminarioides* (Montecinos et al. 2012). *D. antarctica* is a species complex of four deeply divergent taxa, of which only two are present along the SEP coast (i.e. "central Chile" and "subantarctic", Fraser et al. 2010a, b).Within each morphospecies, molecular markers were congruent in revealing the occurrence of cryptic species distributed in parapatry along the SEP coast (Figs. 10.2 and 10.3). The location of phylogeographic discontinuities are specific to each morphospecies: 30°S for *L. nigrescens*, 33°S and 38°S for *M. laminarioides*, and in-between 44°S and 49°S for *D. antarctica*. Interestingly, speciation processes seem to be tightly linked to the processes driving the biogeographic discontinuities. In *L. nigrescens* and *D. antarctica* and between *M. laminarioides* "north" and "center" lineages, genetic discontinuities broadly match the biogeographic boundaries (Meneses and Santelices 2000; Camus 2001; Thiel et al. 2007), which is also the case for several invertebrate species (Brante et al. 2012 Varela and Haye 2012; Haye et al. 2014). However, the genetic break between *M. laminarioides* "center" and "south" lineages is located between 37°S and 39°S, a region where neither a biogeographic boundary nor a phylogeographic break has been described. The absence of a precise estimate of mutation rates makes it difficult to use molecular clocks to assess the time of divergence among lineages of seaweeds. An exploratory estimation of the historical events at the origin of the divergence among lineages has however been attempted. The separation between *M. laminarioides* "north" and "center" and between *M. laminarioides* "center" and "south" were estimated to be between 1.0 and 12.1 Myr and between 0.5 and 5.8 Myr, respectively (Montecinos et al. 2012). The separation between the cryptic species of *L. nigrescens* (i.e. *L. berteroana* and *L. spicata*; González et al. 2012) is estimated to have occurred more recently, between 0.2 and 1.7 Myr (Tellier et al. 2009) and 2.0 and 3.1 Myr (Martin and Zuccarello 2012) depending on the method of estimation. Within *D. antarctica*, most of the genetic diversity is found around New Zealand where a late Miocene/Pliocene radiation occurred between 1.3 and 9.7 Myr (Fraser et al. 2010b). Even if substantial uncertainty is associated with the timing of lineage splits in the three study cases, the authors agree that they predate Pleistocene glaciations. Because the location and putative timing of the phylogeographic breaks are not fully congruent between taxa, it is not possible to identify specific historical events that may have driven speciation in these intertidal seaweeds (Avise et al. 1987).

In the case of *D. antarctica*, phylogenetic reconstruction indicates an ancient split of the "central Chile" lineage (Fraser et al. 2009, 2010b) that could have occurred after the first transoceanic dispersal event of the algae from New Zealand, where the genus originates. This allopatric divergence took place well before the diversification of New Zealand taxa from which evolved the "subantarctic" lineage present in Chile (Fraser et al. 2010b). The presence of the "subantarctic" lineage in Chile has been explained by recent, post-glacial transoceanic dispersal from New Zealand and colonization of Patagonia soon after melting of the coastal ice sheet (Fraser et al. 2009, 2010a, b). Therefore, in the case of *D. antarctica*, the genetic break observed in the SEP around 44–49°S results from a secondary contact between two lineages that have diverged in allopatry on either side of the Pacific Ocean.

Conversely, allopatric speciation has not been proposed as the cause of lineage splitting in either *M. laminarioides* or *L. nigrescens*. The genus *Lessonia* is distributed throughout both New Zealand and Chile, but studies have clearly indicated that divergence between *L. spicata* and *L. berteroana* occurred in Chile (Tellier et al. 2009; Martin and Zuccarello 2012). A similar situation has been suggested for *M. laminarioides*, which is endemic to Chile. While incomplete lineage sorting due to the presence of ancestral haplotypes has been revealed using slow evolving molecular markers, reciprocal monophyly has been systematically observed using faster evolving markers for cryptic lineages of *M. laminarioides* and for both *L. berteroana* and *L. spicata* (Fig. 10.3). The topology of phylogenetic trees built from slow evolving markers was similar for both *Mazzaella* and *Lessonia*, with monophyletic derived lineages embedded within a basal polytomy formed by the more ancestral haplotypes (Fig. 10.3). These slow evolving marker tree topologies have been described as a typical signature of budding/peripatric speciation rather than vicariant speciation (Funk and Omland 2003; Crawford 2010). For both the lineage splitting within *M. laminarioides* and the divergence between *L. berteroana* and *L. spicata*, authors have suggested the existence of budding speciation with a sudden expansion at the range limit of the ancestral lineages (Tellier et al. 2009; Montecinos et al. 2012). However, there is a difference in the direction of the range expansion of these ancestral lineages. A range expansion of *L. spicata* in the area

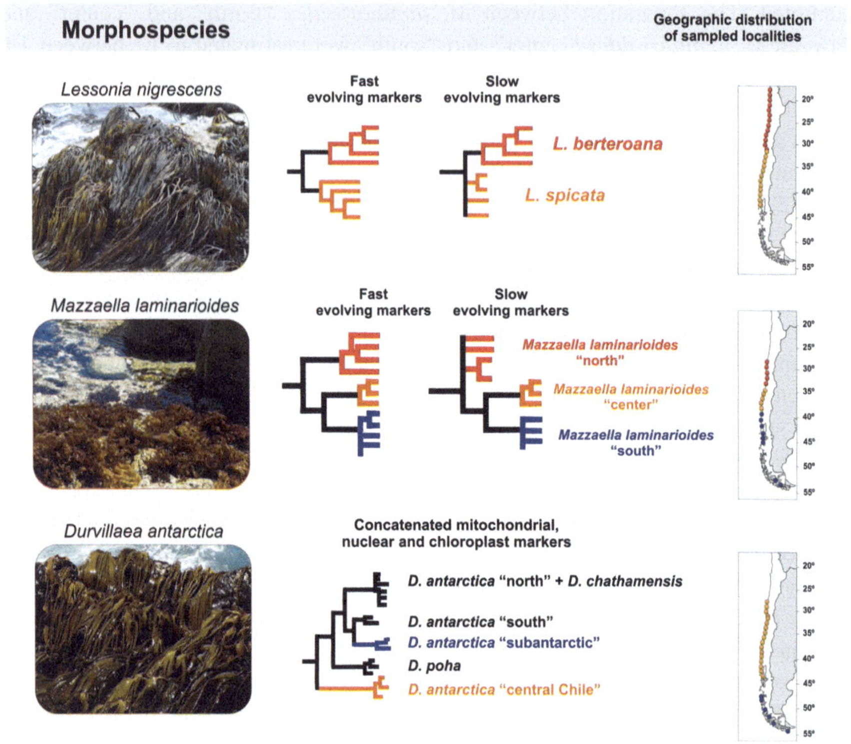

Fig. 10.3 Growth habits, schematic representation of published phylogenetic trees and range distribution of *Lessonia berteroana* and *L. spicata, Durvillaea antarctica* "central Chile" and *D. antarctica* "subantarctic" and *Mazzaella laminarioides* "north", *M. laminarioides* "center" and *M. laminarioides* "south." For *L. berteroana* and *L. spicata* phylogenetic trees were retrieved from Tellier et al. (2009) using fast evolving *atp8/trnS* (mitochondrial) and ITS1 (nuclear) markers and slow evolving ITS2 (nuclear) and RuBisCo spacer (chloroplast) markers. For *D. antarctica* a phylogenetic tree for all four concatenated markers (mitochondrial: COI, chloroplast: *rbc*L and nuclear: 18S rRNA and 28S rRNA) was retrieved from Fraser et al. (2010b). For *M. laminarioides* phylogenetic trees were retrieved from Montecinos et al. (2012) using the fast evolving COI (mitochondrial) marker and slow evolving *rbc*L (chloroplast) marker. Distributions of cryptic species or lineages are based on the work of Tellier et al. (2009, 2011a) for *L. berteroana* and *L. spicata*, on the work of Fraser et al. (2009, 2010a, b) for *D. antarctica* "central Chile" and *D. antarctica* "subantarctic" and Montecinos et al. (2012) for *M. laminarioides* "north", *M. laminarioides* "center" and *M. laminarioides* "south". Photos represent, from top to bottom, thalli of *L. berteroana* growing in the wave-swept intertidal zone in Los Verdes, Iquique (20°25′S, photo E. Macaya), thalli of *D. antarctica* "central Chile" in Mar Brava (41°52′S, photo E. Macaya) and thalli of *M. laminarioides* "south" growing in the high intertidal zone in Caleta Hiuro, Valdivia (39°57′S, photo M-L. Guillemin)

located north of the 30°S could be at the origin of *L. berteroana.* In contrast, range expansion of *M. laminarioides* "north" in the area located south of the 33°S could be at the origin of *M. laminarioides* "center" and "south" (Fig. 10.3; see Tellier et al. 2009; Montecinos et al. 2012 for more details).

Budding speciation (also known as peripatric speciation) was first defined by Mayr (1954) as a speciation process by which an initially small colonizing population becomes reproductively isolated from a species with a larger range. In a paper entitled "Rethinking classic examples of recent speciation in plants", Gottlieb (2004) developed the argument that this process of speciation in which species 'bud off' from ancestral species via small, locally isolated peripheral populations is probably common in plants. In addition, Gottlieb (2004) highlights that recently diverged sister species, for which the overall genetic distance is minimal and the direction of evolution is clear are particularly relevant cases to study this mode of speciation. Many authors have thus argued that budding speciation has a unique signature that, early in the speciation process, sister species should have overlapping or adjacent ranges with very different sizes (i.e. asymmetric ranges) and different realized niche breadths (Funk and Omland 2003; Gottlieb 2004; Grossenbacher et al. 2014; Anacker and Strauss 2014). The prediction of a greater range asymmetry between younger versus older sister pairs was verified recently in two large-scale analyses in plants (114 species of *Mimulus* sampled in North America, Grossenbacher et al. 2014; 71 sister species from 12 families sampled in the California Floristic Province, Anacker and Strauss 2014). Supporting a scenario of budding speciation, strong ecological niche differences were observed between *L. berteroana* and *L. spicata*, including temperature responses (Oppliger et al. 2011, 2012; Vieira et al. 2015), tolerance to air exposure (López-Cristoffanini et al. 2013) and differences in phlorotannin pigments and chlorophyll *a* fluorescence of PSII (Koch et al. 2015). Niche preferences have yet to be studied for the three lineages of *M. laminarioides*. The comparison of range sizes of derived and ancestral cryptic species does not seem to fit the prediction of range asymmetry for sister lineages and cryptic species along the SEP. Indeed, *L. berteroana* and *L. spicata* have a very similar range size of about 1500 km which overlaps less than 200 km between 29°S and 30°S (Tellier et al. 2009, 2011a) (Table 10.1; Figs. 10.2 and 10.3). Similarly, the range size of the ancestral northern lineage of *M. laminarioides* (415 km) is smaller than that of the derived lineages; particularly, the southern lineage is much larger ("central" lineage range of 730 km; "south" lineage range of 2400 km; Table 10.1; Figs. 10.2 and 10.3; Montecinos et al. 2012). The unexpectedly large size of the recently diverged lineages in *Mazzaella* and *Lessonia* could be explained by recent post-speciation range shifts that do not reflect the historical range sizes of the lineages during the budding speciation process. Indeed, in the case of *M. laminarioides*, the "south" lineage has clearly been subjected to successive periods of range expansion/contraction during the glacial periods, which was not the case for the other two lineages that were located north of the glacial coverings (Montecinos et al. 2012). On the other hand, in plants for which clear evidence of range asymmetries support a scenario of budding speciation, the time of divergence between those sister species was determined to be very recent (less than 1 Myr, see for example Baldwin 2005 and Crawford et al. 2006). This does not seem to be the case for the studied seaweeds, and therefore current sister lineage distributions do not necessarily represent the range at budding speciation time for *Mazzaella* or *Lessonia*.

10.4.2 Genetic Diversity and Structure Within SEP Endemic Taxa

In order to study patterns of intraspecific genetic variation, we calculated average per population gene diversity (H) and population-pairwise Φ_{ST} (Fig. 10.4) using the data published for *L. berteroana*, *L. spicata* (Tellier et al. 2009), *M. laminarioides* "north", *M. laminarioides* "center", *M. laminarioides* "south" (Montecinos et al. 2012) and *D. antarctica* "central Chile" (Fraser et al. 2010a). Two contrasting patterns of genetic differentiation were found (Fig. 10.4). On one hand, *M. laminarioides* "center" and "south" displayed the lowest values of Φ_{ST} ($\Phi_{ST} = 0.49$ and $\Phi_{ST} = 0.36$, respectively), and were composed of haplotypes shared among neighbouring populations. On the other hand, the other four taxa were characterized by low diversity within populations (mean $H < 0.2$) and high levels of genetic differentiation (mean pairwise $\Phi_{ST} > 0.80$). Within these four taxa each haplotype was restricted to a single population or at most to a few close-by populations. The extreme case of such a patchy distribution of haplotypes with almost every population composed of a unique haplotype (mean H close to 0, and Φ_{ST} close to 1) was found in the northernmost species *L. berteroana* (Fig. 10.4). In the same way, no haplotypes were shared between populations of *M. laminarioides* "north" (Montecinos et al. 2012).

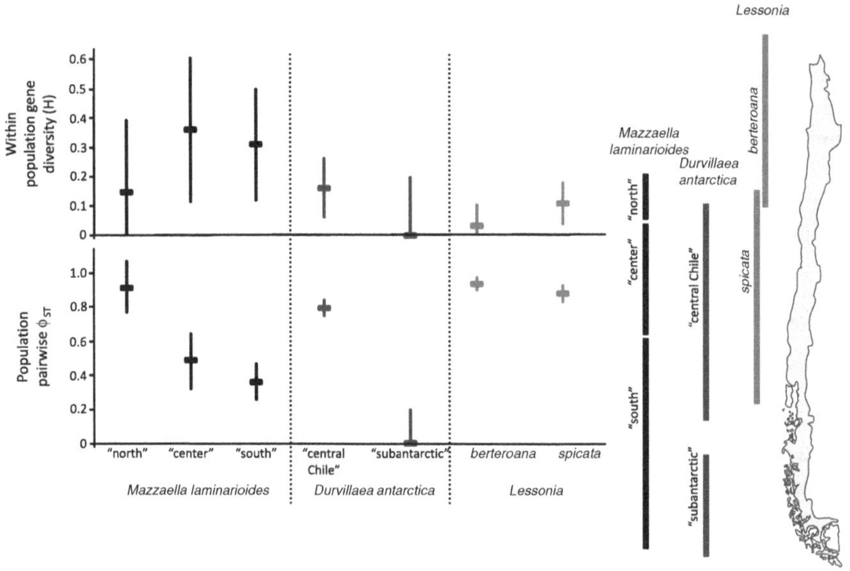

Fig. 10.4 Average gene diversity per population (H) and population-pairwise Φ_{ST} calculated within each taxon of *Mazzaella laminarioides* and for *Durvillaea antarctica* "central Chile", *Lessonia berteroana* and *L. spicata*. Mean ±95 % confidence intervals are shown. Range distribution is given for each taxon (for more information about latitudinal limits of each taxon see Table 10.1, Montecinos et al. 2012; Fraser et al. 2010a; Tellier et al. 2009, 2011b)

Regardless of the taxa studied, genetic differentiation was significant, confirming that seaweeds are generally poor dispersers compared to other marine organisms. Genetic estimates of dispersal distance in seaweeds are generally considered to be lower than 10 to 50 km (see reviews by Kinlan and Gaines 2003; Valero et al. 2011; Durrant et al. 2014). The minimum values of pairwise Φ_{ST}, observed for *M. laminarioides* "south", were similar to values observed along the SEP coast for invertebrates with low dispersal capabilities (Haye et al. 2014). If dispersal is the main cause of the observed patterns of genetic differentiation, we would expect to find a significant relationship between genetic distances and geographic distances. Surprisingly, this was not the case for the three taxa of *M. laminarioides* studied, while isolation-by-distance (IBD) was significant for the three other taxa. For *L. berteroana*, *L. spicata*, *M. laminarioides* "north" and *D. antarctica* "central Chile" there is almost complete genetic isolation among populations separated by at most a few kilometres or tens of kilometres. It is probable that isolation-by-distance occurs at much shorter spatial scales than the species considered for the phylogeographic studies in these taxa.

The occurrence of haplotypes restricted to a few neighbouring populations and distributed in strict parapatry suggests that mechanisms other than limited dispersal may have contributed to such a patchy genetic diversity pattern. The mid to low intertidal habitat occupied by each of the studied species is intrinsically patchy since rocky shores along the SEP coast are interrupted by influential barriers, such as estuaries or sandy beaches (Thiel et al. 2007). Moreover, strong genetic drift is expected when populations are small and/or experience highly dynamic demographics and pass through periods of reduced population size. The combination of isolation and drift is strongly correlated with the level of genetic differentiation between pairs of populations in *D. antarctica* (Fraser et al. 2010a). In addition, strong demographic bottlenecks or local extinctions often occur in the region of the SEP located between 30°S and 40°S due to the recurrent effects of El Niño (Martínez et al. 2003; Thiel et al. 2007) and tectonic activity, which cause sudden changes to the elevation of the coast (Castilla 1988; Castilla et al. 2010; Jaramillo et al. 2012). If a local bottleneck is sufficiently strong, population recovery may be achieved by a reduced number of reproductively successful survivors or founders (in the case of local extinction). Such population dynamics typically reinforce the effects of genetic drift and may significantly contribute to patterns of genetic diversity and structure. The combined effects of reduced population size and high population turnover have been shown to reduce local genetic diversity and increase population genetic differentiation (Walser and Haag 2012). These effects can be observed even in species with high dispersive abilities such as floating seaweeds capable of rafting or species with long-lived spores. Considering gene flow restrictions when population densities reach a maximum, the Monopolization Hypothesis (De Meester et al. 2002; Waters et al. 2013) predicts that space will be saturated with local propagules soon after a bottleneck or a founder event, and this will give little chance for settlement of future immigrants. Under this hypothesis, dispersal is less efficient when patches of organisms are dense (Tellier et al. 2011a; Neiva et al. 2012 and for review see Waters et al. 2013). Moreover, the monopolization effect limits immigration therefore reinforcing local adaptation even if environmental heterogeneity in the system is minimal (De Meester et al. 2002). Under this scenario, historical phylogeographic disjunction patterns may last for long periods of time and can lead to speciation.

10.5 Post-glacial Histories: Distinguishing Between Local Population Recovery and Trans-oceanic Introductions

The phylogeographic literature on Patagonian species (i.e. south of 42°S) indicates complex and unexpected evolutionary histories. For terrestrial species, the high genetic diversity often observed in Patagonia and the strong divergence among populations suggests ancient evolutionary processes that took place long before the LGM. Generally, it is proposed that populations persisted in different refugia in or around the glaciated region during the LGM (see Sérsic et al. 2011 for a review of 33 terrestrial plant and vertebrate species). Freshwater and marine animals seem to follow the same pattern. The use of periglacial refugia is hypothesized for frogs (Nuñez et al. 2011), fish (Ruzzante et al. 2008) and marine gastropods (Sánchez et al. 2011; González-Wevar et al. 2012) whereas glacially embedded refugia are hypothesized for fish (Zemlak et al. 2010), crabs (Xu et al. 2009) and river otters (Vianna et al. 2011).

In contrast, all the seaweed species studied to date show clear phylogeographic signatures of recent, post-glacial expansions in Patagonia. While haplotype networks are generally diversified and complex north of 42°S, only one or very few haplotypes have been detected in the region covered by the Patagonian ice sheet during the LGM (Fig. 10.2). A single haplotype is found in this region for the giant kelp *Macrocystis pyrifera* (Macaya and Zuccarello 2010a, b) and for the bull kelp *D. antarctica* (Fraser et al. 2009, 2010a, b), and a star-like haplotype network is observed in the red algae *Gigartina skottsbergii* (Billard et al. 2015) and *M. laminarioides* "south" (Montecinos et al. 2012). It has generally not been possible to precisely estimate the timing of these demographic expansions due to a lack of accurate mutation rates for most molecular markers in seaweeds. However, the single haplotype shared by *D. antarctica* individuals found among all of the sub-antarctic islands and continental region of South America matches the latitudinal limits of coastal ice sheet coverage during the LGM (Fraser et al. 2009). This strongly suggests that the species colonized these areas only after the ice melted. In addition, the observation of a single haplotype strongly suggests that the sub-antarctic colonization is sufficiently recent that no new mutations have appeared in any of these regions. The large-scale spread of genetic variants has been observed during the recolonization of newly available empty habitats for both buoyant (*D. antarctica*, Fraser et al. 2010a; *M. pyrifera*, Alberto et al. 2010, 2011) and non-buoyant seaweeds (*M. laminarioides*, Faugeron et al. 2001; *G. skottsbergii*, Faugeron et al. 2004). During spatial expansion, gene surfing effects may contribute to the reduction of diversity in the recolonized region (Klopfstein et al. 2006; Hallatschek et al. 2007; Excoffier and Ray 2008; also see chapter by Neiva et al. (2016) in this volume). Indeed, gene surfing could lead to the presence of one or a few haplotypes over large recently colonized geographic areas (Klopfstein et al. 2006; Hallatschek et al. 2007; Excoffier and Ray 2008; Neiva et al. 2010). And finally, after the colonization of the region affected by ice, a density-blocking effect could have limited the establishment of new haplotypes allowing for the persistence of founder effect signatures long after spatial expansion is completed (De Meester et al. 2002; Waters et al. 2013). A better sampling coverage of the Magellan province and more polymorphic markers would be necessary to specifically test these various hypotheses.

Interestingly, there are two different hypothesized source populations for these post-glacial demographic expansions. One seems to be local, on the South American continent. In both *M. laminarioides* "south" and *G. skottsbergii*, population expansion apparently started from a single South American refuge. This conclusion is supported by the existence of a single central haplotype, shared by all populations, in the star-like networks of these two taxa (Fig. 10.2, Montecinos et al. 2012; Billard et al. 2015). Whereas it seems reasonable to think that the refugium was north of 42° S for *M. laminarioides* where most of the haplotype diversity is present, it is difficult to propose an exact location for the refuge of *G. skottsbergii*. For *G. skottsbergii*, the same central haplotype was found near the 42°S range limit, on the Southern tip of South America, and on the Falkland Islands, and no clear difference in haplotype diversity was observed over the whole species distribution range (41°–55°S). On the other hand, the source population of *D. antarctica* "subantarctic" for the post-glacial expansion in Chile was located in the New Zealand Subantarctic Islands (Fraser et al. 2009, 2010a, b). A similar scenario of recent transoceanic dispersal from New Zealand can also be inferred from the genetic diversity of the red alga *Gracilaria chilensis* (Guillemin et al. 2014). For ITS2, the only shared haplotype between both sides of the Pacific Ocean for *G. chilensis* was the central haplotype in the star-like network in Chile. This gives evidence for a recent population expansion after a founder effect due to the migration to Chile from New Zealand (Guillemin et al. 2014). One other seaweed shares its single Patagonian haplotype with subantarctic islands (South Georgia, Marion and Macquire), the giant kelp *M. pyrifera* (Macaya and Zuccarello 2010a, b). Because little diversity was found in mitochondrial sequences of *M. pyrifera,* it is difficult to infer the original source population of this subantarctic demographic expansion based on mtDNA phylogeography alone. Other genetic markers are required to make further inferences. Further discussion of trans-Pacific connectivity can be found in Chapter by Fraser (2016 in this volume). Moreover, more species should be studied to understand the history of the Magellan Province, which hosts high species richness despite the regular perturbations by glacial cycles.

10.6 Prospects and Challenges Ahead

Most seaweed phylogeographic studies show that the SEP coast houses a rich cryptic diversity. Of the 12 studies listed in Table 10.1, six have revealed the existence of cryptic species or subspecies: in *Lessonia* (Tellier et al. 2009), *Durvillaea* (Fraser et al. 2009, 2010a, b), *Adenocystis* (Fraser et al. 2013), *Mazzaella* (Montecinos et al. 2012) and *Nothogenia* (Lindstrom et al. 2015). These results suggest that species diversity is underestimated in the region (see Peters et al. 2010; Fraser et al. 2013). Because the existence of high cryptic diversity could obscure biogeographic patterns (Riddle and Hafner 1999), future phylogeographic studies may lead to new biogeographic paradigms (for an example see the case of the Canadian Arctic marine flora in Saunders and McDevit 2013 where molecular data have allowed the authors to revisit hypotheses about the evolutionary origins and migration of algal taxa in this region). However, strong patterns have emerged from this first comparative study. First, different studies suggest parapatric and/or peripatric speciation occurs along the linear coast of SEP. Some speciation events seem to have taken place at the limit of the Intermediate Area and the Peruvian province (30–33°S). Yet other events of divergence do not correspond with any biogeographic limit or clear ecological barriers to gene flow. In this last case, the role of stochasticity associated with climatic and tectonic catastrophes needs to be further explored as a source of phylogeographic discontinuity. Moreover, in some morphospecies the location of phylogeographic discontinuities have not been clearly pinpointed due to logistical difficulties in accessing sites, which lead to large gaps in sampling. These gaps are most apparent in the Magellan province (between 44°S and 52°S, where access to the benthic habitat is impossible from land and very difficult by boat) and in the Peruvian province (i.e. between 18°S and 24°S, a region where few roads give access to the coast). Access to samples from these areas is essential to better assess species diversity along the SEP and test speciation or recolonization scenarios. Second, if population turnover is indeed high enough to promote genetic differentiation through founder and monopolization effects, effective dispersal is expected to be low. Dispersal and gene flow should be assessed at appropriate spatial scales to test this hypothesis. In the case of strong reductions in gene flow, even slight environmental differences may allow directional selection to promote local adaptation. This is an interesting perspective that future phylogeographic studies could include either through eco-physiological experimentation or by combining analyses of both neutral and selected markers using NGS technology and population genomics approaches (Beaumont and Balding 2004; Luikart et al. 2003). Third, several (cryptic) species actually colonized the SEP from New Zealand soon after the LGM. In this context, it could be interesting to investigate whether species diversity present along the SEP is the result of elevated local speciation rates (Center of Origin hypothesis) or rather due to the accumulation of species formed elsewhere (Center of Accumulation hypothesis; Briggs 2000; Barber 2009).

From a more applied point of view, these new phylogeographic results should lead to changes in management strategies and conservation of natural resources. Small fishing villages along the SEP sometimes heavily rely on algal communities as economically important resources. Both red (*G. chilensis*, *Chondracanthus chamissoi*, *Pyropia* spp., *Sarcothalia crispata*, *M. laminarioides* and *G. skottsbergii*) and brown (*Lessonia* spp., *D. antarctica* and *M. pyrifera*) seaweeds are harvested along the coast (Buschmann et al. 2001; Guillemin et al. 2008; Vásquez 2008; Tellier et al. 2011b). Morpho-species with cryptic diversity should be taken into account when conservation and/or management programs are developed (Tellier et al. 2011b) if differences in population demographics or ecological niches lead to differing susceptibility to environmental threats. For example, ecological differences between *L. berteroana* and *L. spicata* in temperature and desiccation sensitivity (Oppliger et al. 2011, 2012; López-Cristoffanini et al. 2013) suggest that the species will experience different susceptibilities to events like ENSO. Finally, several species of algae are cultivated in Chile. Breeding strategies are emerging (Buschmann et al. 2014) and might face difficulties if reproductive barriers exist between cryptic species. Such reproductive isolation exists between *L. spicata* and *L. berteroana* in their overlapping distribution ranges (Tellier et al. 2011a). Taking into account the complex evolutionary processes occurring at large spatial and temporal scales is a necessity for these applied issues, and offers interesting perspectives of interactions among multiple research lines and disciplines.

Acknowledgments This research was supported by CONICYT FONDECYT/REGULAR No. 1130797 and Centro FONDAP IDEAL No. 15150003 to MLG and FONDECYT/INICIACION No. 11121504 to FT. Additional support came from CONICYT PCCI/Proyectos de Intercambio No. PCCI130047, The Nucleo Milenio "Centro de Conservación Marina"—RC130004, the French Research Agency ANR-ECOKELP (ANR 06 BDIV 012) and International Research Network "Diversity, Evolution and Biotechnology of Marine Algae" (GDRI No. 0803).

References

Alberto F, Raimondi PT, Reed DC, Coelho NC, Leblois R, Whitmer A, Serrão EA. Habitat continuity and geographic distance predict population genetic differentiation in giant kelp. Ecology. 2010;91:49–56.

Alberto F, Raimondi PT, Reed DC, Watson JR, Siegel DA, Mitarai S, Coelho NC, Serrão EA. Isolation by oceanographic distance explains genetic structure for *Macrocystis pyrifera* in the Santa Barbara Channel. Mol Ecol. 2011;20:2543–54.

Anacker BL, Strauss SY. The geography and ecology of plant speciation: range overlap and niche divergence in sister species. Proc R Soc B. 2014;281:20132980.

Avise JC, Arnold J, Ball RM, Bermingham E, Lamb T, Neigel JE, Reeb CA, Saunders NC. Intraspecific phylogeography: the mitochondrial DNA bridge between population genetics and systematics. Ann Rev Ecol Sys. 1987;18:489–522.

Baldwin BG. Origin of the serpentine-endemic herb *Layia discoidea* from the widespread *L. glandulosa* (Compositae). Evolution. 2005;59(11):2473–9.

Barber PH. The challenge of understanding the Coral Triangle biodiversity hotspot. J Biogeogr. 2009;36:1845–6.

Beaumont MA, Balding DJ. Identifying adaptive genetic divergence among populations from genome scans. Mol Ecol. 2004;13:969–80.
Belkin IM, Gordon AL. Southern Ocean fronts from the Greenwich meridian to Tasmania. J Geophys Res. 1996;101(C2):3675–96.
Billard E, Reyes J, Mansilla A, Faugeron S, Guillemin ML. Deep genetic divergence between austral populations of the red alga *Gigartina skottsbergii* reveals a cryptic species endemic to the Antarctic continent. Pol Biol. 2015;. doi:10.1007/s00300-015-1762-4.
Boo GH, Mansilla A, Nelson W, Bellgrove A, Boo SM. Genetic connectivity between trans-oceanic populations of *Capreolia implexa* (Gelidiales, Rhodophyta) in cool temperate waters of Australasia and Chile. Aquat Bot. 2014;119:73–9.
Brante A, Fernández M, Viard F. Phylogeography and biogeography concordance in the marine gastropod *Crepipatella dilatata* (Calyptraeidae) along the southeastern Pacific coast. J Hered. 2012;103(5):630–7.
Brattström H, Johanssen A. Ecological and regional zoogeography of the marine benthic fauna of Chile. Sarsia. 1983;68:289–339.
Briggs JC. Centrifugal speciation and centres of origin. J Biogeogr. 2000;27:1183–8.
Broitman BR, Navarrete SA, Smith F, Gaines SD. Geographic variation of southeastern Pacific intertidal communities. Mar Ecol Prog Ser. 2001;224:21–34.
Buschmann AH, Correa JA, Westermeier R, Hernández-González MC, Norambuena R. Red algal farming in Chile: a review. Aquaculture. 2001;194:203–20.
Buschmann AH, Prescott S, Potin P, Faugeron S, Vasquez JA, Camus C, Infante J, Hernández-González MC, Gutierrez A, Varela DA. The status of kelp exploitation and marine agronomy, with emphasis on *Macrocystis pyrifera*, in Chile. Adv Bot Res. 2014;71:161–88.
Camus PA. Procesos regionales y fitogeografía en el Pacífico Suroriental: el efecto de 'El Niño-Oscilación del Sur'. Rev Chil Hist Nat. 1990;63:11–7.
Camus PA. Biogeografía marina de Chile continental. Rev Chil Hist Nat. 2001;74(3):587–617.
Cárdenas L, Castilla JC, Viard F. A phylogeographical analysis across three biogeographical provinces of the south-eastern Pacific: the case of the marine gastropod *Concholepas concholepas*. J Biogeogr. 2009;36:969–81.
Castilla JC. Earthquake-caused coastal uplift and its effects on rocky intertidal kelp communities. Science. 1988;242(4877):440–3.
Castilla JC, Manriquez PH, Camaño A. Effects of rocky shore coseismic uplift and the 2010 Chilean mega-earthquake on intertidal biomarker species. Mar Ecol Prog Ser. 2010;418:17–23.
Crawford DJ. Progenitor-derivative species pairs and plant speciation. Taxon. 2010;59(5): 1413–23.
Crawford DJ, Landolt E, Les DH, Kimball RT. Speciation in duckweeds (Lemnaceae): phylogenetic and ecological inferences. Aliso. 2006;22(1):229–40.
Darwin C. Voyages of the adventure and Beagle, vol. III. London: H. Colburn; 1839. p. 617.
De Meester L, Gómez A, Okamura B, Schwenk K. The Monopolization Hypothesis and the dispersal–gene flow paradox in aquatic organisms. Acta Oecol. 2002;23:121–35.
Durrant HMS, Burridge CP, Kelaher BP, Barrett NS, Edgar GJ, Coleman MA. Implications of macroalgal isolation by distance for networks of marine protected areas. Cons Biol. 2014; 28(2):438–45.
Excoffier E, Ray N. Surfing during population expansions promotes genetic revolutions and structuration. Trends Ecol Evol. 2008;23(7):347–51.
Faugeron S, Valero M, Destombe C, Martínez EA, Correa JA. Hierarchical spatial structure and discriminant analysis of genetic diversity in the red alga *Mazzaella laminarioides* (Gigartinales, Rhodophyta). J Phycol. 2001;37:705–16.
Faugeron S, Martínez EA, Correa JA, Cardenas L, Destombe C, Valero M. Reduced genetic diversity and increased population differentiation in peripheral and over-harvested populations of *Gigartina skottsbergii* (Rhodophyta, Gigartinales) in southern Chile. J Phycol. 2004;40:454–62.

Faugeron S, Martínez EA, Correa JA, Billot C. Long-term copper mine waste disposal in northern Chile associated with gene flow disruption of the intertidal *Lessonia nigrescens*. Mar Ecol Prog Ser. 2005;288:129–40.

Fenberg PB, Menge BA, Raimondi PT, Rivadeneira MM. Biogeographic structure of the northeastern Pacific rocky intertidal: the role of upwelling and dispersal to drive patterns. Ecography. 2015;38(1):83–95.

Fernández M, Jaramillo E, Marquet PA, Moreno CA, Navarrete SA, Ojeda FP, Valdovinos CR, Vasquez JA. Diversity, dynamics and biogeography of Chilean benthic nearshore ecosystems: an overview and guidelines for conservation. Rev Chil Hist Nat. 2000;73:797–830.

Fraser CI. Change in Southern Hemisphere intertidal communities through climate cycles: the role of dispersing algae. In: Hu ZM, Fraser CI, editors. Seaweed phylogeography: adaptation and evolution of seaweeds under environmental change. Berlin: Springer; 2016.

Fraser CI, Nikula R, Spencer HG, Waters JM. Kelp genes reveal effects of subantarctic sea ice during the Last Glacial Maximum. Proc Nat Acad Sci USA. 2009;106:3249–53.

Fraser CI, Thiel M, Spencer HG, Wasters JM. Contemporary habitat discontinuity and historic glacial ice drive genetic divergence in Chilean kelp. BMC Evol Biol. 2010a;10:203.

Fraser CI, Winter DJ, Spencer HG, Waters JM. Multigene phylogeny of the southern bull-kelp genus *Durvillaea* (Phaeophyceae: Fucales). Mol Phylogenet Evol. 2010b;57:1301–11.

Fraser CI, Zuccarello GC, Spencer HG, Salvatore LC, Garcia GR, Waters JM. Genetic affinities between trans-oceanic populations of non-buoyant macroalgae in the high latitudes of the Southern Hemisphere. PLoS ONE. 2013;8(7):e69138.

Fuentes K, Brante A. Megadisturbances and ecological recovery in the intertidal zone: the role of interspecific competition between bioengineer mussel species. Mar Ecol Prog Ser. 2014;515:161–72.

Funk DJ, Omland KE. Species-level paraphyly and polyphyly: frequency, causes, and consequences, with insights from animal mitochondrial DNA. Ann Rev Ecol Evol Syst. 2003;34:397–423.

González A, Beltrán J, Hiriart-Bertrand L, Flores V, de Reviers B, Correa JA, Santelices B. Identification of cryptic species in the *Lessonia nigrescens* complex (Phaeophyceae, Laminariales). J Phycol. 2012;48:1153–65.

González-Wevar CA, Hüne M, Cañete JI, Mansilla A, Nakano T, Poulin E. Towards a model of postglacial biogeography in shallow marine species along the Patagonian Province: lessons from the limpet *Nacella magellanica* (Gmelin, 1791). BMC Evol Biol. 2012;12:139.

Gottlieb LD. Rethinking classic examples of recent speciation in plants. New Phytol. 2004;161:71–82.

Grossenbacher DL, Veloz SD, Sexton JP. Niche and range size patterns suggest that speciation begins in small, ecologically diverged populations in North American monkeyflowers (*Mimulus spp.*). Evolution. 2014;68:1270–80.

Guillemin M-L, Faugeron S, Destombe C, Viard F, Correa JA, Valero M. Genetic variation in wild and cultivated populations of the haploid-diploid red alga *Gracilaria chilensis*: how farming practices favor asexual reproduction and heterozygosity. Evolution. 2008;62:1500–19.

Guillemin M-L, Valero M, Faugeron S, Nelson W, Destombe C. Tracing the trans-Pacific evolutionary history of a domesticated seaweed (*Gracilaria chilensis*) with archaeological and genetic data. PLoS ONE. 2014;9(12):e114039.

Hallatschek O, Hersen P, Ramanathan S, Nelson DR. Genetic drift at expanding frontiers promotes gene segregation. Proc Nat Acad Sci USA. 2007;104(50):19926–30.

Haye PA, Muñoz-Herrera NC. Isolation with differentiation followed by expansion with admixture in the tunicate *Pyura chilensis*. BMC Evol Biol. 2013;13:252.

Haye PA, Salinas P, Acuña E, Poulin E. Heterochronic phenotypic plasticity with lack of genetic differentiation in the southeastern Pacific squat lobster *Pleuroncodes monodon*. Evol Dev. 2010;12(6):628–34.

Haye PA, Varela AI, Thiel M. Genetic signatures of rafting dispersal in algal-dwelling brooders *Limnoria* spp. (Isopoda) along the SE Pacific (Chile). Mar Ecol Prog Ser. 2012;455:111–22.

Haye PA, Segovia NI, Muñoz-Herrera NC, Gálvez FE, Martinez A, Meynard A, Pardo-Gandarillas MC, Poulin E, Faugeron S. Phylogeographic structure in benthic marine invertebrates of the Southeast Pacific coast of Chile with differing dispersal potential. PLoS ONE. 2014;9(2):e88613.

Hernández-Miranda E, Cisterna J, Díaz-Cabrera E, Veas R, Quinones RA. Epibenthic macrofaunal community response after a mega-earthquake and tsunami in a shallow bay off central-south Chile. Mar Biol. 2014;161(3):681–96.

Hormazabal S, Shaffer G, Leth O. Coastal transitions zone off Chile. J Geophys Res. 2004; 109:1–13.

Hulton N, Sugden D, Payne A, Clapperton C. Glacier modeling and the climate of Patagonia during the last glacial maximum. Quat Res. 1994;42(1):1–19.

Hulton NRJ, Purves RS, McCulloch RD, Sugden DE, Bentley MJ. The last glacial maximum and deglaciation in southern South America. Quat Sci Rev. 2002;21:233–41.

Jaramillo E, Dugan JE, Hubbard DM, Melnick D, Manzano M, Duarte C, Campos C, Sanchez R. Ecological implications of extreme events: footprints of the 2010 earthquake along the Chilean coast. PLoS ONE. 2012;7(5):e35348.

Kaiser J, Lamy F, Hebbeln D. A 70-kyr sea surface temperature record off southern Chile (Ocean Drilling Program Site 1233). Paleoceanography. 2005;20:PA4009.

Keith SA, Kerswell AP, Connolly SR. Global diversity of marine macroalgae: environmental conditions explain less variation in the tropics. Gobal Ecol Biogeogr. 2014;23:517–29.

Kerswell AP. Global biodiversity patterns of benthic marine algae. Ecology. 2006;87:2479–88.

Kinlan BP, Gaines SD. Propagule dispersal in marine and terrestrial environments: a community perspective. Ecology. 2003;84(8):2007–20.

Klopfstein S, Currat M, Excoffier L. The fate of mutations surfing on the wave of a range expansion. Mol Biol Evol. 2006;23(3):482–90.

Knowlton N. Sibling species in the sea. Ann Rev Ecol Syst. 1993;24:189–216.

Knowlton N. Molecular genetic analyses of species boundaries in the sea. Hydrobiologia. 2000;420(1):73–90.

Koch K, Thiel M, Tellier F, Hagen W, Graeve M, Tala F, Laeseke P, Bischof K. Species separation within the *Lessonia nigrescens* complex (Phaeophyceae, Laminariales) is mirrored by ecophysiological traits. Bot Mar. 2015;58(2):81–92.

Lachkar Z, Gruber N. A comparative study of biological production in eastern boundary upwelling systems using an artificial neural network. Biogeosciences. 2012;9(1):293–308.

Leliaert F, Verbruggen H, Vanormelingen P, Steen F, Lopez-Bautista JM, Zuccarello GC, De Clerck O. DNA-based species delimitation in algae. Eur J Phycol. 2014;49:179–96.

Lindstrom SC, Gabrielson PW, Hughey JR, Macaya EC, Nelson WA. Sequencing of historic and modern specimens reveals cryptic diversity in *Nothogenia* (Scinaiaceae, Rhodophyta). Phycologia. 2015;54(2):97–108.

Locarnini RA, Mishonov AV, Antonov JI, Boyer TP, Garcia HE, Branova HA, Zweng MM, Johnson DR. World Ocean Atlas 2009, vol. 1. Temperature, NOAA Atlas NESDIS. 2010, p. 68.

López-Cristoffanini C, Tellier F, Otaíza R, Correa JA, Contreras-Porcia L. Tolerance to air exposure: a feature driving the latitudinal distribution of two sibling kelp species. Bot Mar. 2013;56(5–6):431–40.

Luikart G, England PR, Tallmon D, Jordan S, Taberlet P. The power and promise of population genomics: from genotyping to genome typing. Nat Rev Genet. 2003;4:981–94.

Macaya EC, Zuccarello GC. DNA barcoding and genetic divergence in the giant kelp *Macrocystis* (Laminariales). J Phycol. 2010a;46:736–42.

Macaya EC, Zuccarello GC. Genetic structure of the giant kelp *Macrocystis pyrifera* along the southeastern Pacific. Mar Ecol Prog Ser. 2010b;420:103–12.

Macaya EC, Lopez B, Tala F, Tellier F, Thiel M. Flot and raft: role of buoyant seaweeds in the phylogeography and genetic structure of non-buoyant associated flora. In: Hu ZM, Fraser CI, editors. Seaweed phylogeography: adaptation and evolution of seaweeds under environmental change. Berlin: Springer; 2016.

Martin P, Zuccarello GC. Molecular phylogeny and timing of radiation in *Lessonia* (Phaeophyceae, Laminariales). Phycol Res. 2012;60(4):276–87.

Martínez EA, Cárdenas L, Pinto R. Recovery and genetic diversity of the intertidal kelp *Lessonia nigrescens* (Phaeophyceae) 20 years after El Niño 1982/1983. J Phycol. 2003;39(3):504–8.

Mayr E. Change of genetic environment and evolution. In: Huxley J, Hardy AC, Ford E, editors. Evolution as a process. London: Allen and Unwin; 1954. p. 157–80.

McCulloch RD, Bentley MJ, Purves RS, Hulton NRJ, Sugden DE, Clapperton CM. Climatic inferences from glacial and palaeoecological evidence at the last glacial termination, southern South America. J Quat Sci. 2000;15:409–17.

Meneses I, Santelices B. Patterns and breaking points in the distribution of benthic algae along the temperate Pacific coast of South America. Rev Chil Hist Nat. 2000;73(4):615–23.

Mohtadi M, Hebbeln D. Mechanisms and variations of the paleoproductivity off northern Chile (24°S–33°S) during the last 40,000 years. Paleoceanography. 2004;19:20–3.

Montecinos A, Broitman BR, Faugeron S, Haye PA, Tellier F, Guillemin ML. Species replacement along a linear coastal habitat: phylogeography and speciation in the red alga *Mazzaella laminarioides* along the South East Pacific. BMC Evol Biol. 2012;12(1):97.

Moy CM, Seltzer GO, Rodbell DT, Anderson DM. Variability of El Niño/Southern Oscillation activity at millennial timescales during the Holocene epoch. Nature. 2002;420(6912):162–5.

Neiva J, Pearson GA, Valero M, Serrão EA. Surfing the wave on a borrowed board: range expansion and spread of introgressed organellar genomes in the seaweed *Fucus ceranoides* L. Mol Ecol. 2010;19(21):4812–22.

Neiva J, Pearson GA, Valero M, Serrão EA. Fine-scale genetic breaks driven by historical range dynamics and ongoing density-barrier effects in the estuarine seaweed *Fucus ceranoides* L. BMC Evol Biol. 2012;12:78.

Neiva J, Serrão EA, Assis J, Pearson GA, Coyer JA, Olsen JL, Hoarau G, Valero M. Climate oscillations, range shifts and phylogeographic patterns of North Atlantic Fucaceae. In: Hu ZM, Fraser CI, editors. Seaweed phylogeography: adaptation and evolution of seaweeds under environmental change. Berlin: Springer; 2016.

Nelson WA. *Pyropia plicata* sp. nov. (Bangiales, Rhodophyta): naming a common intertidal alga from New Zealand. PhytoKeys. 2013;21:17–28.

Nuñez JJ, Wood NK, Rabanal FE, Fontanella FM, Sites JW Jr. Amphibian phylogeography in the Antipodes: Refugia and postglacial colonization explain mitochondrial haplotype distribution in the Patagonian frog *Eupsophus calcaratus* (Cycloramphidae). Mol Phylogenet Evol. 2011;58(2):343–52.

Oppliger VL, Correa JA, Faugeron S, Beltran J, Tellier F, Valero M, Destombe C. Sex ratio variation in the *Lessonia nigrescens* complex (Laminariales, Phaeophyceae): effect of latitude, temperature, and marginality. J Phycol. 2011;47(1):5–12.

Oppliger LV, Correa JA, Engelen AH, Tellier F, Vieira V, Faugeron S, Valero M, Gomez G, Destombe C. Temperature effects on gametophyte life-history traits and geographic distribution of two cryptic kelp species. PLoS ONE. 2012;7(6):e39289.

Ortega KJ, Sáez CA, Macaya EC. Changes in invertebrate assemblages inhabiting *Lessonia spicata* (Phaeophyceae) holdfasts after the 2010 earthquake-mediated coastal uplift in Chile. Rev Biol Mar Oceanogr. 2014;49(1):129–34.

Pardo C, Lopez L, Peña V, Hernández-Kantun J, Gall LL, Bárbara I, Barreiro R. A multilocus species delimitation reveals a striking number of species of coralline algae forming maerl in the OSPAR maritime area. PLoS ONE. 2014;9(8):e104073.

Payo DA, Leliaert F, Verbruggen H, D'hondt S, Calumpong HP, De Clerck O. Extensive cryptic species diversity and fine-scale endemism in the marine red alga *Portieria* in the Philippines. Proc R Soc B. 2013;280:20122660.

Peters AF, Mann AD, Córdova CA, Brodie J, Correa JA, Schroeder DC, Cock JM. Genetic diversity of *Ectocarpus* (Ectocarpales, Phaeophyceae) in Peru and northern Chile, the area of origin of the genome-sequenced strain. New Phytol. 2010;188:30–41.

Ponce JF, Rabassa J, Coronato A, Borromei AM. Palaeogeographical evolution of the Atlantic coast of Pampa and Patagonia from the last glacial maximum to the Middle Holocene. Biol J Linn Soc. 2011;103(2):363–79.

Pyron RA, Burbrink FT. Can the tropical conservatism hypothesis explain temperate species richness patterns? An inverse latitudinal biodiversity gradient in the NewWorld snake tribe Lampropeltini. Global Ecol Biogeogr. 2009;18:406–15.

Ramírez ME. Flora marina bentónica de la región Austral de Sudamérica y la Antártica. Anales Instituto Patagonia. 2010;38:57–71.

Ramírez ME, Contreras-Porcia L, Guillemin ML, Brodie J, Valdivia C, Flores-Molina MR, Nunez A, Contador CB, Lovazzano C. *Pyropia orbicularis* sp. nov. (Rhodophyta, Bangiaceae) based on a population previously known as *Porphyra columbina* from the central coast of Chile. Phytotaxa. 2014;158(2):133–53.

Riddle BR, Hafner DJ. Species as units of analysis in ecology and biogeography: time to take the blinders off. Global Ecol Biogeogr. 1999;8(6):433–41.

Ruzzante DE, Walde SJ, Gosse JC, Cussac VE, Habit E, Zemlak TS, Adams EDM. Climate control on ancestral population dynamics: insight from Patagonian fish phylogeography. Mol Ecol. 2008;17:2234–44.

Sánchez R, Sepúlveda RD, Brante A, Cristina CTL. Spatial pattern of genetic and morphological diversity in the direct developer *Acanthina monodon* (Gastropoda: Mollusca). Mar Ecol Prog Ser. 2011;434:121–31.

Santelices B. Phytogeographic characterization of the temperate coast of Pacific South America. Phycologia. 1980;19(1):1–12.

Santelices B. Patterns of reproduction, dispersal and recruitment in seaweeds. Oceanogr Mar Bio. 1990;128:177–276.

Santelices B, Marquet P. Seaweeds, latitudinal diversity patterns, and Rapoport's rule. Divers Distrib. 1998;4(2):71–5.

Santelices B, Meneses I. A reassessment of the phytogeographic characterization of temperate Pacific South America. Rev Chil Hist Nat. 2000;73:605–14.

Saunders GW, McDevit DC. DNA barcoding unmasks overlooked diversity improving knowledge on the composition and origins of the Churchill algal flora. BMC Ecol. 2013;13:9.

Sérsic AN, Cosacov A, Cocucci AA, Johnson LA, Pozner R, Avila LJ, Sites JW Jr, Morando M. Emerging phylogeographical patterns of plants and terrestrial vertebrates from Patagonia. Biol J Linn Soc. 2011;103(2):475–94.

Strub PT, Mesías JM, Montecino V, Rutlant J, Salinas S. Coastal ocean circulation off western South America. In: Brink KH, Robinson AR, editors. The global coastal ocean. New York: Wiley; 1998. p. 273–313.

Tapia FJ, Largier JL, Castillo M, Wieters EA, Navarrete SA. Latitudinal discontinuity in thermal conditions along the nearshore of Central-Northern Chile. PLoS ONE. 2014;9(10):e110841.

Tarazona J, Arntz W. The Peruvian coastal upwelling system. In: Seeliger U, Kjerfve B, editors. Coastal marine ecosystems of Latin America. Ecological studies, vol. 144. Berlin: Springer; 2001. p. 229–44.

Tellier F, Meynard AP, Correa JA, Faugeron S, Valero M. Phylogeographic analyses of the 30°S south-east Pacific biogeographic transition zone establish the occurrence of a sharp genetic discontinuity in the kelp *Lessonia nigrescens*: Vicariance or parapatry? Mol Phylogenet Evol. 2009;53(3):679–93.

Tellier F, Tapia J, Faugeron S, Destombe C, Valero M. The *Lessonia nigrescens* complex (Laminariales, Phaeophyceae) shows strict parapatry and complete reproductive isolation in a secondary contact zone. J Phycol. 2011a;47:894–903.

Tellier F, Vega JMA, Broitman BR, Vasquez JA, Valero M, Faugeron S. The importance of having two species instead of one in kelp management: the *Lessonia nigrescens* species complex. Cah Biol Mar. 2011b;52:455–65.

Thiel M, Macaya EC, Acuña E, Arntz WE, Bastias H, Brokordt K, Camus P, Castilla JC, Castro L, Cortes M, Dumont CP, Escribano R, Fernandez M, Gajardo JA, Gaymer CF, Gomez I, Gonzalez AE, Gonzalez HE, Haye PA, Illanes JE, Iriarte JL, Lancellotti DA, Luna-Jorquera G,

Luxoro C, Manriquez PH, Marin V, Munoz P, Navarrete SA, Perez E, Poulin E, Sellanes J, Hito Sepulveda H, Stotz W, Tala F, Thomas A, Vargas C, Vasquez JA, Vega JMA. The Humboldt Current System of Northern and Central Chile: oceanographic processes, ecological interactions and socioeconomic feedback. Oceanogr Mar Biol. 2007;45:195–344.

Tronholm A, Leliaert F, Sansón M, Afonso-Carrillo J, Tyberghein L, Verbruggen H, De Clerk O. Contrasting geographical distributions as a result of thermal tolerance and long-distance dispersal in two allegedly widespread tropical brown algae. PLoS ONE. 2012;7:e30813.

Valero M, Destombe C, Mauger S, Ribout C, Engel CR, Daguin-Thiebaut C, Tellier F. Using genetic tools for sustainable management of kelps: a literature review and the example of *Laminaria digitata*. Cah Biol Mar. 2011;52(4):467–83.

Varela AI, Haye PA. The marine brooder *Excirolana braziliensis* (Crustacea: Isopoda) is also a complex of cryptic species on the coast of Chile. Rev Chil Hist Nat. 2012;85:495–502.

Vasquez. Production, use and fate of Chilean brown seaweeds: re-sources for a sustainable fishery. J App Phycol. 2008;20:457–67.

Verleye TJ, Louwye S. Late Quaternary environmental changes and latitudinal shifts of the Antarctic Circumpolar Current as recorded by dinoflagellate cysts from offshore Chile (41°S). Quat Sci Rev. 2010;29(7):1025–39.

Vianna JA, Medina-Vogel G, Chehébar C, Sielfeld W, Olavarria C, Faugeron S. Phylogeography of the Patagonian otter *Lontra provocax*: adaptive divergence to marine habitat or signature of southern glacial refugia? BMC Evol Biol. 2011;11:53.

Vidal R, Meneses I, Smith M. Phylogeography of the genus *Spongites* (Corallinales, Rhodophyta) from Chile. J Phycol. 2008;44:173–82.

Vieira C, D'hondt S, De Clerck O, Payri CE. Toward an inordinate fondness for stars, beetles and *Lobophora*? Species diversity of the genus *Lobophora* (Dictyotales, Phaeophyceae) in New Caledonia. J Phycol. 2014;50(6):1101–19.

Vieira VM, Oppliger LV, Engelen AH, Correa JA. A new method to quantify and compare the multiple components of fitness—A study case with Kelp niche partition by divergent microstage adaptations to temperature. PLoS ONE. 2015;10(3):e0119670.

Walser B, Haag CR. Strong intraspecific variation in genetic diversity and genetic differentiation in *Daphnia magna*: the effects of population turnover and population size. Mol Ecol. 2012;21:851–61.

Waters JM, Fraser CI, Hewitt GM. Founder takes all: density-dependent processes structure biodiversity. Trends Ecol Evol. 2013;28(2):78–85.

Weinberger F, Guillemin M-L, Destombe C, Valero M, Faugeron S, Correa JA, Pohnert G, Pehlke C, Kloareg B, Potin P. Defense evolution in the Gracilariaceae (Rhodophyta): substrate-regulated oxidation of agar oligosaccharides is more ancient than the oligoagar-activated oxidative burst. J Phycol. 2010;46(5):958–68.

Xu J, Pérez-Losada M, Jara CG, Crandall KA. Pleistocene glaciation leaves deep signature on the freshwater crab *Aegla alacalufi* in Chilean Patagonia. Mol Ecol. 2009;18:904–18.

Zakas C, Binford J, Navarrete SA, Wares JP. Restricted gene flow in Chilean barnacles reflects an oceanographic and biogeographic transition zone. Mar Ecol Prog Ser. 2009;394:165–77.

Zemlak TS, Habit EM, Walde SJ, Carrea C, Ruzzante DE. Surviving historical Patagonian landscapes and climate: molecular insights from *Galaxias maculatus*. BMC Evol Biol. 2010;10:67.

Chapter 11
Climate Oscillations, Range Shifts and Phylogeographic Patterns of North Atlantic Fucaceae

João Neiva, Ester A. Serrão, Jorge Assis, Gareth A. Pearson, James A. Coyer, Jeanine L. Olsen, Galice Hoarau and Myriam Valero

Abstract Members of the seaweed family Fucaceae have been recurrent models in North Atlantic phylogeographic research; numerous studies have been published since 2000, and this review synthesizes their major findings. Fucoid species exhibited diverse responses to glacial–interglacial cycles, but evidence indicates there were a few common refugial areas such as north-western Iberia, the Celtic Sea (Brittany/Ireland) region and the North-west Atlantic. In genetically rich refugial areas, pervasive genetic breaks confirmed presently limited gene flow between adjacent distinct genetic groups. In contrast with the maintenance of sharp genetic breaks, most species experienced extensive migration during post-glacial expansion. Poleward migrations in the North-east Atlantic followed routes along north-western Ireland and the transgressing English Channel. These patterns support the role of density-blocking in maintaining sharp genetic breaks at contact zones, and of long-distance dispersal from range edges in mediating expansion into uninhabited regions. The data also indicate that expansions involve mostly the genetic groups located at range edges rather than the entire species' gene pool, both poleward during interglacials and toward warmer regions during glacial periods. Fucoid expansions

J. Neiva (✉) · E.A. Serrão · J. Assis · G.A. Pearson
CCMAR—Centro de Ciências do Mar, Universidade do Algarve, Gambelas, 8005-139 Faro, Portugal
e-mail: jmeiva@ualg.pt

J.A. Coyer
Shoals Marine Laboratory, Cornell University, Portsmouth, NH 03801, USA

J.L. Olsen
Groningen Institute for Evolutionary Life Sciences, 9747 AG, Groningen, The Netherlands

G. Hoarau
Faculty of Biosciences and Aquaculture, University of Nordland, 8049 Bodø, Norway

M. Valero
Evolutionary Biology and Ecology of Algae, UMI 3614, UPMC University Paris VI, CNRS, Sorbonne Universités, Station Biologique de Roscoff, CS 90074, Place G. Teissier, 29680 Roscoff, France
e-mail: myriam.valero@sb-roscoff.fr

Z.-M. Hu and C. Fraser (eds.), *Seaweed Phylogeography*,
DOI 10.1007/978-94-017-7534-2_11

have also been linked to introgressive recombination of genomes at (and beyond) contact zones and to gene surfing leading to present large-scale dominance by alleles that were located at the expanding edge. Phylogeographic approaches have also proven useful to identify and track the sources of introductions linked to marine traffic. The integration of environmental niche models with molecular data have further allowed hindcasting southern distributions during glaciation and predicting the potentially negative effects of future climate warming, including the loss of vulnerable, unique trailing-edge lineages, as species' ranges are predicted to continue shifting northward. Collectively, these studies have contributed greatly to elucidating the links between past and ongoing climatic shifts, range dynamics and geographical patterns of genetic variability in the North Atlantic.

Keywords North-Atlantic intertidal · Climatic refugia · Comparative phylogeography · Fucoid seaweeds · Genetic diversity · Latitudinal range shifts · Marine introductions · Glacial–interglacial cycles · Ongoing climate change · Pathways of range expansion · Rafting · Restricted dispersal

11.1 Introduction

The large climatic shifts experienced across the Quaternary glacial–interglacial cycles (2.5 Ma to 12 ka) have played a major role in shaping the modern distributions and genetic make-ups of extant species (Hofreiter and Stewart 2009). In the Northern Hemisphere, climatic, paleontological and molecular evidence show that glacial advances commonly pressed temperate organisms into periglacial or more southern refugial areas (Stewart et al. 2010), sometimes in the form of small and scattered populations (Stewart and Lister 2001). During interglacials, such as the current Holocene period (12 ka–present), warming climate has eventually allowed subsets of species (and specifically subsets of populations within species) to vastly expand their ranges northward (Petit et al. 2002; Bennett and Provan 2008; Provan 2013), sometimes associated with the loss (Hampe and Petit 2005) or displacement (e.g. Davis and Shaw 2001) of the southern range margins.

Coastal organisms inhabiting temperate latitudes of the North Atlantic (*sensu* Spalding et al. 2007) have been particularly impacted by these global climatic shifts. Understanding responses of these species has been limited by the absence and/or inaccessibility of adequate fossil records (particularly from colder periods, when sea levels were lower), but recent phylogeographic studies (typically based on mtDNA and often complemented with nuclear microsatellite loci) have provided new insights. Indeed, a diverse range of taxa including coastal invertebrates (Roman and Palumbi 2004; Kelly et al. 2006; Remerie et al. 2009; Handschumacher et al. 2010), fishes (Wilson and Eigenmann Veraguth 2010; Kettle et al. 2011; Woodall et al. 2011), seagrasses (Coyer et al. 2004a; Olsen et al. 2004; Alberto et al. 2008) and seaweeds (red algae: Provan et al. 2005b; Hu et al. 2010, 2011; Provan and Maggs

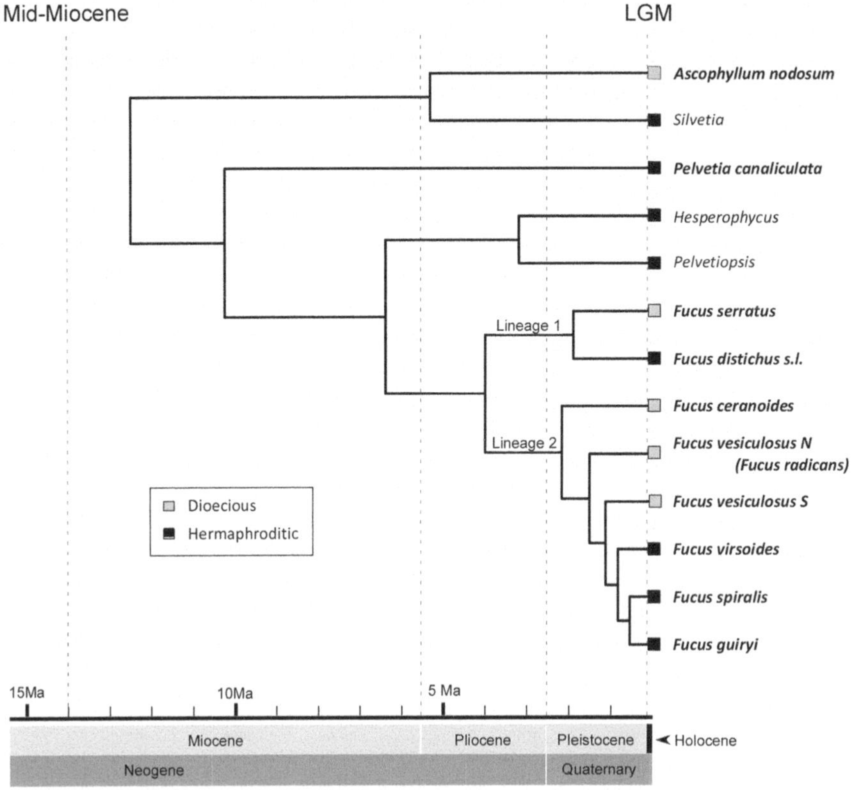

Fig. 11.1 Simplified Bayesian phylogenetic tree of Fucaceae based on 13 coding loci, with branches proportional to divergence time and with Atlantic taxa (and respective mating system) highlighted. Adapted from Cánovas et al. (2011), where detailed reconstruction methods and node age estimates are discussed. *Fucus radicans* is part of the clade *F. vesiculosus* N.; *Fucus distichus s.l.* includes *F. gardneri* and *F. evanescens*

2012; brown algae: Assis et al. 2013) have been surveyed in the last decade for phylogeographic patterns in the North Atlantic alone, advancing considerably our understanding of past distributions and (in some cases) demographic histories of species in this region (Maggs et al. 2008). Members of the brown algal family Fucaceae (Fucales, Phaeophyceae, Stramenopiles; Fig. 11.1) and several red algal taxa (Rhodophyceae; see chapter by Li et al. (2016) in this volume) have been particularly fruitful research models in marine phylogeography of the North Atlantic, contributing to elucidating the links between climatic history, range dynamics and their consequences for genetic variability across species ranges in the region. This prominence stems from their ecology, life history and dispersal capacities.

Fucoids (here focused on the family Fucaceae) play a fundamental and important ecological role in intertidal habitats, where they form conspicuous ecosystem-structuring assemblages on rocky shores and in some estuarine and salt

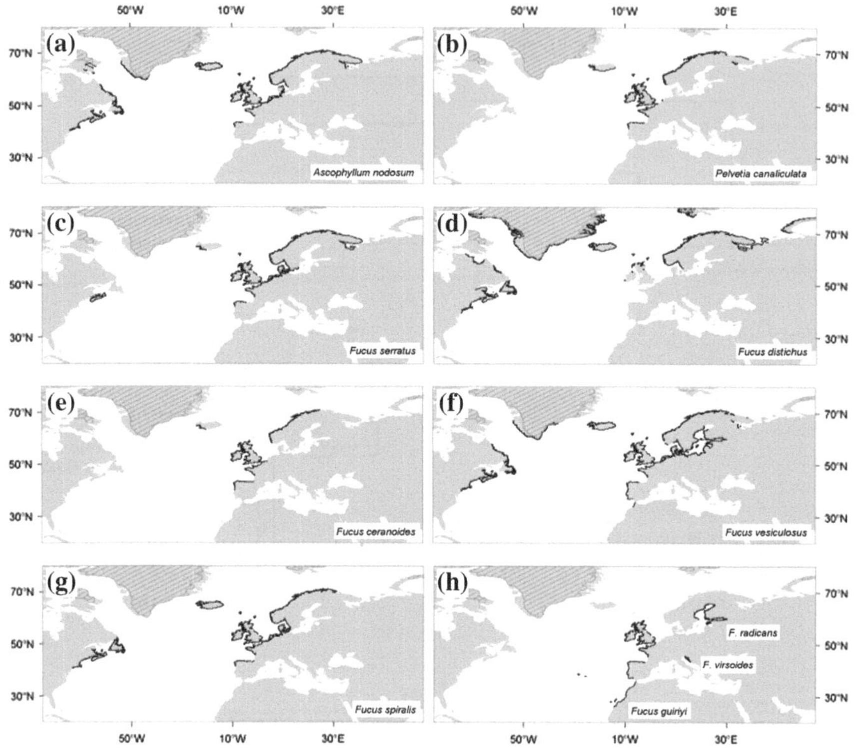

Fig. 11.2 Distribution of Fucaceae species in the North Atlantic. *Dashed areas* depict permanent ice. The distributions of *Fucus distichus* (native all around the Arctic) and *F. spiralis* also include the North Pacific, although the latter appear to have been recently introduced there

marsh environments (Lüning 1990; Chapman 1995). Like kelps (Laminariales and ecologically equivalent members of the order Tilopteridales) and the related subtidal fucoid forests (e.g. Cystoseiraceae, Sargassaceae), intertidal fucoid beds provide habitat structure, energy, food and shelter for other organisms, as well as relief against tidal/seasonal variations in thermal and desiccation stress (Schiel and Foster 2006; Christie et al. 2009; Dijkstra et al. 2012). In the North Atlantic, three genera and *c.* 10 species are present and all but one (*F. distichus sensu lato*) are endemic to the region (Fig. 11.2). *Ascophyllum nodosum* and *Pelvetia canaliculata* are representatives of monotypic genera, whereas *Fucus* comprises *ca.* eight species originating in a dynamic radiation (Serrão et al. 1999; Coyer et al. 2006a; Cánovas et al. 2011) that continues in contemporary times [e.g. speciation of Baltic *F. radicans* in the past 2000–400 years (Pereyra et al. 2009)]. *Fucus distichus*, comprising multiple regional entities of uncertain taxonomic validity, is the only species naturally occurring in both the Pacific and Atlantic, and is also the species extending deepest into the Arctic (Coyer et al. 2011b). It can be hypothesized that the Atlantic monotypic genera (*Ascophyllum*, *Pelvetia*) are the only surviving lineages of past radiations followed by extinctions, similar to the monotypic *Pelvetiopsis* and *Hesperophycus* in the Pacific. Lineage 2 of *Fucus* (Serrão et al. 1999) has greatly diversified along eastern Atlantic shorelines, as has the still poorly studied *Fucus distichus sensu lato* along the Pacific American coast. It can be hypothesized that the newly available European Atlantic habitats might have favoured diversification after the split from the most recent common ancestor with its sister Pacific lineage (Coyer et al. 2006a; Cánovas et al. 2011).

In sharp contrast with marine organisms that exhibit extensive dispersal by planktonic stages, fucoids are characterized by spatially restricted dispersal. They are sessile and perennial diplonts with direct development; free-living stages are restricted to short-lived gametes liberated in gametangia (antheridia and oogonia) that immediately sink and are preferentially released under low water motion (Serrão et al. 1996). Gametes typically do not disperse more than a few metres from the broadcasting parent (Chapman 1995; Serrão et al. 1997; Dudgeon et al. 2001). Rafting by floating thalli may extend the dispersal ranges by several orders of magnitude, and although in general its relevance for genetic connectivity among established populations is perceived to be modest (Fraser et al. 2009; Neiva et al. 2012b, c; Waters et al. 2013), rafting is nonetheless likely to have played an important role in large-scale range expansions, such as post-glacial northern recolonizations. Short-distance dispersal (SDD), as a reproductive trait, implies that fucoid populations persist to a significant extent via local recruitment (Pearson and Serrão 2006), making them particularly susceptible to inbreeding and intra-/inter-specific hybridization (Coyer et al. 2002, 2007, 2011a; Engel et al. 2005; Moalic et al. 2011). SDD coupled with coastal patchiness also causes isolation, small effective population size (Coyer et al. 2008) and density-barrier effects (Neiva et al. 2012c), and facilitates genetic bottlenecks, allelic surfing (spread of particular alleles following chance increase or even fixation at the leading edge of expansions) and introgression during range expansions (Neiva et al. 2010). In short, SDD accentuates population structure, providing an historical footprint of change that can be easily traced (e.g. Hoarau et al. 2007; Assis et al. 2014). In addition, the strong high- to low-intertidal zonation at the local scale (habitat replication) and the natural warm temperate to cold temperate gradients of sea surface temperature (SST) (a phylogeographic cline), provide a natural laboratory for studying ecological speciation and the impacts of large-scale climatic changes for species exhibiting similar/contrasting biogeographical affinities (species "replication") (Lüning 1990; Chapman 1995).

In this review we summarize and synthesize the major findings of phylogeographic studies of Atlantic Fucaceae. We: (1) highlight North Atlantic paleoclimatic history, identify refugia and discuss the principal pathways of post-LGM recolonization that have shaped current distribution patterns; (2) discuss (un)intentional historical introductions of fucoids; and (3) discuss current climate change and modelled dynamics of the projected future ranges of North Atlantic fucoids.

11.2 Climate-Driven Range Dynamics of North Atlantic Fucoids

11.2.1 Glaciations Affect SST, Sea level and Habitat Availability

The last of many Quaternary glaciations, the Last Glacial Maximum (LGM, 0.026–0.019 Ma) was characterized by massive ice sheets and permafrost belts covering large parts of what is now temperate Eurasia and North America (Fig. 11.3a). On the Atlantic European side, the Eurasian ice sheet reached as far south as the Brittany coast of France and the global marine regression (sea levels were lower by as much as 130 m; Lambeck et al. 2015) resulted in the complete emersion of many shallow seas, including the North, Celtic and Irish Seas, and also the English Channel. The latter was a land bridge between continental Europe and the British Isles and was crossed by a large paleo-river draining most of the regional rivers (e.g. Seine, Thames and Rhine) as well as large quantities of melting ice produced by the declining Eurasian ice sheet at the onset of the present interglacial (Gibbard and Lautridou 2003; Ménot et al. 2006). The huge shoreline displacements (hundreds of km in some cases) between glacial and interglacial conditions regularly transfigured the coastal geography of this region, and have been associated with major ecological shifts elsewhere (Graham et al. 2003).

On the Atlantic North American side, the Laurentide ice sheet covered almost all of the Canadian shoreline and stretched as far south as Cape Cod (USA), although small coastal regions (present offshore seamounts) remained relatively ice-free (Carlson and Winsor 2012). Because of compressed SST isotherms along the American coast and the absence of rocky coastlines south of Cape Cod, substantial extinction of the pre-glacial biota undoubtedly occurred, differing from the splayed

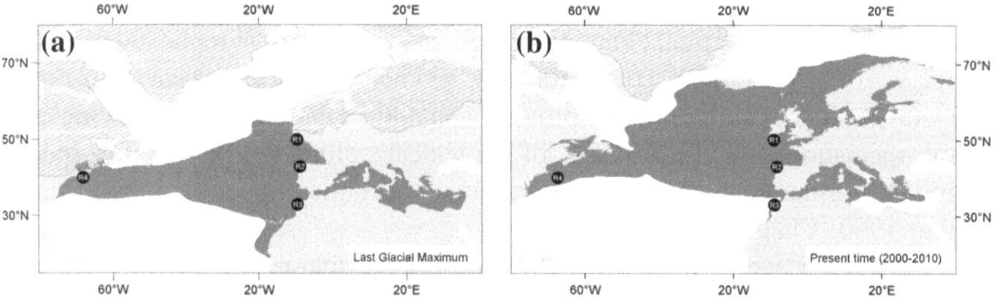

Fig. 11.3 Schematic representation of the North Atlantic during (**a**) the LGM, and (**b**) the present, depicting the Laurentide, Greenland and Eurasian ice sheets (stripped areas), landmasses (*light grey*) and "temperate" oceanographic areas (*dark grey*). The oceanographic area corresponds to average winter SSTs temperatures below 18 °C and average summer temperatures above 8 °C. *Note* the broader latitudinal range of temperate conditions along the North-east Atlantic/European coast as compared with the narrow isotherms on the North-west Atlantic/North American coast. Also note the northward shift since the LGM (*c.* 20 ka ago). R1: Brittany/Ireland refugia; R2: northern Iberia; R3: south-western Iberia/Morocco; R4: North-west Atlantic

SST isotherms and plentiful rocky shore habitat along the European coastlines (Fig. 11.3). The lower species diversity in the North-west Atlantic is supported by molecular studies on numerous benthic organisms and further suggests recent post-glacial (re)colonizations from Europe by some (but not all) amphi-Atlantic organisms (Wares and Cunningham 2001; Ilves et al. 2010; Waltari and Hickerson 2013). In any case, due to the tempering effect of the Gulf Stream, suitable areas for temperate rocky intertidal species have been latitudinally wider along the North-east versus the North-west Atlantic across both full glacial and interglacial conditions and have migrated northwards between the two periods (Fig. 11.3).

11.2.2 Southern Glacial Ranges, Post-Glacial Range Shifts and Glacial Pockets

Seaweed species are especially sensitive and responsive to climatic perturbations at their warmer trailing edges (Wernberg et al. 2011a; Bartsch et al. 2012; Duarte et al. 2013; Nicastro et al. 2013; Oppliger et al. 2014). The lack of a fossil record and the absence of the species today make it difficult to confirm the extent of the southern ranges of fucoids during the last glacial period. Given the southern shift of the isotherms, North-west Africa and some of the Atlantic islands (e.g. Canaries, Azores) could have been colonized during past colder periods, although evidence today is only available through hindcasting using ecological niche models (ENMs) for *Pelvetia canaliculata* (Neiva et al. 2014) and *Fucus vesiculosus* (Assis et al. 2014) (Fig. 11.4). Likewise, evidence for more extensive northern ranges can also be inferred where land masses protruded from the iceless coastal pockets (e.g. offshore Newfoundland). Southern expansions during cold periods have also been inferred in other coastal seaweeds and invertebrates (Kettle et al. 2011; Provan and Maggs 2012; Waltari and Hickerson 2013) and may well be representative of temperate organisms in general.

Long-term biogeographic distributions of Atlantic fucoid seaweeds are expected to consist of: (1) previously favourable (during glacial periods) but presently unsuitable or increasingly marginal southern areas, where species have become extirpated or currently persist as isolated, trailing edge climatic relicts; (2) more central regions where suitable climatic conditions have allowed a more stable presence across glacial/interglacial cycles (i.e. climatic refugia); and (3) previously glaciated/emerged regions that were colonized post-glacially following ice sheet retreat and sea level rise. For low-dispersal marine species such as fucoids, higher regional genetic diversity and uniqueness (as a proxy for long-term persistence and isolation) is expected where glacial and interglacial distributions overlap (Assis et al. 2014). In the North Atlantic, several of these putative long-term climatic refugia have been identified, including distinct areas within the larger paleo-Celtic Sea/Channel area, north-western Iberia, and for a smaller set of species, central-southern Iberia and North-west Africa (Fig. 11.3). Evidence also suggests

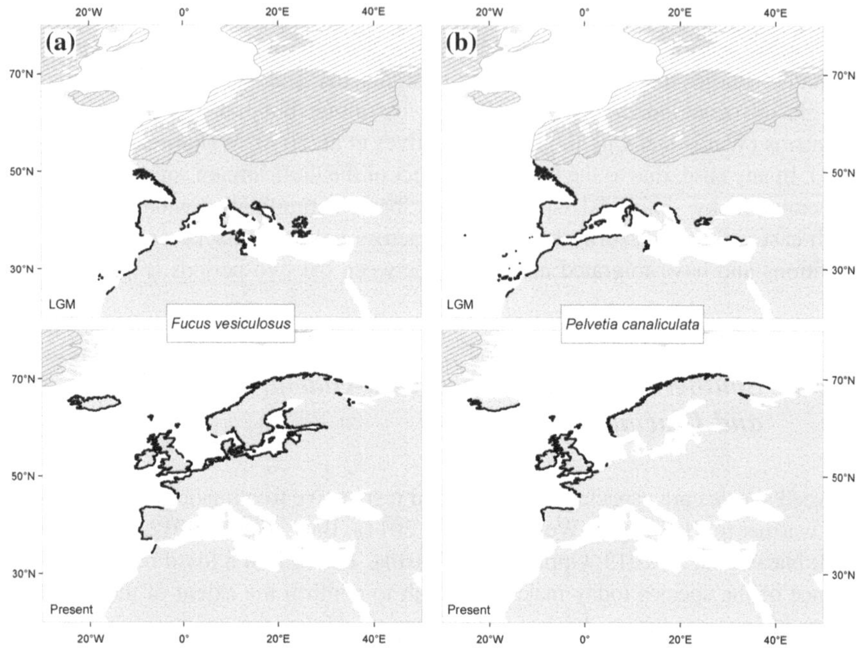

Fig. 11.4 Modelled distributions and latitudinal range shifts experienced by (**a**) *Fucus vesiculosus* [adapted from Assis et al. (2014)] and (**b**) *Pelvetia canaliculata* [adapted from Neiva et al. (2014)], between the LGM (*top*) and the present (*bottom*)

that Iceland and the Andøya region of northern Norway are and were refugia for the colder adapted *F. distichus* (Coyer et al. 2011b).

11.2.3 Glacial Refugia

11.2.3.1 Brittany and South-Western Ireland

One of the earliest suggestions of a refugium for the Fucaceae was the exceptionally high level of microsatellite allelic diversity found in populations of *Fucus serratus* in the Brittany region of France relative to other areas throughout the North Atlantic (Coyer et al. 2003). Subsequent work indicated that the ancient Hurd Deep (a canyon and river in the present-day English Channel), and an inland sea between Brittany and south-western Ireland were glacial refugia for marine species such as seaweeds (Provan et al. 2005b), invertebrates (Jolly et al. 2006) and fish (Finnegan et al. 2013). In addition to *F. serratus*, several other fucoids that have high genetic diversity in the Brittany and south-western Ireland periglacial refugia are *A. nodosum* (Olsen et al. 2010), *F. vesiculosus* (Coyer et al. 2011a; Assis et al. 2014) and *P. canaliculata* (Neiva et al. 2014). For these two last species, as well as for *F. ceranoides*

(Neiva et al. 2010) and *F. spiralis* (Coyer et al. 2011a), these refugia harboured several well-differentiated phylogeographic groups (Figs. 11.5, 11.6 and 11.7). For *F.* serratus, two phylogeographic groups have contributed to distinct phases (via different routes) of the global post-glacial expansion (Hoarau et al. 2007) (Fig. 11.5a). Patterns are more complex for the other species. For *F. vesiculosus*, ENMs show that the combined refugia formed the largest area of persistent habitat in the Atlantic. This proxy for long-term size of favourable habitat explained the high diversity of *F. vesiculosus* in this region (Assis et al. 2014). Brittany was clearly a refugium for *A. nodosum*, but unlike the other fucoid species, there is no evidence for additional refugia in Iberia (Fig. 11.6). In general, genetic data confirm that the

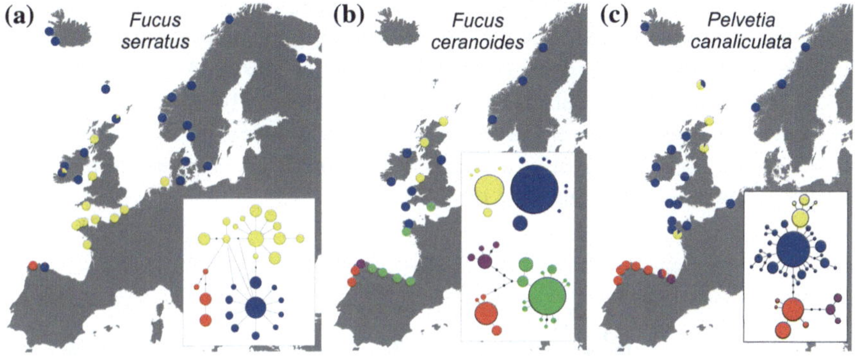

Fig. 11.5 MtIGS haplotype genealogies and geographical distribution of major inferred phylogroups in (**a**) *Fucus serratus*, (**b**) *F. ceranoides* and (**c**) *Pelvetia canaliculata*. Modified from Hoarau et al. (2007) and Neiva et al. (2010, 2014)

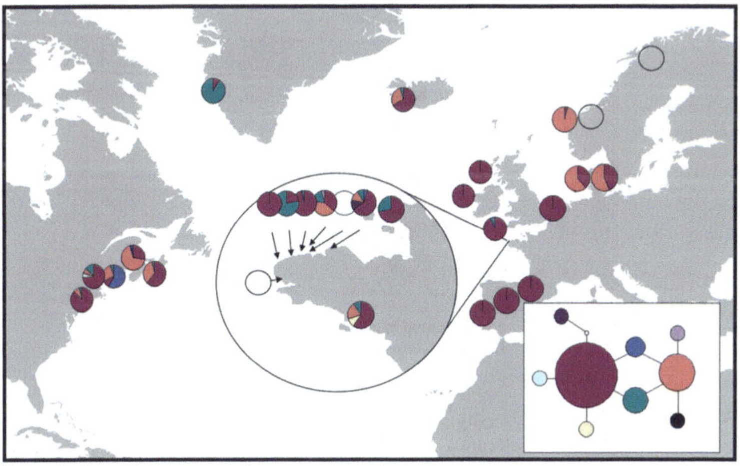

Fig. 11.6 Haplotype distribution and haplotype network for the combined IGS–*trn*W locus of *Ascophyllum nodosum. Grey-shaded pie* charts indicate that samples were missing from that location. Modified from Olsen et al. (2010)

colonization of coastlines north of Brittany and south-western Ireland relied on populations originating in this general periglacial region, with little contribution from Iberia.

The Celtic Sea/Brittany area is a biogeographical transition zone between the Northern European Seas and Lusitanian provinces within the "temperate Northern Atlantic" realm (Spalding et al. 2007). In this contact zone, higher genetic diversity could also be explained by complex patterns of secondary contact and admixture of populations expanding from distinct refugia (Maggs et al. 2008). The occurrence of tension zones with mosaically distributed hybrid zones has been documented for different *Mytilus* species in this area (Bierne et al. 2003). For several *Fucus* species, hybridization has also profoundly shaped patterns of gene diversity in this contact zone. Conflicting nuclear and organellar genomes were observed and explained by organellar sweeps caused by past hybridization and introgression between *F. vesiculosus*, *F. ceranoides* and species of the *F. spiralis/F. guiryi* complex (Engel et al. 2005; Coyer et al. 2006a, 2011a; Neiva et al. 2010; Moalic et al. 2011). For these, patterns of genetic diversity are more complex because post-glacial recolonization did not involve a single expansion from southern refugia and because of inter-specific gene flow (Fig. 11.7).

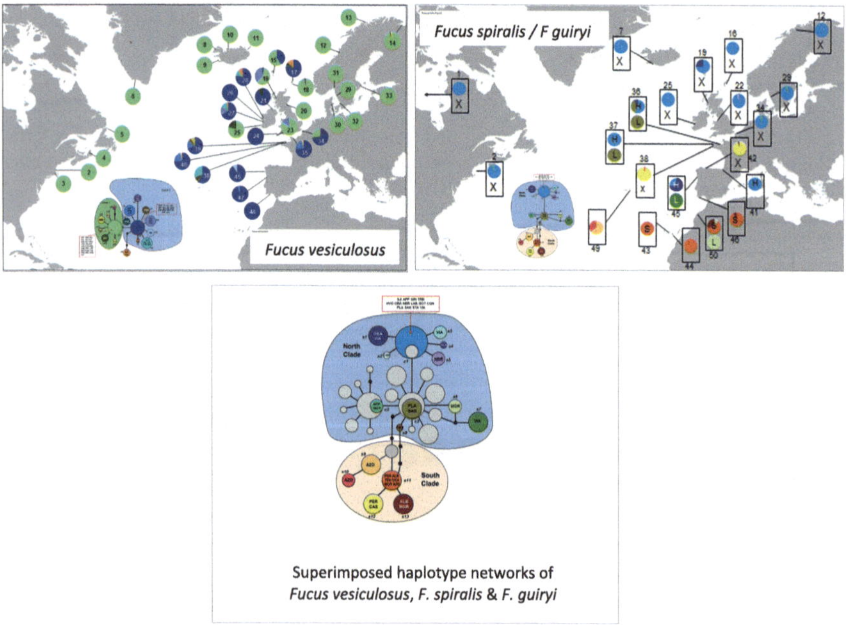

Fig. 11.7 MtDNA haplotype diversity and inset haplotype network for *F. vesiculosus* (*left*) and *F. spiralis/F. guiryi* (*right*). Haplotypes within boxes represent haplotypes collected in low and high intertidal zones at a single site, X indicates not collected/not found. Introgression of haplotypes (*bottom*) for *F. spiralis* (*blue* and *lilac colours*), introgressed *F. guiryi* (*brown*, *green*), pure *F. guiryi* ("South clade", *yellow*, *orange*, *red*), and *F. vesiculosus* (ghosted in *grey*) also based on microsatellite identification. Modified from Coyer et al. (2011a)

11.2.3.2 North-western Iberia

North-western Iberia represents the southern range limit (in some cases already as isolated climatic relicts) for a large number of cold-temperate seaweeds characteristic of present northern European shores. Examples include rhodophytes (e.g. *Palmaria palmata*), kelps (e.g. *Laminaria hyperborea*, *Saccharina latissima*) and fucoids (e.g. *Himanthalia elongata*, *Halidrys siliquosa*), including also the fucoids *A. nodosum*, *P. canaliculata*, *F. serratus*, *F. ceranoides* and *F. spiralis* (Fig. 11.2; Bárbara et al. 2005; Araújo et al. 2009). This is also the southern limit of the rocky shore lineage of *F. vesiculosus* (Ladah et al. 2003); farther south it becomes a distinct genetic lineage found only in estuaries and lagoons. In addition to this southern boundary, Iberian kelps and wrack are physically separated from Brittany by the Bay of Biscay, where a combination of adverse conditions interact to create a shared distributional gap, including exposed shorelines in northern Spain, large stretches of soft substrate in south-western France, and increasing summer SSTs warming towards the innermost part of the gulf (Lüning 1990; Gorostiaga et al. 2004). The scale of this distributional gap varies between species and, for some of them, also in time (Duarte et al. 2013; Fernández 2013).

ENM hindcasting suggests that during the LGM, Iberia was a central climatic optimum for *P. canaliculata* (Neiva et al. 2014) and *F. vesiculosus* (Assis et al. 2014), and probably for cold-temperate seaweeds in general (Fig. 11.4). Notwithstanding its present marginality and isolation, north-western Iberian fucoids harbour high levels of genetic endemism and/or diversity: distinct haplotype lineages in *F. serratus* (Hoarau et al. 2007), *F. ceranoides* (Neiva et al. 2010) and *P. canaliculata* (Neiva et al. 2014) (Fig. 11.5), and genetically differentiated groups (based on microsatellites) in *A. nodosum* (Olsen et al. 2010) and in *F. vesiculosus* (Assis et al. 2014) (Fig. 11.8). The endemism of *F. ceranoides* stands out in this region; concordant mtDNA and microsatellite data show that Iberian populations

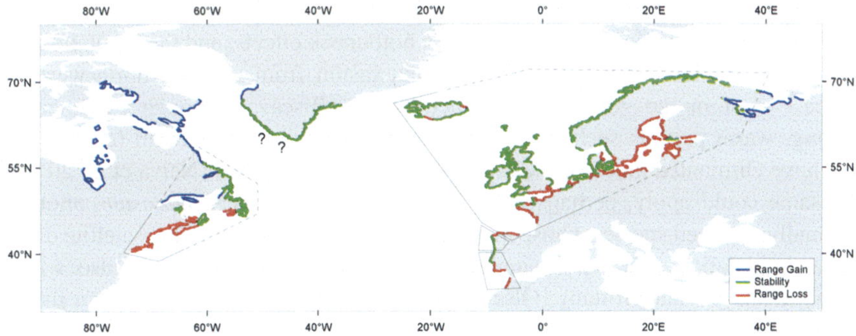

Fig. 11.8 Predicted range changes of *F. vesiculosus* due to near-future (2090–2100) climatic change. Gains and losses are superimposed over four recognized intra-specific genetic clusters (microsatellite data). *Dashed lines* depict the areas where genetic data is not available. Adapted from Assis et al. (2014) and Vuorinen et al. (2015)

are grouped hierarchically in small shoreline sectors (<200 km) showing extremely limited admixture, each defined by a distinct mtDNA lineage and many private alleles (Neiva et al. 2012c). Regional haplotypic diversity was striking, particularly within the ranges of the westernmost, Iberian-endemic phylogroups. For instance, more haplotypes were found in eight estuaries in western Galicia than in the entire northern range, where the species is introgressed with *F. vesiculosus* organelles (Neiva et al. 2010, 2012c). When compared to more central, former periglacial areas, Iberia also differs in terms of microsatellite diversity and intra-regional population differentiation (Neiva et al. 2012b). These genetic data confirm that Iberia still represents a suitable climatic region for *F. ceranoides*.

In contrast, Iberian lineages in other fucoids are much less diverse than extra-Iberian phylogroups or Iberian lineages of *F. ceranoides* (Hoarau et al. 2007; Neiva et al. 2014). In *F. serratus*, microsatellite data further revealed that the two sampled localities (separated by only 140 km) were highly differentiated from each other and from the remaining extra-Iberian range and were also the least diverse (Coyer et al. 2003), suggesting that extant Iberian populations are the remnants of a larger, and ecologically and geographically more central, glacial range. Contemporary marginality is demonstrated by historical distributional records revealing regular expansions and contractions of *F. serratus* along the northern coast of Spain during the past century, apparently associated with decadal-scale oscillations in oceanographic conditions (Arrontes 1993; Duarte et al. 2013). Current low levels of genetic diversity and diminished functional responses at the very rear-edge (see Pearson et al. 2009) in otherwise highly differentiated and isolated populations clearly reflect these recurrent extinction–recolonization cycles, and increased genetic drift in small and demographically unstable populations (Coyer et al. 2003).

Only a single local extinction of Iberian *P. canaliculata* has been documented (Berlengas Island off central Portugal, Neiva et al. 2014). This documentation, coupled with nineteenth century herbarium records describing unattached specimens south of Lisbon suggest that *P. canaliculata* may have occurred hundreds of kilometres south of its current southern edge. The contemporary genetic depletion of the two unique Iberian lineages indicates bottleneck effects, and the homogeneity of most of the Iberian range suggests re-expansion from a single north-western source except in the genetically distinct Bay of Biscay. Although evidence is missing, warm periods such as the mid-Holocene Climatic Optimum (*c.* 9–5 ka) may have eliminated much of the pre-existing Iberian variation (Neiva et al. 2014). The same could apply, perhaps to an even larger extent, to *A. nodosum*, another regionally depleted species whose much differentiated (based on microsatellite data) Iberian populations harbour a single mitochondrial haplotype that is also widespread in and beyond Brittany (Olsen et al. 2010) (Fig. 11.6). North-western Iberia is also the contact zone between the southern species *F. guiryi* and its sister species *F. vesiculosus* (open-shore northern lineage) and *F. spiralis*, both of which find here their southernmost limit. The nuclear genome of *F. guiryi* is introgressed with these species in this sympatric region (but not further south in allopatry), as demonstrated by microsatellite evidence (Moalic et al. 2011) and by 13 protein-coding loci (Zardi et al. 2011) (see also Fig. 11.7).

11.2.3.3 Southern Iberia/Morocco

ENMs suggest that suitable habitat for cold-temperate fucoids (Assis et al. 2014; Neiva et al. 2014) and a range of other coastal organisms (Kettle et al. 2011; Provan and Maggs 2012; Waltari and Hickerson 2013) was present in North-west Africa and the Mediterranean during glacial periods. Discounting the restricted Mediterranean relict *F. virsoides*, only two fucoids presently extend their distributions beyond the north-western Iberian biogeographic boundary (Fig. 11.2): the stress-tolerant high/mid-shore *F. guiryi* (Zardi et al. 2011) and a vanishing lineage of *F. vesiculosus* (Nicastro et al. 2013).

Among all members of the genus, *F. vesiculosus* has the broadest latitudinal distribution, occurring above the Arctic Circle in the north to warm temperate shores in southern Iberia and Morocco. It also has the widest habitat range, from soft sediments in estuaries and coastal lagoons, to the brackish Baltic Sea, in addition to extensive stands on marine rocky shores. Some of this variability might include cryptic species, as was the case of the recently evolved *F. radicans* in the Baltic Sea (Pereyra et al. 2009). South of 41°N latitude in the North-east Atlantic (central Portugal), *F. vesiculosus* is absent from open coastal rocky shores, but is found exclusively in coastal lagoons and estuarine habitats (Ladah et al. 2003), possibly a consequence of the more moderate thermal environment offered by estuarine and lagoon habitats in warmer southern regions (Ladah et al. 2003; Zardi et al. 2013). Concordant with the shift in habitat occupancy, phylogenetic and phylogeographic evidence support, a genetically distinct southern lineage within *F. vesiculosus* (based on multiple protein-coding nuclear genes Cánovas et al. 2011) and microsatellite markers (Nicastro et al. 2013; Assis et al. 2014). This lineage, coinciding with one climate refugium along central Iberia and north-western Morocco, where the species was hindcast to have persisted during both extreme cold (LGM) and warm (mid-Holocene) periods (Assis et al. 2014), is undergoing a regressive trend (Nicastro et al. 2013).

Phylogenetic analyses using multiple protein-coding regions successfully resolved the phylogenetic position of *F. guiryi* as a unique species sister to *F. spiralis* and *F. vesiculosus* (Cánovas et al. 2011; Coyer et al. 2011a; Zardi et al. 2011) (Fig. 11.1). However, clear phylogenetic resolution is restricted to allopatric southern populations (i.e. Portugal and Morocco), while inclusion of individuals from the sympatric range results in unresolved relationships between the three species. This suggests that genetic boundaries in the northern range of *F. guiryi* have been blurred by introgression (Zardi et al. 2011) despite limited contemporary gene flow (Billard et al. 2005, 2010; Coyer et al. 2011a). Further support for introgression comes from mitochondrial data, which show distinct and divergent haplotypes in allopatry in the south (previously referred to as *F. spiralis* South), while northern sympatric populations (previously designated *F. spiralis* Low) share haplotypes with *F. vesiculosus and F. spiralis* (previously designated *F. spiralis* High) (Coyer et al. 2011a) (Fig. 11.7). All these studies indicate that *F. guiryi* represents a unique genetic lineage in the south (central-southern Iberia, Morocco and the Canary Islands), distinct from the introgressed regions farther north. Although the complex history of inter-specific genetic exchange complicates phylogeographic inferences at more northern latitudes, mtDNA shows that the pure and southern allopatric *F. guiryi* exhibits differentiated populations, particularly distinct in the Azores (Fig. 11.7). This pattern of high and structured mtDNA diversity reflects the selfing mating system and poor dispersal abilities of *F. guiryi*, but high diversity also implies a relatively stable existence in the region during glacial and ongoing warmer climatic phases (Coyer et al. 2011a).

11.2.3.4 Northern Europe

A refugium for *F. distichus*, a species with colder climatic affinities, may have existed off the coastal island of Andøya, northern Norway where it presently forms a highly resolved microsatellite cluster (Coyer et al. 2011b). As such, this southern refugium may overlap with the northern refugia identified in Norway for a range of terrestrial and freshwater organisms (e.g. Alm and Birks 1991; Brunhoff et al. 2006). Although an Icelandic refugium has been inferred in several floristic and phylogeographic studies of plants, birds and marine invertebrates, some of these data and their interpretation are controversial (Ægisdóttir and Þórhallsdóttir 2004; Ingólfsson 2009) and no evidence of an Icelandic refugium was found for the closely related species *F. spiralis* and *F. vesiculosus* (Coyer et al. 2011a). Nevertheless, the presence of high diversity, private alleles and an intergenic spacer (IGS) divergence date of 0.06–0.60 Ma suggest that Iceland is a present-day southern refugium for *F. distichus* as the species is not found (naturally) in the British Isles or along the southern half of the west Norwegian coast.

11.2.3.5 Canadian Maritimes

Until phylogeographic data became available in the early 1990s, it was widely believed that the North-west Atlantic could not have been a refugium for rocky shore invertebrates or seaweeds due to a lack of suitable substrate in combination with ice scour and very narrow isotherms for cold-adapted taxa. For example, the presence of *A. nodosum* and *F. vesiculosus* along the Atlantic Canadian coast was thought to be the result of post-LGM recolonization from Europe, possibly via Iceland (e.g. Lüning 1990). Determining the presence of *F. distichus* was more problematic, given its distribution in both the eastern Atlantic and North Pacific. Subsequent phylogeographic studies, however, demonstrated that glacial refugia did exist in the Canadian Maritimes for some marine species (Cunningham and Collins 1998; Wares and Cunningham 2001). Among North-west Atlantic fucoids, *A. nodosum* (Olsen et al. 2010) (Fig. 11.6) and the cold-adapted *F. distichus*

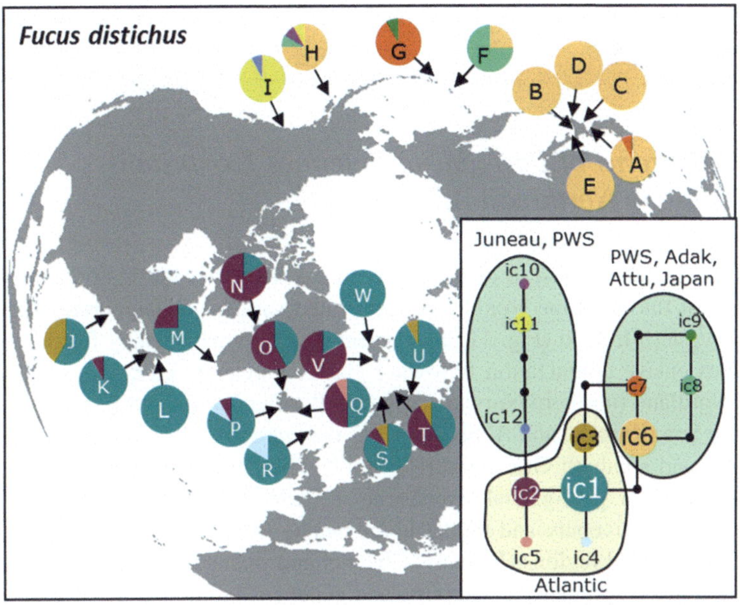

Fig. 11.9 Combined haplotype distribution of mtIGS and COI for *F. distichus*, showing inset of haplotype network. Atlantic haplotypes are derived from two Pacific regions. Modified from Coyer et al. (2011b)

(Coyer et al. 2011b) (Fig. 11.9) have differentiated lineages with unique alleles in the West Atlantic, supporting a scenario of glacial persistence.

In *F. vesiculosus*, distinguishing between in situ survival and post-glacial colonization scenarios has been more challenging. Muhlin and Brawley (2009) found low microsatellite diversity and differentiation in populations spanning Canada to North Carolina, and the most widespread haplotype was identical to a common haplotype in Europe, implying that the populations of *F. vesiculosus* in the North-west Atlantic originated from European colonizers from the south-western Ireland and/or Brittany refugia (Muhlin and Brawley 2009; Coyer et al. 2011a). Another study combining ENMs and microsatellite data (Assis et al. 2014), however, suggested ample habitat availability during the LGM and also important genetic differentiation, more consistent with long-term persistence of *F. vesiculosus* in North-west Atlantic (Fig. 11.8). Introgression could explain conflicting mtDNA signatures of a demographic sweep. Indeed, low mtDNA diversity characterizes the entire distribution of *F. vesiculosus* (Coyer et al. 2011a) and its haplotypes are more similar to *F. spiralis* and *F. guiryi* ("*F. spiralis* south") than among themselves (Coyer et al. 2011a), a finding in contradiction with nuclear evidence of phylogenetic relationship (Cánovas et al. 2011) (Figs. 11.1 and 11.7). MtDNA evidence suggests that *F. spiralis sensu stricto* in the West Atlantic was colonized from the East Atlantic (Coyer et al. 2011a). *F. serratus* did not colonize the West Atlantic by natural means, but was introduced about 150 years ago via human activities (Brawley et al. 2009).

11.2.4 Patterns of Post-glacial Poleward Expansions in the North-east Atlantic

During the LGM, expansion of the Eurasian ice sheet and the regional westward migration of the coastline would have prevented survival of temperate organisms north of western Ireland (Fig. 11.3a). Present distributions of many fucoids clearly indicate extensive colonization following glacier retreat: *F. ceranoides* extends at least to Nordland (northern Norway), whereas *P. canaliculata* and *F. serratus* reach the White Sea (Russia), and *A. nodosum*, *F. vesiculosus* and *F. spiralis* are present off Russia and southern Greenland (Fig. 11.2). Globally, molecular signatures are consistent with a post-glacial expansion scenario, as evidenced by the lower regional genetic diversity and higher homogeneity of colonized areas versus more southern periglacial regions (Brittany, Channel and western Ireland; Fig. 11.10). In a few species, comprehensive sampling and genetic resolution allowed post-glacial colonization sources and routes along the north-eastern Atlantic to be tentatively inferred by combining paleogeographic reconstitutions and the spatial distribution of widespread northern haplotypes.

In *F. serratus*, the distribution of cluster 1 (blue in Fig. 11.5a) suggests a first expansion wave originating in the genetically diverse area of south-eastern Ireland and proceeding via north-western Ireland and Scotland to Scandinavia, eventually reaching the White Sea to the north and the Kattegat and the lower Baltic Seas to

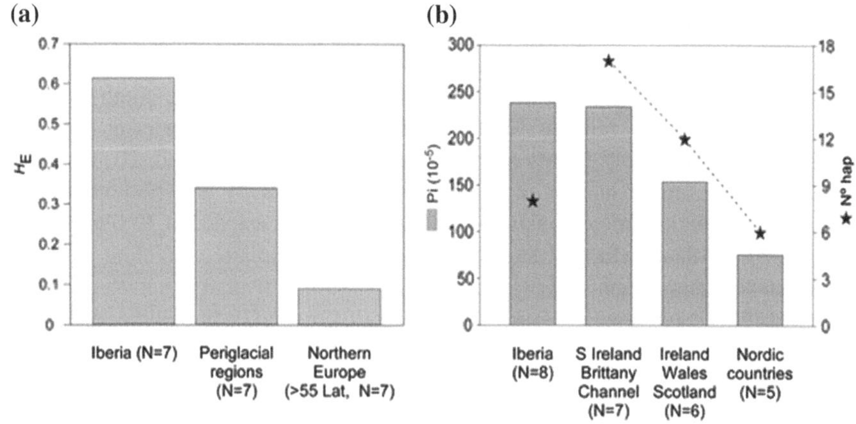

Fig. 11.10 Latitudinal trends in regional genetic diversity in (**a**) *Fucus ceranoides* (microsatellite data, from Neiva et al. (2012b) and (**b**) *Pelvetia canaliculata*. Haplotype data from Neiva et al. (2014). *Note* the common reduction in regional diversity north of periglacial regions

the south (Hoarau et al. 2007). The latter region could only be colonized after its establishment as a marine environment ca. 7500 years bp (Björck 1995), and its populations apparently descended from a single founder event, as implied by mtDNA (*nad*11) heteroplasmy unique to the area (Coyer et al. 2004b). The modern distribution of the cluster 2 haplotypes (yellow in Fig. 11.5a) suggest that a second independent wave, originating from the Hurd Deep sea/Brittany region, proceeded northwards and eastwards following the progressive transgression of the Irish Sea and the English Channel (at the time land bridges connecting Ireland and Great Britain and these to continental Europe). Secondary contact between the "Irish" front and descendants of the first wave was eventually established in the central Irish Sea following the fall of the terrestrial bridge connecting Ireland and South-west England (Hoarau et al. 2007). The westward front eventually expanded into the south of the North Sea following the breach of Dover Straits that resulted in the connection of the transgressed Channel and the Southern Bight some time between 8000 and 7500 BP. The absence of samples throughout much of the eastern part of the North Sea makes it difficult to determine where along eastern England/Scotland secondary contact with cluster 1 has been established.

In *F. vesiculosus*, the greatest demographic expansion occurred well before the LGM, but post-LGM recolonization originated in the south-western Ireland-Brittany area (Coyer et al. 2011a). Two dominant haplotypes were present in this refugium; both expanded north and are presently found in the Shetlands and the Faeroes. One haplotype expanded throughout Scandinavia, the Baltic, and Iceland, and possibly crossed the Atlantic colonizing the Canadian Maritimes and eastern USA [but see Assis et al. 2014; the other spread to France and areas south (Coyer et al. 2011a) (Fig. 11.7)].

In *F. ceranoides* and in *P. canaliculata*, the colonization routes are less clear given the existence of only one phylogroup throughout both periglacial and post-glacially colonized regions (Neiva et al. 2012b, 2014) (Fig. 11.5b, c). In both species, Scottish populations (and nearby areas) are fixed for specific alleles (yellow haplotypes), apparently excluding the region as the source of Scandinavian populations. Alternatively, these distinct haplotypes may simply reflect the arrival there of later waves from the Irish Sea (*F. ceranoides*) or from eastern England (*P. canaliculata*).

Unlike the previously discussed fucoids, *F. distichus* is cold-adapted and of North Pacific origin, having a unique distribution along Arctic ice margins, including northern Canada, central Greenland, Iceland, Svalbard, and Novaya Zemlya (Lüning 1990; Adey and Hayek 2005). Tracing post-glacial pathways in the North Atlantic is, therefore, more speculative, as the populations from which recolonization expansions emerged are no longer present in areas likely to have been populated during glacial periods. Nevertheless, ENMs and extant haplotype distributions suggest that *F. distichus* was restricted to the North-west Atlantic coastal areas between Brittany and central Portugal and in the North-west Atlantic to the Canadian Maritimes (Coyer et al. 2011b) (Fig. 11.9). As *F. distichus* began a post-glacial poleward recolonization, it could no longer survive in the warming refugia along the southern European coast and consequently, the Icelandic/Faeroes region may now represent a southern interglacial refugium (Coyer et al. 2011b).

11.3 Long-Distance Range Expansions but Poor or no Gene Flow Across Genetic Breaks

The apparent absence of barriers in the oceans and the speed of ocean currents, are expected to allow marine organisms to be connected over broad spatial scales (Hellberg 2009). Rafting of reproductive individuals is known to play an important role in long-distance dispersal (LDD) of many seaweed species including kelps and fucoids with pneumatocysts (Norton 1992; Coyer et al. 2001; Thiel and Haye 2006; McKenzie and Bellgrove 2008). LDD, therefore, may explain the extensive post-glacial range expansions experienced by Atlantic fucoids, the colonization of remote mid-Atlantic islands such as Iceland and the Azores, the swift spread of some fucoid invaders (Kraan 2008), and fucoid ubiquity despite large habitat discontinuities (e.g. estuarine species or those circumscribed to scattered sheltered stations in otherwise exposed shores).

Paradoxically, Atlantic fucoids maintain sharp genetic breaks (i.e. fixed or nearly fixed haplotypic and allelic differences) throughout their ranges (Fig. 11.5), suggesting that while LDD may be common, successful gene flow is not. Only SDD maintains populations at seemingly small spatial scales. Regionally, genetic discontinuities are seldom spatially coincident between different species, which are often evenly distributed across contact zones (Hoarau et al. 2007; Neiva et al. 2012b, c, 2014). The origin, stability and pervasiveness of (neutral) genetic breaks thus appear to reflect historical contingencies and intrinsically restricted dispersal (stemming from fucoid life history traits), although dispersal barriers (including large habitat discontinuities) may play a role at the regional level.

Taken together, the phylogeographic patterns of Atlantic fucoids indicate that typical dispersal rates (including rafting) enable colonization of empty habitat patches and range shifts, but are much less effective in neutralizing local to regional founder events or secondary contact, with recurrent genetic drift contributing to population differentiation. The contrasting dispersal and gene flow effects of dispersal into vacant (colonization) and populated (immigration) habitats suggest an important role for density-barriers such as priority-colonization or "founder takes all" effects (De Meester et al. 2002; Tellier et al. 2011; Waters et al. 2013). These are expected when infrequent but successful recruitment occurs into already dense populations and when available habitat patches are saturated (see also Guillemin et al., this book). In newly colonized habitats, exponential population growth ensures a disproportionate contribution of early settlers for the genetic make-up of establishing populations. Conversely, in demographically mature habitats, increased competition and numerical resident/immigrant disparity acts as a demographic buffer against changes in allele frequencies, since new arriving alleles will likely be rare and prone to be lost by genetic drift, rather than magnified by unconstrained growth.

Rare dispersal further constrains the dynamics of the colonization front(s), particularly in narrow intertidal habitats. Several Atlantic fucoids exhibit clear parapatric phylogeographic sectors (Fig. 11.5) thought to have been formed during expansion and secondary contact of differentiated source populations (Hoarau et al. 2007; Neiva et al. 2012b, c, 2014; Assis et al. 2014). In addition, commonly observed poleward reductions in genetic diversity suggest genetic bottlenecks and allelic surfing at species' leading edges, which contributes to the progressive erosion of refugial gene pools in post-glacially colonized regions (Excoffier and Ray 2008). Fucoid research has further shown that allelic surfing may also facilitate introgression between hybridizing species, as revealed by the massive spread of introgressed *F. vesiculosus* organellar genomes throughout the central and northern range of *F. ceranoides* associated with the post-glacial expansion of the latter (Neiva et al. 2010). Other outcomes are possible, such as newly formed hybrid zones between *F. serratus* and *F. distichus* (Coyer et al. 2002; Hoarau et al. 2015). In contrast to other fucoid species, the phylogeographic pattern of *A. nodosum* was typical of long lifespan species with large population size, high fecundity, high within population diversity and weak large-scale differentiation (Olsen et al. 2010).

11.4 Human Impacts—Species Introductions and Climate Change

11.4.1 Unintentional Introductions Linked to Maritime Traffic Can Affect Phylogeography

Anthropogenically mediated LDD of coastal species, including Atlantic fucoids, is of increasing concern because it can potentially affect local ecosystem functioning and native biodiversity. Common vectors include shipping, mariculture, aquarium trade and intentional introductions (Miller et al. 2004; Provan et al. 2005a; Hewitt et al. 2007; Kim et al. 2010; Riosmena-Rodríguez et al. 2012). Documented examples of human-mediated introductions of Atlantic fucoids include *F. distichus* into the Oslofjord (southern Norway) in the mid-1890s (Schueller and Peters 1994), *F. serratus* to Iceland and the Faroes in the mid-1800s and 1900s respectively (Coyer et al. 2006b) and Nova Scotia (North-west Atlantic) around 1868 (Brawley et al. 2009), and of *A. nodosum* to San Francisco Bay, first detected in 2001 (Miller et al. 2004). Another putative human-mediated introduction (the unlikely alternative being a trans-Arctic crossing) is that of *F. spiralis* in the Pacific, first reported by Harvey in 1862 (Norris and Conway 1974), who did not identify the species, but reported it to be similar to the *Fucus* species from the Canary Islands. Its single haplotype is identical to a dominant haplotype in northern Atlantic populations, indicating a recent origin from an Atlantic source (Coyer et al. 2011a).

Several fucoid introductions have proven successful with both intercalation and expansion beyond the original introduction point. For example, *F. distichus* rapidly expanded south of its introduction site in Oslofjorden in the mid-1890s, appearing in the Skagerrak Sea in 1924, the central Kattegat Sea in 1933 and in the Kiel Bight (western Baltic) by 1992 (Schueller and Peters 1994). Similarly, the original distribution of *F. spiralis* from the Aleutian Islands to northern Washington State (Norris and Conway 1974) has expanded to include Oregon (Neiva et al. 2012a) and California (EAS and GAP personal observations 2014). *Fucus serratus* has expanded over 1500 km along North American shorelines, by both natural and further human-mediated transport (Johnson et al. 2011). As an exception, alien *A. nodosum* was eradicated from San Francisco Bay in 2002, but the number and size of the individuals removed suggest that it had been present, but undetected for several years (Miller et al. 2004). It was probably introduced as discarded seafood or bait worm packing material, and as it presently continues to appear in seafood restaurants of San Francisco Bay (JLO personal observations), re-establishment is likely.

Molecular techniques have been used to determine the native or introduced status of species, and to determine putative population sources. Microsatellite-based approaches indicated that Icelandic populations of *F. serratus* originated from the Oslofjorden (southern Norway), an introduction that had to have occurred sometime between 900 AD, when the first Icelandic settlers arrived from Norway, and 1900, when it was first noted in a phycological survey (Coyer et al. 2006b). Despite anthropogenic passages to Iceland from numerous areas in Norway, the UK and Germany during the intervening years, historical records and microsatellite assignment tests show that the most parsimonious source of the introduced population was via logs from the Oslofjorden logging area during the mid-nineteenth century (Coyer et al. 2006b). Further analysis indicated that *F. serratus* was introduced to the eastern Faeroes Islands from the Reykjavik area, probably via fishing activities, during the late twentieth century (Coyer et al. 2006b).

Human emigration and shipping activities also played a role in the introduction of *F. serratus* to Nova Scotia, where a similar combination of historical records, microsatellite genotypes and assignment tests revealed two introductions in the mid-nineteenth century. One stemmed from Galway (Ireland) to Pictou and the other from Greenock (Scotland) to Western Cape Breton Island (Brawley et al. 2009). In both cases, the introduction vector was probably ballast rocks; ballasted ships arrived in Nova Scotia with passengers and the ballast was discarded before departing with log cargo (Brawley et al. 2009). Ballast rocks taken from the shallow intertidal at the source inevitably were inhabited by a thriving community of invertebrates and algae, which could easily survive in the dark and damp holds for the weeks-long transatlantic voyages. There has been at least one unsuccessful intentional introduction of *F. serratus* from Europe to North America. In 1902, the New York Botanical Gardens transplanted several individual *F. serratus* from Pictou, Nova Scotia to Pelham Bay and Hunters Island near New York City as part of a programme to increase local diversity (Anonymous 1992), but the transplant presumably was unsuccessful. Through unknown means, *F. serratus* was found at

Newburyport, Massachusetts in 1852, but in 1902, was reported to have "long ceased to exist there" (Anonymous 1992).

In addition to potential competitive interactions with the local flora, a further consequence of inadvertent or intentional fucoid introductions may be hybridization with closely related native species. Molecular data revealed signatures of introgressive hybridization between the recently introduced *F. serratus* and its native sister species *F. distichus* (*F. evanescens* in Coyer et al. 2006b) in Iceland, and vice versa in Oslofjorden (Coyer et al. 2002, 2007) that were an order of magnitude higher than between the two species in northern Norway where they have been in sympatry for several thousands of years (Hoarau et al. 2015). The reductions in hybridizations and introgressions with increasing time of sympatry strongly suggest rapid evolution of pre-zygotic reinforcement mechanisms (Hoarau et al. 2015), as shown also when comparing southern allopatric *F. guiryi* and a contact zone with *F. vesiculosus* and *F. spiralis* farther north (Moalic et al. 2011).

11.4.2 Climate Change and Predicted Loss of Southern Ranges and Associated Endemic Diversity

Species' responses to climate change range from shifts in phenology and productivity to local extinctions associated with physiological tipping points. Such responses can be a function of adaptation, maladaptation, acclimation history of the individuals and their parents, or even speciation. For example, the embryos of *F. vesiculosus* descendent from warm-acclimated individuals survive for longer periods when exposed to thermal stress (Li and Brawley 2004). Local adaptation has been reported for southern range populations of *F. serratus* (Jueterbock et al. 2014), but at the southernmost limit, the populations show no local adaptation, coping instead by restricting their vertical range in the intertidal (Pearson et al. 2009). In another example, involving *Laminaria digitata* (which reaches its southern limit in southern Brittany) and *L. hyperborea* (which extends its distribution farther south to northern Spain), reduced genetic diversity in southern Brittany was only observed for *L. digitata* (Robuchon et al. 2014) and was best explained by local maladaptation through altered sexual reproduction (Oppliger et al. 2014). Under strong selective pressures, fucoid species may even undergo speciation (e.g. *F. radicans*, Pereyra et al. 2009). Nevertheless, the rapid pace of recent climate change seems to largely outpace the potential of species to acclimate or adapt, typically producing more detrimental effects and leading to extinction of seaweed populations at regional scales (Wernberg et al. 2011b; Duarte et al. 2013; Nicastro et al. 2013).

Local extinctions are probably the most common outcome of climate change, particularly at lower latitude trailing edges, where environmental conditions generally border the thermal tolerance of species and small microhabitat differences can make an enormous difference (e.g. Mota et al. 2015 for *F. vesiculosus*). For instance, the progressive warming trend (due in part to intense heat waves) reported for the Iberian Peninsula and the Mediterranean Sea (Belkin 2009; Lima and Wethey 2012), contracted southern ranges by hundreds of kilometres for many fucoid species such as *F. serratus* and *H. elongata* (Martínez et al. 2012; Duarte et al. 2013), *F. vesiculosus* (Nicastro et al. 2013), *F. virsoides* (Mačić 2006) and *Cystoseira* spp. (Perkol-Finkel and Airoldi 2010). Distinct model-based approaches using contrasting scenarios of greenhouse gas emissions showed that several fucoid species (e.g. *F. serratus*, *F. vesiculosus* and *A. nodosum*) may shift ranges northward as a unit, likely colonizing the Arctic shores in Canada, Greenland and Svalbard within the next 50–75 years, while severely contracting ranges south of 45° N (Jueterbock et al. 2013; Assis et al. 2014) (Figs. 11.8 and 11.11).

Extinction of canopy-forming seaweeds causes loss of essential habitats, but a major concern is impoverishment of intra-specific diversity (Provan 2013). This is particularly important for the low-dispersing fucoids, which presently harbour high unique diversity at vulnerable lower latitude trailing edges due to long-term

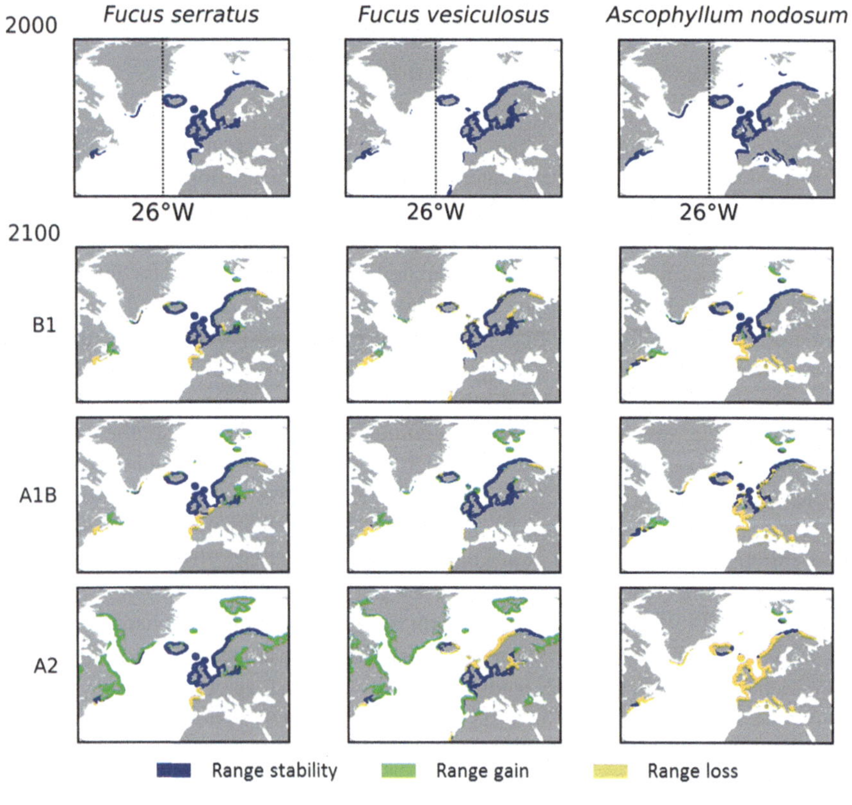

Fig. 11.11 Predicted range changes of *F. serratus*, *F. vesiculosus* and *Ascophyllum nodosum* in the coming century based on increasingly severe IPCC model predictions (B1 < A1B < A2) of climate warming. The *dotted line* delimits North-west from North-east Atlantic. From Jueterbock et al. (2013)

persistence in southern climatic refugia (Hoarau et al. 2007; Diekmann and Serrão 2012; Neiva et al. 2012b, c; Provan and Maggs 2012) (Figs. 11.5 and 11.8). In *F. vesiculosus*, a distinct southern genetic lineage has regressed more than 1000 km (Nicastro et al. 2013) and is predicted to continue towards extinction (Fig. 11.8) (Assis et al. 2014). Because unique genetic lineages are not morphologically distinct, the loss of gene pools may occur unnoticed in many species (i.e. shifting genetic baselines; Assis et al. 2013).

11.5 General Conclusions

Fucoid seaweeds are the most extensively investigated group of algae in Atlantic phylogeographic research. Their life history, ecology and rapid speciation are well studied making it possible to gain insights into dispersal, population structure and distribution at multiple spatial resolutions. Fucoid species with cold-to-warm temperate affinities vary in responses to climatic shifts that overlap for some species along a few common persistence zones (i.e. refugia). Most studies found persistent climatic refugia in north-western Iberia, as inferred from the distinct lineages that evolved and persisted. For some, the region has become progressively marginal in the course of the present (warming) interglacial. For *F. guiryi*, Iberia appears to be an introgressive contact zone, likely associated with expansion from the non-introgressed south (north-western Morocco), another long-term climatic refugium for a reduced set of more warm-tolerant species. The large Irish to Brittany paleocoastline is the most impressive refugium as its large size allowed persistence of populations with high genetic diversity for most fucoid species. The few fucoids (*F. distichus*, *A. nodosum* and *F. vesiculosus*) occurring in the North-west Atlantic (south of the Laurentide ice sheet) exhibit distinct genetic variability, supporting a hypothesis of glacial persistence. The eurythermal *F. vesiculosus* is noteworthy as the only species having distinct genetic lineages supporting persistence in all these refugia, which is further supported by hindcasting ENMs.

For most fucoids, the scale of Atlantic post-glacial expansions was notable, with several reaching Russia and Greenland. Waves of colonization involved routes along north-western Scotland into Scandinavia, and simultaneously along the transgressing English Channel into the North Sea. These created large geographical areas along the recolonized northern ranges where populations are genetically homogeneous and less diverse. Range expansions have also changed genetic diversity and structure by means of introgressive recombination of genomes at contact zones. Phylogeographic tools have also proven useful to identify cryptic and non-cryptic introductions, which have had some consequences for native biodiversity in terms of introgression and reduced genetic variation.

Collectively, North Atlantic fucoids demonstrate how limited dispersal modes can have contrasting effects at the scales of marine meta-population (connectivity) versus range dynamics (habitat tracking), promoting both sharp genetic divergence between refugial gene pools and large-scale homogeneity along recently colonized areas, depending on the past demographic conditions of populations. Climatic refugia typically harbour differentiated gene pools, but for several species the present highest diversity hotspots are not the southernmost species range limits, but rather in the large, stable Brittany–Celtic Sea refugium. This might result from the past loss of southernmost diversity due to population bottlenecks, drift and local extinctions at species' rear edges, and is likely to intensify in the coming decades of climatic change.

In summary, Atlantic fucoids are excellent models to investigate the responses of species to global environmental change. Putative climatic refugia occurred across several main regions (broadly North-west Africa, north-western Iberia, Celtic Sea, Grand Banks), but each species survived only in a subset of these regions and their relative size or importance was variable for individual fucoid species with contrasting physiological affinities and ranges. Their many range shifts, including southern contractions and northward recolonizations also provide useful insights about how climate warming has affected and likely will affect rocky intertidal communities. Future scenarios for changing fucoid distributions do not include species extinctions, but do include loss of unique, rear-edge diversity as poleward range expansions are counteracted by southern contractions. Future research challenges should focus on how specific gene pools and genetic structuring (e.g. intertidal zonation) drive adaptation to local conditions. Ecological and functional genomic approaches applied in an already robust phylogeographic framework will open further possibilities to understand the complex dynamics of community genetic/genomic interactions. As the "canary in a mine", fucoid algae are only a sign of the major changes expected on rocky shores in the coming decades.

References

Adey WH, Hayek LAC. The biogeographic structure of the western North Atlantic rocky intertidal. Cryptogam Algologie. 2005;26:35–66.

Ægisdóttir HH, Þórhallsdóttir ÞE. Theories on migration and history of the North-Atlantic flora: a review. Jokull. 2004;54:1–16.

Alberto F, Massa S, Manent P, Diaz-Almela E, Arnaud-Haond S, Duarte CM, Serrão EA. Genetic differentiation and secondary contact zone in the seagrass *Cymodocea nodosa* across the Mediterranean-Atlantic transition region. J Biogeogr. 2008;35:1279–94.

Alm T, Birks HH. Late Weichselian flora and vegetation of Andøya, Northern Norway-macrofossil (seed and fruit) evidence from Nedre Æråsvatn. Nord J Bot. 1991;11:465–76.

Anonymous (1992) Increasing New York's Sea Flora. New York Times.

Araújo R, Bárbara I, Tibaldo M, Berecibar E, Tapia PD, Pereira R, Santos R, Pinto IS, Ba I, Dı P. Checklist of benthic marine algae and cyanobacteria of northern Portugal. Bot Mar. 2009;52:24–46.

Arrontes J. Nature of the distributional boundary of *Fucus serratus* on the north shore of Spain. Mar Ecol Prog Ser. 1993;93:183–93.

Assis J, Coelho NC, Alberto F, Valero M, Raimondi P, Reed D, Serrão EA. High and distinct range-edge genetic diversity despite local bottlenecks. PLoS ONE. 2013;8(7):e68646.

Assis J, Serrão EA, Claro B, Perrin C, Pearson GA. Climate-driven range shifts explain the distribution of extant gene pools and predict future loss of unique lineages in a marine brown alga. Mol Ecol. 2014;23:2797–810.

Bárbara I, Cremades J, Calvo S, López-Rodriguez MC, Dosil J. Checklist of the benthic marine and brackish Galician algae (NW Spain). Anales del Jardín Botánico de Madrid. 2005; 62:69–100.

Bartsch I, Wiencke C, Laepple T. Global seaweed biogeography under a changing climate: the prospected effects of temperature. In: Wiencke C, Bischof K, editors. Seaweed biology. Berlin Heidelberg: Springer; 2012. p. 383–406.

Belkin IM. Rapid warming of large marine ecosystems. Prog Oceanogr. 2009;81:207–13.

Bennett K, Provan J. What do we mean by "refugia"? Quat Sci Rev. 2008;27:2449–55.

Bierne N, Borsa P, Daguin C, Jollivet D, Viard F, Bonhomme F, David P. Introgression patterns in the mosaic hybrid zone between *Mytilus edulis* and *M. galloprovincialis*. Mol Ecol. 2003;12:447–61.

Billard E, Daguin C, Pearson GA, Serrão EA, Engel C, Valero M. Genetic isolation between three closely related taxa: *Fucus vesiculosus*, *F. spiralis*, and *F. ceranoides* (Phaophyceae). J Phycol. 2005;41:900–5.

Billard E, Serrão EA, Pearson GA, Destombe C, Valero M. Fucus vesiculosus and spiralis species complex: a nested model of local adaptation at the shore level. Mar Ecol Prog Ser. 2010;405:163–74.

Björck S. A review of the history of the Baltic Sea, 13.0–8.0 ka BP. Quat Int. 1995;27:19–40.

Brawley SH, Coyer JA, Blakeslee AMH, Hoarau G, Johnson LE, Byers JE, Stam WT, Olsen JL. Historical invasions of the intertidal zone of Atlantic North America associated with distinctive patterns of trade and emigration. Proc Nat Acad Sci USA. 2009;106:8239–44.

Brunhoff C, Yoccoz NG, Ims RA, Jaarola M. Glacial survival or late glacial colonization? Phylogeography of the root vole (*Microtus oeconomus*) in north-west Norway. J Biogeogr. 2006;33:2136–44.

Cánovas FG, Mota CF, Serrão EA, Pearson GA. Driving south: a multi-gene phylogeny of the brown algal family Fucaceae reveals relationships and recent drivers of a marine radiation. BMC Evol Biol. 2011;11:371.

Carlson AE, Winsor K. Northern hemisphere ice-sheet responses to past climate warming. Nat Geosci. 2012;5:607–13.

Chapman ARO. Functional ecology of fucoid algae: twenty-three years of progress. Phycologia. 1995;34:1–32.

Christie H, Norderhaug KM, Fredriksen S. Macrophytes as habitat for fauna. Mar Ecol Prog Ser. 2009;396:221–33.

Coyer JA, Diekmann OE, Serrão EA, Procaccini G, Milchakova N, Pearson GA, Stam WT, Olsen JL. Population genetics of dwarf eelgrass *Zostera noltii* throughout its biogeographic range. Mar Ecol Prog Ser. 2004a;281:51–62.

Coyer JA, Hoarau G, Costa JF, Hogerdijk B, Serrão EA, Billard E, Valero M, Pearson GA, Olsen JL. Evolution and diversification within the intertidal brown macroalgae *Fucus spiralis/F. vesiculosus* species complex in the North Atlantic. Mol Phylogenet Evol. 2011a;58:283–96.

Coyer JA, Hoarau G, Oudot-Le Secq MP, Stam WT, Olsen JL. A mtDNA-based phylogeny of the brown algal genus *Fucus* (Heterokontophyta; Phaeophyta). Mol Phylogenet Evol. 2006a;39:209–22.

Coyer JA, Hoarau G, Van Schaik J, Luijckx P, Olsen JL. Trans-Pacific and trans-Arctic pathways of the intertidal macroalga *Fucus distichus* L. reveal multiple glacial refugia and colonizations from the North Pacific to the North Atlantic. J Biogeogr. 2011b;38:756–71.

Coyer JA, Hoarau G, Sjøtun K, Olsen JL. Being abundant is not enough: a decrease in effective over eight generations in a Norwegian population of the seaweed *Fucus serratus*. Biol Lett. 2008;4:755–7.

Coyer JA, Hoarau G, Skage M, Stam WT, Olsen JL. Origin of *Fucus serratus* (Heterokontophyta; Fucaceae) populations in Iceland and the Faroes: a microsatellite-based assessment. Eur J Phycol. 2006b;41:235–46.

Coyer JA, Hoarau G, Stam WT, Olsen JL. Geographically specific heteroplasmy of mitochondrial DNA in the seaweed *Fucus serratus* (Heterokontophyta: Phaeophyceae, Fucales). Mol Ecol. 2004b;13:1323–6.

Coyer JA, Hoarau G, Stam WT, Olsen JL. Hybridization and introgression in a mixed population of the intertidal seaweeds *Fucus evanescens* and *F. serratus*. J Evol Biol. 2007;20:2322–33.

Coyer JA, Jason Smith G, Andersen RA. Evolution of *Macrocystis spp*. (Phaeophyceae) as determined by ITS1 and ITS2 sequences. J Phycol. 2001;37:574–85.

Coyer JA, Peters AF, Hoarau G, Stam WT, Olsen JL. Hybridization of the marine seaweeds Fucus serratus and *Fucus evanescens* (Heterokontophyta: Phaeophyceae) in a 100-year-old zone of secondary contact. Proc R Soc B. 2002;269:1829–34.

Coyer JA, Peters AF, Stam WT, Olsen JL. Post-ice age recolonization and differentiation of *Fucus serratus* L. (Phaeophyceae; Fucaceae) populations in Northern Europe. Mol Ecol. 2003;12:1817–29.

Cunningham CW, Collins TM. Beyond area relationships: extinction and recolonization in molecular marine biogeography. In: Desalle R, Schierwater B, editors. Molecular approaches to ecology and evolution. Basel: Birkhauser Verlag; 1998. p. 297–321.

Davis MB, Shaw RG. Range shifts and adaptive responses to Quaternary climate change. Science. 2001;292:673–9.

De Meester L, Gómez A, Okamura B, Schwenk K. The monopolization hypothesis and the dispersal–gene flow paradox in aquatic organisms. Acta Oecologia. 2002;23:121–35.

Diekmann OE, Serrão EA. Range-edge genetic diversity: locally poor extant southern patches maintain a regionally diverse hotspot in the seagrass *Zostera marina*. Mol Ecol. 2012; 21:1647–57.

Dijkstra JA, Boudreau J, Dionne M. Species-specific mediation of temperature and community interactions by multiple foundation species. Oikos. 2012;121:646–54.

Duarte L, Viejo RM, Martínez B, DeCastro M, Gómez-Gesteira M, Gallardo T. Recent and historical range shifts of two canopy-forming seaweeds in North Spain and the link with trends in sea surface temperature. Acta Oecol. 2013;51:1–10.

Dudgeon S, Kübler JE, Wright WA, Vadas RL, Petraitis PS. Natural variability in zygote dispersal of *Ascophyllum nodosum* at small spatial scales. Funct Ecol. 2001;15:595–604.

Engel CR, Daguin C, Serrão EA. Genetic entities and mating system in hermaphroditic Fucus spiralis and its close dioecious relative *F. vesiculosus* (Fucaceae, Phaeophyceae). Mol Ecol. 2005;14:2033–46.

Excoffier L, Ray N. Surfing during population expansions promotes genetic revolutions and structuration. Trends Ecol Evol. 2008;23:347–51.

Fernández C. The retreat of large brown seaweeds on the north coast of Spain: the case of *Saccorhiza polyschides*. Eur J Phycol. 2013;46:352–60.

Finnegan AK, Griffiths AM, King RA, Machado-Schiaffino G, Porcher JP, Garcia-Vazquez E, Bright D, Stevens JR. Use of multiple markers demonstrates a cryptic western refugium and postglacial colonisation routes of Atlantic salmon (*Salmo salar* L.) in Northwest Europe. Heredity. 2013;111:34–43.

Fraser CI, Nikula R, Spencer HG, Waters JM. Kelp genes reveal effects of subantarctic sea ice during the Last Glacial Maximum. Proc Nat Acad Sci USA. 2009;106:3249–53.

Gibbard PL, Lautridou JP. The Quaternary history of the English Channel: an introduction. J Quat Sci. 2003;18:195–9.

Gorostiaga JM, Santolaria A, Secilla A, Casares C, Diez I. Check-list of the Basque coast benthic algae (North of Spain). Anales del Jardín Botánico de Madrid. 2004;61:155–80.

Graham MH, Dayton PK, Erlandson JM. Ice ages and ecological transitions on temperate coasts. Trends Ecol Evol. 2003;18:33–40.

Hampe A, Petit RJ. Conserving biodiversity under climate change: the rear edge matters. Ecol Lett. 2005;8:461–7.

Handschumacher L, Steinarsdóttir MB, Edmands S, Ingólfsson A. Phylogeography of the rock-pool copepod *Tigriopus brevicornis* (Harpacticoida) in the northern North Atlantic, and its relationship to other species of the genus. Mar Biol. 2010;157:1357–66.

Hellberg ME. Gene flow and isolation among populations of marine animals. Ann Rev Ecol Evol Syst. 2009;40:291–310.

Hewitt C, Campbell M, Schaffelke B. Introductions of seaweeds: accidental transfer pathways and mechanisms. Bot Mar. 2007;50:326–37.

Hoarau G, Coyer JA, Giesbers MCWG, Jueterbock A, Olsen JL. Pre-zygotic isolation in the macroalgal genus *Fucus* from four contact zones a tale of reinforcement? R Soc Open Sci. 2015;2:140538.

Hoarau G, Coyer JA, Veldsink JH, Stam WT, Olsen JL. Glacial refugia and recolonization pathways in the brown seaweed *Fucus serratus*. Mol Ecol. 2007;16:3606–16.

Hofreiter M, Stewart J. Ecological change, range fluctuations and population dynamics during the Pleistocene. Curr Biol. 2009;19:R584–94.

Hu ZM, Guiry MD, Critchley AT, Duan DL. Phylogeographic patterns indicate trans-Atlantic migration from Europe to North America in the red seaweed *Chondrus crispus* (Gigartinales, Rhodophyta). J Phycol. 2010;46:889–900.

Hu ZM, Li W, Li JJ, Duan DL. Post-Pleistocene demographic history of the North Atlantic endemic Irish moss *Chondrus crispus*: glacial survival, spatial expansion and gene flow. J Evol Biol. 2011;24:505–17.

Ilves KL, Huang W, Wares JP, Hickerson MJ. Colonization and/or mitochondrial selective sweeps across the North Atlantic intertidal assemblage revealed by multi-taxa approximate Bayesian computation. Mol Ecol. 2010;19:4505–19.

Ingólfsson A. A marine refugium in Iceland during the last glacial maximum: fact or fiction? Zool Sci. 2009;38:663–5.

Johnson LE, Brawley SH, Adey WH. Secondary spread of invasive species: historic patterns and underlying mechanisms of the continuing invasion of the European rockweed *Fucus serratus* in eastern North America. Biol Invasions. 2011;14:79–97.

Jolly MT, Viard F, Gentil F, Thiébaut E, Jollivet D. Comparative phylogeography of two coastal polychaete tubeworms in the Northeast Atlantic supports shared history and vicariant events. Mol Ecol. 2006;15:1841–55.

Jueterbock A, Kollias S, Smolina I, Fernandes JM, Coyer JA, Olsen JL, Hoarau G. Thermal stress resistance of the brown alga *Fucus serratus* along the North-Atlantic coast: acclimatization potential to climate change. Mar Genomics. 2014;13:27–36.

Jueterbock A, Tyberghein L, Verbruggen H, Coyer JA, Olsen JL, Hoarau G. Climate change impact on seaweed meadow distribution in the North Atlantic rocky intertidal. Ecol Evol. 2013;3:1356–73.

Kelly DW, MacIsaac HJ, Heath DD. Vicariance and dispersal effects on and speciation in a widespread estuarine invertebrate. Evolution. 2006;60:257–67.

Kettle AJ, Morales-Muñiz A, Roselló-Izquierdo E, Heinrich D, Vøllestad LA, Rosell E. Refugia of marine fish in the northeast Atlantic during the last glacial maximum: concordant assessment from archaeozoology and palaeotemperature reconstructions. Clim Past. 2011;7:181–201.

Kim SY, Weinberger F, Boo SM. Genetic data hint at a common donor region for invasive atlantic and pacific populations of *Gracilaria vermiculophylla* (Gracilariales, Rhodophyta). J Phycol. 2010;46:1346–9.

Kraan S. *Sargassum muticum* (Yendo) Fensholt in Ireland: an invasive species on the move. J Appl Phycol. 2008;20(5):825–32.

Ladah L, Bermudez R, Pearson GA, Serrão EA. Fertilization success and recruitment of dioecious and hermaphroditic fucoid seaweeds with contrasting distributions near their southern limit. Mar Ecol Prog Ser. 2003;262:173–83.

Lambeck K, Rouby H, Purcell A, Sun Y, Sambridge M. Sea level and global ice volumes from the Last Glacial Maximum to the Holocene. PNAS. 2015;111:15296–303.

Li JJ, Hu ZM, Duan DL. Survival in glacial refugia vs. postglacial dispersal in the North Atlantic: the cases of red seaweeds. In: Hu ZM, Fraser CI, editors. Seaweed: adaptation and evolution of seaweeds under environmental change. Springer: Berlin Heidelberg; 2016.

Li R, Brawley SH. Improved survival to heat stress in intertidal embryos simultaneously exposed to hypersalinity and the effect of parental thermal history. Mar Biol. 2004;144:205–13.

Lima FP, Wethey DS. Three decades of high-resolution coastal sea surface temperatures reveal more than warming. Nat Commun. 2012;3:704.

Lüning K (1990) Seaweeds: their environment, biogeography, and ecophysiology. Wiley.

Mačić M. Distribution of seaweed *Fucus virsoides* J. Agardh in Boka Kotorska Bay (South Adriatic Sea). Ann Ser Hist Nat. 2006;16:1–4.

Maggs CA, Castilho R, Foltz D, Henzler C, Jolly MT, Kelly J, Olsen J, Perez KE, Stam W, Väinölä R, Viard F, Wares J. Evaluating signatures of glacial refugia for north atlantic benthic marine taxa. Ecology. 2008;89:108–22.

Martínez B, Arenas F, Rubal M, Burgués S, Esteban R, García-Plazaola I, Figueroa FL, Pereira R, Saldaña L, Sousa-Pinto I, Trilla a, Viejo RM. Physical factors driving intertidal macroalgae distribution: physiological stress of a dominant fucoid at its southern limit. Oecologia. 2012;170:341–53.

McKenzie PF, Bellgrove A. Dispersal of *Hormosira banksii* (Phaeophyceae) via detached fragments: reproductive viability and longevity. J Phycol. 2008;44:1108–15.

Ménot G, Bard E, Rostek F, Weijers JWH, Hopmans EC, Schouten S, Damsté JSS. Early reactivation of European rivers during the last deglaciation. Science. 2006;313:1623–5.

Miller AW, Chang AL, Cosentino-Manning N, Ruiz GM. A new record and eradication of the Northern Atlantic alga *Ascophyllum nodosum* (Phaeophyceae) from San Francisco Bay, California, USA. J Phycol. 2004;40:1028–31.

Moalic Y, Arnaud-Haond S, Perrin C, Pearson GA, Serrão EA. Travelling in time with networks: revealing present day hybridization versus ancestral polymorphism between two species of brown algae, *Fucus vesiculosus* and *F. spiralis*. BMC Evol Biol. 2011;11:33.

Mota CF, Engelen AH, Serrão EA, Pearson GA. Some don't like it hot: microhabitat-dependent thermal and water stresses in a trailing edge population. Funct Ecol. 2015;29:640–9.

Muhlin JF, Brawley SH. Recent versus Relic: discerning the genetic signature of *Fucus vesiculosus* (Heterokontophyta; Phaeophyceae) in the Northwestern Atlantic. J Phycol. 2009;45:828–37.

Neiva J, Assis J, Fernandes F, Pearson GA, Serrão EA. Species distribution models and mitochondrial DNA phylogeography suggest an extensive biogeographical shift in the high-intertidal seaweed *Pelvetia canaliculata*. J Biogeogr. 2014;41:1137–48.

Neiva J, Hansen GI, Pearson GA, Van De Vliet MS, Maggs CA, Serrão EA. *Fucus cottonii* (Fucales, Phaeophyceae) is not a single genetic entity but a convergent salt-marsh morphotype with multiple independent origins. Eur J Phycol. 2012a;47:461–8.

Neiva J, Pearson GA, Valero M, Serrão EA. Surfing the wave on a borrowed board: range expansion and spread of introgressed organellar genomes in the seaweed *Fucus ceranoides* L. Mol Ecol. 2010;19:4812–22.

Neiva J, Pearson GA, Valero M, Serrão EA. Drifting fronds and drifting alleles: the genetic architecture of the estuarine seaweed *Fucus ceranoides* L. J Biogeogr. 2012b;39:1167–78.

Neiva J, Pearson GA, Valero M, Serrão EA. Fine-scale genetic breaks driven by historical range dynamics and ongoing density-barrier effects in the estuarine seaweed *Fucus ceranoides* L. BMC Evol Biol. 2012c;12:78.

Nicastro KR, Zardi GI, Teixeira S, Neiva J, Serrão EA, Pearson GA. Shift happens: trailing edge contraction associated with recent warming trends threatens a distinct genetic lineage in the marine macroalga *Fucus vesiculosus*. BMC Biol. 2013;11:6.

Norris RE, Conway E. *Fucus spiralis* L. in the northeast Pacific. Syesis. 1974;7:79–81.

Norton TA. Dispersal by macroalgae. Brit Phycol J. 1992;27:293–301.

Olsen JL, Stam WT, Coyer JA, Reusch TBH, Billingham M, Boström C, Calvert E, Christie H, Granger S, la Lumière R, Milchakova N, Oudot-Le Secq MP, Procaccini G, Sanjabi B, Serrão EA, Veldsink J, Widdicombe S, Wyllie-Echeverria S. North Atlantic phylogeography and large-scale population differentiation of the seagrass *Zostera marina* L. Mol Ecol. 2004;13:1923–41.

Olsen JL, Zechman FW, Hoarau G, Coyer JA, Stam WT, Valero M, Åberg P. The phylogeographic architecture of the fucoid seaweed *Ascophyllum nodosum*: an intertidal "marine tree" and survivor of more than one glacial-interglacial cycle. J Biogeogr. 2010;37:842–56.

Oppliger LV, Von Dassow P, Bouchemousse S, Robuchon M, Valero M, Correa JA, Mauger S, Destombe C. Alteration of sexual reproduction and genetic diversity in the kelp species *Laminaria digitata* at the southern limit of its range. PLoS ONE. 2014;9(7):e102518.

Pearson GA, Lago-Leston A, Mota C. Frayed at the edges: selective pressure and adaptive response to abiotic stressors are mismatched in low diversity edge populations. J Ecol. 2009;97:450–62.

Pearson GA, Serrão EA. Revisiting synchronous gamete release by fucoid algae in the intertidal zone: fertilization success and beyond? Integr Comp Biol. 2006;46:587–97.

Pereyra RT, Bergström L, Kautsky L, Johannesson K. Rapid speciation in a newly opened postglacial marine environment, the Baltic Sea. BMC Evol Biol. 2009;9:70.

Perkol-Finkel S, Airoldi L. Loss and recovery potential of marine habitats: an experimental study of factors maintaining resilience in subtidal algal forests at the Adriatic Sea. PLoS ONE. 2010;5(5):e10791.

Petit J, Brewer S, Cheddadi R, Coart E, Cottrell J, Csaikl UM, Van Dam B, Deans JD, Espinel S, Fineschi S, Finkeldey R, Glaz I, Goicoechea PG, Lowe AJ, Svejgaard J, Ko AO, Munro RC, Flemming S, Ma G, Popescu F, Slade D, Tabbener H, De Vries SGM, Ziegenhagen B, De Beaulieu J, Kremer A. Identification of refugia and post-glacial colonisation routes of European white oaks based on chloroplast DNA and fossil pollen evidence. For Ecol Manag. 2002;156:49–74.

Provan J. The effects of past, present and future climate change on range-wide genetic diversity in northern North Atlantic marine species. Front Biogeogr. 2013;5:60–6.

Provan J, Maggs CA. Unique genetic variation at a species' rear edge is under threat from global climate change. Proc R Soc B. 2012;279:39–47.

Provan J, Murphy S, Maggs CA. Tracking the invasive history of the green alga *Codium fragile ssp. tomentosoides*. Mol Ecol. 2005a;14:189–94.

Provan J, Wattier RA, Maggs CA. Phylogeographic analysis of the red seaweed *Palmaria palmata* reveals a Pleistocene marine glacial refugium in the English Channel. Mol Ecol. 2005b;14:793–803.

Remerie T, Vierstraete A, Weekers PHH, Vanfleteren JR, Vanreusel A. Phylogeography of an estuarine mysid, *Neomysis integer* (Crustacea, Mysida), along the north-east Atlantic coasts. J Biogeogr. 2009;36:39–54.

Riosmena-Rodríguez R, Boo GH, López-Vivas JM, Hernández-Velasco A, Sáenz-Arroyo A, Boo BM. *Sargassum filicinum* (Fucales, Phaeophyceae) is on the move along the Mexican Pacific coastline. Bot Mar.2012;55:547–551.

Robuchon M, Le Gall L, Mauger S, Valero M. Contrasting genetic diversity patterns in two sister kelp species co-distributed along the coast of Brittany, France. Mol Ecol. 2014;23:2669–85.

Roman J, Palumbi SR. A global invader at home: population structure of the green crab, *Carcinus maenas*, in Europe. Mol Ecol. 2004;13:2891–8.

Schiel DR, Foster MS. The population biology of large brown seaweeds: ecological consequences of multiphase life histories in dynamic coastal environments. Ann Rev Ecol Evol Syst. 2006;37:343–72.

Schueller G, Peters A. Arrival of *Fucus evanescens* (Phaeophyceae) in Kiel Bight (Western Baltic). Bot Mar. 1994;37:471–477.

Serrão EA, Alice LA, Brawley SH. Evolution of the Fucaceae (Phaeophyceae) inferred from nrDNA-ITS. J Phycol. 1999;35:382–94.

Serrão EA, Kautsky L, Lifvergren T, Brawley SH. Gamete dispersal and mortality in baltic *Fucus vesiculosus*. Phycologia. 1997;36:101–2.

Serrão EA, Pearson GA, Kautsky L, Brawley SH. Successful external fertilization in turbulent environments. Proc Nat Acad Sci USA. 1996;93:5286–90.

Spalding MD, Fox HE, Allen GR, Davidson N, Ferdaña ZA, Finlayson MAX, Halpern BS, Jorge MA, Lombana AL, Lourie SA, Martin KD, Manus MC, Molnar J, Recchia CA, Robertson J. Marine ecoregions of the world: a bioregionalization of coastal and shelf areas. Bioscience. 2007;57:573–83.

Stewart JR, Lister AM. Cryptic northern refugia and the origins of the modern biota. Trends Ecol Evol. 2001;16:608–13.

Stewart JR, Lister AM, Barnes I, Dalén L. Refugia revisited: individualistic responses of species in space and time. Proc R Soc B. 2010;277:661–71.

Tellier F, Tapia J, Faugeron S, Destombe C, Valero M. The *Lessonia nigrescens* species complex (Laminariales, Phaeophyceae) shows strict parapatry and complete reproductive isolation in a secondary contact zone. J Phycol. 2011;47:894–903.

Thiel M, Haye PA. The ecology of reafting in the marine environment. III. Biogeographical and evolutionary consequences. Oceanogr Mar Biol Ann Rev. 2006;44:323–429.

Vuorinen I, Hänninen J, Rajasilta M, Laine P, Eklund J, Montesino-Pouzols F, Corona F, Junker K, Meier MHE, Dippner JW. Scenario simulations of future salinity and ecological consequences in the Baltic Sea and adjacent North Sea areas–implications for environmental monitoring. Ecol Indic. 2015;50:196–205.

Waltari E, Hickerson MJ. Late Pleistocene species distribution modelling of North Atlantic intertidal invertebrates. J Biogeogr. 2013;40:249–60.

Wares JP, Cunningham CW. Phylogeography and historical ecology of the North Atlantic intertidal. Evolution. 2001;55:2455–69.

Waters JM, Fraser CI, Hewitt GM. Founder takes all: density-dependent processes structure biodiversity. Trends Ecol Evol. 2013;28:78–85.

Wernberg T, Russell BD, Moore PJ, Ling SD, Smale DA, Campbell A, Coleman MA, Steinberg PD, Kendrick GA, Connell SD. Impacts of climate change in a global hotspot for temperate marine biodiversity and ocean warming. J Exp Mar Biol Ecol. 2011a;400:7–16.

Wernberg T, Russell BD, Thomsen MS, Gurgel CFD, Bradshaw CJ, Poloczanska ES, Connell SD. Seaweed communities in retreat from ocean warming. Curr Biol. 2011b;21:1828–32.

Wilson AB, Eigenmann Veraguth I. The impact of Pleistocene glaciation across the range of a widespread European coastal species. Mol Ecol. 2010;19:4535–53.

Woodall LC, Koldewey HJ, Shaw PW. Historical and contemporary population genetic connectivity of the European short-snouted seahorse *Hippocampus hippocampus* and implications for management. J Fish Biol. 2011;78:1738–56.

Zardi GI, Nicastro KR, Canovas F, Costa JF, Serrão EA, Pearson GA. Adaptive traits are maintained on steep selective gradients despite gene flow and hybridization in the intertidal zone. PLoS ONE. 2011;6:e19402.

Zardi GI, Nicastro KR, Ferreira Costa J, Serrão EA, Pearson GA. Broad scale agreement between intertidal habitats and adaptive traits on a basis of contrasting population genetic structure. Estu Coast Shelf Sci. 2013;131:140–8.

Chapter 12
Survival in Glacial Refugia Versus Postglacial Dispersal in the North Atlantic: The Cases of Red Seaweeds

Jing-Jing Li, Zi-Min Hu and De-Lin Duan

Abstract The past two decades have witnessed the North Atlantic used as a model system to explore how climate changes and environmental conditions shaped the spatiotemporal distributions and biogeographic processes of intertidal seaweed species. The tectonic reconfigurations of the Northwestern and Northeastern Atlantic shores caused by the Quaternary ice ages allow us to examine two evolutionarily contrasting scenarios: survival in local glacial refugia versus postglacial trans-Atlantic dispersal. In this chapter, we collected comparative data from the red algae *Chondrus crispus*, *Mastocarpus stellatus*, *Palmaria palmata,* and *Porphyra umbilicalis* across the North Atlantic to illustrate the effects of paleoclimatic oscillations on historical demography, lineage divergence, and genetic connectivity. The genetic signals detected in *C. crispus* and *P. palmata* are consistent with the hypothesis that they survived in situ on each side of the North Atlantic during the Quaternary glaciations, while the phylogeographic evidence for *M. stellatus* and *P. umbilicalis* indicates postglacial trans-Atlantic dispersals. Bayesian coalescent analysis detected signals of demographic expansions in the four algal species, during the late Pliocene to the middle Pleistocene. In addition, the dated genetic splits between lineages were compatible with the expansion times for each species. In summary, our comparative analysis revealed contrasting biogeographic processes in these seaweeds despite their similar contemporary distributional ranges in the North Atlantic. These results also highlight the importance of comparative phylogeographic surveys in exploring dynamic evolutionary patterns and phylogeographic histories of intertidal marine organisms.

Keywords Demographic history · Divergence time · Genetic diversity · Genealogical isolation · The quaternary ice age · Postglacial recolonization

J.-J. Li · Z.-M. Hu (✉) · D.-L. Duan
Key Laboratory of Experimental Marine Biology, Institute of Oceanology, Chinese Academy of Sciences, Qingdao 266071, China
e-mail: huzimin9712@163.com

J.-J. Li · Z.-M. Hu · D.-L. Duan
Qingdao National Laboratory for Marine Science and Technology, Qingdao 266071, China

Z.-M. Hu and C. Fraser (eds.), *Seaweed Phylogeography*,
DOI 10.1007/978-94-017-7534-2_12

12.1 Introduction

The North Atlantic Ocean has undergone major geological and climatic changes since its establishment in the Mesozoic (Roth 1986). Elucidating the ecological and evolutionary impacts of these changes can help us to understand the processes underpinning contemporary marine coastal biodiversity and community structure. Nevertheless, it was not until the 1980s that scientists started to unravel how intertidal organisms in the North Atlantic respond to climate change and environmental shifts, over the decadal to centennial timescales to understand the processes shaping species and genetic diversity.

Seaweeds are a fascinating and diverse group of organisms. They constitute key components of many marine ecosystems and play important ecological roles in rocky intertidal communities by providing animals with food and shelter. Seaweeds have been demonstrated to be excellent models for investigating the impacts of environmental shifts and climate change on population genetic differentiation and the evolutionary dynamics of intertidal marine organisms.

The effects of past geological and climate changes in the North Atlantic can be observed in the composition and floristic relationships of seaweed assemblages. During 1982 and 1986, phycologists organized three rounds of the Seaweed Biogeography Workshop at Hamburg, Groningen, and Helgoland, and relevant theme articles were published in *Helgoländer Meeresunters* (38:225–417; 41:233–383). In 1989, the 4th Seaweed Biogeography Workshop, held in St. Andrews, Canada, produced a book *Evolutionary Biogeography of the Marine Algae in the North Atlantic*. These studies build a fundamental framework of biogeography and evolution of seaweeds in the North Atlantic in terms of integrative paleogeography, taxonomy, and phylogenetics.

Research on population genetics and evolutionary biology of seaweeds has made rapid progress since the mid-1990s, aided by advanced DNA-sequencing technologies (see Chapter by Hu et al. (2016) in this volume). In this chapter, we summarize the main findings of recent research on the phylogeographic dynamics and evolutionary histories of red seaweeds in the North Atlantic in a comparative evolutionary context.

12.2 Climate Shifts and Environmental Variables in the North Atlantic

Among the broad range of environmental and ecological factors that, singly or in combination, shaped the phylogeographic structure and geographic distributions of North Atlantic intertidal organisms, paleoclimatic oscillations during the Quaternary ice ages have been increasingly considered critically important. During the last major Ice Age (0.12–0.1 Ma), much of the northern European and northern North American land mass was covered by ice sheets (Dyke et al. 2002; Clark et al.

2009). The transfer of water from the oceans to land-based ice sheets caused sea level to drop dramatically to a brief lowstand of *c.* 130 m at the Last Glacial Maximum (LGM, 0.026–0.019 Ma) (Lambeck and Chappell 2001; Lambeck et al. 2002), making the coastlines of North Atlantic quite different from the present (Fig. 12.1). Along European coasts, the ice sheets reached as far south as the British Isles, and the dramatic drop in sea level caused intertidal species to be restricted to southern glacial or cryptic periglacial refugia (Hewitt 1996; Provan and Bennett 2008). Consequently, the littoral organisms currently found in these areas, including some intertidal seaweeds, could expand from periglacial or more southern refugia as the ice sheet retreated after the LGM (Provan et al. 2005; Hoarau et al. 2007; Hu et al. 2010; Neiva et al. 2012; Provan and Maggs 2012; Neiva et al. 2014; Li et al. 2015).

In contrast, the North American-Atlantic littoral species may have been confronted with relatively hasher living conditions during Pleistocene ice ages (Riggs et al. 1996; Wares and Cunningham 2001; Krebes et al. 2011). Although both sides of the North Atlantic were at times covered by huge ice sheets (Fig. 12.1), the European marginal areas still retained suitable rocky habitats for littoral organisms in comparison with the limited rocky habitats along the ice-free North American-Atlantic coasts (Olsen et al. 2010; Hu et al. 2011; Krebs et al. 2011). The loss of large amounts of coastal rocky habitat during glacial periods may account for intertidal species' extinctions, low genetic diversity, and weak population

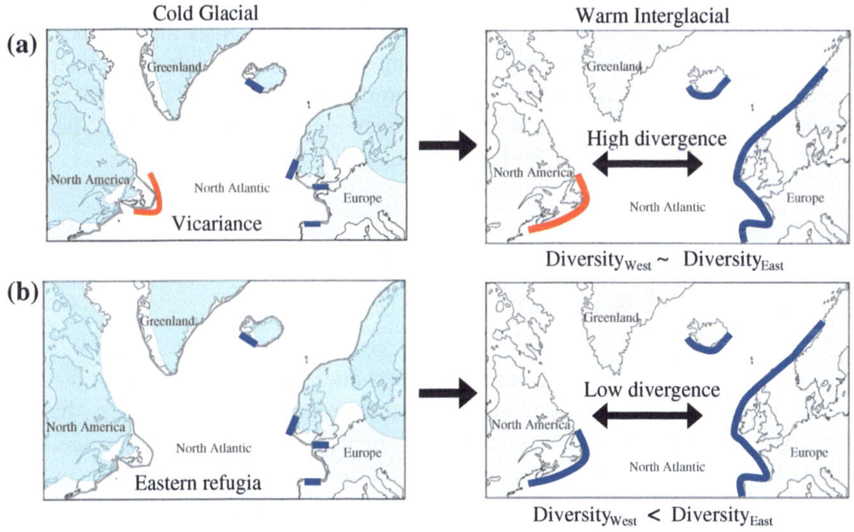

Fig. 12.1 Two biogeographically contrasting scenarios illustrate the evolutionary patterns of North Atlantic seaweeds. The *light-blue* and *gray lines* indicate the approximate boundaries of the Laurentide (North America), Greenland and Eurasian ice sheets, and the coastlines during the Last Glacial Maximum (LGM), respectively (synthesized from Charbit et al. 2007; Olsen et al. 2010). *Blue* and *red bars* show the distributions of species along North American and European coasts, respectively

structure (Wares and Cunningham 2001; Teasdale and Klein 2010). Genetic characters induced by biogeographic processes during glacial periods can be tracked in the present-day organisms inhabiting the North Atlantic coasts.

On the other hand, the sea surface temperature (SST) isotherms (4–12 °C) during the LGM were more compressed than the present (Charbit et al. 2007; Olsen et al. 2010). The present-day 4–12 °C isotherms in summer cover most of the European coasts (Charbit et al. 2007). However, the 4–12 °C summer isotherms during the LGM only spread along the Iberian Peninsula (Charbit et al. 2007). In particular, the isotherms along North American coasts were more compressed than those along European coasts (Charbit et al. 2007). The 6 °C isotherm during the LGM was located even farther south to the present 12 °C isotherm (Olsen et al. 2010). The different distribution patterns of SST isotherms between North American and European coasts during the LGM have been postulated as important factors contributing to the contrasting range shifts and biogeographic dynamics of intertidal species on the two sides of the North Atlantic (Teasdale and Klein 2010; Hu et al. 2010; Waltari and Hickerson 2013).

Ecologically, present-day harsh winters may pose a big challenge for intertidal organisms to maintain large population sizes along the northern North American coasts (McKindsey and Bourget 2001; Saucier et al. 2003; Scrosati and Heaven 2007). During winter, ice can be either fast ice or free-floating in the sea along the northern coasts. The attached shore ice can be over a meter thick and completely cover the intertidal habitat (Dyke et al. 2002). Upon thawing, the ice can rip plants and sessile invertebrates from the substrate (Mathieson et al. 1982). Free-floating ice in the sea can also affect shoreline organisms when it comes into contact with the shore and scours the substrate, removing plants and animals from intertidal habitats (Minchinton et al. 1997). These environmental conditions may give rise to distinct ecological stresses, driving particular adaptive responses in benthic organisms along the northern North American coast.

12.3 The North Atlantic as a Model for Phylogeographic Studies of Red Seaweeds

In the North Atlantic, most phylogeographic studies to date have concentrated on the influence of basin-wide or regional marine environmental shifts (e.g., habitat discontinuities, ocean currents, salinity gradients, and geographic barriers) on population genetic structure and spatiotemporal distributions of intertidal seaweeds (Table 12.1).

All contemporary marine organisms occurring in the Baltic Sea must have immigrated after the opening of the Danish Straits 8000 years ago and responded to the development of the steep Skagerrak–Baltic salinity gradient (Björck 1995). Therefore, the Baltic Sea provides a unique model system for exploring genetic imprints of postglacial colonization and ecological differentiation. Gabrielsen et al. (2002) applied mtDNA *cox*2-3 spacer and RAPDs markers to investigate ecological

Table 12.1 A chronological list of phylogeographic studies of North Atlantic red seaweeds

References	Species	Marker	Region	Main results
Gabrielsen et al. (2002)	*Ceramium tenuicorne*	*cox*2-3, RAPD	The Baltic Sea	Conspicuous genetic variation in growth and reproduction along the salinity gradient
Gurgel et al. (2004)	*Gracilaria tikvahiae*	ITS, *rbc*L	Eastern North America	Eastern North American *G. tikvahiae* has evolved from a common ancestry with four phylogeographic groups detected
Zuccarello et al. (2005)	*Mastocarpus stellatus*	*cox*2-3, RuBisCo spacer	Eastern North Atlantic	Two types of life history correspond to genetic isolation of haplotype groups with a north–south distribution pattern
Provan et al. (2005)	*Palmaria palmata*	ITS, 16S-*trn*I-*trn*A-23S-5S, *rbc*L-*rbc*S, *rpl*12-*rps*31-*rpl*9 regions, *cox*2-3	Eastern North Atlantic	*P. palmata* persisted throughout the last glacial maximum at three glacial refugia, the English Channel (Hurd Deep), northwestern Ireland, and Norway/Iceland
Zuccarello et al. (2006)	*Bostrychia radicans/ B. moritzuana*	RuBisCo spacer, *cox*2-3	Eastern North America	The split of two genetic lineages in eastern North America may be related to the formation of a strait between the western Gulf of Mexico and southern Georgia in the Miocene/Pliocene
Bouza et al. (2006)	*Gelidium canariense*	RAPD	The Canary Islands	Genetic structure of *G. canariense* from the Canary Island agrees with an isolation-by-distance model, with gene flow occurred from eastern to western islands
Hu et al. (2007)	*Chondrus crispus*	ITS, 5.8S rRNA gene	North Atlantic	Genetic data indicated a single phylogenetic Atlantic origin for *C. crispus* and two main clades identified within *C. crispus*

(continued)

Table 12.1 (continued)

References	Species	Marker	Region	Main results
Hu et al. (2010)	*Chondrus crispus*	ITS-2, *cox*1	North Atlantic	Survived on both sides of the North Atlantic during the late Pleistocene ice ages, and the English Channel population played a predominant contribution to postglacial recolonization along the European coasts
Teasdale and Klein (2010)	*Porphyra umbilicalis*	ITS, a ribosomal DNA group-I intron, *cox*2-3	North Atlantic	North American populations were extirpated during the last glacial maximum and subsequently recolonized from European donor populations
Couceiro et al. (2011)	*Ahnfeltiopsis pusilla*	AFLP, *cox*2-3	Northwest Iberian Peninsula	Genetic diversity within the northwest Iberian populations was low, and *A. pusilla* is a very poor disperser
Krueger-Hadfield et al. (2011)	*Chondrus crispus*	SSR	The Brittany	High genetic divergence between high and low intertidal populations which may promote local adaptation to different physical conditions
Provan and Maggs (2012)	*Chondrus crispus*	SNP, SSR, two single-copy nuclear regions	North Atlantic	Iberian populations ("rear edge") were characterized by unique variance, and did not contribute to postglacial recolonization along the European coasts. North American refugium did have present-day rear-edge populations

(continued)

Table 12.1 (continued)

References	Species	Marker	Region	Main results
Provan et al. (2013)	*Chondrus crispus*	SSR, SNP	The British Islands	Isolation-by-distance (IBD) only was detected at intermediate scales (<200 m). It worked with long-term processes (e.g., postglacial recolonization) and structured genetic variation
Li et al. (2015)	*Palmaria palmata*	*rpl*12-*rps*31-*rpl*9 regions, *cox*2-3	North Atlantic	Deep genetic lineage split and contrasting phylogeographic structure between European and North American populations
Li et al. (2016)	*Mastocarpus stellatus*	*cox*2-3, RuBisCo spacer	North Atlantic	North American populations were extirpated during the last glacial maximum and subsequently recolonized from North European donors

differentiation of the red alga *Ceramium tenuicorne* sampled in three salinity regions: Oslofjorden (high), Kattegat (medium), and Baltic (low). They found that *C. tenuicorne* has conspicuous variation in growth and reproduction along this gradient, and genetic divergence was detected, suggesting that *C. tenuicorne* lineages may come from different gene pools (Gabrielsen et al. 2002).

The coast along the southeastern USA presents another ideal system for surveying how ecological factors led to genetic divergence in red seaweeds. The strait between the Gulf of Mexico and southern Georgia was closed in the Miocene/Pliocene, which may have acted as a geographic barrier for seaweed species. For example, Zuccarello et al. (2006) used mtDNA *cox*2-3 and cpDNA Rubisco spacer to investigate the genetic structure of *Bostrychia moritziana*/*Bostrychia radicans* samples from southern coasts of the USA. The genetic divergence between the western Gulf of Mexico and eastern North American samples coincided with late Pleistocene geographic barriers.

The local hydrodynamics in the North Atlantic also played an important role in structuring population genetic differentiation of red seaweeds. Bouza et al. (2006) detected a significant relationship between $F_{ST}/(1 - F_{ST})$ and geographical distance for the red seaweed *Gelidium canariense* from the Canary Islands. They concluded that the observed pattern was more likely the result of continuous disturbance provoked by the action of local hydrodynamics than of selection. Similarly, Couceiro et al. (2011) detected a pattern of isolation by distance (IBD) for the red alga

Ahnfeltiopsis pusilla along the northwestern Iberian coasts at large and regional scales based on AFLP and mtDNA *cox*2-3 spacer. This pattern suggests long-term migration-drift equilibrium incompatible with a recent introduction because *A. pusilla* is a poor disperser. Marine macroalgae with limited dispersal potential tend to show a pattern of IBD, where populations exhibiting geographical proximity are more genetically similar than those separated by long distances (Faugeron et al. 2001).

A recent study of the red alga *Chondrus crispus* also identified a potential IBD pattern at a fine micro-geographic scale (<100 m) (Krueger-Hadfield et al. 2011), but at larger geographic scales (>200 km), no significant correlation was detected between genetic differentiation and geographic distance (Provan et al. 2013). These results imply that other factors such as rapid postglacial recolonization from isolated refugia may play a fundamental role in shaping range-wide genetic connectivity of North Atlantic red algae at macro-geographic scales, e.g., *Porphyra umbilicalis* (Teasdale and Klein 2010), *Palmaria palmata* (Provan et al. 2005; Li et al. 2015), *C. crispus* (Hu et al. 2007, 2010; Provan and Maggs 2012), and *Mastocarpus stellatus* (Zucarrello et al. 2005; Li et al. 2016). For example, the Pleistocene glacial–interglacial cycles have periodically imposed significant changes on the diversity and distribution of littoral seaweeds (Hewitt 2000; Maggs et al. 2008). Coalescent analysis and model-based phylogeographic simulations have revealed genetic exchange between regions for *P. palmata* (Li et al. 2015), *M. stellatus* (Li et al. 2016), and *C. crispus* (Hu et al. 2010; Hu 2013; Provan and Maggs 2012). Interestingly, the genetic exchange between *P. palmata* populations on both sides of the North Atlantic was restricted, leading to deep divergent genetic lineages (Li et al. 2015).

Although the impacts of glacial advances and the role of glacial refugia in poleward recolonization are well documented, range dynamics at the rear edges (the southern distributional edges in the Northern Hemisphere) have largely been ignored. Climate amelioration eventually allowed species to expand their ranges poleward, often associated with genetic erosion of rear-edge populations (Provan and Maggs 2012; Neiva et al. 2014). Provan and Maggs (2012) examined the genetic diversity of red seaweed *C. crispus* across the species' range based on mtDNA single-nucleotide polymorphisms (SNPs), two nuclear markers, and microsatellites. They discovered unique genetic variation in the rear-edge populations in Iberia, which they linked to long-term persistence in the Iberian refugium, with some genotypes not contributing to the recolonization of Europe after the Ice Age.

12.4 Two Contrasting Hypotheses

Biodiversity differs markedly between European and North American coasts, with the former having higher species diversity and more endemics than the latter. Given the especially harsh LGM conditions for rocky intertidal species on North American coasts (Ingólfsson 1992, Riggs et al. 1996, Wares and Cunningham 2001), some researchers have argued that the present-day North American populations may have come from Europe after the last glacial period (van Oppen et al. 1995; Wares and Cunningham 2001; Teasdale and Klein 2010). However, others have suggested that there were cryptic North American refugia in which intertidal species could survive the glacial period. Fortunately, different demographic responses to climatic changes should leave behind characteristic patterns of genetic diversity and phylogeographic structure (Bermingham et al. 1992; Maggs et al. 2008), and thus we can reconstruct the evolutionary histories of intertidal organisms with a trans-Atlantic distribution based on genetic data collected from contemporary populations. If an ancestral amphi-Atlantic species was isolated into eastern and western glacial refugia during ice ages (e.g., the LGM), present-day populations across the North Atlantic should show strong genetic differentiation (Fig. 12.1a). Alternatively, if the glacial period led to restriction to one side of the North Atlantic (often the East), postglacial trans-Atlantic migration should be reflected in weak genetic differentiation (Fig. 12.1b). Here, we collected genetic data from the red seaweeds, *C. crispus*, *P. palmata*, *M. stellatus*, and *P. umbilicalis* and conducted comparative phylogeographic analysis to test the contrasting evolutionary hypotheses in the North Atlantic Ocean: survival in local glacial refugia versus postglacial trans-Atlantic dispersal.

12.5 Glacial Refugia in the North Atlantic

Paleoecological and genetic studies have identified several possible glacial refugia in the North Atlantic during the LGM (Hewitt 1996, 2004; Maggs et al. 2008; Provan and Bennett 2008). For the study cases of marine algae, there are four well-surveyed red seaweeds, *C. crispus*, *P. palmata*, *M. stellatus,* and *P. umbilicalis* (Fig. 12.2), which present important population-level phylogeographic insights into potential glacial refugia in the North Atlantic (also see Chapter by Neiva et al. (2016) in this volume).

12.5.1 Eastern North Atlantic Refugia

At the LGM, the Scadinavian Ice Sheet reached the European northern continental margin and covered most of Scotland (Fig. 12.2a). The southern limit of sea ice along the European coasts is controversial, possibly extending as far south as the Bay of Biscay (Zaragosi et al. 2001).

(i) The northwest Iberian Peninsula

The Iberian Peninsula is the westernmost of the three major southern European peninsulas (the Iberian, Italian, and Balkan), bordering on the southeast and east by the Mediterranean Sea, and on the north, west, and southwest by the Atlantic Ocean. The coastline of the Iberian Peninsula on the Mediterranean side has been

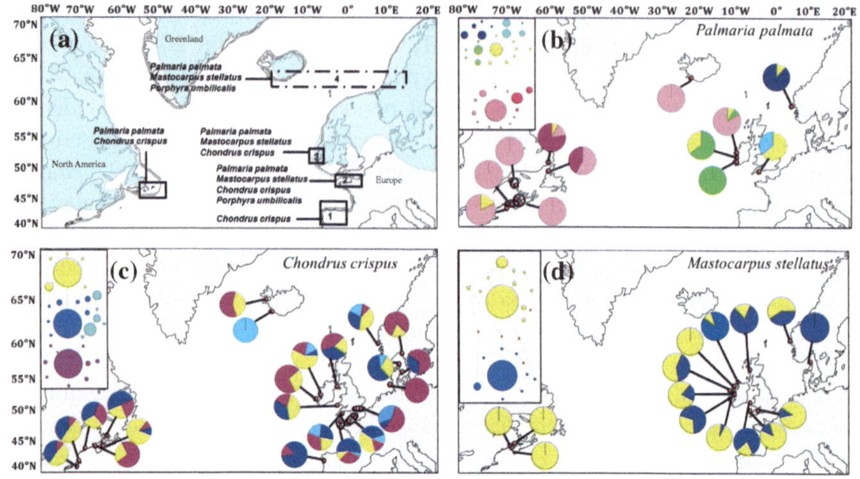

Fig. 12.2 **a** Map of the North Atlantic shows the position of proposed glacial refugia in numbered boxes and the red seaweeds that support each one. *1* the Iberian Peninsula; *2* the English Channel; *3* Galway Bay; *4* Norway/Iceland/the Faroe Islands; *5* Grand Banks. mtDNA haplotypes genealogies and geographical distribution of major inferred phylogroups in *Palmaria palmata* (**b**), *Chondrus crispus* (**c**), and *Mastocarpus stellatus* (**d**)

inundated by sea-level rise from a minimum of 115–120 m since the LGM, and the coastal shelf remains submerged. However, the continental shelf on the Atlantic side drops steeply, indicating a potential glacial refugium in the north (e.g., the Galicia Bank, formed as an extension between the continent shelf and the abyss in the Bay of Biscay).

The northwest Iberian Peninsula (the Galicia) was ice-free during the LGM (Fig. 12.2a) and was first proposed as a glacial refugium by phylogeographic study of terrestrial species (Petit et al. 2003). Afterward, many phylogeographic studies of intertidal species also supported this inference, such as the brown alga *Fucus serratus* (Hoarau et al. 2007) and the green crab *Carcinus maenas* (Roman and Palumbi 2004). High genetic diversity and unique haplotypes of the red alga *C. crispus* were found in the northwest Iberian Peninsula (Fig. 12.2c), thus indicating that *C. crispus* may have survived in a late Pleistocene glacial refugium in the Northeastern Atlantic (Hu et al. 2010, 2011; Provan and Maggs 2012).

(ii) The English Channel (The Hurd Deep)

During the LGM, the English Channel was dry land due to lower sea levels (about 135 m below present), but the existence of an enigmatic trench (Hurd Deep), about 100 m deeper than the adjoining sea floor (Lambeck et al. 2002), provided a suitable place for the survival of intertidal species during glacial periods (Provan et al. 2005). High microsatellite allelic diversity in northern Brittany and the Cornwall area was first found in *Ascophyllum nodosum* (Stam et al. 2000) and *F. serratus* (Coyer et al. 2003), and Provan et al. (2005) proposed the Hurd Deep in the English Channel as a Pleistocene marine glacial refugium to explain phylogeographic patterns in the endemic red alga *Palmaria palamta* (Fig. 12.2b). Other red seaweeds that have high levels of genetic diversity in this region include *M. stellatus* (Fig. 12.2d) (Zuccarello et al. 2005; Li et al. 2016), *C. crispus* (Fig. 12.2c) (Hu et al. 2010, 2011; Provan and Maggs 2012), and *P. umbilicalis* (Teasdale and Klein 2010). For *C. crispus* and *M. stellatus*, the English Channel populations made a prominent contribution to postglacial recolonization along European coasts (Provan and Maggs 2012). Evidence for the Hurd Deep as a glacial refugium during the Quaternary ice ages has also been detected in the brown algae *F. serratus* (Hoarau et al. 2007), *A. nodosum* (Olsen et al. 2010), and periwinkle *Littorina saxatilis* (Doellman et al. 2011).

(iii) Southwestern Ireland (Galway Bay)

Phylogeographic studies on terrestrial fauna and flora with a Lusitanian distribution first demonstrated the existence of the southwestern Irish glacial refugium (Sinclair et al. 1998; Rowe et al. 2006). Although the Quaternary ice sheets in the Northern Hemisphere covered most of Ireland (Fig. 12.2a), geomorphology and fossil evidence indicated that southern Ireland was partly unglaciated (Lambeck and Chappell 2001). In particular, genetic evidence from other marine intertidal organisms favors the existence of a Galway Bay refugium, including the soft-bottom polychaete worms *Pectinaria koreni* and *Owenia fusiformi* (Jolly et al. 2006), periwinkle (Doellman et al. 2011), the brown alga *F. serratus* (Hoarau et al. 2007), and the red algae *P. palmata* (Provan et al. 2005; Li et al. 2015), *M. stellatus* (Zuccarello et al. 2005; Li et al. 2016), and *C. crispus* (Hu et al. 2010, 2011; Provan and Maggs 2012). In *P. palmata*, large numbers of private haplotypes and small haplotype radiations indicate long-term persistence in this region (Hu et al. 2010, 2011; Provan and Maggs 2012; Li et al. 2015). Our comparative phylogeographic survey of the red algae *M. stellatus* and *C. crispus* further confirmed this hypothesis, as the colonization of northern Europe mostly relied on populations from this region (Fig. 12.3).

(iv) Northern European refugia (Norway/Iceland/The Faroe Islands)

Although northern Europe was covered by ice caps during the Quaternary ice ages (Fig. 12.2), the marginal coastal areas may have been suitable for the survival of intertidal species (Sejrup 2005). Norway, Iceland, and the Faroe Islands were key areas for intertidal species to survive through glacial periods. Terrestrial fossils (Larsen et al. 1987; Alm and Birks 1991) lend credence to the existence at the LGM of terrestrial refugia in northern Norway, but there is as yet no convincing evidence for marine species. Phylogeographically, a divergent Icelandic population of the isopod *Idotea balthica* is consistent with the hypothesis that there was a refugium in Iceland or the nearby Faroes (Wares 2001). Similar biogeographic patterns have also been observed in the green crab *C. maenas* (Roman and Palumbi 2004), supporting the notion that the Faroes were a glacial refugium during the Pleistocene ice ages. Phylogeographic studies on seaweeds have provided strong genetic evidence for the existence of northern European refugia, with high genetic diversity

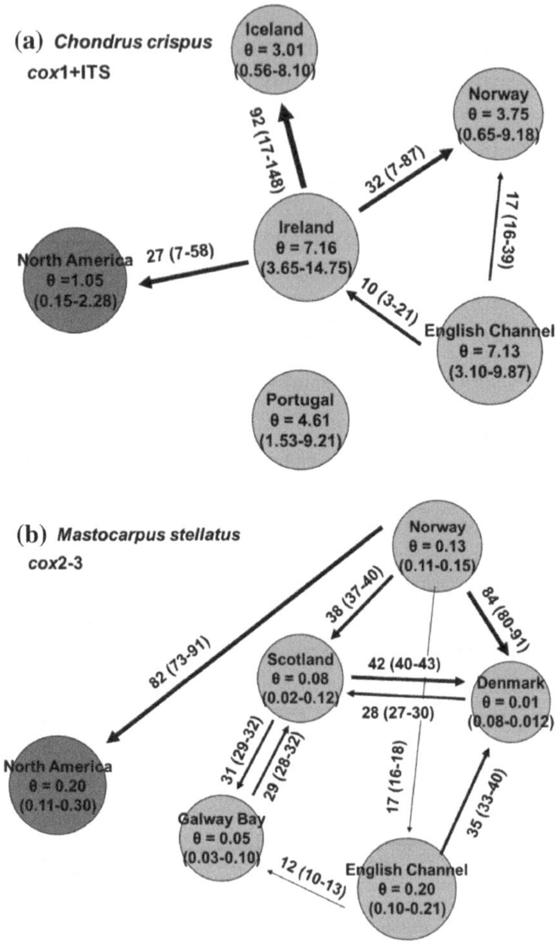

Fig. 12.3 Estimated historical effective population size (θ) and the number of migrants per generation (*N*m) between representative regions (95 % confidence intervals in parentheses) for *Chondrus crispus* (**a**) and *Mastocarpus stellatus* (**b**). For clarity of presentation *N*m values less than 10 are not shown. The North American population is shown in *dark gray* and the European populations are shown in *light gray*

and unique haplotypes detected in this region (*P. palmata*, Provan et al. 2005; Li et al. 2015; *M. stellatus*, Zuccarello et al. 2005; Li et al. 2016, and *P. umbilicalis*, Teasdale and Klein 2010), but the exact geographic sites of refugia are still uncertain. Our comparative coalescence analyses of *M. stellatus* and *C. crispus* confirmed that the colonization of northern Europe and North America mostly relied on populations originating in these periglacial regions (Fig. 12.3).

12.5.2 Western North Atlantic Refugia

To date there are at least three regions in North America, Grand Banks, southern Nova Scotia shelf, and George's Bank, which have been proposed as potential glacial refugia for intertidal species (Young et al. 2002; Maggs et al. 2008). For intertidal red algae, the most likely area was Grand Banks, because recent studies revealed that the Laurentide ice sheet extended to the edge of the continental shelf in the Gulf of Maine (Miller et al. 2002; Charbit et al. 2007; Uchupi and Bolmer 2008), and the area south of the southern limit of the ice sheet lacked suitable rocky substrate (Riggs et al. 1996; Wares and Cunningham 2001). Previous phylogeographic studies have also revealed that the Grand Banks were unglaciated during the LGM (Brochmann et al. 2003; Maggs et al. 2008), thus making it the most likely area for western Atlantic refugia. Deep genetic divergence was detected between European and North American *P. palmata* populations, indicating long isolation between the populations in these regions (Li et al. 2015) (Fig. 12.2). High levels of genetic diversity and large numbers of private haplotypes were detected in North American populations of *C. crispus* (Fig. 12.2c). The genetic evidence supports the presumption that there were glacial refugia for these seaweeds to survive in the western North Atlantic (Provan et al. 2005; Hu et al. 2010, 2011; Provan and Maggs 2012; Li et al. 2015).

12.6 Vicariant Isolation and Trans-Atlantic Dispersal

We used two organellar markers (*cox*2-3 and *rps*) to assess the phylogeographic history of *P. palmata* (Li et al. 2015). A genealogical split that cleanly distinguishes the western North Atlantic populations from the eastern (see Fig. 12.1 in Li et al. 2015) probably stems from a vicariant separation during the Pleistocene ice ages that drove isolation between the western and eastern North Atlantic populations of *P. palmata*. Genetic substructure was detected in European populations and the division was likely related to survival in scattered glacial refugia along the European coasts (Li et al. 2015). It seems that vicariant isolation has played a predominant role in shaping the genetic structure of *P. palmata*, because IMA analysis revealed quite restricted genetic introgression among regions (Li et al. 2015).

Phylogeographic surveys also showed that the red alga *C. crispus* likely survived in multiple refugia in the North Atlantic despite no genetically distinct groups/populations detected within this species. Molecular diversity statistics using two-locus scnDNA data indicated that *C. crispus* in Europe recolonized via a stepping-stone model (Provan and Maggs 2012). Here we reassessed the amounts of genetic exchange between nearby regions using IMA based on combined ITS and *cox*1 data (Nielsen and Wakeley 2001; Hey and Nielsen 2007; Hey 2010) (Fig. 12.3a). We found that the direction of trans-Atlantic dispersal was mainly from south to north and the reverse dispersal was nearly negligible (<10) (Fig. 12.3a). High levels of genetic exchange after glacial periods may have led to genetic homogeneity of *C. crispus* along European coasts. Interestingly, the populations along the Iberian Peninsula possessed a particularly high proportion of endemic haplotypes and alleles across both markers analyzed (ITS, *cox*1) (Provan and Maggs 2012). The detected variance may reflect the failure of the Iberian refugium (Portugal) to contribute to postglacial recolonization to the same extent as the English Channel populations, due to the general lack of suitable substrate in the Bay of Biscay (Hoarau et al. 2007). In addition, we did not detect genetic exchange between Iberian (Portugal) and other populations of *C. crispus* (Fig. 12.3a).

The red alga *M. stellatus* showed similar genetic homogeneity with two main *cox*2-3 haplotypes identified throughout its distributional range, but the mechanistic processes for the weak genetic structure seem more complicated. Previous research showed that *M. stellatus* has two inter-sterile breeding groups which tend to have a north–south distribution pattern (Guiry and West 1983), and this laboratory hybridization-based hypothesis has later been demonstrated by organellar genetic variation (Zucarello et al. 2005). Our recent analysis based on cpDNA RuBisCo spacer also presented population-level phylogeographic evidence of the north–south distribution of two lineages, but mtDNA *cox*2-3 analysis showed that 'southern' haplotypes were also widely distributed in northern Europe (Li et al. 2016). One possible explanation is that asexual plants in the north hybridized with direct or asexual-type from the south, and consequently generated a 'mixed' mitochondrial genome while the plastid genome remained intact (Johnson et al. 2012; Li et al. 2016). Intraspecific genetic surveys based on *cox*2-3 indicated that large amounts of trans-Atlantic dispersal occurred between the northern and southern regions (Fig. 12.3b), which may also have contributed to the admixture of the two diverged lineages (Li et al. 2016).

12.6.1 Demographic Histories

Based on the previous work on *P. palmata* and *C. crispus*, population divergence and demographic history were calculated over regional scales (the codes for each geographical region are shown in Table 12.2). However, we calculated divergence times only between lineages in *M. stellatus* (Li et al. 2016), because the datasets do not fit for the isolation-by-migration model used in IMA (Nielsen and Wakeley 2001; Hey and Nielsen 2007; Hey 2010). The divergence times between regions/lineages in the North Atlantic red seaweeds are shown in Fig. 12.4. It seems that most of the region-level divergence times in *C. crispus* occurred during the last glacial and interglacial periods (0.013–0.128 Ma) (Fig. 12.4). The population expansion times for *C. crispus* also happened during the similar period (Fig. 12.5), indicating that the Illinoian glacial period (0.191–0.120 Ma) may have played an important role in shaping the phylogeographic history of *C. crispus*. Nevertheless, the divergence times between Portugal (POR) and other populations were much older with the estimates ranging from 0.116 to 0.312 Ma (Fig. 12.4). As discussed above, the likely cause may be that the Portugal population did not contribute to postglacial recolonization along European coasts due to the appearance of a geographical barrier in the Bay of Biscay (i.e., lack of rocky habitat). This phylogeographic pattern has also been detected in other seaweed species and marine invertebrates in the North Atlantic

Table 12.2 Regional codes and corresponding areas

Species	Code	Region	Code	Region
Palmaria palmata	NWY	Norway	ICE	Iceland
	GAL	Galway Bay	BFD	Bay of Fundy
	ENC	English Channel	GSL	Gulf of St. Lawrence
Chondrus crispus	IC	Iceland	POR	Portugal
	NOR	Norway	NA	North America
	ISI	Ireland		
	EC	English Channel		

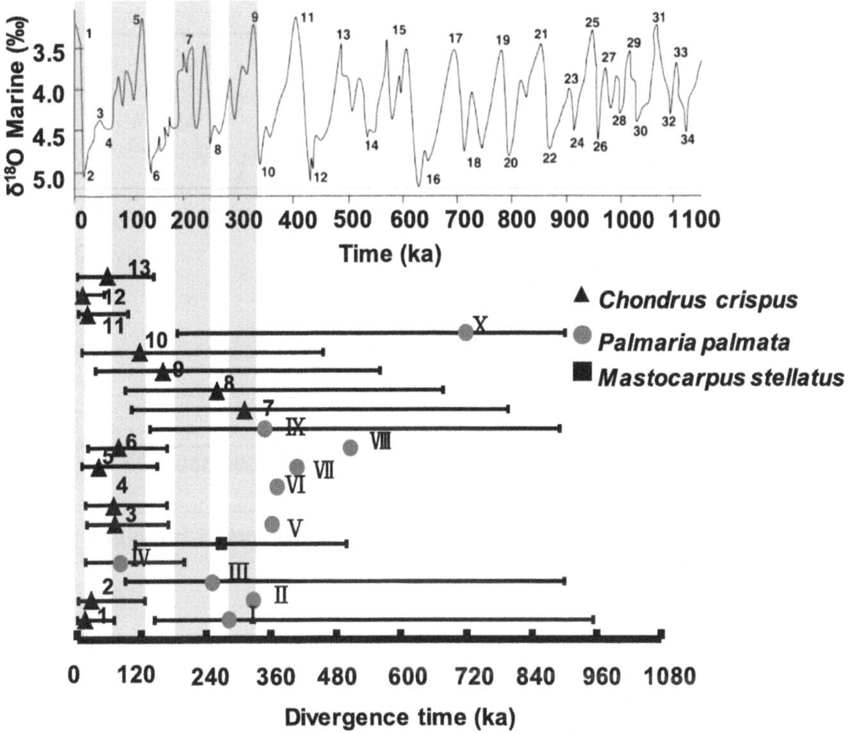

Fig. 12.4 Estimated timing of divergence for pairwise intraspecific lineages in the three red seaweeds (*Chondrus crispus*, *Palmaria palmata*, and *Mastocarpus stellatus*) and their relation to Pleistocene glacial maxima inferred from 57 globally distributed $\delta^{18}O$ records (after Lisiecki and Raymo 2005). *Gray bands* represent interglacial periods. *Chondrus crispus*: *1* IC-NOR; *2* IC-ISI; *3* NOR-ISI; *4* EC-IC; *5* EC-NOR; *6* ISI-EC; *7* ISI-POR; *8* EC-PO; *9* POR-NOR; *10* POR-IC; *11* IC-NA; *12* NOR-NA; *13* ISI-NA; *Palmaria palmata*: *I* ENC-GAL; *II* ENC-NWY; *III* GAL-NWY; *IV* GSL-BFD; *V* NWY-BFD; *VI* GAL-GSL; *VII* NWY-GSL; *VIII* ENC-BFD; *IX* GAL-BFD; *X* ENC-GSL. See Table 12.2 for site code explanation

(Hoarau et al. 2007; Maggs et al. 2008). In contrast, the genetic divergence between North American and European populations of *C. crispus* occurred much more recently, between 0.02 and 0.056 Ma, which may be driven by high levels of post-glacial trans-Atlantic dispersal (Fig. 12.3a).

In comparison with *C. crispus*, the intraspecific genetic divergence of the red alga *P. palmata* was quite ancient, between 0.080 and 0.716 Ma (Fig. 12.4). *M. stellatus* underwent a genetic lineage split at *c.* 0.267 Ma, corresponding to the Elster glacial period (*c.* 0.23–0.30 Ma). The extended Bayesian skyline plots (EBSPs) analysis suggests that European *P. palmata* populations expanded at *c.* 0.169–0.322 Ma (Fig. 12.5), indicating that the two interglacial periods, Holstein and Cromer in the late Pleistocene (*c.* 0.180–0.350 Ma; Coyer et al. 2011), may

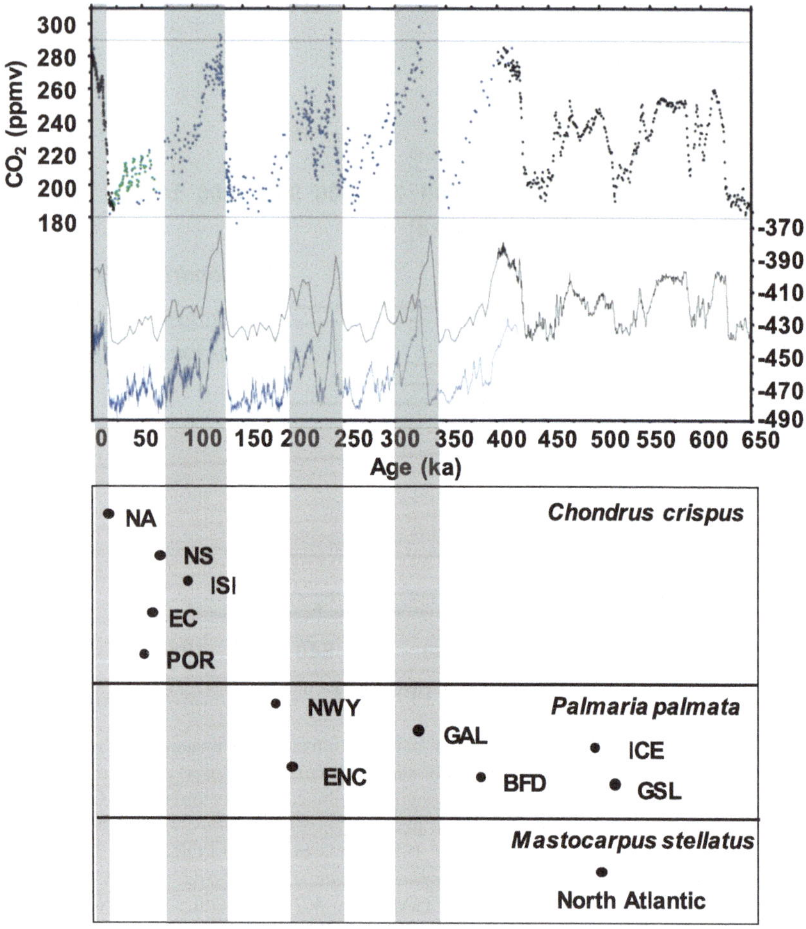

Fig. 12.5 Expansion times estimated from extended Bayesian skyline plots analysis using BEAST. The results overlap with Pleistocene interglacial (*gray bands*) and glacial periods (*white bands*). The composite CO_2 record (the graph above) was adopted from Siegenthaler et al. (2005)

have had a dominant effect on the demographic history of *P. palmata*. Such an expansion time frame resembles that of the brown alga *Fucus vesiculosus* (0.200–0.300 Ma; Coyer et al. 2011). Although the expansion times for North American and Icelandic *P. palmata* populations were much earlier (*c*. 0.400–0.537 Ma) than those for European populations, they are closer to estimates for *M. stellatus* (*c*. 0.520 Ma) (Fig. 12.5). Our comparative phylogeographic analyses imply that the demographic history of the three red seaweeds was likely structured by the mid-to-late Pleistocene (*c*. 0.84–0.13 Ma) glacial episodes.

12.7 Survival in Glacial Refugia Versus Postglacial Dispersal

Accumulating evidence indicates that the three red algal species reviewed in this chapter have inhabited North Atlantic coasts for a long time, and survived the LGM in the Northern Hemisphere. Nevertheless, genetic dynamic analyses indicate that they have probably undergone different trans-Atlantic dispersal processes. We detected contrasting genetic variation patterns between North American and European populations of *M. stellatus*, and the European side showed significantly higher genetic diversity than North America. The organellar haplotypes detected in North America were not only shared with European populations, but also exhibited a range-wide distribution in the North Atlantic (Li et al. 2016). We thus can hypothesize that North American *M. stellatus* populations derived from a trans-Atlantic dispersal from Europe after the LGM. More importantly, our *cox*2-3 data indicate dispersal of *M. stellatus* from Europe to North America (Fig. 12.3b). Interestingly, our identified postglacial trans-Atlantic dispersal pathways are similar to those found in *P. umbilicalis*: the Norway population underwent a bi-directional dispersal, with southward movement resulting in colonization of other parts of Europe while the westward movement seeded populations in North America (Fig. 12.3b) (Teasdale and Klein 2010; Li et al. 2016). The extremely weak genetic differentiation between North American *M. stellatus* populations suggests that the trans-Atlantic dispersal might have occurred only in the past few hundred years, mediated by anthropogenic activities (Johnson et al. 2012). Similar dispersal has also been inferred for the brown alga *F. serratus*, which was introduced to Canadian Maritime Provinces from western Ireland on rocks used as ballast in shipping (Brawley et al. 2009).

The existence of North American refugia is important for the second trans-Atlantic distribution scenario. Phylogeographic studies of red seaweeds, *C. crispus,* and *P. palmata* consistently indicate that the marginal area of the Grand Banks may have been a potential refugium for Northwestern Atlantic populations (Hu et al. 2010; 2011; Li et al. 2015). In particular, the presence of private haplotypes of *C. crispus* and *P. palmata* in North America is strongly indicative of long-term survival along the North American coasts. Ware (2001) proposed that the <20,000 years since the LGM is too short to account for significant genetic divergence detected in marine coastal invertebrates. The phylogeographic patterns observed in *C. crispus* and *P. palmata* provide further genetic evidence of glacial survival on both sides of the North Atlantic, because the divergence rates of organellar genes in seaweeds are much slower than those in invertebrates. The calculated divergence times between North American and European populations (Fig. 12.4) show that the trans-Atlantic separation occurred just before the LGM in *C. crispus* (*c.* 0.02–0.056 Ma), more recently than those in *P. palmata* (*c.* 0.348–0.716 Ma). This evolutionary scenario has also been confirmed in a growing body of literature, including on seaweeds (Olsen et al. 2010), sea urchins (Addison and Hart 2005), and mussels (Riginos and Henzler 2008).

12.8 Concluding Remarks

Phylogeographical studies of the red algae, *Chondrus crispus*, *Palmaria palmata,* and *Mastocarpus stellatus* in the North Atlantic uncovered the potential effects of marine environments (e.g., habitat discontinuities, ocean currents, salinity gradients, and geographic barriers) on these species' spatiotemporal distributions and population connectivity. Population genetic structure, historical demography, and gene flow are important molecular estimates to elucidate trans-Atlantic evolutionary processes and biogeographic histories of marine coastal flora. Despite the three algal species exhibiting similar distributional ranges in the North Atlantic, they are characterized by distinct phylogeographic histories and modes. These results suggest that the evolutionary effects of climatic changes on contemporary populations and species are far more complicated than previously expected. In other words, marine intertidal organisms with distinct biological features (e.g., life-history and dispersal capability) may leave behind distinct genetic imprints in extant populations. The comparative phylogeographic surveys demonstrate that the western and eastern North Atlantic shores are an excellent model system to distinguish evolutionary scenarios for marine organisms with a tran-Atlantic distribution: survival in glacial refugia versus postglacial trans-Atlantic migration. Methodologically, single-copy nuclear markers may provide important insights into speciation, evolution, and diversification patterns of marine algae, although organellar genetic markers (mainly mtDNA and cpDNA) are still the most popular tools for elucidating the evolutionary histories of seaweeds. We propose to use model-based phylogeographic inferences (e.g., Approximate Bayesian computation, ABC) to test evolutionarily contrasting scenarios in seaweeds.

Mutation rate is the key parameter to estimate divergence times and demographic histories for molecular phylogeographies. However, several studies show that genes' recent mutation rates are time dependent and for many species are an order of magnitude larger than mutation rates estimated from phylogenetic divergences (Ho et al. 2005, 2007). Hence, the use of mutation rates derived from divergences between species separated more than one million years tends to overestimate the timeframe for molecular estimates (Ho et al. 2008). In other words, it may be not appropriate to date population-level divergence events using species-level mutation rates because of the lack of a good fossil record for most seaweed groups (see Chapter by Grant (2016) in this volume).

Acknowledgments We would like to thank Ceridwen Fraser and two anonymous reviewers for valuable comments on the early version of this chapter. This study was partially supported by National Natural Science Foundation of China (31370264) granted to Z.M. Hu.

References

Addison JA, Hart MW. Colonization, dispersal, and hybridization influence phylogeography of North Atlantic sea urchins (*Strongylocentrotus droebachiensis*). Evolution. 2005;59:532–43.

Alm T, Birks HH. Late Weichselian flora and vegetation of Andøya, Northern Norway-Macro fossil (seed and fruit) evidence from Nedre Aerasvatn. Nord J Bot. 1991;11:465–76.

Bermingham E, Rohwer S, Freeman S, Wood C. Vicariance biogeography in the Pleistocene and speciation in North American wood warblers: Atest of Mengel's model. Proc Nat Acad Sci USA. 1992;89:6624–8.

Björck S. A review of the history of the Baltic Sea, 13.0–8.0 ka BP. Quat Int. 1995;27:19–40.

Bouza N, Caujapé-Castells J, González-Pérez MA, Sosa PA. Genetic structure of natural populations in the red algae *Gelidium canariense* (Gelidiales, Rhodophyta) investigated by random amplified polymorphic DNA (RAPD) markers. J Phycol. 2006;42:304–11.

Brawley SH, Coyer JA, Blakeslee AMH, Hoarau G, Johnson LE, Byers JE, Stam WT, Olsen JL. Historical invasions of the intertidal zone of Atlantic North America associated with distinctive patterns of trade and emigration. Proc Nat Acad Sci USA. 2009;106:8239–44.

Brochmann C, Gabrielsen TM, Nordal I, Landvik JY, Elven R. Glacial survival or tabula rasa? The history of North Atlantic biota revisited. Taxon. 2003;52:417–50.

Charbit S, Ritz C, Philippon G, Peyaud V, Kageyama M. Numerical reconstructions of the Northern Hemisphere ice sheets through the last glacial-interglacial cycle. Clim Past. 2007;3:15–37.

Clark PU, Dyke AS, Shakun JD, Carlson AE, Clark J, Wohlfarth B, Mitrovica JX, Hostetler SW, McCabe AM. The Last Glacial Maximum. Science. 2009;325:710–4.

Couceiro L, Maneiro I, Ruiz JM, Barreiro R. Multiscale genetic structure of an endangered seaweed *Ahnfeltiopsis pusilla* (Rhodophyta): implications for its conservation. J Phycol. 2011;47:259–68.

Coyer JA, Peters AF, Stam WT, Olsen JL. Post-ice age recolonization and differentiation of *Fucus serratus* L. (Phaeophyceae; Fucaceae) populations in Northern Europe. Mol Ecol. 2003;12:1817–29.

Coyer JA, Hoarau G, Costa JF, Hogerdijk B, Serrão EA, Billard E, Valero M, Pearson GA, Olsen JL. Evolution and diversification within the intertidal brown macroalgae *Fucus spiralis*/ *F. vesiculosus* species complex in the North Atlantic. Mol Phylogenet Evol. 2011;58:283–96.

Doellman MM, Trussell GC, Grahame JW, Vollmer SV. Phylogeographic analysis reveals a deep lineage split within North Atlantic *Littorina saxatilis*. Proc R Soc B. 2011;278:3175–83.

Dyke AS, Andrews JT, Clark PU, England JH, Miller GH, Shaw J, Veillette JJ. The Laurentide and Innuitian ice sheets during the Last Glacial Maximum. Quat Sci Rev. 2002;21:9–31.

Faugeron S, Valero M, Destombe C, Martinez EA, Correa JA. Hierarchical spatial structure and discriminant analysis of genetic diversity in the red alga *Mazzaella laminarioides* (Gigartinales, Rhodophyta). J Phycol. 2001;37:705–16.

Gabrielsen TM, Brochmann C, Rueness J. The Baltic Sea as a model system for studying postglacial colonization and ecological differentiation, exemplified by the red alga *Ceramium tenuicorne*. Mol Ecol. 2002;11:2083–95.

Grant WS. Paradigm shifts in the analysis of seaweeds. In: Hu ZM, Fraser CI, editors. Seaweed phylogeography: adaptation and evolution of seaweeds under environmental change. Springer, Heidelberg; 2016.

Guiry MD, West JA. Life history and hybridization studies on *Gigartina stellata* and *Petrocelis cruenta* (Rhodophyta) in the North Atlantic. J Phycol. 1983;19:474–94.

Gurgel CFD, Fredericq S, Norris JN. Phylogeography of *Gracilaria tikvahiae* (Gracilariaceae, Rhodophyta): A study of genetic discontinuity in a continuously distributed species based on molecular evidence. J Phycol. 2004;40:748–58.

Hewitt GM. Some genetic consequences of ice ages, and their role in divergence and speciation. Biol J Linn Soc. 1996;58:247–76.

Hewitt GM. The genetic legacy of the Quaternary ice ages. Nature. 2000;405:907–13.

Hewitt GM. Genetic consequences of climatic oscillations in the Quaternary. Phil Trans R Soc B. 2004;359:183–95.

Hey J. Isolation with migration models for more than two populations. Mol Biol Evol. 2010;27:905–20.

Hey J, Nielsen R. Integration within the Felsenstein equation for improved Markov chain Monte Carlo methods in population genetics. Proc Nat Acad Sci USA. 2007;104:2785–90.

Ho SYW, Phillips MJ, Drummond AJ, Cooper A. Accuracy of rate estimation using relaxed-clock models with a critical focus on the early metazoan radiation. Mol Biol Evol. 2005;22:1355–63.

Ho SYW, Shapiro B, Phillips MJ, Copper A, Drummond AJ. Evidence for time dependency of molecular rate estimates. Syst Biol. 2007;56:515–22.

Ho SYW, Saarma U, Barnett R, Haile J, Shapiro B. The effect of inappropriate calibration: three case studies in molecular ecology. PLoS ONE. 2008;3(2):e1615.

Hoarau G, Coyer JA, Veldsink JH, Stam WT, Olsen JL. Glacial refugia and recolonization pathways in the brown seaweed *Fucus serratus*. Mol Ecol. 2007;16:3606–16.

Hu ZM. Intertidal population genetic dynamics at a microgeographic seascape scale. Mol Ecol. 2013;22:3191–4.

Hu ZM, Duan DL, Lopez-Bautista J. Seaweed phylogeography from 1994 to 2014: an overview. In: Hu ZM, Fraser CI, editors. Seaweed phylogeography: adaptation and evolution of seaweeds under environmental Change. Springer, Heidelberg; 2016.

Hu ZM, Zeng XQ, Critchley AT, Morrell SL, Duan DL. Phylogeography of the Northern Atlantic species *Chondrus crispus* (Gigartinales, Rhodophyta) inferred from nuclear rDNA internal transcribed spacer sequences. Hydrobiologia. 2007;575:315–27.

Hu ZM, Guiry MD, Critchley AT, Duan DL. Phylogeographic patterns indicate transatlantic migration from Europe to North America in the red seaweed *Chondrus crispus* (Gigartinales, Rhodophyta). J Phycol. 2010;46:889–900.

Hu ZM, Li W, Li JJ, Duan DL. Post-Pleistocene demographic history of the North Atlantic endemic Irish moss *Chondrus crispus*: glacial survival, spatial expansion and gene flow. J Evol Biol. 2011;24:505–17.

Ingólfsson A. The origin of the rocky shore fauna of Iceland and the Canadian Maritimes. J Biogeogr. 1992;19:705–12.

Johnson LE, Brawley SH, Adey WH. Secondary spread of invasive species: historic patterns and underlying mechanisms of the continuing invasion of the European rockweed *Fucus serratus* in eastern North America. Biol Invasions. 2012;14:79–97.

Jolly MT, Viard F, Gentil F, Thiebaut E, Jollivet D. Comparative phylogeography of two coastal polychaete tubeworms in the Northeast Atlantic supports shared history and vicariant events. Mol Ecol. 2006;15:1841–55.

Krebes L, Blank M, Bastrop R. Phylogeography, historical demography and postglacial colonization routes of two amphi-Atlantic distributed amphipods. Syst Biodivers. 2011;9:259–73.

Krueger-Hadfield SA, Collen J, Daguin-Thiebaut C, Valero M. Genetic population structure and mating system in *Chondrus crispus* (Rhodophyta). J Phycol. 2011;47:440–50.

Lambeck K, Chappell J. Sea level change through the last glacial cycle. Science. 2001;292: 679–86.

Lambeck K, Esat TM, Potter EK. Links between climate and sea levels for the past three million years. Nature. 2002;419:199–206.

Larsen E, Gulliksen S, Lauritzen SE, Lie R, Lovlie R, Mangerud J. Cave stratigraphy in western Norway: multiple Weischselian glaciations and interstadial vertebrate fauna. Boreas. 1987;16:267–92.

Li JJ, Hu ZM, Duan DL. Genetic data from the red alga *Palmaria palmata* reveal a mid-Pleistocene deep genetic split in the North Atlantic. J Biogeogr. 2015;42(5):902–13.

Li JJ, Hu ZM, Liu RY, Zhang J, Liu SL, Duan DL. Phylogeographic surveys and apomictic genetic connectivity in the North Atlantic red seaweed *Mastocarpus stellatus*. Mol Phylogenetic Evol. 2016;94: 463–72.

Lisiecki LE, Raymo ME. A Pliocene-Pleistocene stack of 57 globally distributed benthic $\delta^{18}O$ records. Paleoceanography. 2005;20:17.

Maggs CA, Castilho R, Foltz D, Henzler C, Jolly MT, Kelly J, Olsen J, Perez KE, Stam W, Väinölä R, Viard F, Wares J. Evaluating signatures of glacial refugia for North Atlantic benthic marine taxa. Ecology. 2008;89:S108–22.

Mathieson AC, Penniman CA, Busse PK, Tvetergallagher E. Effects of ice on *Ascophyllum nodosum* within the Great Bay Estuary System of New Hampshire-Maine. J Phycol. 1982;18:331–6.

McKindsey CW, Bourget E. Diversity of a northern rocky intertidal community: the influence of body size and succession. Ecology. 2001;82:3462–78.

Miller GH, Wolfe AP, Steig EJ, Sauer PE, Kaplan MR, Briner JP. The Goldilocks dilemma: big ice, little ice, or "just-right" ice in the Eastern Canadian Arctic. Quat Sci Rev. 2002;21:33–48.

Minchinton TE, Scheibling RE, Hunt HL. Recovery of an intertidal assemblage following a rare occurrence of scouring by sea ice in Nova Scotia, Canada. Bot Mar. 1997;40:139–48.

Neiva J, Serrão EA, Assis J, Pearson GA, Coyer JA, Olsen JL, Hoarau G, Valero M. Climate oscillations, and phylogeographic patterns of North Atlantic Fucaceae. In: Hu ZM, Fraser CI, editors. Seaweed phylogeography: adaptation and evolution of seaweeds under environmental change. Springer, Heidelberg; 2016.

Neiva J, Pearson GA, Valero M, Serrão EA. Drifting fronds and drifting alleles: range dynamics, local dispersal and habitat isolation shape the population structure of the estuarine seaweed *Fucus ceranoides*. J Biogeogr. 2012;39:1167–78.

Neiva J, Assis J, Fernandes F, Pearson GA, Serrão EA. Species distribution models andmmitochondrial DNA phylogeography suggest an extensive biogeographical shift in the. J Biogeogr. 2014;41:1137–48.

Nielsen R, Wakeley J. Distinguishing migration from isolation: a Markov chain Monte Carlo approach. Genetics. 2001;158:885–96.

Olsen JL, Zechman FW, Hoarau G, Coyer JA, Stam WT, Valero M, Åberg P. The phylogeographic architecture of the fucoid seaweed *Ascophyllum nodosum*: an intertidal 'marine tree' and survivor of more than one glacial-interglacial cycle. J Biogeogr. 2010;37:842–56.

Petit RJ, Aguinagalde I, de Beaulieu JL, Bittkau C, Brewer S, Cheddadi R, Ennos R, Fineschi S, Grivet D, Lascoux M, Mohanty A, Muller-Starck GM, Demesure-Musch B, Palme A, Martin JP, Rendell S, Vendramin GG. Glacial refugia: hotspots but not melting pots of genetic diversity. Science. 2003;300:1563–5.

Provan J, Bennett KD. Phylogeographic insights into cryptic glacial refugia. Trends Ecol Evol. 2008;23:564–71.

Provan J, Maggs CA. Unique genetic variation at a species' rear edge is under threat from global climate change. Proc R Soc B. 2012;279:39–47.

Provan J, Wattier RA, Maggs CA. Phylogeographic analysis of the red seaweed *Palmaria palmata* reveals a Pleistocene marine glacial refugium in the English Channel. Mol Ecol. 2005;14: 793–803.

Provan J, Glendinning K, Kelly R, Maggs CA. Levels and patterns of population genetic diversity in the red seaweed *Chondrus crispus* (Florideophyceae): a direct comparison of single nucleotide polymorphisms and microsatellites. Biol J Linn Soc. 2013;108:251–62.

Riggs SR, Snyder SW, Hine AC, Mearns DL. Hardbottom morphology and relationship to the geologic framework: mid-Atlantic continental shelf. J Sediment Res. 1996;66:830–46.

Riginos C, Henzler CM. Patterns of mtDNA diversity in North Atlantic populations of the mussel *Mytilus edulis*. Mar Biol. 2008;155:399–412.

Roman J, Palumbi SR. A global invader at home: population structure of the green crab, *Carcinus maenas*, in Europe. Mol Ecol. 2004;13:2891–8.

Roth PH. Mesozoic paleoceanography of the North Atlantic and Tethys Oceans. In: Summerhayes CP, Shackleton NJ, editors. North Atlantic Paleoceanography, vol. 21. Geological Society of America Special Publication 1986; pp. 299–320.

Rowe G, Harris DJ, Beebee TJC. Lusitania revisited: a phylogeographic analysis of the natterjack toad Bufo calamita across its entire biogeographical range. Mol Phylogenet Evol. 2006;39:335–46.

Saucier FJ, Roy F, Gilbert D, Pellerin P, Ritchie H. Modeling the formation and circulation processes of water masses and sea ice in the Gulf of St. Lawrence, Canada. J Geophys Res Oceans. 2003;108:3269.

Scrosati R, Heaven C. Spatial trends in community richness, diversity, and evenness across rocky intertidal environmental stress gradients in eastern Canada. Mar Ecol Prog Ser. 2007;342:1–14.

Sejrup HP, Hjelstuen BO, Dahlgren KIT, Haflidason H, Kuijpers A, Nygard A, Praeg D, Stoker MS, Vorren TO. Pleistocene glacial history of the NW European continental margin. Mar Pet Geol. 2005;22:1111–29.

Siegenthaler U, Stocker TF, Monnin E, Luthi D, Schwander J, Stauffer B, Raynaud D, Barnola JM, Fischer H, Masson-Delmotte V, Jouzel J. Stable carbon cycle-climate relationship during the late Pleistocene. Science. 2005;310:1313–7.

Sinclair WT, Morman JD, Ennos RA. Multiple origins for Scots pine (*Pinus sylvestris* L) in Scotland: evidence from mitochondrial DNA variation. Heredity. 1998;80:233–40.

Stam WT, Olsen JL, Coyer JA. Post-glacial recolonization and biogeographic patterns in the North Atlantic. Phycologia. 2000;40s:46.

Teasdale BW, Klein AS. Genetic variation and biogeographical boundaries within the red alga *Porphyra umbilicalis* (Bangiales, Rhodophyta). Bot Mar. 2010;53:417–31.

Uchupi E, Bolmer ST. Geologic evolution of the Gulf of Maine region. Earth Sci Rev. 2008;91:27–76.

van Oppen MJH, Draisma SGA, Olsen JL, Stam WT. Multiple trans-Arctic passages in the red alga *Phycodrys rubens*: evidence from nuclear rDNA ITS sequences. Mar Biol. 1995;123: 179–88.

Waltari E, Hickerson MJ. Late Pleistocene species distribution modelling of North Atlantic intertidal invertebrates. J Biogeogr. 2013;40:249–60.

Wares JP. Intraspecific variation and geographic isolation in *Idotea balthica* (Isopoda: Valvifera). J Crustac Biol. 2001;21:1007–13.

Wares JP, Cunningham CW. Phylogeography and historical ecology of the North Atlantic intertidal. Evolution. 2001;55:2455–69.

Young AM, Torres C, Mack JE, Cunningham CW. Morphological and genetic evidence for vicariance and refugium in Atlantic and Gulf of Mexico populations of the hermit crab *Pagurus longicarpus*. Mar Biol. 2002;140:1059–66.

Zaragosi S, Eynaud F, Pujol C, Auffret GA, Turon JL, Garlan T. Initiation of the European deglaciation as recorded in the northwestern Bay of Biscay slope environments (Meriadzek Terrace and Trevelyan Escarpment): a multi-proxy approach. Earth Planet Sci Lett. 2001;188:493–507.

Zuccarello GC, Schidlo N, McIvor L, Guiry MD. A molecular re-examination of speciation in the intertidal red alga *Mastocarpus stellatus* (Gigartinales, Rhodophyta) in Europe. Eur J Phycol. 2005;40:337–44.

Zuccarello GC, Buchanan J, West JA. Increased sampling for inferring phylogeographic patterns in *Bostrychia radicans*/*B. moritziana* (Rhodomelaceae, Rhodophyta) in the eastern USA. J Phycol. 2006;42:1349–52.

Chapter 13
Comparative Population Genetics of Red Alga Occupying Different Salinity Conditions

Helena Korpelainen

Abstract Osmotic stress is one of the major abiotic stresses in seaweeds, which may physiologically acclimate to changing osmolarity by altering their transcriptome under hyper- or hypo-osmotic conditions. Also local adaptations leading to genetic differentiation between populations may develop. DNA-based genetic markers and sequencing provide the essential tools to investigate genetic diversity and differentiation and signs of selection. Whole-genome or transcriptome sequencing facilitates marker development and allows in-depth investigations on the population genetic structures of organisms. In this chapter, a special attention is paid on a set of genetic studies conducted on the marine red alga *Furcellaria lumbricalis*, including populations in the Atlantic Ocean (35 psu) and the brackish Baltic Sea (3.8-8 psu). The amount of genetic variation did not differ between ocean and brackish populations, but differences were observed between marker types. The expressed sequence tag (EST)—derived microsatellites possessed less variation and showed greater differentiation than the putatively neutral microsatellites, whereas the EST-derived SNP markers contained even less variation and showed even more differentiation. Yet, the most distinct result was that *F. lumbricalis* showed definite differentiation between the ocean and brackish populations in exprcssed genomic regions, while such differentiation was not detected by presumably neutral loci. Thus, suboptimal salinity is a stress factor that affects population genetic structures. However, such differentiation would have been missed if the investigations on *F. lumbricalis* had relied on the analysis of only neutral markers.

Keywords *Furcellaria* · Microsatellites · Population genetics · Salinity tolerance · SNP markers · Transcription profiling

---

H. Korpelainen (✉)
Department of Agricultural Sciences, Viikki Plant Science Centre,
University of Helsinki, P.O. Box 27, 00014 Helsinki, Finland
e-mail: helena.korpelainen@helsinki.fi

Z.-M. Hu and C. Fraser (eds.), *Seaweed Phylogeography*,
DOI 10.1007/978-94-017-7534-2_13

13.1 Introduction

High tolerance, phenotypic plasticity and local adaptations may allow species to occur across a range of environments. Adaptation to a combination of environmental factors presumes genetic differentiation, but tolerance depends on physiological responses that usually limit productivity outside the optimal environment (Kamer and Fong 2000; Oetjen and Reusch 2007). Osmotic stress is one of the major abiotic stresses to which many seaweeds are exposed. Yet, osmoregulatory processes are important to all living organisms, since the maintenance of intracellular osmotic pressure or chemical potential of metabolites is of fundamental importance for cell survival (Hoffmann 1987).

High osmolarity caused by elevated salinity lowers the external water potential. To cope with hyperosmolarity, seaweeds may increase the uptake of ions (K^+ and Cl^-), removal of Na^+, water loss, and the synthesis of osmotically active carbohydrates (Karsten et al. 1992; Mostaert et al. 1995). The amount of soluble carbohydrates can be adjusted to maintain osmobalance in cells during salinity stress (Lüning et al. 1990). Hypo-osmotic stress due to decreased salinity increases the external water potential, which may induce water uptake that results in increased cell volume and turgor pressure and in the loss of ions and organic solutes, thus leading to osmotic adjustment. Species exposed to such stresses may show lowered performance due to inefficient cellular metabolism, changed cellular ultrastructure and defective ion metabolites in the cell (Kirst 1990). Furthermore, osmotic stress can induce intracellular generation of reactive oxygen species (ROS) that cause oxidative damage to lipids, proteins, and nucleic acids (Teo et al. 2009).

In seaweeds, adaptation and acclimation to osmotic stresses are of key importance, and it is important to understand the specific mechanisms by which osmotic stress impacts those organisms and what kind of stress-related consequences arise in nature. The distribution of algal species under suboptimal salinity may also be influenced by the impairment of the reproductive system due to reduced performance of gametes or unsuccessful fertilization (Raven 1999; Serrão et al. 1999). It has been shown that extended exposure to higher or lower than optimal salinities inhibits cell division and may result in stunted growth (Graham and Wilcox 2000). It has been discovered that many marine species occurring along the salinity gradient of the brackish Baltic Sea in the northern Europe show at least some degree of physiological adaptation to the brackish water conditions (Kristiansen et al. 1994; Düwel 2001; Bergström and Kautsky 2005). Furthermore, some marine species show marked genetic differentiation between populations living in the Baltic Sea and in marine habitats with higher salinity levels (Luttikhuizen et al. 2003; Olsen et al. 2004; Johannesson and André 2006; Kostamo et al. 2012; Olsson and Korpelainen 2013), although in some cases the diversity and differentiation may be the result of invasions by different evolutionary lineages instead of local adaptation processes (Röhner et al. 1997; Väinölä 2003; Nikula et al. 2007).

13.2 Life Histories and Reproduction of Red Algae

A combination of sexual and asexual modes of reproduction is common in red algae, and asexual reproduction tends to be prevalent in the marginal regions of distribution (Hawkes 1990; Maggs 1998). The occurrence of asexual reproduction has been discovered in the hyposaline waters of the Baltic Sea in several macroalgal species (Bergström et al. 2005; Tatarenkov et al. 2005). Since red algae lack motile sperm, they rely on water currents to transport gametes to female organs or on vegetative fragments to propagate new locations (Lee 2008). Red algae display an alternation of generations. In addition to the gametophyte generation, many red algae have two sporophyte generations, (i) carposporophytes that produce carpospores, which germinate into (ii) tetrasporophytes, which then produce tetraspores that germinate into gametophytes (Lee 2008). Carpospores may also germinate directly into thalloid gametophytes, or carposporophytes may produce tetraspores without going through the tetrasporophyte phase. Thus, the life histories of red algae are complex and varied.

For instance, the marine red alga *Furcellaria lumbricalis* (Hudson) Lamoroux has a triphasic life cycle consisting of a haploid sexual phase (the gametophyte) and two diploid phases: the carposporophyte, which grows parasitically on the female gametophyte, and the tetrasporophyte (Austin 1960a, b). In the Atlantic Ocean, *F. lumbricalis* grows in subtidal habitats on sheltered to moderately expose rocky shores (Schwenke 1971) where its distribution depends on water turbidity, competition and the presence of a suitable growing substrate (Taylor 1975; Holmsgaard et al. 1981). In the low-salinity conditions of the brackish Baltic Sea—a marginal habitat for the species—*F. lumbricalis* grows under the brown alga *Fucus vesiculosus* in the bladder wrack belt, but also in the red algal belt among other red algae (Rosenvinge 1917; Wærn 1952: Mäkinen et al. 1988). Kostamo and Mäkinen (2006) have shown that *F. lumbricalis* is unlikely to reproduce sexually in the northern and eastern parts of the hyposaline Baltic Sea where the salinity is below 7 psu and that the spores produced in the Baltic Sea populations are smaller and more often deformed than those in the Atlantic Ocean populations. This finding indicates that low salinity creates a stressful environment and suggests that there may be consequences on the genetic structure of populations.

13.3 Genetic Tools for Population Genetic Analyses on Red Algae

DNA-based genetic markers and sequencing provide essential tools to measure genetic diversity within and differentiation among populations of red algae. Recent developments in high-throughput sequencing technologies enable the effective discovery of single nucleotide polymorphisms (SNPs, i.e., variation occurring when a single nucleotide, A, T, C, or G, differs) and other polymorphic DNA markers, such as microsatellites (i.e., repeats of 1–6 base pair long sequences. Consequently, it has become easier to develop markers for, e.g., molecular ecological and conservation genetic research on natural populations. Along with better availability of markers, there has been an increasing interest to compare information of EST (expressed sequence tag)—derived markers that represent portions of expressed genes with data based on noncoding markers. Studies on expressed genomic regions allow the discovery of variation involved in adaptation and make it possible to link patterns of adaptive variation to environmental factors. The presence of adaptive genetic variation is necessary for the survival of organisms when exposed to environmental changes.

Microsatellite repeat sequences are known to be ubiquitous in prokaryotic and eukaryotic genomes and present in both coding and noncoding regions. However, the distribution of microsatellites is not homogeneous within a genome, and the frequency of microsatellite sequences also varies across taxa, in terms of both absolute numbers of microsatellite loci and repeat motifs (Hancock 1999). The results of Korpelainen et al. (2007) suggest that fewer microsatellite regions are present in the red algal genome than in land plants. Even so, microsatellite markers have been successfully developed for many red algal species (Wattier et al. 1997; Luo et al. 1999; Guillemin et al. 2005; Andreakis et al. 2007; Xie et al. 2009, 2013; Krueger-Hadfield et al. 2011, 2013; Kostamo et al. 2012; Choi et al. 2013). In addition, some genetic studies on red algae have successfully used mitochondrial markers, such as cytochrome oxidase subunit 1 (*cox*1) and intergenic spacer between the cytochrome oxidase subunits 2 and 3 (*cox*2-3 spacer) (Yow et al. 2013).

Although whole-genome or transcriptome sequencing and the utilization of resulting libraries greatly facilitate marker development and allow the effective discovery of both neutral and adaptive genetic markers in any organism, the majority of such studies on plant genomes have concentrated on flowering plants. EST projects on red algae include *Gracilaria gracilis* (Lluisma and Ragan 1997), *Porphyra yezoensis* (Lee et al. 2000; Nikaido et al. 2000; Asamizu et al. 2003; Kitade et al. 2008), *Chondrus crispus* (Collén et al. 2006), *Saccharina japonica* (Liu et al. 2010), *F. lumbricalis* (Kostamo et al. 2011), *Pyropia haitanensis* (Xie et al. 2013) and *Pyropia tenera* (Choi et al. 2013). On average, one SNP can be expected every 500–1000 bp of coding sequence, and the mutation rates of SNPs (often about 10^{-8}–10^{-9}) are low when compared with the mutation rates of microsatellites (10^{-4}) (Brumfield et al. 2003). The results by Olsson and Korpelainen (2013) on the red alga *F. lumbricalis* are congruent with this information on the frequency of SNPs (on average, one SNP per 558 bp within the sequenced genomic region).

Kostamo et al. (2012) and Olsson and Korpelainen (2013) developed both putatively neutral and EST-derived microsatellite markers and SNP markers and used them to conduct comparative population genetic analyses of *F. lumbricalis* populations located in geographical locations with different salinity conditions in Northern and Western Europe. Population genetic information obtained from SNP markers (Olsson and Korpelainen 2013) was compared to the results of analyses of *F. lumbricalis* based on putatively neutral and adaptive (EST-derived) microsatellite

Table 13.1 Mean genetic diversity and genetic differentiation (F_{ST}) among Atlantic Ocean and Baltic Sea populations, and mean F_{ST} values between Atlantic Ocean and Baltic Sea populations of *Furcellaria lumbricalis* based on putatively neutral and EST-derived microsatellite markers, and SNP markers

Variable	Atlantic Ocean	Baltic Sea	Atlantic Ocean versus Baltic Sea
Genetic diversity			
Putatively neutral microsatellites	0.734	0.716	
EST-derived microsatellites	0.285	0.272	
SNP markers	0.132	0.111	
Genetic differentiation (F_{ST})			
Putatively neutral microsatellites	0.158	0.124	0.136
EST-derived microsatellites	0.230	0.108	0.401
SNP markers	0.353	0.278	0.535

Data from Kostamo et al. (2012) and Olsson and Korpelainen (2013). Genetic diversity is measured as expected heterozygosity for microsatellites and as Nei's (1987) gene diversity for SNP markers

markers (Kostamo et al. 2012). Although the new primer pairs were designed for *F. lumbricalis*, the SNP markers may have utility in population genetic and phylogenetic studies across red algal species, since SNP marker regions possess lower levels of variation than typically hypervariable microsatellite regions (Table 13.1).

13.4 Challenges of Red Algal Populations Occupying Different Salinity Conditions

In nature, many plants are adversely affected and challenged by various environmental factors that have negative effects on survival, development, and reproduction. Natural selection in the wild is largely created by environmental stress, which is most intensively evoked in an organism at the edges of its ecological niche. The extent to which an organism is able to deal with stresses determines the limits of its ecological amplitude. The possibility to discover large amounts of expressed (i.e., protein coding) sequence information offers a unique chance to screen and detect molecular variation of genes at a genome-wide level, and to discover the polymorphisms that affect the success of organisms in natural environments.

Osmotic stress is one of the most significant natural abiotic stresses of seaweeds. At suboptimal salinities, the growth of red algae is reduced, as shown, e.g., in *Gracilaria* species (Bird and McLachlan 1986; Choi et al. 2006; Guillemin et al. 2013) and *Dixoniella grisea*, in which also mannitol levels increased considerably when salinity increased from the optimal level of 10–60 psu, indicating its role as an osmolyte (Eggert et al. 2007). Seaweeds may physiologically acclimate to changing osmolarity by altering their transcriptome under hyper- or hypo-osmotic conditions, as shown, e.g., in *Gracilaria changii* (Teo et al. 2009), *F. lumbricalis*

(Kostamo et al. 2011), *Kappaphycus alvarezii* (Liu et al. 2011) and *Porphyra yezoensis* (Uji et al. 2012), all of which have shown changes in gene expression levels or in the proportions of ESTs representing different functional categories.

The Baltic Sea provides a unique model system for the study of genetic effects of postglacial colonization followed by ecological differentiation. The entire Baltic Sea was covered by the Northern European ice cap during the last glaciation (Andersen and Borns 1994). The melting of the continental icecap was followed by several freshwater and marine phases, and resulted in the opening of the current connection to the Atlantic Ocean about 8000 years ago. Consequently, a new colonization route to marine organisms was established (Björck 1995). Since then, the channel connecting the brackish Baltic Sea to the North Sea and the Atlantic Ocean has been reduced, which has resulted in a cline of decreasing salinity toward its inner parts and a relatively stable salinity regime on a local scale (Kullenberg 1981). The present salinity gradient of the Baltic Sea ranges from c. 30 psu in the channel to 2 psu in its most northern and eastern parts. Brackish water is a stressful environment for marine organisms, and only marine species capable of survival and reproduction in reduced salinity can remain in the Baltic Sea, resulting in a reduction in species numbers in all major taxa (Middleboe et al. 1997).

For instance, the red alga *Ceramium tenuicorne* possesses considerable variation in growth and reproduction along the salinity gradient of the Baltic Sea (Gabrielsen et al. 2002) and its genetic variation shows the presence of a continuous cline corresponding the salinity gradient, even though the used marker type (RAPDs) unlikely represents adaptive genetic variation. These results may still be indicative of limited successful migration of genotypes possibly adapted to local salinity conditions. The genetic variation pattern of *C. tenuicorne* also corresponds to the previously demonstrated ecotypic differentiation among its populations sampled along this gradient (Düwel 2001). Another red alga, *F. lumbricalis*, occurs in the cold waters of the North Atlantic and Arctic Ocean (Holmsgaard et al. 1981; Bird et al. 1991). The species is well known among larger red algae for its tolerance of low salinity (Bird et al. 1991). An experimental assessment of the effects of salinity on growth has shown that the increase in the biomass is maximal at the salinity of 20 psu under favorable light and temperature conditions (Bird et al. 1979). The distribution of *F. lumbricalis* in the Baltic Sea extends as far as the 4 psu isohaline regions of the Gulf of Bothnia and Gulf of Finland (Zenkevitch 1963; Bergström and Bergström 1999). Only recently, extensive genetic studies have been conducted on *F. lumbricalis* (Kostamo et al. 2011, 2012; Olsson and Korpelainen 2013).

13.5 Genetic Diversity and Differentiation of Red Algal Populations Occupying Different Salinity Conditions

Understanding how environmental factors influence the spatial distribution of genetic variation provides insight into microevolutionary processes. Through studies on *F. lumbricalis*, Kostamo et al. (2011, 2012) and Olsson and Korpelainen

(2013) have tried to develop a better understanding of the genetic adaptation potential of red algae living on the edge of their habitat range in the brackish Baltic Sea, which provides a unique, evolutionary relatively young environment. Genetic studies on *F. lumbricalis* may aid in the planning of conservation measures for also other species living in this vulnerable ecosystem. Thus far, only a low number of marine organisms have successfully adapted to the low-salinity waters of the Baltic Sea during the postglacial period (Middleboe et al. 1997).

13.5.1 Gene Expression Patterns

In order to identify genes with potential roles in the salinity tolerance and other stress responses, and to gain knowledge for further studies on the mechanisms behind the physiological and ecological stress responses of *F. lumbricalis*, Kostamo et al. (2011) conducted a small-scale transcriptome analysis and constructed an expressed sequence tag (EST) library for algal material originating from the marine environment along the coast of Northern Ireland (35 psu) when subjected to extremely low salinity (6 psu). These sequences were compared with EST sequences originating from algal material growing naturally at 6-psu salinity in the northern Baltic Sea along the southern coast of Finland in order to generate new markers for population genetic and stress tolerance adaptation studies. In all, 28 % of annotated ESTs (26 and 30 % in the Atlantic Ocean and Baltic Sea, respectively) played a role in general or specific abiotic stress responses, while 4.3 % of annotated ESTs were similar to genes with known roles in specific salinity stress responses (4.9 and 3.8 % in the Atlantic Ocean and Baltic Sea samples, respectively). Some differences between the two sequence datasets were observed in the proportions of ESTs representing different functional categories indicating moderate functional divergence between the ocean and brackish water populations of *F. lumbricalis* (Kostamo et al. 2011).

13.5.2 Genetic Diversity Patterns

Genetic diversity patterns of *F. lumbricalis* populations occurring in different salinity conditions, ranging from 35 psu of the Atlantic regions to 3.6 psu in eastern Gulf of Finland in the Baltic Sea, were investigated with three types of markers: putatively neutral microsatellites, and possibly adaptive microsatellite and SNP markers developed from EST library sequences (Kostamo et al. 2011, 2012; Olsson and Korpelainen 2013). Information from presumably neutral microsatellite markers was compared with data of EST-derived microsatellite and SNP markers to reveal genetic variation patterns in genomic regions potentially subjected to different evolutionary forces. The hypothesis was that the divergence pattern among *F. lumbricalis* populations differs between the putatively neutral and adaptive markers and that greater amounts of differentiation will be detected based on EST-derived and SNP markers.

Despite presumed asexual propagation in the Baltic Sea populations, the amount of genetic variation did not differ between the Atlantic Ocean and Baltic Sea populations, e.g., expected heterozygosities equalled 0.734 and 0.716, respectively, based on the neutral microsatellites, 0.285 and 0.272, respectively, based on EST-derived microsatellites, and 0.132 and 0.111, respectively, based on SNP markers (Table 13.1). The sole transition zone population from Sweden living at mid-range but variable salinity conditions possessed the highest amount of variability at EST-derived marker loci (0.420), which may be due to migration from multiple directions or adaptation to a wide range of environmental conditions (i.e., salinity). No evidence of recent bottlenecks was found in the combined Atlantic Ocean samples or in the combined Baltic Sea samples. Although marginal habitats, such as the brackish Baltic Sea, are often expected to have reduced population sizes Primary>Population size that would result in some loss of diversity over time through genetic drift (Johannesson and André 2006), there was no such evidence in relation to *F. lumbricalis*. The question of strict asexuality in *F. lumbricalis* in low-salinity conditions still remains open, as no multilocus genotypes (MLG) were shared by more than one sample in each population. In general, asexual species or populations often possess considerable amounts of genetic variation (Ellstrand and Roose 1987; Bengtson 2003), resulting from new mutations as well as from remnant sexuality and/or multiple origins. Therefore, populations with a predominantly asexual mode of population regeneration can display almost any pattern of genotypic structure.

Although the amount of genetic variation did not differ between the ocean and brackish populations, there was a great difference in variability between the marker types, EST-derived markers possessing considerably less variation than neutral microsatellites, and SNP markers showing even less variation (Table 13.1). It is evident that the expressed regions are subject to selection, unlike neutral microsatellite regions, and that leads to lower amounts of variability. However, genetic drift, gene flow and reproductive patterns affect genetic variation at all loci to the same extent. The influence of drift and/or reproductive patterns was visible as frequent heterozygote deficiencies were detected at neutral microsatellite loci: 52 % of tests showed significant heterozygote deficiencies and 19 % excesses (Kostamo et al. 2012). At EST-derived microsatellite loci, selection counteracts the effects of drift and reproductive patterns, and that shows as fewer heterozygote deficiencies (Kostamo et al. 2012).

13.5.3 Differentiation Along a Salinity Gradient

Selection is implicated when alleles (allele frequencies) at a given locus vary along with a specific environmental factor that creates an environmental cline. Therefore, it is justified to presume that genetic characteristics of algae varying along a salinity gradient would indicate the presence of selection and resulting adaptation. Studies have previously demonstrated that red algal ecotypes react differently to different salinities, as shown, e.g., in *Bostrychia radicans* and *Caloglossa leprieurii* (Yarish et al. 1979), *Ceramium strictum* (Rueness and Kornfeldt 1992) and *Phycodrys rubens* (Van Oppen et al. 1995).

Based on the analysis of molecular variance (AMOVA), the genetic differentiation pattern of *F. lumbricalis* populations varied depending on the marker type (Kostamo et al. 2012; Olsson and Korpelainen 2013). Genetic differentiation (F_{ST}) based on putatively neutral microsatellites showed similar moderate values among both Atlantic Ocean and Baltic Sea populations and between Atlantic Ocean and Baltic Sea populations, 0.158, 0.124, and 0.136, respectively (Table 13.1). A comparable differentiation estimate (0.164) has been obtained for *P. haitanensis* based on putatively neutral microsatellites (Bi et al. 2014). However, the differentiation pattern of *F. lumbricalis* was different based on EST-derived microsatellites (0.230, 0.108, and 0.401, respectively) and SNP markers (0.353, 0.278, and 0.535, respectively) (Table 13.1). Thus, the results for *F. lumbricalis* indicated definite differentiation between the ocean and brackish populations in expressed genomic regions, while such differentiation was not detected with presumably neutral loci.

Besides AMOVA, the Bayesian Structure analysis, when conducted for expressed marker data, clearly resolved two main clusters, the Atlantic Ocean and the Baltic Sea (Fig. 13.1). Indeed, some of the Baltic Sea populations included in the study were from extreme habitats at the edge of the species' distribution range. The only transition zone population from Sweden clustered to the Atlantic Ocean group (Kostamo et al. 2012; Olsson and Korpelainen 2013). Among the brackish populations, there was a clear transition in the genetic pattern from the higher salinity conditions of the western-central Baltic Sea toward the extreme low-salinity conditions of the Gulf of Finland in the eastern Baltic Sea, which was especially distinct based on EST-derived data (Kostamo et al. 2012).

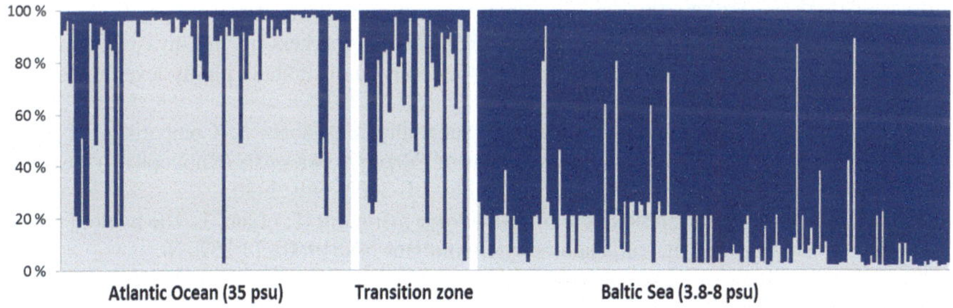

Fig. 13.1 Assignment of individual samples of *Furcellaria lumbricalis* to different pools as inferred by Bayesian clustering analysis based on EST-derived microsatellite markers. Samples represent four populations from the Atlantic Ocean (35 psu), one transition zone population from Sweden (15–30 psu) and six brackish populations from the Baltic Sea (3.8–8 psu) (data from Kostamo et al. 2012). Each color represents one of the two clusters formed in the analysis

Although definite trends were detected in the genetic structure of *F. lumbricalis* in the overall analysis based on EST-derived and SNP markers, it was shown that divergence patterns vary considerably among the loci (Kostamo et al. 2012; Olsson and Korpelainen 2013). Differentiation, apparently resulting from strong selection was most notable at the EST-derived microsatellite loci FI2143 and FI2838, which showed considerable differences in allele frequencies and possessed F_{ST} values equaling 0.389 and 0.354, respectively (for comparison, mean F_{ST} for all SNP markers 0.522, for all EST-derived microsatellite loci 0.267 and for all presumably neutral loci 0.095). These two loci were also among the most variable EST-derived marker loci (PIC values 0.239 and 0.374, respectively). Differentiation reflects differences in allele frequencies, which was especially clear in the frequency of allele 63 at locus FI2143: rare or not detected in the Atlantic Ocean populations but with a frequency range 0.217–0.656 in the Baltic Sea populations (Kostamo et al. 2012). A definite pattern was also detected for allele 213 at locus FI2838: a frequency range of 0.150–196 in the Atlantic Ocean and 0.125–0.978 in the Baltic Sea, the frequency increasing strongly toward extremely low salinity conditions. Although the neutrality tests provided no convincing evidence for selection in the studied genes, it is notable that deviations from neutrality were found only in the brackish Baltic Sea and transition zone populations of *F. lumbricalis*. Low salinity is a stress factor, which appears in the population genetic structures. However, such differentiation would have been missed if the investigations on *F. lumbricalis* had relied on neutral marker information only. Foregoing studies emphasize the importance of studying population genetic structures across geographic gradients using different genetic markers. Contrasts between neutral and adapatively important markers can potentially reveal the effects of natural selection.

References

Andersen BG, Borns HW Jr. The ice age world. Oslo: Scandinavian Press; 1994.

Andreakis N, Kooistra WH, Procaccini G. Microsatellite markers in an invasive strain of *Asparagopsis taxiformis* (Bonnemaisoniales, Rhodophyta): insights in ploidy level and sexual reproduction. Gene. 2007;406:144–51.

Asamizu E, Nakajima M, Kitade Y, Saga N, Nakamura Y, Tabata S. Comparison of RNA expression profiles between the two generations of *Porphyra yezoensis* (Rhodophyta) based on expressed sequence tag frequency analysis. J Phycol. 2003;39:923–30.

Austin AP. Life history and reproduction of *Furcellaria fastigiata* (L.) Lam. 1. The haploid plants and the development of the carposporophyte. Ann Bot NS. 1960a;24:257–76.

Austin AP. Life history and reproduction in *Furcellaria fastigiata* (L.) Lam. 2. The tetrasporophyte and reduction division in the tetrasporangium. Ann Bot NS. 1960b;24:296–312.

Bengtson BO. Genetic variation in organisms with sexual and asexual reproduction. J Evol Biol. 2003;16:189–99.

Bergström L, Bergström U. Species diversity and distribution of aquatic macrophytes in the Northern Quark, Baltic Sea. Nord J Bot. 1999;19:375–83.

Bergström L, Kautsky L. Local adaptation of *Ceranium tenuicorne* (Ceramiales, Rhodophyta) within the Baltic Sea. J Phycol. 2005;42:36–42.

Bergström L, Tatarenkov A, Johannesson K, Jönsson RB, Kautsky L. Genetic and morphological identification of *Fucus radicans* sp. nov. (Fucales, Phaeophyceae) in the brackish Baltic Sea. J Phycol. 2005;41:1025–38.

Bi YH, Wu YY, Zhou ZG. Genetic diversity of wild populations of *Pyropia haitanensis* based on SSR analysis. Biochem Syst Ecol. 2014;54:307–12.

Bird CJ, McLachlan J. The effect of salinity on distribution of species of *Gracilaria* Grev. (Rhodophyta, Gigartinales): an experimental assessment. Bot Mar. 1986;29:231–8.

Bird CJ, Chen LC-M, McLachlan J. Effects of light, temperature and salinity in culture of *Chondrus crispus*, *Furcellaria lumbricalis*, *Gracilaria tikvahiae* (Gigartinales, Rhodophyta) and *Fucus serratus* (Fucales, Phaeophyta). Bot Mar. 1979;22:521–7.

Bird NL, Saunders GW, McLachlan J. Biology of *Furcellaria lumbricalis* (Huds.) Lamoroux (Rhodophyta: Gigartinales), a commercial carrageenophyte. J Appl Phycol. 1991;3:61–82.

Björck S. A review of the history of the Baltic Sea, 13.0–8.0 ka BP. Quat Int. 1995;27:19–40.

Brumfield R, Beerli P, Nickerson D, Edwards S. The utility of single nucleotide polymorphisms in inferences of population history. Trends Ecol Evol. 2003;18:249–56.

Choi S, Hwang MS, Im S, Kim N, Jeong WJ, Park EJ, Gong YG, Choi DW. Transcriptome sequencing and comparative analysis of the gametophyte thalli of *Pyropia tenera* under normal and high temperature conditions. J Appl Phycol. 2013;25:1237–46.

Choi HG, Kim YS, Kim JH, Lee SJ, Park EJ, Ryu J, Nam KW. Effects of temperature and salinity on the growth of *Gracilaria verrucosa* and *G. chorda*, with the potential for mariculture in Korea. J Appl Phycol. 2006;18:269–77.

Collén J, Roeder V, Rousvoal S, Collin O, Kloareg B, Boyen C. An expressed sequence tag analysis of thallus and regenerating protoplasts of *Chondrus crispus* (Gigartinales, Rhodophyceae). J Phycol. 2006;42:104–12.

Düwel L. Experimental studies on macroalgae along the salinity gradient in the Baltic Sea area. PhD Thesis, University of Copenhagen; 2001.

Eggert A, Raimund S, Michalik D, West J, Karsten U. Ecophysiological performance of the primitive red alga *Dixoniella grisea* (Rhodellophyceae) to irradiance, temperature and salinity stress: growth responses and the osmotic role of mannitol. Phycologia. 2007;46:22–8.

Ellstrand NC, Roose ML. Patterns of genotypic diversity in clonal plant species. Am J Bot. 1987;74:123–31.

Gabrielsen TM, Brochmann C, Rueness J. The Baltic Sea as a model system for studying postglacial colonization and ecological differentiation, exemplified by the red alga *Ceramium tenuicorne*. Mol Ecol. 2002;11:2083–95.

Graham LD, Wilcox L. Algae. NJ: Prentice-Hall Inc.; 2000.

Guillemin M-L, Destombe C, Faugeron S, Correa JA, Valero M. Development of microsatellite DNA markers in the cultivated seaweed, *Gracilaria chilensis* (Gracilariales, Rhodophyta). Mol Ecol Notes. 2005;5:155–7.

Guillemin M-L, Sepúlveda RD, Correa JA, Destombe C. Differential ecological responses to environmental stress in the life history phases of the isomorphic red alga *Gracilaria chilensis* (Rhodophyta). J Appl Phycol. 2013;25:215–24.

Hancock JM. Microsatellites and other simple sequences: genomic context and mutational mechanisms. In: Goldstein DB, Schlötterer C, editors. Microsatellites, evolution and applications. Oxford: Oxford University Press; 1999. p. 1–9.

Hawkes MW. Reproductive strategies. In: Cole KM, Sheath RG, editors. Biology of the red algae. Cambridge: Cambridge University Press; 1990. p. 305–47.

Hoffmann EK. Volume regulation in cultured cells. Curr Topics Membr Transp. 1987;30:125–80.

Holmsgaard MH, Greenwell M, McLachlan J. Biomass and vertical distribution of *Furcellaria lumbricalis* and associated algae. Proc Int Seaweed Symp. 1981;10:309–14.

Johannesson K, André C. Life on the margin: genetic isolation and diversity loss in a peripheral marine ecosystem, the Baltic Sea. Mol Ecol. 2006;15:2013–29.

Karsten U, West JA, Mostaert AS, King RJ, Barrow KD, Kirst GO. Mannitol in the red algal genus *Caloglosia* (Harvey) J. Agardh. J Plant Physiol. 1992;140:292–7.

Kirst GO. Salinity tolerance of eukaryotic marine algae. Ann Rev Plant Physiol Mol Biol. 1990;41:21–53.

Kitade Y, Asamizu E, Fukuda S, Nakajima M, Ootsuka S, Endo H, Tabata S, Saga N. Identification of genes preferentially expressed during asexual sporulation in *Porphyra yezoensis* gametophytes (Bangiales, Rhodophyta). J Phycol. 2008;44:113–23.

Korpelainen H, Kostamo K, Virtanen V. Microsatellite marker identification using genome screening and restriction-ligation. Biotechniques. 2007;42:479–86.

Kostamo K, Korpelainen H, Olsson S. Comparative study on the population genetics of the red algae *Furcellaria lumbricalis* occupying different salinity conditions. Mar Biol. 2012;159: 561–71.

Kostamo K, Mäkinen A. Observations on the mode and seasonality of reproduction in *Furcellaria lumbricalis* (Gigartinales, Rhodophyta) populations in the northern Baltic Sea. Bot Mar. 2006;49:304–9.

Kostamo K, Olsson S, Korpelainen H. Search for stress-responsive genes in the red alga *Furcellaria lumbricalis* (Rhodophyta) by expressed sequence tag analysis. J Exp Mar Biol Ecol. 2011;404:21–5.

Kramer K, Fong P. A fluctuating salinity regime mitigates the negative effects of reduced salinity on the estuarine macroalga, *Enteromorpha intestinalis* (L) Iink. J Exp Mar Biol Ecol. 2000;254:53–69.

Kristiansen AA, Pedersen PM, Moseholm L. Salinity-temperature effects on growth and reproduction of *Scytosiphon lomentaria* (Fucophyceae) along the salinity gradient in Danish waters. Phycologia. 1994;33:444–54.

Krueger-Hadfield SA, Collen J, Daguin-Thiiebaut C, Valero M. Genetic population structure and mating system in *Chondrus crispus* (Rhodophyta). J Phycol. 2011;47:440–50.

Krueger-Hadfield SA, Roze D, Mauger S, Valero M. Intergametophytic selfing and microgeographic genetic structure shape populations of the intertidal red seaweed *Chondrus crispus*. Mol Ecol. 2013;22:3242–60.

Kullenberg G. Physical oceanography. In: Voipio A, editor. The Baltic Sea. Amsterdam: Elsevier; 1981. p. 135–81.

Lee EK, Seo SB, Kim TH, Sung SK, An G, Lee CH, Kim YJ. Analysis of expressed sequence tags of *Porphyra yezoensis*. Mol Cells. 2000;10:338–42.

Lee RE. Phycology. 4th ed. Cambridge: Cambridge University Press; 2008.

Liu FL, Wang XL, Yao JT, Fu WD, Duan DL. Development of expressed sequence tag derived microsatellite markers for *Saccharina (Laminaria) japonica*. J Appl Phycol. 2010;22:109–11.

Liu CL, Wang XL, Huang XH, Liu J. Identification of hypo-osmotically induced genes in *Kappachycus alvarezii* (Solieriaceae, Rhodophyta) through expressed sequence tag analysis. Bot Mar. 2011;54:557–62.

Lluisma AO, Ragan MA. Expressed sequence tags (ESTs) from the marine red alga *Gracilaris gracilis*. J Appl Phycol. 1997;9:287–93.

Lüning K, Charles Y, Kirkman H. Seaweeds: their environment, biogeography and ecophysiology. New York: Wiley; 1990.

Luo H, Mörchen M, Engel CR, Destombe C, Epplen JT, Saumitou-Laprade P, Valero M. Characterization of microsatellite markers in the red alga *Gracilaria gracilis*. Mol Ecol. 1999;8:700–2.

Luttikhuizen PC, Drent J, Baker AJ. Disjunct distribution of highly diverged mitochondrial lineage clade and population subdivision in a marine bivalve with pelagic larval dispersal. Mol Ecol. 2003;12:15–29.

Maggs CA. Life history variation in *Dasya ocellata* (Dasyaceae, Rhodophyta). Phycologia. 1998;37:100–5.

Mäkinen A, Kääriä J, Rajasilta M. Factors controlling the occurrence of *Furcellaria lumbricalis* (Huds.) Lamour. and *Phyllophora truncata* (Pallas) Zinova in the upper littoral of the archipelago of SW Finland. Kieler Meeresforsch Sonderheft. 1988;6:1404–46.

Middleboe AL, Sand-Jensen K, Brodersen K. Patterns for macroalgal distribution in the Kattegatt-Baltic region. Phycologia. 1997;36:208–19.

Mostaert AS, Karsten U, King RJ. Physiological responses of *Caloglossa leprieurii* (Ceramiales, Rhodophyta) to salinity stress. Phycol Res. 1995;43:215–22.

Nei M. Molecular evolutionary genetics. New York: Columbia University Press; 1987.

Nikaido I, Asamizu E, Nakajima M, Nakamura Y, Saga N, Tabata S. Generation of 10,154 expressed sequence tags from a leafy gametophyte of a marine red alga, *Porphyra yezoensis*. DNA Res. 2000;7:223–7.

Nikula R, Strelkov P, Väinölä R. Diversity and trans-Arctic invasion history of mitochondrial lineages in the North Atlantic *Macoma balthica* complex (Bivalvia: Tellinidae). Evolution. 2007;61:928–41.

Oetjen K, Reusch TBH. Genome scans detect consistent divergent selection among subtidal vs. intertidal populations of the marine angiosperm *Zostera marina*. Mol Ecol. 2007;16:5156–67.

Olsen JL, Stam WT, Coyer JA, Reusch TBH, Billingham M, Boström C, Calvert E, Christie H, Granger S, La Lumiere R, Milchakova N, Oudot-Le Secq MP, Pocaccini G, Sanjabi B, Serrão E, Veldsink J, Widdicombe S, Wyllie-Echeverria S. North Atlantic phylogeography and large-scale population differentiation of the seagrass *Zostera marina*. Mol Ecol. 2004;13: 1923–41.

Olsson S, Korpelainen H. Single nucleotide polymorphism found in the red alga *Furcellaria lumbricalis* (Gigartinales): new markers for population and conservation genetic analyses. Aquat Conser Mar Freshw Ecosyst. 2013;23:460–7.

Raven JA. Constraints of marine algal inversion of low-salinity environments: sex in the Baltic. J Phycol. 1999;35:210–2.

Röhner MR, Bastorp R, Jürss K. Genetic differentiation in *Hediste diversicolor* (Polychaetea: Neredididae) for the North Sea and the Baltic Sea. Mar Biol. 1997;130:171–80.

Rosenvinge LK. The marine algae of Denmark. Contributions to their natural history II. Rhodophyceae II (Cryptonemiales). Kongelige Danske Videnskabernes Selskab Skr; 1917.

Rueness J, Kornfeldt R-A. Ecotypic differentiation in salinity responses of *Ceramium strictum* (Rhodophyta) from Scandinavian waters. Sarsia. 1992;77:207–12.

Schwenke H. Water movements 5.2. plants. In: Kinne O, editor. Marine ecology I. Environmental factors. Wiley-Interscience: London; 1971. p. 1091–121.

Serrão EA, Brawley SH, Hedman J, Kautsky L, Samuelson G. Reproductive success in *Fucus vesiculosus* (Phaeophyceae) in the Baltic Sea. J Phycol. 1999;35:1–2.

Tatarenkov A, Bergström L, Jönsson RB, Serrão EA, Kautsky L, Johannesson K. Intriguing asexual life in marginal populationa of the brown seaweed *Fucus vesiculosus*. Mol Ecol. 2005;14:647–51.

Taylor ARA. The *Chondrus crispus–Furcellaria fastigiata* community at Campbell's Cove, Prince Edward Island. Technical Report Series 88. Industry Development Branch, Fisheries and Marine Services, Environment Canada, Ottawa; 1975.

Teo SS, Ho CL, Teoh S, Rahim RA, Phang SM. Transcriptomic analysis of *Gracilaria changii* (Rhodophyta) in response to hyper- and hypoosmotic stresses. J Phycol. 2009;45:1093–9.

Uji T, Hirata R, Mikami K, Mizuta H, Saga N. Molecular characterization and expression analysis of sodium pump genes in the marine red alga *Porphyra yezoensis*. Mol Biol Rep. 2012;39:7973–80.

Väinölä R. Repeated trans-Arctic invasions of littoral bivalves: molecular zoogeography of the *Macoma balthica* complex. Mar Biol. 2003;143:935–46.

Van Oppen MJH, Olsen JL, Stam WT. Genetic variation within and among North Atlantic and Baltic populations of the benthic alga *Phycodrys rubens* (Rhodophyta). Eur J Phycol. 1995;30:251–60.

Wærn M. Rocky-shore algae in the Öregrund Archipelago. Acta Phytogeogr Suecia. 1952; 30:1–298.

Wattier R, Dallas JF, Destombe C, Saumitou-Laprade P, Valero M. Single locus microsatellites in Gracilariales (Rhodophyta): high level of genetic variability within *Gracilaria gracilis* and conservation in related species. J Phycol. 1997;33:868–80.

Xie CT, Chen CS, Ji DH, Xu Y. Characterization, development and exploitation of EST-derived microsatellites in Porphyra haitanensis Chang et Zheng (Bangiales, Rhodophyta). J Appl Phycol. 2009;21:367–74.

Xie CT, Li B, Xu Y, Ji D, Chen CS. Characterization of the global transcriptome for *Pyropia haitanensis* (Bangiales, Rhodophyta) and development of cSSR markers. BMC Genom. 2013;14:107.

Yarish C, Edwards P, Casey S. A culture study of salinity responses in ecotypes of two estuarine red alga. J Phycol. 1979;15:341–6.

Yow YY, Lim PE, Phang SM. Assessing the use of mitochondrial cox1 gene and co2–3 spacer for genetic diversity study of Malaysian Gracilaria changii (Gracilariaceae, Rhodophyta) from Peninsular Malaysia. J Appl Phycol. 2013;25:831–8.

Zenkevitch L. Biology of the seas of the U.S.S.R. New York: Interscience Publishers; 1963.

Chapter 14
Phylogeography of Macroalgal Species Distributed in Brackish Water: *Ulva prolifera* (Ulvophyceae) and *Pyropia tenera* (Bangiophyceae)

Satoshi Shimada, Kensuke Ichihara, Yuka Masakiyo, Riko Kawaguchi and Norio Kikuchi

Abstract Comparative studies of closely related species or populations in contrasting environments can potentially provide insights into adaptive mechanisms. We review the phylogeography and population diversity of brackish water species derived from marine species, *Ulva prolifera* Müller (Ulvophyceae) and *Pyropia tenera* (Kjellman) N. Kikuchi, M. Miyata, M.S. Hwang & H.G. Choi (Bangiophyceae). Brackish *U. prolifera* and marine *Ulva linza* L. have been resolved as closely related species based on phylogenetic analysis of moleclar markers, with *U. linza* apparently parental to *U. prolifera*. Our analyses of 5S rDNA spacer region in samples from an Ulvalean bloom in Qingdao, China, indicate that the species appear to be derived from Japanese *U. prolifera*. Hybridization tests suggest that *U. linza* and the Qingdao bloom samples are probably distinct species, but gene flow is possible between them. The threatened brackish water species, *P. tenera*, is morphologically and phylogenetically related to the coastal species, *Pyropia yezoensis* (Ueda) M.S. Hwang & H.G. Choi. One form, *P. yezoensis* Ueda f. *narawaensis* Miura (new combination "*Pyropia yezoensis* f. *narawaensis*" has not yet been proposed), is the largest aquaculture source of "Nori" in Japan. Hybridization between these species has been reported, especially between male *P. tenera* and female *P. yezoensis*. Sequences of the nrITS1 region and *rbc*L gene, and PCR-RFLP (ARP4 gene) analyses suggested that *P. tenera* is distributed across 15 prefectures from Kyushu to Tohoku in Japan; but is

---

S. Shimada (✉) · Y. Masakiyo · R. Kawaguchi
Division of the Natural/Applied Sciences, Graduate School of Humanities and Sciences, Ochanomizu University, 2–1–1 Otuka, Bunkyo-ku 112–8610, Tokyo, Japan
e-mail: shimada.satoshi@ocha.ac.jp

K. Ichihara
Department of Integrated Biosciences, Graduate School of Frontier Sciences, University of Tokyo, 5-1-5 Kashiwanoha, Kashiwa 277-8562, Chiba, Japan

N. Kikuchi
Coastal Branch of Natural History Museum and Institute, Chiba, Yoshio 123, Katsuura 299-5242, Chiba, Japan

Z.-M. Hu and C. Fraser (eds.), *Seaweed Phylogeography*,
DOI 10.1007/978-94-017-7534-2_14

restricted to estuarine and brackish water habitats. Based on SSR analysis on the genetically identified *P. tenera* samples, we conclude that this species spread from Kyushu to Tohoku through the Kanto region.

Keywords Endangered species · Green algae · Green tide · Japan · NrITS1 · Red algae · SSR

14.1 Introduction

Macroalgae comprise three major groups: Ulvophyceae (green seaweeds), Phaeophyceae (brown seaweeds) and Rhodophyceae (red seaweeds), which are found on seashores all over the world (van den Hoek et al. 1995). Comparative studies of related species in different environments can potentially elucidate mechanisms of adaptation. In addition to marine species, each group includes brackish and freshwater species, e.g., *Dichotomosiphon tuberosus* (Braum) Ernst (Bryopsidales, Ulvophyceae), *Heribaudiella fluviatilis* (Areschoug) Svedelius (Sphacelariales, Phaeophyceae) and *Caloglossa continua* (Okamura) King et Puttock (Ceramiales, Rhodophyceae) (Guiry et al. 2015). Molecular phylogenetic studies have revealed that many of these brackish/freshwater species evolved from marine ancestors (Lam and Zechman 2006; Krayesky et al. 2012; Silberfeld et al. 2014). Brackish water species have acquired molecular, functional and physiological characteristics required for adaptation to low-salinity conditions. However, many questions about the biological evolution of brackish and freshwater algae are yet to be resolved. Which genes are responsible for this new environmental adaptation? When and where did the brackish/freshwater species separate from marine ancestral species? What are their distributions relative to environmental gradients?

Phylogeography and population genetics are among the most powerful tools used to understand the provenance, genetic structure, distributional expansion and population diversity of organisms (e.g., dinoflagellates, Nagai et al. 2009; crayfish, Koizumi et al. 2012; humans, Jinam et al. 2012; butterflies, Jeratthitikul et al. 2013; hares, Nunome et al. 2014a, b). Land plants have been particularly well studied (van Inghelandt et al. 2010; Kaya et al. 2013; Xu et al. 2014). For instance, the probable distributions of woody plants at the Last Glacial Maximum (LGM, 0.026–0.019 Ma) has been estimated with palaeodistribution modeling (Sakaguchi et al. 2012), endangered plants were interpreted using population diversity and conservation units (Matsuda and Setoguchi 2012), and extinction risks for wild pollinated conifer populations were results from Simple Sequence Repeat (SSR) marker analyses (Iwasaki et al. 2013). These research techniques are also useful for examining the evolutionary history of seaweeds and phylogenetically related brackish and freshwater macroalgae (Perevra et al. 2009). Phylogeographic research can provide important clues to help understand molecular evolution, as it relates to environmental adaptation through all three aquatic environments (freshwater, brackish water, and seawater).

In this chapter, we review the phylogeography and population diversity of brackish water species derived from marine species in two case studies: *Ulva prolifera* Müller (Ulvophyceae) and *Pyropia tenera* (Kjellman) N. Kikuchi, M. Miyata, M.S. Hwang & H.G. Choi (Bangiophyceae).

14.2 *Ulva* (Ulvophyceae, Chlorophyta)

The genus *Ulva* is one of the most common seaweeds (Canter-Lund and Lund 1995; van den Hoek et al. 1995; Shimada et al. 2007). New species from the genus have been described year after year (Horimoto et al. 2011; Masakiyo and Shimada 2014; Matsumoto and Shimada 2015). This genus includes brackish species (*U. prolifera*) (Hiraoka and Shimada 2004) and freshwater species (*Ulva limnetica* Ichihara et Shimada) (Ichihara et al. 2009a). Due to the euryhaline nature of *Ulva*, low-salinity tolerance and adaptation can be found, as reported in many studies (Reed and Russell 1979; Martins et al. 1999; Ohno et al. 1999; Kamer and Fong 2000; McAvoy and Klung 2005; Ichihara et al. 2009b, 2011, 2013).

Ulva prolifera (Fig. 14.1a), which was originally described from Nebbelund, Lolland Island, Denmark (Müller 1778), has branched tubular thalli and is commonly distributed in estuarine and brackish waters (Burrow 1991; Shimada et al. 2008; Hiraoka et al. 2003, 2011). This species is of major importance to the

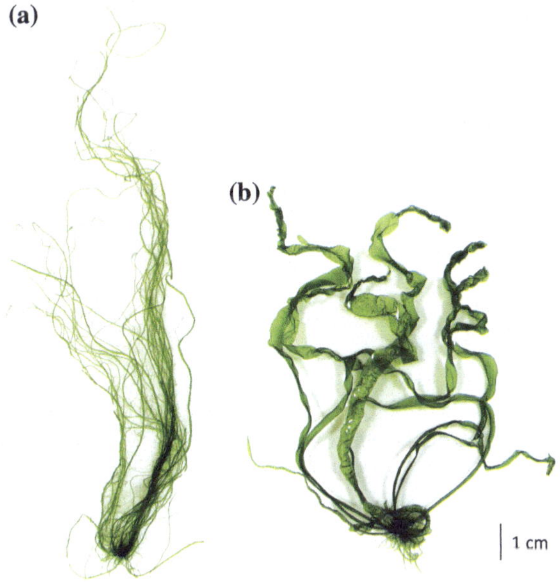

Fig. 14.1 Cultured thalli of molecular-identified *Ulva prolifera* (**a**) and *Ulva linza* (**b**) under 20 °C, 16:8 h LD, 15–25 μE m^{-2} s^{-1} photon flux with PES medium (Provasoli 1968)

Japanese fisheries industry, as it is a food for human consumption and has a regulated IQ (import quota) under Japanese domestic law (Kawashima et al. 2014). Due to an increasing demand, the supply of *U. prolifera* is also supplemented by land-based culture techniques in tanks using seawater (Hiraoka and Oka 2008). This species and *Ulva linza* L. are closely related based on phylogenetic studies (Shimada et al. 2003).

Ulva linza (Fig. 14.1b) is distributed in marine habitats, and possesses unbranched distromatic folious thalli without a stipe or margin (Brodie et al. 2007). In freshwater conditions, the cell viability of *U. linza* dropped to approximately 20 % after 7 days, while that of *U. prolifera* remained at almost 100 % (Ichihara et al. 2013). These physiological differences seem to be the source of their distinct distributions, with *U. prolifera* inhabiting estuaries with fluctuations in salinity, while *U. linza* occurs in marine environments (Ohno et al. 1999; Hiraoka and Shimada 2004). *Ulva linza* has higher intraspecific diversity than *U. prolifera*, suggesting that the latter brackish species is derived from marine *U. linza* (Shimada et al. 2008).

The "green tide" phenomenon occurs under conditions when an extensive biomass of free-floating green algae accumulates on shallow beaches. Free-floating *Ulva* causes green tide at several locations around the world (Fletcher 1996; Ohno 1999). In June 2008, it was reported worldwide that a vast algal bloom had occurred in Qingdao, China. This massive green-tide covered about 600 km^2 along the coast of Qingdao, the host city for the Olympic sailing regatta. An army of more than 10,000 recruits was deployed for a period of one week to help remove the algal bloom. Qingdao city reported that the total mass of the bloom was about 10,000 ton wet weight (BBC NEWS 2008).

Based on satellite imaging data, Liu et al. (2009) proposed the possibility that nonlocal sources were responsible for the Qingdao bloom. According to these authors, on the 15th of May 2008, some small green patches covering a total surface area of about 80 km^2 were observed near the coasts of Yancheng and Lianyungang, Jiangsu province. Soon, they grew rapidly and moved into the middle of the Yellow Sea. Finally, a large quantity of biomass formed the famous green tide along the coasts of Qingdao.

The world's-largest super floating macroalgal bloom of *U. prolifera* has lasted eight years so far, reoccurring every summer in the Yellow Sea (Xing et al. 2014). The inter-annual variability in human-induced nutrient pollution from 2001 to 2012 was assessed, and a significant increase in nutrient uptake was found in the macroalgal bloom phase (2007–2012). Annual in situ nutrient concentrations increased rapidly from 2000 to 2011 in the Jiangsu Shoal, which was the origin of the drifting macroalgae (Xing et al. 2014).

Shimada et al. (2010) carried out a phylogeographic study of *U. prolifera* using samples from Japan and the Qingdao bloom. Figures 14.2 and 14.3 show the geographical distribution of sequence types of the 5S rDNA spacer region and the statistical parsimony network of *U. prolifera*, respectively. Type-A (red circle) individuals are widely distributed along the Pacific coast of Japan from Okinawa to the Kanto region. The blue lineage, derived from type-A in Kanto, radiated to the

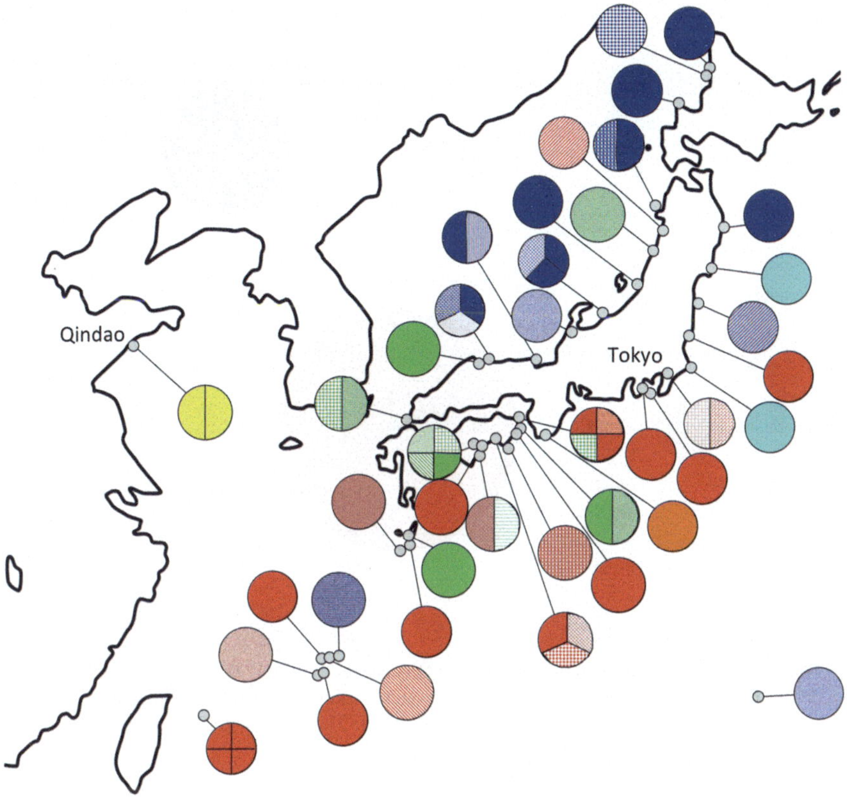

Fig. 14.2 Geographical distributions of sequence types of the 5S rDNA spacer region. Combinations of color and pattern represent different types and are same as Fig. 14.3

northern part of Honshu, and spread to the Sea of Japan through the Tsugaru channel (type-N: blue circle). The Qingdao bloom samples (type-QB: yellow circle) were derived from the blue lineage. There are 54 indels between type-QB and type-N. These results indicate that the individuals causing the Qingdao bloom might have recently derived from the Japanese *U. prolifera,* but are not the same taxon as the latter.

Hiraoka et al. (2011) experimentally hybridized individuals of *U. linza,* Japanese *U. prolifera* and the Qingdao bloom strain. The sexually reproducing Qingdao strains were successfully crossed with *U. prolifera* with no evidence of a reproductive barrier, but could not be crossed with *U. linza* due to gamete incompatibility. However, the results of *U. prolifera* × *U. linza* showed an unusual mating activity: male gametes of *U. prolifera* cannot fuse with female gametes of *U. linza,* but female gametes of *U. prolifera* successfully fuse with male gametes of *U. linza.* The hybrid zygotes can normally develop into sporophytes, and the F1 sporophyte produces zoospores via meiosis. Additionally, the zoospores grow into F1

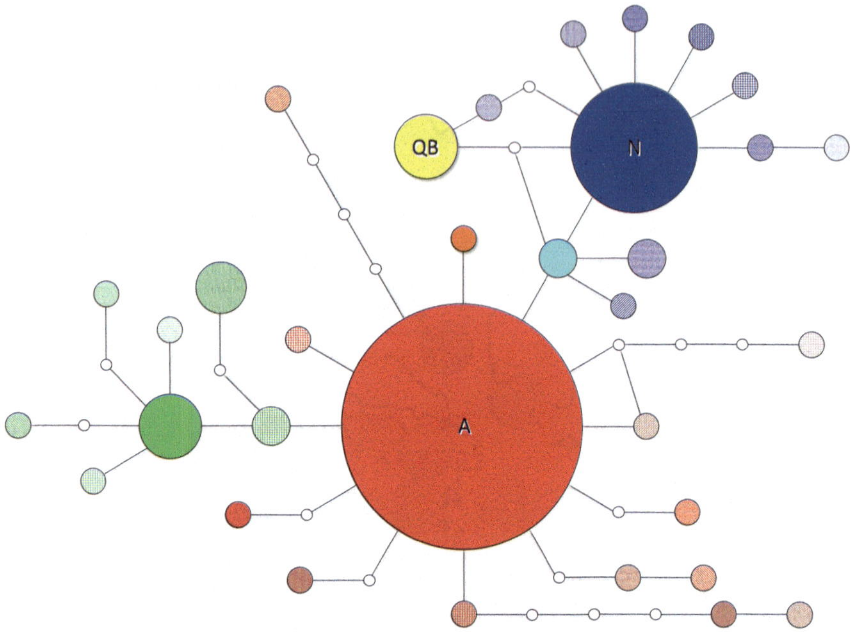

Fig. 14.3 Statistical parsimony network of 5S rDNA spacer region. Indels were treated as missing data. Each line connecting sequence types corresponds to one base mutation. A *white circle* in the parsimony network represents a missing sequence type. The size of circles corresponds to the frequency of each type

gametophytes which can produce normal gametes. These results indicate that *U. linza* and the Qingdao bloom samples are probably different species, but gene flow is nonetheless still possible between them through the Japanese *U. prolifera*.

Microsatellite/SSR markers have been developed for many organisms due to their unique advantages such as being codominant, highly polymorphic, well distributed throughout the genome, and their simplicity of application for analysis and consistency (Varshney et al. 2002; Tanaka et al. 2011). Recently, Expressed Sequence Tag (EST)-SSR markers were developed for *U. prolifera* from the south Yellow Sea (Zhang et al. 2014). The analysis of genetic variation using the SSR markers indicated that most individuals collected at the same site clustered together in a tree produced by the unweighted pair-group mean analysis method (UPGMA). The clusters in the tree showed some consistency with the geographical origins of the samples. In addition, 12 of 13 free-floating samples were grouped as a single clade separated from the attached samples. These free-floating samples were collected from different sites at different dates. These results indicate that the free-floating masses of *U. prolifera* may share the same origin and have been dispersed along coastal areas by wind and ocean currents (Zhang et al. 2014).

Ulva linza and *U. prolifera*, including the Qingdao blooms, is an ideal species-complex to study speciation and environmental adaptation to low (from

U. linza to Japanese *U. prolifera*) and high (from Japanese *U. prolifera* to the Qingdao bloom samples) salinities. Kawashima et al. (2013) reported that the 5S rDNA spacer analysis of Japanese and Chinese *U. prolifera* commercial products detected the potential for intercrossing between *U. linza* and *U. prolifera*. Therefore, hybrids of *U. linza* and *U. prolifera* are likely to appear in nature. We are now analyzing gene flow between field populations in both species to gain an accurate understanding of their taxonomic status. RNA-seq analysis is being conducted using next generation sequencing (NGS) to elucidate the molecular evolution of low-salinity adaptation in *U. prolifera*, by comparing the gene expression levels of *U. prolifera* and *U. linza* cultured in seawater, brackish water and freshwater conditions. NGS allows us to study gene expression in nonmodel organisms without previous genomic resources (e.g., species-specific primers), and RNA-seq is a very promising application for the study of environmental adaptation (Ekblom and Galindo 2011). RAD-seq (Restriction Site Associated DNA Sequence) from NGS analysis is also useful in studies of speciation, population genetics and phylogeography (Davey and Biaxter 2010; Wagner et al. 2013). NGS analyses will undoubtedly provide a better understanding into algal problems such as green tides. These studies also present new insights into speciation and environmental adaptation, which is the major driver of biological evolution.

14.3 *Pyropia* (Bangiophyceae, Rhodophyta)

Pyropia is a foliose red algal genus, consisting of 60 species (Guiry et al. 2015). It is currently recognized as a distinct genus from the closely related genera *Porphyra*, *Boreophyllum*, *Miuraea*, *Lysithea*, *Fuscifolium*, and *Wildemania* (Sutherland et al. 2011). “Nori” (laver), used in sushi, is an edible seaweed product made using thalli of *Pyropia* species obtained from the field and/or aquaculture, including *Pyropia yezoensis* (Ueda) M.S. Hwang & H.G. Choi, *Pyropia haitanensis* (T.J. Chang & B.F. Zheng) N. Kikuchi, M. Miyata, *Pyropia pseudolinearis* (Ueda) N. Kikuchi, M. Miyata, M.S. Hwang & H.G. Choi and others (Park et al. 2003; Touhata et al. 2013; Xia et al. 2013).

Pyropia yezoensis was originally described from Hokkaido, Japan (Ueda 1932) (Fig. 14.4a). Its type locality is “Muroran?” based on the label of the holotype specimen (Sutherland et al. 2011). One form of this species, *P. yezoensis* Ueda f. *narawaensis* Miura (new combination “*Pyropia yezoensis* f. *narawaensis*” has not yet been proposed), is the largest aquaculture source of laver in Japan (Miura 1988). Recently, Nakamura et al. (2013) determined a symbiont-free genome sequence of this species using NGS of purified protoplasts. The *P. yezoensis* genome could represent a model genome for examining red algal life history, and will provide insights into Nori aquaculture in the near future (Nakamura et al. 2013).

Pyropia tenera (Kjellman) N. Kikuchi, M. Miyata, M.S. Hwang & H.G. Choi was described in Japan based on dried Nori sheets (Kjellman 1897) (Fig. 14.4b). Although this species was used for Nori cultivation from the Edo period

Fig. 14.4 Cultured young thalli of molecular-identified samples of *Pyropia yezoensis* (**a**) and *Pyropia tenera* (**b**) under 15 °C, 8:16 h LD, 15–25 μE m^{-2} s^{-1} photon flux with PES medium (Provasoli 1968)

(17 century) to the 1960s, *P. yezoensis* is better than *P. tenera* in color, gloss and tolerance of strong waves and windy conditions, and thus almost all cultivators now use *P. yezoensis* in Japanese aquaculture (Miura 1988; Notoya 2004). *Pyropia tenera* has been categorized as an endangered species (CR + EN) by the Ministry of the Environment in Japan (Environment Agency of Japan 2000). The habitat of this species thought to be limited primarily to the vicinity of the mouths of large and small rivers (brackish water), and only eight localities of *P. tenera* in Japan were known in 2002 (Kikuchi et al. 2002).

This threatened species, *P. tenera*, is morphologically and phylogenetically related to *P. yezoensis* (Yoshida 1998; Sutherland et al. 2011). Hybridization between *P. tenera* and *P. yezoensis* has been reported, especially between male *P. tenera* and female *P. yezoensis* (Niwa et al. 2009). The reverse combination (hybridization between female *P. tenera* and male *P. yezoensis*) also exists, but does not lead to normal development, and thus it is thought that there are few occurrences in the field (Niwa et al. 2010; Niwa and Sakamoto 2010). Although hybrids between female *P. tenera* and male *P. yezoensis* showed *P. tenera*-type in both nrITS (nuclear encoded Internal Transcribed Spacer) and *rbc*L (ribulose-1, 5-bisphosphate carboxylase/oxygenase large subunit) because of concerted evolution, all four types of individuals (1: pure *P. tenera,* 2: pure *P. yezoensis*, 3: hybrids between male *P. tenera* and female *P. yezoensis,* and 4: hybrids between female *P. tenera* and male *P. yezoensis*) can be clearly distinguished by a combination of Niwa's methods using the direct sequence of nrITS and *rbc*L, and PCR-RFLP (Restriction Fragment Length Polymorphism) of the ARP4 gene (Niwa et al. 2009, 2010; Niwa and Sakamoto 2010).

We first determined the distribution of *P. tenera* and the intercross of *P. tenera-yezoensis* in 165 samples identified morphologically as *P. tenera* collected at 46 localities in 17 prefectures around Japan (Ohnishi et al. 2013). Sequences of the nrITS1 region and *rbc*L gene, and PCR-RFLP (ARP4 gene) indicated that *P. tenera*

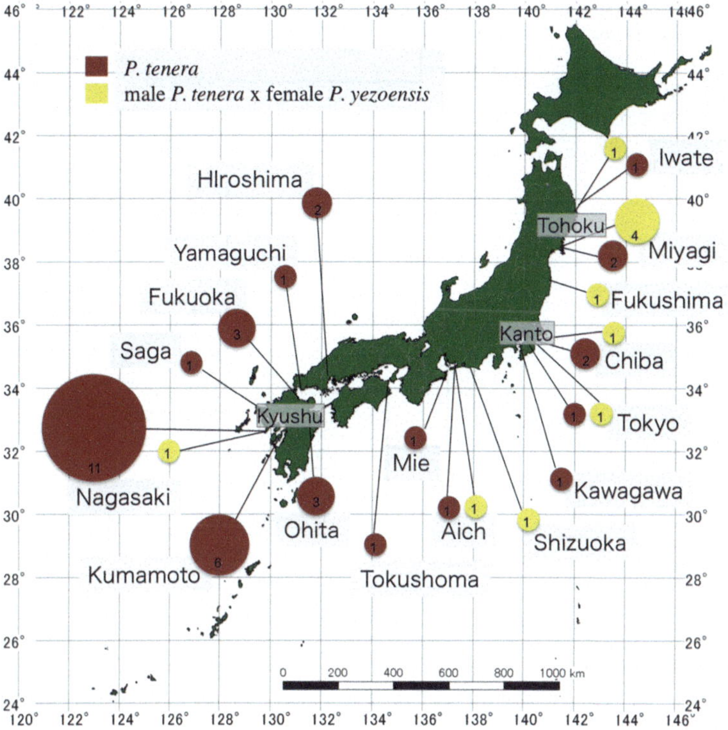

Fig. 14.5 Distribution map of *Pyropia tenera* (*red*) and hybridization between male *P. tenera* × female *P. yezoensis* (*yellow*) based on molecular identification. The number in the *circle* indicates locality numbers in each prefecture. The size of circles corresponds to the frequency of each type

occurred at 38 localities in 15 prefectures (Fig. 14.5). *Pyropia tenera* was restricted to estuarine and brackish water habitats, mainly in the Kyushu region (western part of Japan) and in the Tohoku region, but not in the Sea of Japan. *Nemalionopsis tortuosa* Yoneda et Yagi (Thoreales, Rhodophyta), designated as an endangered species (CR + EN), is also found in many localities around Kyushu (Shimada et al. 2012). Since several threatened algal species have been reported from Kyushu (Environment Agency of Japan 2000), it is one of the most important regions for the conservation of macroalgae in Japan.

Hybrids between male *P. tenera* and female *P. yezoensis* were detected at 11 localities in 8 prefectures, mostly located in Tokai, and Kanto to Tohoku regions (Fig. 14.5). Hybridization was restricted to *P. tenera*'s brackish water habitat. The original distribution of *P. yezoensis* is thought to have been from Kanto to Hokkaido (Niwa et al. 2009), thus it is reasonable that the hybrids are most common in the eastern to northern parts of Honshu. As these species are cultivated on the Pacific coast of Japan from Kyushu to Tohoku region (Miura 1988), hybrids in

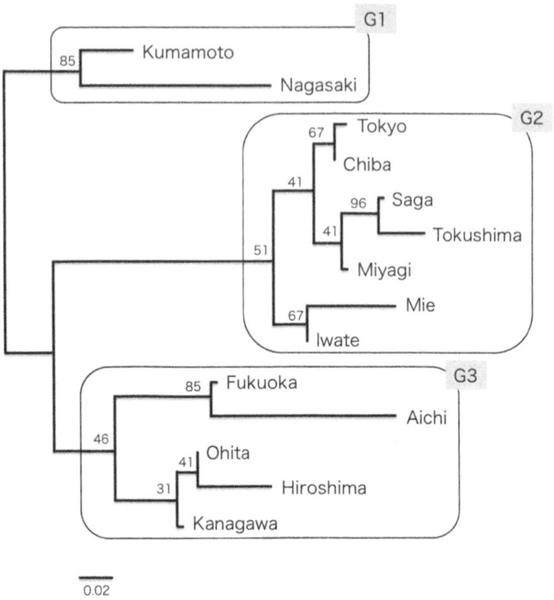

Fig. 14.6 Population tree of Japanese *Pyropia tenera* constructed by neighbor-joining method with D_A data from three SSR markers. Modified from Ohnishi et al. (2013)

the Nagasaki Prefecture may be the result of crosses between wild *P. tenera* and cultivated *P. yezoensis*. Crossing between the two species apparently occurs relatively easily.

Next, from SSR analysis of the molecular-identified *P. tenera* samples, we determined genotypes of three microsatellite markers (Pye13, Pye41, Pye53 of Niwa et al. 2010). Three genetic groups (G1, G2 and G3) appeared in neighbor-joining trees of inter-population genetic distances (Nei and Tajima 1983) computed with POPULATION 1.2.31 (http://bioinformatics.org/~tryphon/populations/) (Fig. 14.6). G1 comprised Kumamoto and Nagasaki populations (Kyushu region) with a high bootstrap value (85 %). Populations of G2 (51 % bootstrap value) were mainly distributed along the Pacific coast from the Kanto to Tohoku region. G3 was supported by a bootstrap value of only 46 %, and the populations in this cluster were geographically scattered (Fig. 14.6).

The cluster analysis generated using STRUCTURE (Pritchard et al. 2000) and Structure Harvester v. 2.3.1 (Earl and von Holdt 2012) revealed four genetically differentiated clusters in the Japanese *P. tenera* (Fig. 14.7). The clustering patterns for $K = 4$ were highly consistent for 10 independent runs. The result was consistent with the previous POPULATION analysis. Cluster 1 (yellow) was present in only Kumamoto and Nagasaki populations, which is consistent with the G1 POPULATION analysis. Cluster 2 (red) appeared mostly in G2 populations. Clusters 3 (green) and 4 (blue) were detected in almost all populations, indicating that these lineages in these clusters might be ancestral. G3 populations (Fukuoka, Aichi, Ohita, Hiroshima, and Kanagawa populations) overlapped at least 50 % with clusters 3 and 4 and were absent cluster 1.

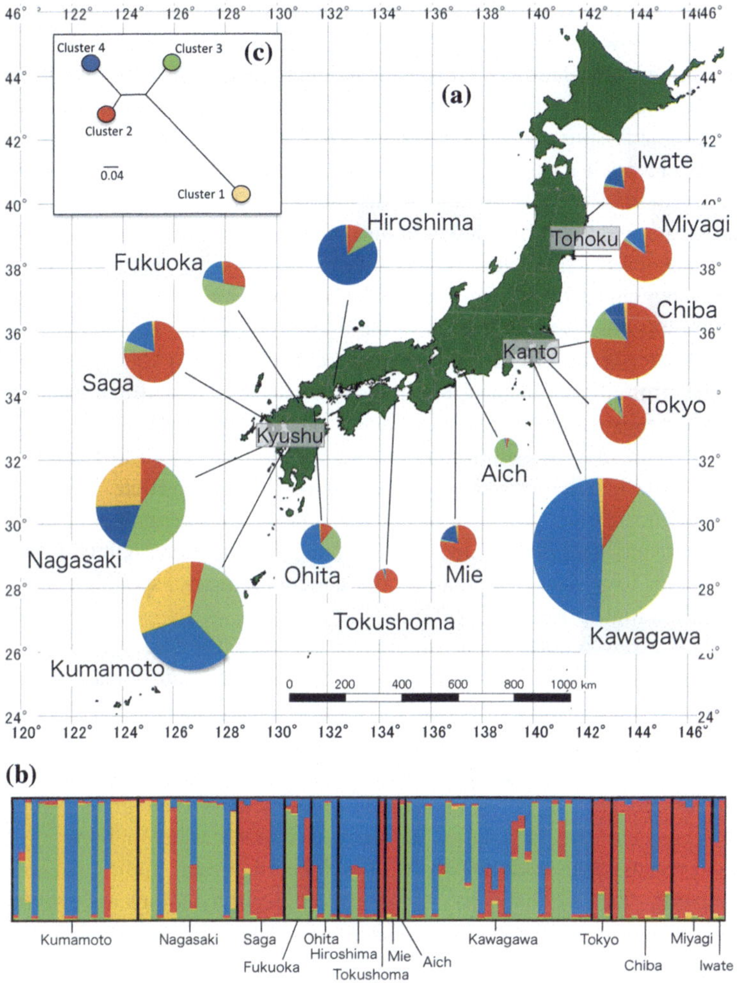

Fig. 14.7 Best clustering result ($K = 4$ clusters) for the 14 populations of the Japanese *Pyropia tenera* based on the STRUCTURE method. **a** Cluster distribution in each prefecture, **b** bar plot showing each individual. Each individual is represented by a *vertical line* partitioned into *colored segments*; the segment length is proportional to the individual's estimated K cluster membership coefficient. The sampled populations are separated by *black vertical lines*. **c** Neighbor-joining tree of the four clusters. Modified from Ohnishi et al. (2013)

The populations of Kumamoto Prefecture had higher levels of genetic diversity (Allele Richness (AR): 1.542, H_E (expected heterozygosity): 0.508, F_{IS} (inbreeding coefficient of an individual with respect to the local subpopulation): 0.556) and Nagasaki (AR: 1.615, H_E: 0.582, F_{IS}: 0.575) than the average of all the other populations (AR: 1.444, H_E: 0.278, F_{IS}: 0.239). This large genetic diversity may indicate a long stable history during which diversity has accumulated. However, the high F_{IS} indicated inbreeding or subpopulations. On the other hand, the populations in the region extending from Kanto to Tohoku region consisted largely of cluster 2 (red) and had moderately low levels of genetic diversity (AR: 1.291–1.389, H_E: 0.167–0.323) (Table 6 in Ohnishi et al. 2013). This result and the direction of the Kuroshio Current, flowing from the southern to the northern Pacific coast of Japan, indicate that *P. tenera* probably spread from Kyushu to Tohoku through the Kanto region.

Recently, *P. tenera* has been rediscovered at the Tamagawa River (Kanagawa Prefecture, Kanto Region), on the border between Tokyo and Kanagawa, near Haneda International Airport, (Kikuchi and Niwa 2006). In Tokyo Bay, the mouth of the Tamagawa River, *P. tenera* was farmed from the seventeenth century until the 1950s (Okamura 1909; Miyashita 2003). However, since the 1950s, aquaculture of *P. tenera* has not been carried out in the vicinity of Tokyo Bay due to increased landfill and deterioration of farming environments (Kawasaki City Museum 1995). The samples from the Tamagawa River site often showed ancestral clusters 3 and 4 in our SSR analysis. These results indicate that the populations of *P. tenera* at Tamagawa River may include remnant descendants from the Edo Period. Also, a population-specific allele has been observed in this area (allele 229 of pye53) (Table 7 in Ohnishi et al. 2013). In recent years, however, the number of individuals in this region has declined, increasing the risk of extinction (Kikuchi and Niwa 2006). In order to protect this area-specific genetic diversity, specific conservation activities are a high priority.

There are a few reports about differences between the two species: *P. tenera* blades are softer in texture and considered better in taste and flavor (Ueda 1932), and it easier to infect with a virus than *P. yezoensis* (Tatyana et al. 2012). However, it is very difficult to identify *P. tenera, P. yezoensis* and their hybrids by gross morphology alone (Niwa et al. 2005; Kikuchi 2014). By using molecular-identified samples, researchers can better investigate differences in their gross morphology and color, growth rate, adaptation to marine/brackish water habitats, taste and flavor, and resistance to viral infections. It is hoped that such studies can contribute to elucidating the mechanisms of morphological/physiological evolution of seaweeds.

Acknowledgements We are deeply indebted to Dr. Takaya Iwasaki of Tokyo Univ., for helpful discussions. We greatly appreciate the stimulating discussions, constructive criticism and valuable suggestions on an early version of the manuscript by Senior Lecturer Alecia Bellgrove and Casual Academic Vanessa Skrzypczyk of Marine Biologist at Deakin University Warrnamboolfor. We also thank the Japanese Society of Phycology for the permission of the reproduction of the figures.

References

BBC NEWS. 2008. http://news.bbc.co.uk/2/hi/asia-pacific/7486814.stm.

Brodie J, Maggs CA, John DM. Green seaweeds of Britain and Ireland. London: British Phycological Society; 2007. p. 242.

Burrow EM (1991) Seaweeds of the British Isles, vol 2: chlorophyta. London: British Museum (Natural History), p. 238.

Canter-Lund H, Lund JWG. Freshwater algae. Their microscopic world explored. Bristol: Biopress; 1995.

Davey JW, Biaxter ML. RADseq: next-generation population genetics. Brief Funct Genom. 2010;9:416–23.

Earl D, von Holdt B. STRUCTURE HARVESTER: a website and program for visualizing STRUCTURE output and implementing the Evanno method. Conserv Genet Res. 2012;4: 359–61.

Ekblom R, Galindo J. Applications of next generation sequencing in molecular ecology of non-model organisms. Heredity. 2011;107:1–15.

Environment Agency of Japan. In: Threatened wildlife of Japan. Red data book, 2nd ed. 2000.

Fletcher RT. The occurrence of "green tide". In: Schramm W, Nienhuis PH, editors. Marine benthic vegetation—recent changes and the effects of eutrophication. Berlin: Springer; 1996. p. 7–43.

Guiry M, Guiry MD, Guiry GM. AlgaeBase. World-wide electronic publication, National University of Ireland, Galway. 2015. http://www.algaebase.org; Accessed 21 Jan 2015.

Hiraoka M, Oka N. Tank cultivation of *Ulva prolifera* in deep seawater using a new "germling cluster" method. J Appl Phycol. 2008;20:97–102.

Hiraoka M, Shimada S. Biology of a special green laver, *Ulva prolifera* from the Shimanto River. Aquabiology. 2004;26:508–15.

Hiraoka M, Dan A, Shimada S, Hagihira M, Migita M, Ohno M. Different life histories of *Enteromorpha prolifera* (Ulvales, Chlorophyta) from four rivers on Shikoku Island, Japan. Phycologia. 2003;42:275–84.

Hiraoka M, Ichihara K, Zhu W, Ma J, Shimada S. Culture and hybridization experiments on an *Ulva* clade including the Qingdao strains blooming in the Yellow sea. PLoS ONE. 2011;6(5): e19371.

Horimoto R, Masakiyo Y, Ichihara K, Shimada S. *Enteromorpha*-like *Ulva* (Ulvophyceae, Chlorophyta) growing in the Todoroki river, Ishigaki island, Japan, with special reference to *Ulva meridionalis* Horimoto et Shimada sp. nov. Bull Nat Sci Mus. 2011;37:155–67.

Ichihara K, Arai S, Uchimura M, Fay EJ, Ebata H, Hiraoka M, Shimada S. New species of freshwater *Ulva*, *Ulva limnetica* (Ulvales, Ulvophyceae) from Ryukyu archipelago, Japan. Phycol Res. 2009a;57:94–103.

Ichihara K, Arai S, Shimada S. cDNA cloning of a lectin-like gene preferentially expressed in freshwater from the macroalga *Ulva limnetica* (Ulvales, Chlorophyta). Phycol Res. 2009b;57:104–10.

Ichihara K, Mineur F, Shimada S. Isolation and temporal expression analysis of freshwater-induced genes in *Ulva limnetica* (Ulvales, Chlorophyta). J Phycol. 2011;47:584–90.

Ichihara K, Miyaji K, Shimada S. Comparing the low-salinity tolerance of *Ulva* species distributed in different environments. Phycol Res. 2013;61:52–7.

Iwasaki T, Sase T, Takeda S, Ohsawa TA, Ozaki K, Tani N, Ikeda H, Suzuki M, Endo R, Tohei K, Watano Y. Extensive selfing in an endangered population of *Pinus parviflora* var. *parviflora* (Pinaceae) in the Boso hills, Japan. Tree Genet Genomes. 2013;9:693–705.

Jeratthitikul E, Hara T, Yago M, Itoh T, Wang M, Usami S, Hikida S. Phylogeography of Fischer's blue, *Tongeia fischeri*, in Japan: evidence for introgressive hybridization. Mol Phylogenet Evol. 2013;66:316–26.

Jinam T, Nishida N, Hirai M, Kawamura S, Oota H, Umetsu K, Kimura R, Ohashi J, Tajima A, Yamamoto T, Tanabe H, Mano S, Suto Y, Kaname T, Naritomi K, Yanagi K, Niikawa N, Omoto K, Tokunaga K, Saitou N. The history of human populations in the Japanese archipelago inferred from genome-wide SNP data with a special reference to the Ainu and the Ryukyuan populations. J Human Genet. 2012;57:787–95.

Kamer K, Fong P. A fluctuating salinity regime mitigates the negative effects of reduced salinity on the estuarine macroalga, *Enteromorpha intestinalis* (L.) link. J Exp Mar Biol Ecol. 2000;254:53–69.

Kawasaki City Museum. Sea and life. The period when Nori was foaming at Kawasaki. Kawasaki: Kawasaki City Museum; 1995.

Kawashima Y, Akasaki T, Matsumoto Y, Yamazaki Y, Shimada S. Species identification of imported and Japanese commercial green algal products, based on phylogenetic analyses using the nrITS2 and 5S rDNA spacer regions. Fish Sci. 2013;79:521–9.

Kawashima Y, Akasaki T, Matsumoto Y, Shimada S. Development of a rapid and accurate PCR-based detection method for commercially valuable green algal species. Fish Sci. 2014;80:859–67.

Kaya HB, Cetin O, Kaya H, Sahin M, Sefer F, Kahraman A, Tanyolac B. SNP discovery by Illumina-based transcriptome sequencing of the olive and the genetic characterization of Turkish olive genotypes revealed by AFLP, SSR and SNP markers. PLoS ONE. 2013;8(9): e73674.

Kikuchi N. Discrimination between *Pyropia* species and forma by morphological data. In: Fujiyoshi E, Tamaki M, Kobayashi M, Aritaki M, editors. Characteristics of Nori cultivars. Nagasaki: Seikai National Fisheries Research Institute; 2014.

Kikuchi N, Niwa K. Habitat and morphology of the endangered species *Porphyra tenera* (Bangiales, Rhodophyta) at the estuary of Tamagawa River in Tokyo Bay. Jap J Phycol. 2006;54:149–56.

Kikuchi M, Yoshida T, Yoshinaga K. Distribution of some endangered species of *Porphyra*. Ecosophia. 2002;9:112–7.

Kjellman FR. Japanska arter af slägtet *Porphyra*. Bihang Till K. Svenska Vet Akad Handlingar. 1897;23:1–34.

Koizumi I, Usio N, Kawai T, Azuma N, Masuda R. Loss of genetic diversity means loss of geological information: the endangered Japanese crayfish exhibits remarkable historical footprints. PLoS ONE. 2012;7(3):e33986.

Krayesky DM, Norris JN, West JA, Kamiya M, Viguerie M, Wysor BS, Fredericq S. Two new species of *Caloglossa* (Delesseriaceae, Rhodophyta) from the Americas, *C. confusa* and *C. fluviatilis*. Phycologia. 2012;51(5):513–30.

Lam DW, Zechman FW. Phylogenetic analyses of the Bryopsidales (Ulvophyceae, Chlorophyta) based on rubisco large subunit gene sequences. J Phycol. 2006;42:669–78.

Liu D, Keesing JK, Xing Q, Shi P. World's largest macroalgal bloom caused by expansion of seaweed aquaculture in China. Mar Pollut Bull. 2009;58:888–95.

Martins I, Oliveria JM, Flindt MR, Marques JC. The effect of salinity on the growth rate of the macroalgae *Enteromorpha intestinalis* (Chlorophyta) in the Mondego estuary (West Portugal). Acta Oecol. 1999;20:259–65.

Masakiyo Y, Shimada S. Species diversity of the genus *Ulva* (Ulvophyceae, Chlorophyta) in Japanese waters, with special reference to *Ulva tepida* Masakiyo et S. Shimada sp. nov. Bull Nat Mus Nat Sci Ser B. 2014;40:1–13.

Matsuda J, Setoguchi H. Isolation and characterization of microsatellite loci in Asarum leucosepalum (Aristolochiaceae), an endangered plant endemic to Tokunoshima Island in the Ryukyu Archipelago. Conserv Genet Res. 2012;4:579–81.

Matsumoto M, Shimada S. Systematics of green algae resembling *Ulva conglobata*, with a description of *Ulva adhaerens* sp. nov. (Ulvales, Ulvophyceae). Eur J Phycol. 2015;50: 100–11.

McAvoy KM, Klug JL. Positive and negative effects of riverine input on the estuarine green alga *Ulva intestinalis* (syn. *Enteromorpha intestinalis*) (Linnaeus). Hydrobiologia. 2005;545:1–9.

Miura A. Taxonomic studies of *Porphyra* species cultivated in Japan, referring to their transition to the cultivated variety. J Tokyo Univ Fish. 1988;75:311–25.

Miyashita A. Thing and human cultural history. Tokyo: Hosei University Press; 2003.

Müller OF (1778) Icones plantarum. *Florae danicae*, vol. 5, fasc 13. Copenhagem.

Nagai S, Nishitani G, Sakamoto S, Sugaya T, Lee CK, Kim CH, Itakura S, Yamaguchi M. Genetic structuring and transfer of marine dinoflagellate *Cochlodinium polykrikoides* in Japanese and Korean coastal waters revealed by microsatellites. Mol Ecol. 2009;18:2337–52.

Nakamura Y, Sasaki N, Kobayashi M, Ojima N, Yasuike M, Shigenobu Y, Satomi M, Fukuma Y, Shiwaku K, Tsujimoto A, Kobayashi T, Nakayama I, Ito F, Nakajima K, Sano M, Wada T, Kuhara S, Inouye K, Gojobori T, Ikeo K. The first symbiont-free genome sequence of marine red alga, susabi-nori (*Pyropia yezoensis*). PLoS ONE. 2013;8(3):e57122.

Nei M, Tajima F. Maximum likelihood estimation of the number of nucleotide substitutions from restriction sites data. Genetics. 1983;105:207–17.

Niwa K, Sakamoto T. Allopolyploidy in natural and cultivated populations of *Porphyra*. J Phycol. 2010;46:1097–105.

Niwa K, Kikuchi N, Aruga Y. Morphological and molecular analysis of the endangered species *Porphyra tenera* (Bangiales, Rhodophyta). J Phycol. 2005;41:294–304.

Niwa K, Iida S, Kato A, Kawai H, Kikuchi N, Kobiyama A, Aruga Y. Genetic diversity and introgression in two cultivated species (*Porphyra yezoensis* and *Porphyra tenera*) and closely related wild species of *Porphyra* (Bangiales, Rhodophyta). J Phycol. 2009;45:493–502.

Niwa K, Kobiyama A, Sakamoto T. Interspecific hybridization in the haploid blade-forming marine crop *Porphyra* (Bangiales, Rhodophyta): occurrence of allodiploidy in surviving F1 gametophytic blades. J Phycol. 2010;46:693–702.

Notoya M. *Porphyra*. In: Ohno M, editor. Biology and technology of economic seaweeds. Tokyo: Uchida Rokakuho; 2004.

Nunome M, Kinoshita G, Tomozawa M, Torii H, Matsuki R, Yamada F, Matsuda Y, Suzuki H. Lack of association between winter coat colour and genetic population structure in the Japanese hare, *Lepus brachyurus* (Lagomorpha: Leporidae). Biol J Linn Soc. 2014a;111(4):761–76.

Nunome M, Kinoshida G, Motozawa M, Torii H, Matsuki R, Yamada F, Matsuda Y, Suzuki H. Lack of association between winter coat colour and genetic population structure in the Japanese hare, *Lepus brachyurus* (Lagomorpha: Leporidae). Biol J Linn Soc. 2014b;111:761–76.

Ohnishi M, Kikuchi N, Iwasaki T, Kawaguchi R, Shimada S. Population genomic structures of endangered species (CR + ER), *Pyropia tenera* (Bangiales, Rhodophyta). Jpn J Phycol. 2013;61:87–96.

Ohno M. *Ulva* and extensive biomass. In: Notoya M, editor. Utilization of *Ulva* and restoration of the environment. Tokyo: Seizando; 1999. p. 1–15.

Ohno M, Mizutani Y, Taino Takahashi I. Ecology of the edible green alga *Enteromorpha prolifera* in Shimanto River, Southern Japan. Bull Mar Sci Fish Kochi Univ. 1999;19:27–35.

Okamura K. Asakusa Nori (*Porphyra tenera*). Tokyo: Hakubunn-kan; 1909.

Park CS, Hwang EK, Sohn CH. A atable seeding method for *Porphyra pseudolinearis* Ueda (Rhodophyta): developing a new species for cultivation of *Porphyra* in Korea. Aquat Res. 2003;34:895–8.

Perevra RT, Bergström L, Kautsky L, Johannesson K. Rapid speciation in a newly opened postglacial marine environment, the Baltic Sea. BMC Evol Biol. 2009;9:70.

Pritchard JK, Stephens M, Donnelly P. Inference of population structure using multilocus genotype data. Genetics. 2000;155:945–59.

Provasoli L. Media and prospects for the cultivation of marine algae. In: Watanabe A, Hattori A (eds) Cultures and collections of Algae. Proceedings of US Japan conference; September 1966, Hakone, Japan. Tokyo: Japanese Society of Plant Physiologists; 1968. pp 63–75.

Reed RH, Russell G. Adaptation to salinity stress in populations of *Enteromorpha intestinalis* (L.) link. Estu Coast Mar Sci. 1979;8:251–8.

Sakaguchi S, Qiu YX, Liu YH, Qi XS, Kim SH, Han J, Takeuchi Y, Worth JR, Yamasaki M, Sakurai S, Isagi Y. Climate oscillation during the quaternary associated with landscape heterogeneity promoted allopatric lineage divergence of a temperate tree *Kalopanax septemlobus* (Araliaceae) in East Asia. Mol Ecol. 2012;21(15):3823–38.

Shimada S, Hiraoka M, Nabata S, Iima M, Masuda M. Molecular phylogenetic analyses of the Japanese *Ulva* and *Enteromorpha* (Ulvales, Ulvophyceae), with special reference to the free-floating *Ulva*. Phycol Res. 2003;51:99–108.

Shimada S, Yokoyama N, Masuda M. Genus *Ulva* (Ulvophyceae, Chlorophyta) in Hokkaido, Japan. Jpn J Bot. 2007;82:190–204.

Shimada S, Yokoyama N, Arai S, Hiraoka M. Phylogeography of the genus *Ulva* (Ulvophyceae, Chlorophyta), with special reference to the Japanese freshwater and brackish taxa. J Appl Phycol. 2008;20:979–89.

Shimada S, Nagano M, Hiraoka M, Ichihara K, Minerur F, Zhu W. Phylogeographic analysis of the genus *Ulva* (Ulvales, Chlorophyta), including bloom sample in Qingdao, China. Coast Mar Sci. 2010;34:117–22.

Shimada S, Ichihara K, Masakiyo Y, Iima M, Yoshida Y, Kumano S. Threatened species *Nemalionopsis tortuosa* (Thoreales, Rhodophyta) in Japan, new locality and current condition of its all reported habitats. Algal Res. 2012;5:9–16.

Silberfeld T, Rousseeau F, de Reviers B. An updated classification of brown algae (Ochrophyta, Phaeophyceae). Cryptog Algol. 2014;35(2):117–56.

Sutherland JE, Lindstrom SC, Nelson WA, Brodie J, Lynch MD, Hwang MS, Choi HG, Miyata M, Kikuchi N, Oliveira MC, Farr T, Neefus C, Mols-Mortensen A, Milstei D, Müller KM. A new look at an ancient order: generic revision of the Bangiales (Rhodophyta). J Phycol. 2011;47(5):1131–51.

Tanaka N, Demise T, Ishii M, Shji Y, Nakaoka M. Genetic structure and gene flow of eelgrass *Zostera marina* populations in Tokyo Bay, Japan: implications for their restoration. Mar Biol. 2011;158:871–82.

Tatyana A, Shim JB, Hwang MS, Kim GH. Host-parasiteinteractions and host species susceptibility of the marine oomycete parasite, *Olpidiopsis* sp., from Korea that infects red algae. J Appl Phycol. 2012;24:135–44.

Touhata K, Namikoshi A, Suzuki T, Iguchi J, Mizusawa N, Hara T, Imamura S, Yabu T, Yamashita Y, Yamashita M. Origin identification of dried seaweed product "Nori" by PCR–RFLP analysis of *Pyropia yezoensis* in the internal transcribed spacer ITS-1 region. Fish Sci. 2013;79:865–75.

Ueda S. Systematic study of the genus *Porphyra* in Japan. Suiko Kenkyu Kokoku. 1932;28:1–45.

van den Hoek C, Man DG, Jahns HM. Algae. An introduction to phycology. Cambridge: Cambridge University Press; 1995.

van Inghelandt D, Melchinger AE, Lebreton C, Stich B. Population structure and genetic diversity in a commercial maize breeding program assessed with SSR and SNP markers. Theor Appl Genet. 2010;120:1289–99.

Varshney RK, Thiel T, Stein N, Langridge P, Graner A. In silico analysis of frequency and distribution of microsatellites in ESTs of some cereal species. Cell Mol Biol Lett. 2002; 7(2A):537–46.

Wagner CE, Keller I, Wittwer S, Selz OM, Mwaiko S, Greuter L, Sivasundar A, Seehausen O. Genome-wide RAD sequence data provide unprecedented resolution of species boundaries and relationships in the Lake Victoria cichlid adaptive radiation. Mol Ecol. 2013;22(3):787–98.

Xia C, Li B, Ji D, Chen C. Characterization of the global transcriptome for *Pyropia haitanensis* (Bangiales, Rhodophyta) and development of cSSR markers. BMC Genom. 2013;14:107.

Xing Q, Gao M, Gao X, Tosi L, Schmitt FG, Zhang Y, Shi P, Wei J, Luo Y. Progressive eutrophication behind the world-largest super floating macroalgal bloom in the Yellow Sea. Biogeosci Dis. 2014;11:7029–54.

Xu P, Xu SZ, Wu XH, Tao Y, Wang BG, Wang S, Qin DH, Lu ZF, Li GJ. Population genomic analyses from low-coverage RAD-Seq data: a case study on the non-model cucurbit bottle gourd. Plant J. 2014;77:430–42.

Yoshida T. Marine algae of Japan. Tokyo: Uchida Rokakuho; 1998.

Zhang L, Wang G, Liu C, Chi S, Liu T. Development and utility of EST-SSR markers in *Ulva prolifera* of the south Yellow Sea. Acta Oceanol Sin. 2014;33(10):105–13.

Part V
Troubleshooting New Genomic Approaches for Seaweeds

Chapter 15
DNA Extraction Techniques for Genomic Analyses of Macroalgae

Laura J. Wilson, Xénia A. Weber, Tania M. King and Ceridwen I. Fraser

Abstract Extracting high-quality DNA in large quantities from Phaeophyceae (brown), Rhodophyta (red), and Chlorophyta (green) macroalgae presents a substantial obstacle for modern molecular studies. Macroalgal tissues are rich in polysaccharides and polyphenols that are known to interfere with downstream molecular techniques. These compounds are released during DNA extraction procedures and often persist despite purification attempts. A wide range of DNA extraction and purification methods have been developed in attempts to overcome these challenges. Here, we review methods of macroalgal DNA extraction, including commercial kits. We discuss each method's merits and limitations, and examine its potential use in both traditional and high-throughput molecular approaches. Finally, we present our own findings from a range of DNA extraction, purification, and sequencing trials based on the commercially available MoBio PowerPlant® Pro and PowerPlant® Clean kits, carried out on a number of Phaeophyceae species. We found that DNA yield and quality can be improved by lengthy soaking of tissue in extraction buffers, altering the homogenization procedures and modifying solutions used. The DNA extraction protocol presented here can be scaled up for high-throughput analyses and preliminary results suggest suitability for next-generation sequencing.

Keywords Brown algae · Chlorophyta · Green algae · High-throughput · Inhibitors · Phaeophyceae · Polyphenols · Polysaccharides · Red algae · Rhodophyta

---

L.J. Wilson · X.A. Weber · C.I. Fraser (✉)
Fenner School of Environment and Society, Australian National University, Daley Road, Acton, ACT 2601, Australia
e-mail: ceridwen.fraser@gmail.com

T.M. King
Department of Zoology, University of Otago, 340 Great King Street, Dunedin 9016, New Zealand

Z.-M. Hu and C. Fraser (eds.), *Seaweed Phylogeography*,
DOI 10.1007/978-94-017-7534-2_15

15.1 Background

Successfully isolating large quantities of high-quality, high molecular weight DNA from macroalgae is notoriously difficult (Rogers and Bendich 1994; Varela-Alvarez et al. 2011). Macroalgal tissues are rich in polysaccharides and polyphenols, such as alginic acid and fucoidan (Saunders 1993; Hu and Zhou 2001; Wang et al. 2005; Hoarau et al. 2007; Verbruggen et al. 2007; Ivanova et al. 2008; Snirc et al. 2010; Varela-Alvarez et al. 2011; Maeda et al. 2012). These compounds are easily released during DNA extraction processes and often remain present even after purification (Rogers and Bendich 1994; Maeda et al. 2012). Polysaccharides and polyphenols can inhibit the activity of enzymes, including restriction enzymes, polymerases, and ligases (Wang et al. 2005; Varela-Alvarez et al. 2011; Maeda et al. 2012). Incomplete removal of these compounds during DNA extraction can inhibit downstream laboratory techniques, such as PCR amplification, restriction enzyme digestion, and cloning (Phillips et al. 2001; Varela-Alvarez et al. 2011; Maeda et al. 2012). In addition, polyphenols can reduce the solubility of extracted DNA, and the presence of polysaccharides often results in a viscous extraction solution, which impedes activities such as agitating and pipetting (Hoarau et al. 2007; Ivanova et al. 2008; Demeke and Jenkins 2010; Maeda et al. 2012). The composition and concentration of polysaccharides and polyphenolic compounds can vary greatly between macroalgal species, developmental stages, and distributions, making it difficult to develop a ubiquitously effective extraction technique (Hu et al. 2004; Wong et al. 2007; Ivanova et al. 2008; Demeke and Jenkins 2010; Varela-Alvarez et al. 2011; Sahu et al. 2012).

Overall, the removal of polysaccharides and polyphenols during macroalgal DNA extraction has proven to be challenging, time consuming, and expensive (Rogers and Bendich 1994; Varela-Alvarez et al. 2011). Obtaining large amounts of high-quality DNA is particularly important for modern, high-throughput genomic research. Thus, the lack of high-yielding and consistently effective DNA extraction methods is a major limiting factor for large-scale macroalgal molecular studies (Phillips et al. 2001; Wang et al. 2005; Ivanova et al. 2008; McDevit and Saunders 2009; Maeda et al. 2012).

Here, we review the main techniques used to extract DNA from macroalgae, and discuss their effectiveness for traditional (i.e. Sanger sequencing) and high-throughput ('Next Generation Sequencing') approaches. We also present our results from several recent trials that determined DNA extraction protocol modifications suitable for extracting DNA from brown (phaeophycean) algae, which have since been used for genomic analyses.

15.2 Techniques Used for Extracting DNA from Macroalgae

A wide variety of techniques have been used to extract DNA from macroalgae for traditional molecular analyses, ranging from relatively simple, single-step methods to complicated procedures and commercial kits (Tables 15.1 and 15.2). Further, detailed notes on each DNA extraction protocol are provided in the final section of this chapter.

Although some of the more simple DNA extraction methods, such as Chelex (Bio-Rad) resin (Walsh et al. 1991; Casquet et al. 2012) and SDS buffers, typically release good volumes of DNA and have occasionally been used for macroalgal molecular research (e.g. Goff and Moon 1993; Saunders 1993; Billot et al. 1998; Vidal et al. 2002; Cohen et al. 2004; Hu et al. 2004; Fraser et al. 2009a; Torrano-Silva et al. 2014), they also release large amounts of polysaccharides and polyphenols. Such techniques can produce DNA that is unusable for genetic analysis without further extensive purification steps (Jin et al. 1997; Hoarau et al. 2007; Varela-Alvarez et al. 2011; Yang et al. 2013), although some researchers have found that extreme dilution (10× to 100×) can reduce the concentration of inhibitors sufficiently to enable PCR (e.g. Fraser et al. 2009a; Heesch et al. 2009). The unreliable results obtained with these methods have precluded their routine use for macroalgal DNA extraction (Hoarau et al. 2007). Instead, various other DNA extraction methods, often originally intended for use on terrestrial plants, have been employed or modified to address the problems associated with the presence of polysaccharides and polyphenols in macroalgae (Phillips et al. 2001; Varela-Alvarez et al. 2011; Kirkendale et al. 2013). Many researchers have based macroalgal DNA extractions on cetyltrimethylammonium bromide (CTAB) treatments (Doyle 1991; De Jong et al. 1998) and lithium chloride extractions (Hong et al. 1992, 1995). These methods also typically include purification steps, such as caesium chloride and ethidium bromide ultracentrifugation (Phillips et al. 2001; Goff and Coleman 1988), agarose gel electrophoresis purification (Saunders 1993), and/or column purification techniques [e.g. hydroxyapatite columns (Parsons et al. 1990), Sepharose columns (Mayes et al. 1992), or Qiagen purification columns (Hu and Zhou 2001; Hoarau et al. 2007)]. Many of these methods (e.g. ultracentrifugation) are complicated, expensive, and time consuming, and require large volumes of macroalgal tissue (Hu and Zhou 2001). Conversely, the simpler, cheaper methods (e.g. lithium chloride extractions) can give inconsistent results and often require numerous modifications to be transferable among macroalgal species (Hong et al. 1992; Rogers and Bendich 1994; Varela-Alvarez et al. 2011; Kirkendale et al. 2013).

CTAB-based DNA extraction methods have become the most widely adopted method for isolating macroalgal DNA (Siemer et al. 1998; Hu and Zhou 2001; Hu et al. 2004; Vieira-Pinto et al. 2014). CTAB selectively binds nucleic acids under high sodium chloride (NaCl) concentration, and forms a CTAB–DNA precipitate after the NaCl concentration is reduced, allowing the polysaccharides to be

Table 15.1 Non-commercial DNA extraction methods used for macroalgae including comments on pre- and post-extraction processes required, the cost and time associated with these methods, and the potential for scaling up to a 96-well format

DNA extraction method	Macroalgae groups this method has been used on	Pre-extraction processes required?	Post-extraction purifications required?	Toxic organic solvents required?	Cost	Time	Potential for scaling up to a 96-well format
Chelex® Resin	Rodophyta Phaeophyceae Chlorophyta	Yes/No	No	No	Low	Low–medium	Low–medium
SDS buffer	Phaeophyceae Rodophyta	Yes	Yes/No	Yes/No	Low–high	Low–high	Low–high
Proteinase K protocol	Rodophyta	Yes/No	Yes	Yes	Low–medium	Low–medium	Low
CTAB buffer	Rodophyta Phaeophyceae Chlorophyta	Yes	Yes/No	Yes	Low–high	Low–high	Low–medium
Lithium chloride extraction	Rodophyta	No	No	No	Low	Low	High
Extraction buffer 1	Phaeophyceae	Yes	Yes	Yes/No	Low–medium	Low–high	Low–high
Extraction buffer 2	Chlorophyta	Yes	Yes	Yes	High	Medium–high	Low

Further detailed notes about these DNA extraction methods are provided in the final section of this chapter

Table 15.2 Commercial DNA extraction methods used for macroalgae including comments on pre- and post-extraction processes required, the cost and time associated with each method and the potential for scaling up to a 96-well format

DNA extraction kit	Macroalgae groups this method has been used on	Pre-extraction processes required?	Post-extraction purifications required?	Toxic organic solvents required?	Cost	Time	Potential for scaling up to a 96-well format
Plant/seed DNA kit (ZR-96, Zymo)	Chlorophyta	No	No	No	Medium–high	Low	High
DNeasy Plant Mini Kit (QIAGEN)	Rodophyta Phaeophyceae Chlorophyta	No	Yes/No	Yes/No	High	Low–medium	Medium–high
NucleoSpin 96 tissue kit (MACHEREY-NAGEL)	Rodophyta	No	No	No	High	Low	High
ISO-PLANT (Nippon gene)	Phaeophyceae	No	Yes	No	High	Low	High
Plant genomic DNA kit (Tiangen Biotech)	Phaeophyceae	No	No	No	High	Low	Low

Further detailed notes about these DNA extraction methods are provided in the final section of this chapter removed (Doyle 1991; Rogers and Bendich 1994; Coyer et al. 1995; Hu and Zhou 2001; Hughey et al. 2001; Rindi et al. 2007; Maeda et al. 2012. This method has been reported as a relatively fast, cheap, and effective DNA extraction method for samples containing large amounts of polysaccharides (Phillips et al. 2001). However, in most cases modifications to the basic CTAB protocol are required to extract useable DNA from macroalgae (Hu and Zhou 2001; Wang et al. 2005). Such modifications often involve tissue preparation processes prior to extraction, for example, nuclear or organellar isolation (Lane et al. 2006; Alberto et al. 2009; Varela-Alvarez et al. 2011); enzymatic dissociation of tissues (Hu and Zhou 2001); incubation and/or washing tissue in various solutions to soften cell walls [e.g. hydrochloric acid (Maeda et al. 2012)] and solubilize inhibiting compounds [e.g. sodium carbonate (Na_2CO_3) (Maeda et al. 2012)]; addition of chemicals to the extraction buffer which are known to assist the removal of polysaccharides and polyphenols [e.g. mercaptoethanol, diethyldithiocarbamic acid (DIECA) (Phillips et al. 2001; McDevit and Saunders 2009; Maeda et al. 2012)]; and/or post-extraction DNA purification by caesium chloride density gradient ultracentrifugation or column purification (Hu and Zhou 2001; Phillips et al. 2001; Hoarau et al. 2007). However, many of these modifications can be lengthy, costly, and difficult to scale up, posing a significant barrier to their use in large-scale or high-throughout studies. In addition, most CTAB-based DNA extraction protocols use toxic organic solvents, particularly phenol and chloroform. This can make these methods unsuitable, depending on the researcher's personal preference, availability of specialized containment facilities, and appropriate waste removal processes to minimize human and environmental exposure (Niu et al. 2008; Kotchoni and Gachomo 2009).

Commercial extraction kits—usually those developed for terrestrial plants—have also been used for macroalgal DNA isolation. Commercial kits are typically based upon membrane technology that is simpler to use than direct precipitation protocols, do not use toxic organic solvents, and can yield macroalgal DNA suitable for PCR amplification and sequencing, [e.g. red algae *Lithothamnion corallioides*, *Phymatolithon calcareum*, and *Lithophylloideae* spp. (Carro et al. 2014; Richards et al. 2014), brown algae *Sargassum thunbergii* (Liu et al. 2012) and *Laminaria* spp. (Yotsukura et al. 2001), and green algae *Chordaria* spp. (Famà et al. 2002)]. Additionally, commercial kits have been used for post-extraction DNA purification (e.g. Manghisi et al. 2010; Saunders and McDevit 2012). While commercial DNA extraction kits have several benefits over noncommercial extraction methods such as CTAB buffers, their convenience and safety benefits often outweigh high financial cost, restricting their use in studies with a large number of samples and/or limited financial resources (Niu et al. 2008). In addition, commercial kits often show inconsistent and species-specific results. Extensive modifications to the manufacturers' protocols are often required to adapt commercial DNA extraction kits to macroalgal species, including post-extraction purification procedures and repeated extractions (Hoarau et al. 2007; McDevit and Saunders 2009; Snirc et al. 2010; Maeda et al. 2012; Pardo et al. 2014). These factors limit the applicability of commercial kits to the extraction of macroalgal DNA.

15.3 The Next-Generation of DNA Extraction and Sequencing

Despite inconsistent results and a high failure rate, the methods discussed above have produced DNA that is sufficiently free of polyphenolic compounds and polysaccharides to allow the PCR amplification of nuclear, mitochondrial, and chloroplast genes, as well as the amplification of microsatellite loci, restriction enzyme digests, hybridization, and sequencing of PCR products (Hu and Zhou 2001; Phillips et al. 2001; Van der Strate et al. 2002; Varela-Alvarez et al. 2011; Arnaud-Haond et al. 2013; Kirkendale et al. 2013). However, with the development of 'Next Generation Sequencing' (NGS) technologies, molecular studies are moving away from amplification of a single locus, microsatellites, and cloning techniques (Phillips et al. 2001). Whereas traditional methods targeting, for example, a single locus via PCR can be successful with small amounts of DNA, NGS approaches typically require hundreds of nanograms of high-purity, high molecular weight DNA (Davey et al. 2011). The few macroalgal studies that have used NGS technologies have generally focused on a single macroalgal genome, often to design microsatellite primers (e.g. Arnaud-Haond et al. 2013), in which case DNA from multiple individuals can be pooled to acquire enough genetic material for NGS. In contrast, approaches that assess separate individuals, such as population genetic analyses, require each sample to yield large amounts of DNA. Although carrying out multiple extractions of the same individual is possible, such an approach would usually be considered prohibitively expensive and time consuming.

One of the major advantages of NGS is that it enables rapid genomic screening of large numbers of individuals (McDevit and Saunders 2009; Snirc et al. 2010), and thus is ideal for population genetic, phylogeographic, and phylogenetic research. Many macroalgal DNA extraction methods discussed above involve processes which are, however, not easily adaptable to 96-well plate format (e.g. phenol–chloroform steps, caesium chloride ultracentrifugation, nuclei or organelle isolation, enzymatic dissociation of tissues) (Hu and Zhou 2001; McDevit and Saunders 2009; Snirc et al. 2010). Therefore, although these DNA extraction techniques can produce good yields of good quality DNA from relatively small amounts of macroalgae tissue, they are poorly suited to high-throughput techniques (Van der Strate et al. 2000; Phillips et al. 2001; Wang et al. 2005; Hoarau et al. 2007; McDevit and Saunders 2009).

Several attempts have been made to design macroalgal DNA extraction protocols that can be carried out in a standardized 96-well format while still producing suitable yields of good quality DNA. For example, Snirc et al. (2010) designed a modified protocol based on the DNeasy Plant mini kit protocol (QIAGEN Sciences, Germantown USA) which extracted an average of 1940 ng of DNA per 20 mg of dried tissue, from a range of brown macroalgal species. This modified protocol involved the use of an alternative lysis buffer with a high salt concentration to decrease the levels of co-extracted polysaccharides, the addition of antioxidant compounds to the lysis buffer, including polyvinylpyrrolidone (PVP) and bovine serum albumin (BSA) to bind polyphenols, as well as the addition of a chloroform–isoamyl alcohol purification step (Snirc et al. 2010). Although Snirc et al. (2010) state that this method has been designed to be easily scaled up to a 96-well plate format, with potential use of automated devices, it is unclear exactly which steps in the protocol this applies to. Chloroform purifications are often seen to be difficult to carry out in a 96-well plate format, suggesting that only the second half of this DNA extraction protocol is suitable for scaling up (McDevit and Saunders 2009). If this is the case, it would result in a DNA extraction protocol similar to the CTAB-based DNA extraction method published by Hoarau et al. (2007) which involves carrying out tissue pulverization, cell lysis, and chloroform purification steps in single tubes before transferring samples to a 96-well microtitre filtration plate for further purification. Alternatively, McDevit and Saunders (2009) developed a high-throughput DNA extraction protocol for brown macroalgae in which all steps could be carried out in a 96-well format. This modification of an earlier DNA extraction protocol by Lane et al. (2006) introduced an acetone wash step that removed some of the PCR inhibiting compounds, thereby eliminating the need for the time-consuming organelle extractions which prevented Lane et al's (2006) protocol from being scaled up. These methods appear promising for large-scale studies involving Sanger sequencing of PCR-amplified single gene or microsatellite loci, but have yet to be tested using NGS techniques (Hoarau et al. 2007; McDevit and Saunders 2009; Snirc et al. 2010).

15.4 New Trials to Improve Existing Extraction Protocols for Brown Algae

With the goal of obtaining high-yield, high-quality DNA from large numbers of brown algal (phaeophycean) individuals, we have tested several DNA extraction protocols both with and without additional modifications. Here, we present the results of these trials, including a new DNA extraction protocol based on modifications to a commercially available kit, with which we have managed to successfully sequence more than 60 samples on an Illumina HiSeq 2500 platform using the NGS 'Genotyping by Sequencing (GBS)' (Elshire et al. 2011) approach. Our trials indicate that both the quantity and quality of genetic material can be improved by (i) preliminary soaking of algal tissue in extraction buffers, (ii) tissue homogenization via bead beating, and (iii) modifying some solutions. The DNA extraction methods trialled were based on four commercially available kits: the DNeasy Plant Mini kit (QIAGEN Sciences, Germantown, USA); PowerBiofilm™ DNA Isolation kit (MoBio Laboratories Inc., Carlsbad, USA); PowerPlant® Pro DNA Isolation Kit; and PowerClean® Pro DNA Clean-Up Kit (MoBio Laboratories Inc., Carlsbad, USA). These kits, with various protocol modifications, as described below, were tested on four genera of Phaeophyceae: *Ecklonia*, *Sargassum*, *Durvillaea* and

Adenocystis. The specimens had been preserved in 95 % ethanol, dried at 60 °C, and stored for up to several years with silica gel beads.

Various DNA extraction methods were tested on the brown macroalga *Ecklonia radiata*, including the PowerBiofilm™ DNA isolation kit (Mo Bio) following the manufacturer's protocol, a published macroalgae-specific modified protocol using the DNeasy Plant Mini kit (Snirc et al. 2010) and various modifications to the manufacturer's protocols of the MoBio PowerPlant® Pro and PowerClean® Pro kits. The volume and quality of DNA extracted, as well as ease of use, varied considerably between protocols. The modified DNeasy Plant Mini kit protocol (Snirc et al. 2010) produced the highest DNA yield (1210 ng) (Table 15.3); however, this was the only protocol tested which required the use of toxic organic solvents, including phenol and chloroform. Due to a preference not to use such chemicals, further testing of this method was discontinued in favour of the PowerBiofilm™ DNA Isolation and PowerPlant® Pro kits. Of these two DNA extraction methods, the PowerBiofilm™ Isolation kit produced the highest DNA yield (1137 ng); however, a modified PowerPlant® Pro kit DNA extraction method, followed by purification with the PowerClean® Pro kit, extracted better quality DNA with regards to both purity, based on spectrometry absorbance ratios, and molecular weight, as indicated by gel electrophoresis. This extraction protocol involved the following modifications to the PowerPlant® Pro manufacturer's protocol: (i) prior to DNA extraction, dried tissue was ground to a powder in a mortar and pestle with zirconia or silica beads; (ii) in order to avoid over homogenizing the samples and over shearing the DNA, the homogenization step was replaced with a 10-min incubation at 65 °C, after which the samples were vortexed for 30 s; (iii) a DNA precipitate formed after the addition of PD4 and PD6 was removed, washed with 70 % ethanol, and resuspended in PowerPlant® Pro solution PD7.

Although DNA extraction and purification using this modified PowerPlant® Pro and PowerClean® Pro protocol was relatively effective, this method was still highly inefficient for multiple reasons. First, it produced relatively small quantities of

Table 15.3 DNA yield (ng), as measured by spectrophotometry, obtained from brown macroalgae species, *Ecklonia radiata*, using a modified QIAGEN DNeasy Plant Mini kit, a PowerBiofilm™ DNA Isolation kit, and a modified MoBio PowerPlant® Pro & PowerClean® Pro extraction protocol

DNA extraction method	Macroalgae species	Average amount of tissue per sample (mg)	Number of samples pooled	Average DNA yield (ng)
Modified DNeasy Plant Mini kit protocol	*Ecklonia radiata*	20	1	1210
PowerBiofilm™ DNA Isolation kit manufacturer's protocol	*Ecklonia radiata*	15	1	1137.5
Modified PowerClean® Pro & PowerClean® Pro protocols	*Ecklonia radiata*	10	8	463.6

DNA, which would be insufficient for whole-genome sequencing. Second, the majority of high molecular weight DNA appeared to precipitate out of solution upon the addition of the PowerPlant® Pro extraction buffers, PD4 and PD6. This clogged the column filter, inhibiting further column purification steps and prevented completion of the manufacturer's protocol, largely eliminating the benefits associated with using a commercial, column-based DNA extraction kit. To obtain the high molecular weight DNA required for whole-genome sequencing, the precipitate had to be removed, washed, and resuspended, as outlined above. Despite these difficulties, with considerable pooling and concentration of multiple DNA extractions, DNA suitable for 454-sequencing was obtained from *Ecklonia radiata*. These complications make this method unsuitable for large-scale, high-throughput molecular studies. In particular, the removal of the DNA precipitate is time consuming and would be difficult to scale up to a 96-well format. Furthermore, due to pooling of the DNA samples from multiple individuals, this method is also unsuitable for studies requiring separate DNA sequences from large numbers of individuals.

Subsequently, modifications to the PowerPlant® Pro and PowerClean® Pro protocols were tested in an attempt to mitigate these problems. These modifications included changing the amount of dried macroalgal tissue used, adding isopropanol or the Phenolic Separation Solution supplied in the PowerPlant® Pro kit at step 1 to prevent later precipitation of the DNA, soaking tissue in purified H_2O or extraction solutions, adjusting incubation conditions, altering homogenization steps, varying the amount of PowerPlant® Pro solutions added, and additional washing and filtration steps. These modifications were tested on brown macroalgae *Sargassum* sp., *Durvillaea antarctica*, *Durvillaea poha*, and *Adenocystis utricularis*. To compare the effects of these protocol modifications, DNA yield was quantified by fluorometry using an Invitrogen Qubit 2.0 Fluorometer (Life Technologies, Mulgrave, Australia). DNA samples were also visualized using agarose gel electrophoresis to ascertain the size of the extracted DNA molecules.

The greatest DNA yield gains were made by increasing the soaking time of tissue in extraction buffers to at least 24 h, repeating homogenization steps up to three times, milling samples with beads for 5 min, and adding isopropanol during incubation. Other protocol modifications, such as soaking of tissues in extraction buffers for longer than 24 h and the temperature of RNase A incubation, saw no substantial improvement in yield. The most successful extraction protocol involved soaking small pieces of tissue (<2 mm^2 dried material) for at least 24 h in the PowerPlant® Pro bead tubes (MoBio Laboratories Inc., Carlsbad USA) at 65 °C in 350 μL of solution PD1 and 50 μL of solution PD2. After soaking, 100 μL of isopropanol and 3 μL RNase A were added and samples were briefly vortexed, before being incubated for 30 min at 65 °C, vortexing briefly every ten minutes. Samples were then homogenized with a TissueLyser Bead Mill (QIAGEN Sciences Germantown USA) for 5 min at 30 Hz. The remaining DNA extraction steps were carried out following the PowerPlant® Pro manufacturer's protocol. For maximum elution efficiency, the eluted DNA was reloaded into the filter column and centrifuged at 10,000× g for 30 s. Subsequent DNA purification, carried out using the

PowerClean® Pro DNA Clean-up Kit, largely followed the manufacturer's protocol, with the only modification being the inclusion of a 650 μL, 100 % ethanol wash before adding the final elution buffer.

With the above PowerPlant® Pro and PowerClean® Pro protocol modifications DNA yield, when measured by fluorometry, increased from an average of less than 2.5 to 9–487 ng, depending on species (Table 15.4). Agarose gel electrophoresis revealed that samples extracted using the unmodified PowerPlant® Pro manufacturer's protocols displayed a poorly defined faint smear, with an unusual curved

Table 15.4 DNA yield (ng), as measured by fluorometry, obtained from <2 mm^2 tissue sample of brown macroalgae, *Sargassum* sp., *Durvillaea* spp., and *Adenocystis utricularis* using various modified protocols based on the Mo Bio PowerPlant® Pro & PowerClean® Pro manufacturer's protocol

Mo Bio protocol modification	Macroalgae species	DNA yield (ng)
2 h soaking at 65 °C	*Sargassum* sp.	<2.5
4 h soaking at 65 °C	*Sargassum* sp.	<2.5
250 μL PD3 + repeated PD3 incubation	*Sargassum* sp.	<2.5
Rep. homogenization + isopropanol	*Sargassum* sp.	3–9
Soaking in PD1 & PD2	*Sargassum* sp.	10–28
Soaking in PD1 & PD2 + bead milling	*Sargassum* sp.	6–35
24 h soaking in PD1 & PD2	*Sargassum* sp.	14–50
Rep. homogenization + isopropanol	*Durvillaea potatorum*	6.85
24 h soaking in PD1 & PD2 + 5 min bead milling + isopropanol + rep homogenization	*Durvillaea antarctica*	80–152
	Durvillaea potatorum	44–71
	Durvillaea poha	38–51
	Adenocystis utricularis	128–487
	Sargassum sp.	9–74
	Sargassum polycystum	49–115
45 h soaking in PD1 & PD2 + 5 min bead milling + isopropanol + rep. homogenization	*Durvillaea antarctica*	371
	Durvillaea potatorum	71
	Sargassum polycystum	37–92
71 h soaking in PD1 & PD2 + 5 min bead milling + isopropanol + rep. homogenization	*Durvillaea antarctica*	240
	Durvillaea potatorum	208

taper. This distortion was likely due to the presence of phenolic and polysaccharide compounds. Subsequent DNA purification following the PowerClean® Pro manufacturer's protocol produced a less curved band (Fig. 15.1a). The addition of the phenolic separating solution during the PowerPlant® Pro DNA extraction process, as recommended in the manufacturers' protocol for samples high in phenolics, did not appear to alter the curved, tapered shape of the DNA (Fig. 15.1b). In comparison, when DNA was extracted following the aforementioned modified extraction protocol, higher molecular weight and less distorted smears were obtained.

The improvements to DNA yield and quality reported here are likely due to a combination of enhanced mechanical and chemical cell lysis. Extending tissue soaking beyond the typical 15–120 min period may have encouraged chemical cell lysis by softening cell walls without releasing as many structural polysaccharides as traditional grinding techniques (Mayes et al. 1992; Patwary and van der Meer 1994; Hu and Zhou 2001; Lim et al. 2001; Harvey and Goff 2006; Sim et al. 2007). Other than improvements to yield, the modifications also made samples easier to handle. Longer soaking of the macroalgal tissue reduced solution viscosity, and the addition of isopropanol limited the precipitation of DNA out of solution. Intra-specific variation in DNA quality could largely be attributed to differences in specimen age and quality, as DNA extracted from more recently collected specimens tended to be of higher molecular weight.

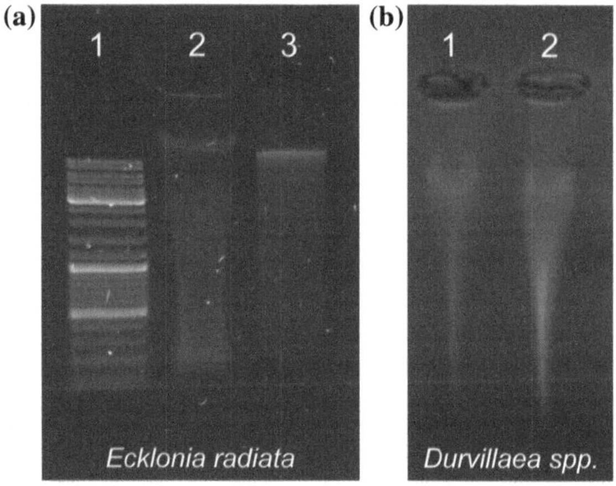

Fig. 15.1 DNA samples that display an unusual curved taper, possibly due to the presence of polysaccharide and polypeholic compounds. **a** DNA extraction from *Ecklonia radiata*. Lane 1, 100 bp/1 kb DNA Ladder (ThermoFisher); Lane 2, genomic DNA extracted using the MoBio PowerPlant® Pro kit, Lane 3, genomic DNA extracted using the MoBio PowerPlant® Pro kit then further purified using the MoBio PowerClean® Pro kit. **b** DNA extracted from *Durvillaea spp.* Lane 1, genomic DNA extracted from *Durvillaea chathamensis* using the MoBio PowerPlant® Pro kit with the provided phenolic separating solution (PSS), tissue was soaked in water at 60 °C prior to extraction; Lane 2, genomic DNA extracted from *Durvillaea antarctica* using the MoBio PowerPlant® Pro kit without the provided phenolic separating solution, tissue was soaked in water at 60 °C prior to extraction

Based on a commercial kit available in both low- and high-throughput formats, this DNA extraction method can be easily adapted to a high-throughput 96-well format, offering substantial time and labour savings. As such, this method has the potential to be used in large-scale NGS studies. Indeed, with a low-throughput version of the above extraction protocol, we obtained 30–100 ng of high-purity, high molecular weight total genomic DNA from individual *Durvillaea* specimens that had been preserved on silica gel for several years. Furthermore, these specimens successfully yielded a dataset comprising more than 100,000 SNPs across more than 60 *Durvillaea* plants, using an Illumina HiSeq 2500 platform (unpublished data). However, extracting DNA from *Sargassum* using the PowerPlant® Pro-htp 96-Well DNA Isolation kit and a self-constructed 96-well version of the PowerClean® Pro kit was less successful, resulting in an extremely low DNA yield which limited GBS library preparation and sequencing results. This discrepancy between results obtained from single-tube and 96-well plate DNA extraction methods may be due to the difficulty of avoiding pelleted inhibitor compounds when using a multichannel pipette. Samples extracted using the 96-well format may thus have higher concentrations of inhibitor compounds than those extracted using the single-tube kits. Alternatively, the discrepancy observed when upscaling the method may be a species-specific issue due to the quantity or type of inhibitory compounds released by *Sargassum* as compared with *Durvillaea*.

15.5 Conclusion

DNA extraction and purification protocols for Phaeophyceae (brown), Rhodophyta (red) and Chlorophyta (green) macroalgae have regularly been improved to remove inhibitory polysaccharide and polyphenolic compounds and to increase the yields of useable, high-quality, high molecular weight DNA. However, these compounds still pose a challenge to macroalgal molecular studies and without further refinement of DNA extraction and purification protocols, polysaccharides and polyphenols will continue to inhibit downstream genetic analyses. The continuing development of next-generation sequencing technologies offers the potential to transform macroalgal population genetic, phylogeographic and phylogenetic research, allowing rapid genomic screening of large numbers of individuals. Unfortunately, many macroalgal-specific DNA extraction protocols are poorly suited to the high-throughput formats required for NGS. As such, there is an urgent need to develop a relatively universal, high-yielding DNA extraction method for macroalgae, which is time efficient and suited to a 96-well plate. The methods and protocol modifications presented here are the promising steps towards developing such a protocol, although their suitability across a wider range of taxa has yet to be determined.

15.6 Catalogue of Reviewed Methods

In this section, the details of each method mentioned above are summarized to facilitate direct comparisons between methods. Specifically, this section provides information about pre-extraction processes, relative costs, time involved, and the potential for scaling the method to high throughput (plate format). These details expand on the information in Tables 15.1 and 15.2.

15.6.1 Noncommercial DNA Extraction Methods

15.6.1.1 Chelex® Resin

Pre-extraction processes: *Yes.* Some DNA extraction methods which use Chelex® Resin involve the isolation of gonimoblast or spore cells prior to DNA extraction (e.g. Goff and Moon 1993) which is thought to produce more consistent PCR amplification results than using vegetative tissues.

Cost: *Low.* Chelex® Resin is a relatively cheap reagent. Standard laboratory equipment is required.

Time: *Low–High.* This is a quick DNA extraction method with little hands-on time for extracting DNA from vegetative tissue. However, excising gonimoblast cells or harvesting spores under a microscope, prior to DNA extraction, as done by Goff and Moon (1993), increases the time required, particularly for large numbers of samples.

Potential for scaling up to a 96-well format: *Medium–High.* If extracting DNA from vegetative tissue, there is a medium to high potential for scaling up, although care must be taken when removing the supernatant with a multiple channel pipette and this may be simpler in single tubes. Scaling up is most likely largely unnecessary as Chelex® Resin is a relatively quick DNA extraction method with limited hands-on time. Even if gonimoblast or spore cells are isolated prior to DNA extraction this process would still have to be carried out one sample at a time, after which, if desired, the process of DNA extraction from these cells could be scaled up.

Additional comments: In some cases extreme dilution (10–100×) of the extracted DNA is required to achieve consistent PCR amplification (e.g. Fraser et al. 2009b; Heesch et al. 2009).

Additional references: Billot et al. (1998); Cohen et al. (2004); Torrano-Silva et al. (2014)

15.6.1.2 SDS Buffer

Pre-extraction processes: *Yes*. Extraction protocols include grinding of tissues in liquid nitrogen (Parsons et al. 1990; Mayes et al. 1992; Hu et al. 2004), organelle (mesiospore) isolation (Mayes et al. 1992), soaking tissues in β-mercaptoethanol to remove polysaccharides (Maeda et al. 2012), and washing tissues in polysaccharide eliminating buffer (Wang et al. 2005).

Post-extraction purification: *Yes*. Post-extraction purification procedures are not always used with a SDS Buffer DNA extraction method (e.g. Hu et al. 2004; Wang et al. 2005); however, when used they include gel filtration using Sepharose columns (Mayes et al. 1992), the GENECLEAN® II (Bio101) kit (Maeda et al. 2012), binding to hydroxyapatite (Parsons et al. 1990), and caesium chloride and ethidium bromide density gradient ultracentrifugation (Parsons et al. 1990).

Toxic organic solvents: *Yes*. SDS buffer-based DNA extraction methods often involve phenol–chloroform or chloroform–isoamyl purification steps, although some methods without such steps have been developed (e.g. Maeda et al. 2012).

Cost: *Low–High*. The cost of this DNA extraction method depends on the post-extraction DNA purification techniques used. Without any post-extraction purification this is a lost cost method, using relatively standard reagents and laboratory equipment, whereas commercial DNA purification kits are more costly, as are methods such as caesium chloride and ethidium bromide density gradient ultracentrifugation as they require specialize equipment (e.g. an ultracentrifuge).

Time: *Low–High*. Some methods involved multiple phenol/chloroform/isoamyl alcohol purifications and ethanol washes which can be time consuming, particularly for large numbers of samples. If organelle extraction were to be carried out prior to DNA extraction, this method would take much longer as each sample would have to be processed individually, and if using the same method as Mayes et al. (1992), by hand, through layers of Miracloth (Calbiochem®) and cheesecloth. Caesium chloride and ethidium bromide density gradient ultracentrifugation also requires time-intensive centrifugation steps, although this is not hands-on time.

Potential for scaling up to a 96-well format: *Low–High*. DNA extraction protocols that involve phenol/chloroform/isoamyl alcohol purification steps, organelle isolation, and/or caesium chloride and ethidium bromide density gradient ultracentrifugation would be difficult and expensive to scale up. Alternatively, methods that do not require phenol/chloroform/isoamyl alcohol purification steps or organelle extraction and use suitable post-extraction purification methods, such as the GENECLEAN® II (Bio101) kit (Maeda et al. 2012), offer greater potential for scaling up.

15.6.1.3 Proteinase K Protocol

Pre-extraction processes: *Yes*. Some protocols (e.g. Manghisi et al. 2010) involve grinding of tissues in liquid nitrogen prior to DNA extraction.

Post-extraction techniques: *Yes*. Post-extraction procedures that have been used with this DNA extraction protocol include gel purification (Saunders 1993) and Wizard® DNA Clean-Up System (Promega) (Manghisi et al. 2010).

Toxic Organic Solvents: *Yes*. Proteinase K Protocol DNA extraction protocols involve phenol/chloroform/isoamyl alcohol purification steps.

Cost: *Low–Medium*. Proteinase K DNA extraction methods use relatively cheap reagents and standard laboratory equipment; therefore, the cost depends on the post-extraction DNA purification technique used. Gel purification is a relatively cheap purification method, whereas purification kits, such as the Wizard® DNA Clean-Up System (Promega), are more expensive.

Time: *Low–Medium*. These methods involve multiple phenol/chloroform purifications, chloroform/isoamyl alcohol purifications, ethanol washes, and/or gel purification steps, which can be time consuming, particularly for large numbers of samples.

Potential for scaling up to a 96-well format: *Low*. The phenol/chloroform/ isoamyl alcohol purification steps in these protocols inhibit scaling up.

Additional Comments: Proteinase K DNA extraction protocols display high levels of variation in success between species.

15.6.1.4 CTAB Buffer

Pre-extraction processes: *Yes*. Protocols include processes to break up macroalgal tissues prior to DNA extraction, such as grinding tissues in liquid nitrogen (Phillips et al. 2001), and bead milling of tissues (Hoarau et al. 2007), and the soaking of tissues in reagents (e.g. β-mercaptoethanol) to help remove polysaccharides. Some protocols also involve enzymatic dissociation of tissues and unicell or nuceli isolation prior to DNA extraction (Hu and Zhou 2001; Alberto et al. 2009; Varela-Alvarez et al. 2011).

Post-extraction purifications: *Yes*. Post-extraction purification procedures are not always used with a CTAB DNA extraction method (e.g. Hu and Zhou 2001; Hoarau et al. 2007; Varela-Alvarez et al. 2011); however, when used they include the GENECLEAN® II (Bio101) kit (Maeda et al. 2012) and caesium chloride and ethidium bromide density gradient ultracentrifugation (Billot et al. 1998; Phillips et al. 2001).

Toxic Organic Solvents: *Yes*. These DNA extraction methods involve phenol/chloroform and/or chloroform/isoamyl alcohol purification steps.

Cost: *Low–High*. The cost of this DNA extraction method depends on the post-extraction DNA purification techniques used. Without any post-extraction purification this is a lost cost method, using relatively standard reagents and laboratory equipment, whereas commercial DNA purification kits are more costly, as are methods such as caesium chloride and ethidium bromide density gradient ultracentrifugation as they require expensive, specialized equipment (e.g. an ultracentrifuge).

Time: *Medium–High*. These methods involve multiple phenol/chloroform purifications, chloroform/isoamyl alcohol purifications, and ethanol washes which can be time consuming, particularly for large numbers of samples. Protocols that involve unicell or nuclei isolation are relatively time consuming, as each sample has to be processed individually, and often by hand. Caesium chloride and ethidium bromide density gradient ultracentrifugation requires time-intensive centrifugation steps, although this is not hands-on time.

Potential for scaling up to a 96-well format: *Low–Medium*. Methods that involve phenol/chloroform and/or chloroform/isoamyl alcohol purification steps, unicell or nuclei, and/or caesium chloride and ethidium bromide density gradient ultracentrifugation would be difficult and expensive to scale up. Hoarau et al. (2007) developed a method which could be partially scaled up by transferring the extracted DNA to a 96-well microtitre plate for purification; however, prior to this, bead milling of the tissues and the CTAB DNA extraction must still be carried out in a single–tube format. While it is worth noting that 96-well plate bead milling devices are available, due to the difficulties associated with scaling up the CTAB DNA extraction process itself, the benefit of scaling up the bead milling steps is most likely negligible.

Additional Comments: DNA extraction success may also be improved by the addition of, or increasing the volumes of, components known to be chelators (e.g. ethylenediaminetetraacetic acid (EDTA)), reductants (e.g. β-mercaptoethanol), or complexors of tannins and polyphenols (e.g. polyvinylpolyrrolidone (PVPP), diethyldithiocarbamic acid (DIECA)) to the extraction buffer (e.g. Phillips et al. 2001).

15.6.1.5 Lithium Chloride Extraction

Cost: *Low*. This method uses relatively cheap reagents and standard laboratory equipment.

Time: *Low*. This is a relatively quick method that involves incubation, precipitation, centrifugation, and ethanol wash steps, but avoids the slightly more time-consuming phenol/chloroform/isoamyl alcohol purification steps.

Potential for scaling up to a 96-well format: *High*. The incubation, precipitation, centrifugation, and ethanol wash steps in these protocols are easily scaled up.

Additional Comments: DNA extracted using these protocols results in inconsistent PCR amplification with highly variable success between species.

Additional references: Hong et al. (1992, 1995); Hong et al. (1997)

15.6.1.6 DNA Extraction Buffer 1*

*Buffer contains 0.1M tris base pH = 8, 0.3M $CaCl_2$, 0.05M disodium ethylenediaminetetraacetic acid (EDTA), 0.2M NaCl, 10 % Tween-20, and 20 mg mL^{-1} Proteinase K.

Pre-extraction processes: *Yes.* Protocols include processes to break up macroalgal tissues prior to DNA extraction, such as grinding tissues in liquid nitrogen and the soaking of tissues in reagents (e.g. acetone) to help remove polysaccharides (McDevit and Saunders 2009). Some protocols also involve organelle isolation prior to DNA extraction (Lane et al. 2006).

Post-extraction purification: *Yes.* The Wizard® DNA Clean-Up (Promega) system has been used for DNA purification following DNA extraction using this buffer (Lane et al. 2006; McDevit and Saunders 2009).

Toxic organic solvents: *Yes.* Some methods involve phenol/chloroform and/or chloroform/isoamyl alcohol purification steps.

Cost: *Medium.* While the DNA extraction buffer uses relatively standard reagents and these protocols require standard laboratory equipment, commercial DNA purification kits such as the Wizard® DNA Clean-Up (Promega) system can be costly.

Time: *Low–High.* Protocols that involve organelle isolation (e.g. Lane et al. 2006) are time consuming as they involve grinding, stirring, and filtration steps, which must be carried out by hand. However, the protocol can be shortened if organelle isolation steps are replaced with washing the tissues in acetone and grinding them in liquid nitrogen (e.g. McDevit and Saunders 2009).

Potential for scaling up to a 96-well format: *Low–High.* Methods which involve phenol/chloroform purifications, chloroform/isoamyl alcohol purifications, and/or organelle isolation would be difficult to scale up; however, the method designed by McDevit and Saunders (2009) is intended to be scaled up and automated.

15.6.1.7 DNA Extraction Buffer 2*

*Buffer contains NaCl, tris-HCl, ethylenediaminetetraacetic acid (EDTA), Triton X-100, β-mercaptoethanol, and dithiothreitol (DTT).

Pre-extraction processes: *Yes.* This protocol involves grinding of tissues in liquid nitrogen prior to DNA extraction.

Post-extraction Purifications: *Yes.* This protocol involves DNA purification by caesium chloride and ethidium bromide density gradient ultracentrifugation.

Toxic organic solvents: *Yes.* This DNA extraction method involves phenol/chloroform and chloroform/isoamyl alcohol purification steps.

Cost: *High.* Ethidium bromide density gradient ultracentrifugation requires expensive, specialized equipment (e.g. a ultracentrifuge).

Time: *Medium–High.* This method involves multiple phenol/chloroform purifications, chloroform/isoamyl alcohol purifications, and ethanol washes which can be time consuming, particularly for large numbers of samples. Caesium chloride and ethidium bromide density gradient ultracentrifugation requires time-intensive centrifugation steps, although this is not hands-on time.

Potential for scaling up to a 96-well format: *Low*. Methods that involve phenol/chloroform purifications, chloroform/isoamyl alcohol purifications, and ethidium bromide density gradient ultracentrifugation would be difficult and/or expensive to scale up.

Additional Reference: Van der Strate et al. (2000).

15.6.2 Commercial DNA Extraction Methods

15.6.2.1 Plant/Seed DNA Kit (ZR-96, Zymo)

Cost: *Medium–High*. Commercial kits are typically more expensive than non-commercial DNA extraction methods, perhaps with the exception of those that involve ethidium bromide density gradient ultracentrifugation.

Time: *Low*. Commercial DNA extraction kits are designed to be relatively efficient, reduce the hands-on time required for DNA extraction, and avoid more time consuming steps such as phenol/chloroform/isoamyl alcohol purifications.

Potential for scaling up to 96-well format: *High*. This is a 96-well kit.

Associated Papers: Kirkendale et al. (2013)

15.6.2.2 DNeasy Plant Mini Kit (QIAGEN)

Post-extraction Purifications: *Yes*. Post-extraction purification procedures are not always used with the DNeasy Plant Mini Kit (e.g. Snirc et al. 2010; Famà et al. 2002; Richards et al. 2014); however, Maeda et al. (2012) used the GENECLEAN® II (Bio101) kit.

Toxic organic solvents: *Yes*. The DNeasy Plant Mini Kit manufacturers' protocol does not use any toxic organic solvents; however, modified versions of the protocol (e.g. Snirc et al. 2010) do include chloroform/isoamyl alcohol purification steps.

Cost: *Medium–High*. Commercial kits are typically more expensive than non-commercial DNA extraction methods, perhaps with the exception of those that involve ethidium bromide density gradient ultracentrifugation.

Time: *Low–Medium*. Commercial DNA extraction kits are designed to be relatively efficient, reduce the hands-on time required for DNA extraction, and to avoid more time-consuming steps such as chloroform/isoamyl alcohol purifications. However, the modified protocol by Snirc et al. (2010) increases the time required due to the addition of chloroform/isoamyl alcohol purification steps.

Potential for scaling up to a 96-well plate: *Low–High*. There is a 96-well version of the DNeasy Plant Mini Kit available (DNeasy 96 Plant Kit, QIAGEN); however, the chloroform/isoamyl alcohol purification steps in the modified protocol developed by Snirc et al. (2010) preclude scaling up. The GENECLEAN® II (Bio101) purification kit also offers the potential to be scaled up to a 96-well format as it does not involve any specialized purification columns or tubes, instead involving the binding of DNA to a silica matrix, wash and centrifugation steps, and elution of the DNA which could be carried out in a 96-well plate.

15.6.2.3 NucleoSpin® 96 Tissue Kit (Macherey-Nagel)

Cost: *Medium–High.* Commercial kits are typically more expensive than non-commercial DNA extraction methods, perhaps with the exception of those that involve ethidium bromide density gradient ultracentrifugation.

Time: *Low.* Commercial DNA extraction kits are designed to be relatively efficient, reduce the hands-on time required for DNA extraction, and avoid more time-consuming steps such as phenol/chloroform/isoamyl alcohol purifications.

Potential for scaling up to a 96-well format: *High.* This is a 96-well kit.

Additional reference: Carro et al. (2014)

15.6.2.4 ISO-PLANT (Nippon Gene)

Post-extraction purification: Yotsukura et al. (2001) used the GENECLEAN® II kit after DNA extraction carried out following the ISO-PLANT kit the manufacturer's protocol.

Cost: *Medium–High.* Commercial kits are typically more expensive than non-commercial DNA extraction methods, perhaps with the exception of those involving caesium chloride–ethidium bromide density gradient ultracentrifugation.

Time: *Low.* Commercial DNA extraction kits are designed to be relatively efficient, to reduce the hands-on time required for DNA extraction and to avoid more time-consuming steps such as phenol/chloroform/isoamyl alcohol purifications.

Potential for scaling up to a 96-well kit: *High.* These kits include only solutions and silica beads and do not require specialized tubes or columns. They, therefore, could be used in a 96-well plate.

15.6.2.5 Plant Genomic DNA Kit (Tiangen Biotech)

Cost: *Medium–High.* Commercial kits are typically more expensive than non-commercial DNA extraction methods, perhaps with the exception of those that involve caesium chloride–ethidium bromide density gradient ultracentrifugation.

Time: *Low.* Commercial DNA extraction kits are designed to be relatively efficient, to reduce the hands-on time required for DNA extraction and to avoid more time-consuming steps such as phenol/chloroform/isoamyl alcohol purifications.

Potential to scale up to a 96-well plate: *Low.* This kit requires specialized spin columns and is not available in a 96-well kit.

Additional reference: Liu et al. (2012)

Acknowledgments We are very grateful to the technical support team of Mo Bio Laboratories incorporated, Niccy Aitken (Australian National University), Duckahul Park (Landcare Research, New Zealand), and Leon Smith (Australian National University) for providing technical advice and assistance. Likewise, we would like to thank Jon Waters (University of Otago) for advice and comments on this work. LJW was funded by a scholarship from the North Australian Marine Research Association and an Australian Research Council Linkage grant led by Graham Edgar (University of Tasmania) (ARC LP100200122), XW was funded by a Capacity-Building scholarship from the Australian Biological Resources Study, TK was funded by the Department of Zoology at the University of Otago, and CIF was funded by an Australian Research Council DECRA grant (DE140101715).

References

Alberto F, Whitmer A, Coelho NC, Zippay M, Varela-Alvarez E, Raimondi PT, Reed DC, Serrao EA. Microsatellite markers for the giant kelp *Macrocystis pyrifera*. Conserv Genet. 2009;10(6):1915–7.

Arnaud-Haond S, Candeias R, Serrao EA, Teixeira SJL. Microsatellite markers developed through pyrosequencing allow clonal discrimination in the invasive alga *Caulerpa taxifolia*. Conserv Genet Resour. 2013;5(3):667–9.

Billot C, Rousvoal S, Estoup A, Epplen JT, Saumitou-Laprade P, Valero M, Kloareg B. Isolation and characterization of microsatellite markers in the nuclear genome of the brown alga *Laminaria digitata* (Phaeophyceae). Mol Ecol. 1998;7(12):1778–80.

Carro B, Lopez L, Pena V, Barbara I, Barreiro R. DNA barcoding allows the accurate assessment of European maerl diversity: a proof-of-concept study. Phytotaxa. 2014;190(1):176–89.

Casquet J, Thebaud C, Gillespie RG. Chelex without boiling, a rapid and easy technique to obtain stable amplifiable DNA from small amounts of ethanol-stored spiders. Mol Ecol Res. 2012; 12(1):136–41.

Cohen S, Faugeron S, Martinez EA, Correa JA, Viard F, Destombe C, Valer M. Molecular identification of two sibling species under the name *Gracilaria chilensis* (Rhodophyta, Gracilariales). J Phycol. 2004;40(4):742–7.

Coyer JA, Steller DL, Alberte RS. A field-compatible method for extraction of fingerprint-quality DNA from *Macrocystis pyrifera* (Phaeophyceae). J Phycol. 1995;31(1):177–80.

Davey JW, Hohenlohe PA, Etter PD, Boone JQ, Catchen JM, Blaxter ML. Genome-wide genetic marker discovery and genotyping using next-generation sequencing. Nat Rev Genet. 2011; 12(7):499–510.

De Jong Y, Van der Wurff A, Stam W, Olsen J. Studies on Dasyaceae. 3. Towards a phylogeny of the Dasyaceae (Ceramiales, Rhodophyta), based on comparative rbcL gene sequences and morphology. Eur J Phycol. 1998;33(3):187–201.

Demeke T, Jenkins GR. Influence of DNA extraction methods, PCR inhibitors, and quantification methods on real-time PCR assay of biotechnology-derived traits. Anal Bioanal Chem. 2010;396(6):1977–90.

Doyle J. DNA protocols for plants. In: Hewitt G, Johnston AB, Young JP, editors. Molecular techniques in taxonomy, vol. 57., NATO ASI seriesBerlin: Springer; 1991. p. 283–93.

Elshire RJ, Glaubitz JC, Sun Q, Poland JA, Kawamoto K, Buckler ES, Mitchell SE. A robust, simple genotyping-by-sequencing (GBS) approach for high diversity species. PLoS ONE. 2011;6(5):e19379.

Famà P, Wysor B, Kooistra WHCF, Zuccarello GC. Molecular phylogeny of the genus *Caulerpa* (Caulerpales, Chlorophyta) inferred from chloroplast tufa gene. J Phycol. 2002;38(5):1040–50.

Fraser CI, Hay CH, Spencer HG, Waters JM. Genetic and morphological analyses of the southern bull kelp *Durvillaea antarctica* (Phaeophyceae: Durvillaeales) in New Zealand reveal cryptic species. J Phycol. 2009a;45(2):436–43.

Fraser CI, Spencer HG, Waters JM. Glacial oceanographic contrasts explain phylogeography of Australian bull kelp. Mol Ecol. 2009b;18(10):2287–96.
Goff LJ, Coleman AW. The use of plastid DNA restriction endonuclease patterns in delineating red algal species and populations. J Phycol. 1988;24(3):357–68.
Goff LJ, Moon DA. PCR amplification of nuclear and plastid genes from algal herbarium specimens and algal spores. J Phycol. 1993;29(3):381–4.
Harvey JBJ, Goff LJ. A reassessment of species boundaries in *Cystoseira* and *Halidrys* (Phaeophyceae, Fucales) along the north american west coast. J Phycol. 2006;42(3):707–20.
Heesch S, Broom JES, Neill KF, Farr TJ, Dalen JL, Nelson WA. *Ulva, Umbraulva* and *Gemina*: Genetic survey of New Zealand taxa reveals diversity and introduced species. Eur J Phycol. 2009;44(2):143–54.
Hoarau G, Ja Coyer, Stam WT, Olsen JL. A fast and inexpensive DNA extraction/purification protocol for brown macroalgae. Mol Ecol Notes. 2007;7(2):191–3.
Hong YK, Coury DA, Polne-Fuller M, Gibor A. Lithium chloride extraction of DNA from the seaweed *Porphyra perforata* (Rhodophyta). J Phycol. 1992;28(5):717–20.
Hong YK, Kim SD, Polne-Fuller M, Gibor A. DNA extraction conditions from *Porphyra perforata* using LiCl. J Appl Phycol. 1995;7(2):101–7.
Hong YK, Sohn CH, Lee KW, Kim HG. Nucleic acid extraction from seaweed tissues for polymerase chain reaction. J Mar Biotechnol. 1997;5:95–9.
Hu YJ, Zhou ZG. Extraction of RAPD-friendly DNA from *Laminaria japonica* (Phaeophyta) after enzymatic dissociation of the frozen sporophyte tissues. J Appl Phycol. 2001;13(5):415–22.
Hu ZM, Zeng XQ, Wang AH, Shi CJ, Duan DL. An efficient method for DNA isolation from red algae. J Appl Phycol. 2004;16(3):161–6.
Hughey JR, Silva PC, Hommersand MH. Solving taxonomic and nomenclatural problems in Pacific Gigartinaceae (Rhodophyta) using DNA from type material. J Phycol. 2001;37(6): 1091–109.
Ivanova NV, Fazekas AJ, Hebert PDN. Semi-automated, membrane-based protocol for DNA isolation from plants. Plant Mol Biol Report. 2008;26(3):186–98.
Jin HJ, Kim JH, Sohn C, DeWreede RE, Choi TJ, Towers GHN, Hudson JB, Hong YK. Inhibition of Taq DNA polymerase by seaweed extracts from British Columbia, Canada, and Korea. J Appl Phycol. 1997;9(4):383–8.
Kirkendale L, Saunders GW, Winberg P. A molecular survey of *Ulva* (Chlorophyta) in temperate Australia reveals enhanced levels of cosmopolitanism. J Phycol. 2013;49(1):69–81.
Kotchoni S, Gachomo E. A rapid and hazardous reagent free protocol for genomic DNA extraction suitable for genetic studies in plants. Mol Biol Rep. 2009;36(6):1633–6.
Lane CE, Mayes C, Druehl LD, Saunders GW. A multi-gene molecular investigation of the kelp (Laminariales, Phaeophyceae) supports substantial taxonomic re-organization. J Phycol. 2006;42(4):493–512.
Lim PE, Thong KL, Phang SM. Molecular differentiation of two morphological variants of *Gracilaria salicornia*. J Appl Phycol. 2001;13(4):335–42.
Liu FL, Sun XT, Wang WJ, Liang ZR, Wang FJ. Development of a female-specific RAPD marker for *Sargassum thunbergii* gender identification using bulked segregant analysis. Aquat Bot. 2012;102:79–81.
Maeda T, Kawai T, Nakaoka M, Yotsukura N. Effective DNA extraction method for fragment analysis using capillary sequencer of the kelp, Saccharina. J Appl Phycol. 2012;25(1):337–47.
Manghisi A, Morabito M, Bertuccio C, Le Gall L, Couloux A, Cruaud C, Genovese G. Is routine DNA barcoding an efficient tool to reveal introductions of alien macroalgae? A case study of *Agardhiella subulata* (Solieriaceae, Rhodophyta) in Cape Peloro lagoon (Sicily, Italy). Cryptog Algol. 2010;31(4):423–33.
Mayes C, Saunders GW, Tan IH, Druehl LD. DNA extraction methods for kelp (Laminariales) tissue. J Phycol. 1992;28(5):712–6.
McDevit DC, Saunders GW. On the utility of DNA barcoding for species differentiation among brown macroalgae (Phaeophyceae) including a novel extraction protocol. Phycol Res. 2009; 57(2):131–41.

Niu C, Kebede H, Auld DL, Woodward JE, Burow G, Wright RJ. A safe inexpensive method to isolate high quality plant and fungal DNA in an open laboratory environment. Afr J Biotechnol. 2008;7:2818–22.

Pardo C, Lopez L, Pena V, Hernandez-Kantun J, Le Gall L, Barbara I, Barreiro R. A multilocus species delimitation reveals a striking number of species of coralline algae forming maerl in the OSPAR maritime area. PLoS ONE. 2014;9(8):e104073.

Parsons TJ, Maggs CA, Douglas SE. Plastid DNA restriction analysis links the heteromorphic phases of an apomictic red algal life history. J Phycol. 1990;26(3):495–500.

Patwary MU, van der Meer JP. Application of rapd markers in an examination of heterosis in *Gelidium vagum* (Rhodophyta). J Phycol. 1994;30(1):91–7.

Phillips N, Smith CM, Morden CW. An effective DNA extraction protocol for brown algae. Phycol Res. 2001;49(2):97–102.

Richards JL, Gabrielson PW, Fredericq S. New Insights into the genus *Lithophyllum* (Lithophylloideae, Corallinaceae, Corallinales) from deepwater rhodolith beds offshore the NW Gulf of Mexico. Phytotaxa. 2014;190(1):162–75.

Rindi F, McIvor L, Sherwood AR, Friedl T, Guiry MD, Sheath RG. Molecular phylogeny of the green algal order Prasiolales (Trebouxiophyceae, Chlorophyta). J Phycol. 2007;43(4):811–22.

Rogers SO, Bendich AJ. Extraction of total cellular DNA from plants, algae, and fungi. In: Gelvin S, Schilperoort R, editors. Plant molecular biology manual. Netherlands: Springer; 1994. p. 183–90.

Sahu SK, Thangaraj M, Kathiresan K. DNA extraction protocol for plants with high levels of secondary metabolites and polysaccharides without using liquid nitrogen and phenol. ISRN Mol Biol. 2012;2012:1–6.

Saunders GW. Gel purification of red algal genomic DNA: an inexpensive and rapid method for the isolation of polymerase chain reaction-friendly DNA. J Phycol. 1993;29(2):251–4.

Saunders GW, McDevit DC. Acquiring DNA sequence data from dried archival red algae (Florideophyceae) for the purpose of applying available names to contemporary genetic species: a critical assessment. Botany. 2012;90(3):191–203.

Siemer BL, Stam WT, Olsen JL, Pedersen PM. Phylogenetic relationships of the brown algal orders Ectocarpales, Chordariales, Dictyosiphonales, and Tilopteridales (Phaeophyceae) based on rubisco large subunit and spacer sequences. J Phycol. 1998;34(6):1038–48.

Sim MC, Lim PE, Gan SY, Phang SM. Identification of random amplified polymorphic DNA (RAPD) marker for differentiating male from female and sporophytic thalli of *Gracilaria changii* (Rhodophyta). J Appl Phycol. 2007;19(6):763–9.

Snirc A, Silberfeld T, Bonnet J, Tillier A, Tuffet S, Sun JS. Optimization of DNA extraction from brown algae (Phaeophyceae) based on a commercial kit. J Phycol. 2010;46(3):616–21.

Torrano-Silva BN, Riosmena-Rodriguez R, Oliveira MC. Systematic position of *Paulsilvella* in the Lithophylloideae (Corallinaceae, Rhodophyta) confirmed by molecular data. Phytotaxa. 2014;190(1):94–111.

Van der Strate HJ, Olsen JL, Van De Zande L, Edwards KJ, Stam WT. Isolation and characterization of microsatellite loci in the benthic seaweed, *Cladophoropsis membranacea* (Cladophorales, Chlorophyta). Mol Ecol. 2000;9(9):1442–3.

Van der Strate HJ, Van de Zande L, Stam WT, Olsen JL. The contribution of haploids, diploids and clones to fine-scale population structure in the seaweed *Cladophoropsis membranacea* (Chlorophyta). Mol Ecol. 2002;11(3):329–45.

Varela-Alvarez E, Glenn TC, Serrao EA, Duarte CM, Martinez-Daranas B, Valero M, Marba N. Dinucleotide microsatellite markers in the genus *Caulerpa*. J Appl Phycol. 2011;23(4):715–9.

Verbruggen H, Leliaert F, Maggs CA, Shimada S, Schils T, Provan J, Booth D, Murphy S, De Clerck O, Littler DS, Littler MM, Coppejans E. Species boundaries and phylogenetic relationships within the green algal genus *Codium* (Bryopsidales) based on plastid DNA sequences. Mol Phylogenet Evol. 2007;44(1):240–54.

Vidal R, Meneses I, Smith M. Enhanced DNA extraction and PCR amplification of SSU ribosomal genes from crustose coralline algae. J Appl Phycol. 2002;14(3):223–7.

Vieira-Pinto T, Oliveira MC, Bouzon J, Sissini M, Richards JL, Riosmena-Rodriguez R, Horta PA. *Lithophyllum* species from Brazilian coast: range extension of *Lithophyllum margaritae* and description of *Lithophyllum atlanticum* sp nov. (Corallinales, Corallinophycidae, Rhodophyta). Phytotaxa. 2014;190(1):355–69.

Walsh PS, Metzger DA, Higuchi R. Chelex 100 as a medium for simple extraction of DNA for PCR-based typing from forensic material. Biotechniques. 1991;10(4):506–13.

Wang GG, Li Y, Xia PP, Duan DL. A simple method for DNA extraction from sporophyte in the brown alga *Laminaria japonica*. J Appl Phycol. 2005;17(1):75–9.

Wong TKM, Ho CL, Lee WW, Rahim RA, Phang SM. Analyses of expressed sequence tags from *Sargassum binderi* (Phaeophyta). J Phycol. 2007;43(3):528–34.

Yang LE, Jin QP, Xiao Y, Xu P, Lu S. Improved methods for basic molecular manipulation of the red alga *Porphyra umbilicalis* (Rhodophyta: Bangiales). J Appl Phycol. 2013;25(1):245–52.

Yotsukura N, Kawai T, Motomura T, Ichimura T. Random amplified polymorphic DNA markers for three Japanese *Laminaria* species. Fish Sci. 2001;67(5):857–62.

Index

Z.-M. Hu and C. Fraser (eds.), *Seaweed Phylogeography*,
DOI 10.1007/978-94-017-7534-2

The manufacturer's authorised representative in the EU is Springer Nature Customer Service Centre GmbH, Europaplatz 3, 69115 Heidelberg, Germany. If you have any concerns regarding our products, please contact ProductSafety@springernature.com

Printed and bound by CPI Group (UK) Ltd, Croydon, CR0 4YY
30/09/2026
02225449-0016